Ilya J. Bakelman

Convex Analysis and Nonlinear Geometric Elliptic Equations

Springer-Verlag
Berlin Heidelberg New York
London Paris Tokyo
Hong Kong Barcelona
Budapest

Ilya J. Bakelman †

Mathematics Subject Classification (1991):
35JXX, 53CXX

ISBN-13:978-3-642-69883-5 **e-ISBN-13: 978-3-642-69881-1**
DOI: 10.1007/978-3-642-69881-1

CIP will follow

Softcover reprint of the hardcover 1st edition 1994

Typesetting: Camera ready copy by author

SPIN 10000094 41/3140- 5 4 3 2 1 – Printed an acid-free paper

In Memoriam

Ilya J. Bakelman died tragically in a highway accident on August 30, 1992. Since 1981 he had been Professor of Mathematics at Texas A&M University, following a long and distinguished career as Chair Professor and Head of the Geometry Section at Leningrad State Pedagogical Institute. He was a leader in the study of nonlinear elliptic partial differential equations by methods of differential and convex geometry. In Russia he is also recognized as a reformer of mathematical education at both school and university levels.

Much of Bakelman's work was devoted to boundary value problems for mean curvature and Monge–Ampere equations in more than two variables and their generalizations. He developed powerful analytic tools to obtain solvability results for these problems, the most useful of which may have been his R-curvature. Of particular importance is his a priori bound for the absolute value of solutions of elliptic equations in nondivergence form. This bound is independent of the moduli of continuity of the coefficients so that it can be applied to nonlinear equations. It has been used extensively and has many important applications.

Ilya Bakelman was born November 30, 1928, in Leningrad, now St. Petersburg. As a youngster he was a member of the mathematical circle of Leningrad Pioneer Palace, the cradle of many well-known mathematicians. As a university student, he in turn led such a circle; later, when some of his pupils were denied admission to higher education because of discrimination, he continued to teach them privately and to help them start scientific careers. Bakelman received the equivalent of a Ph.D. degree from the University of Leningrad in 1954 and the degree of Doctor of Sciences from Leningrad State Pedagogical Institute in 1960. His teachers were A.D. Alexandrov in geometry and V.I. Smirnov, O.A. Ladyzhenskaya and S.G. Mikhlin in partial differential equations and applied mathematics. As a professor in Leningrad he organized a strong research group and totally revised the mathematical education of Soviet teachers, according to his former colleagues. His former students can now be found in educational leadership positions throughout Russia.

He emigrated from Russia in 1979 and spent two years as a visiting professor at the University of Minnesota before coming to Texas A&M. As a teacher in Texas, Bakelman became very popular with students, as his friendliness and infectious enthusiasm for mathematics overcame the remaining barriers of language and culture. His research lectures were models of clarity, followed and enjoyed by mathematicians with no special interest in the subject matter. His research continued actively, despite a scarcity of close collaborators and graduate students. During this last decade Bakelman was a frequent long-term visitor at such centers of research as IHES, MSRI, IMA, IAS, ETH, and the Universities of Bonn and Rome. At the time of his death he was completing this monograph, *Convex Analysis and Nonlinear Geometric Elliptic Equations*. Earlier he had written several monographs and textbooks in Russian, one of which, *Inversions*, was translated into Japanese, English, and German.

Dr. Bakelman's untimely death leaves a void in the mathematical community which may never be filled. He will be greatly missed by his family, colleagues and friends around the world.

William Rundell

We dedicate this book to the beloved memory of our husband and father,

Ilya J. Bakelman,

to whom its publication meant so very much.

Irene and Alexander

Acknowledgements

We would like to express our profound gratitude to Professor Steven Taliaferro of the Texas A&M University Mathematics Department for his inestimable help in finishing the editing of Ilya's book as well as completing its final organization.

Likewise, we would like to express our great thanks to Dr. William Rundell, Head of the Texas A&M Mathematics Department, and everyone of the faculty whose assistance after Ilya's death was so significant to the publication of the book.

Our special gratitude goes to Robin Campbell, the departmental technical secretary, whose expertise provided the impeccably typed manuscript, and to Karola Feltz, the Department's Administrative Assistant, who coordinated the details of this endeavor and helped to carry it to fruition.

Finally, we would like to thank all of Ilya's colleagues and friends in the United States, the many Universities and Research Centers throughout Europe and Russia, who have worked with him these past many decades and who expressed their fond remembrances to the family after his tragic death.

Irene R. Bakelman
Alexander I.R. Bakelman

Editorial Comments

This book, which represents much of Professor Bakelman's work of the last ten years, was nearly finished at the time of his death in 1992. Since I had been helping him with the readability and homogeneity of the text, the Mathematics Department Head, Dr. William Rundell, asked me to put the final touches on the book to get it ready for publication, and Springer Verlag agreed.

The task which was left for me to finish was to insert reference numbers into the text and to shorten the book. The publisher and Prof. Bakelman had previously agreed that the book should be shortened by about 200 pages. In order not to introduce any errors and also to preserve the author's method or presentation, I either completely deleted a section or left it unaltered, except for corrections of obvious typographical errors. I was also careful not to delete any material that would be needed later. Needless to say Prof. Bakelman could have finished this final editing far better and more easily than I. However, he had discussed the final form of this book with me, and in spite of my lack of expertise in his area, I believe this final version is close to what he had envisioned.

I would like to thank James Serrin and Neil Trudinger for their generous help with the references and Robin Campbell of our Department for the impeccably typed manuscript.

Steven D. Taliaferro
Texas A&M University

Introduction

Investigations in modern nonlinear analysis rely on ideas, methods and problems from various fields of mathematics, mechanics, physics and other applied sciences. In the second half of the twentieth century many prominent, exemplary problems in nonlinear analysis were subject to intensive study and examination. The united ideas and methods of differential geometry, topology, differential equations and functional analysis as well as other areas of research in mathematics were successfully applied towards the complete solution of complex problems in nonlinear analysis.

It is not possible to encompass in the scope of one book all concepts, ideas, methods and results related to nonlinear analysis. Therefore, we shall restrict ourselves in this monograph to nonlinear elliptic boundary value problems as well as global geometric problems. In order that we may examine these problems, we are provided with a fundamental vehicle: The theory of convex bodies and hypersurfaces. In this book we systematically present a series of centrally significant results obtained in the second half of the twentieth century up to the present time. Particular attention is given to profound interconnections between various divisions in nonlinear analysis.

The theory of convex functions and bodies plays a crucial role because the ellipticity of differential equations is closely connected with the local and global convexity properties of their solutions. Therefore it is necessary to have a sufficiently large amount of material devoted to the theory of convex bodies and functions and their connections with partial differential equations.

The structure and summary of this book is as follows. The material is divided into three main parts. Part I contains the foundations of the theory of convex bodies (Chapter 1) and applications to the Minkowski problem for convex closed hypersurfaces (Chapter 2). The Minkowski problem can be reduced to certain Monge–Ampere equations on the unit sphere. The topics in Chapter 2 are decidedly more advanced than those in Chapter 1; the multidimensional Minkowski problem in particular was actively expounded upon through the mid 1980's (Chapter 6, Part II). In Chapter 2, we present basic properties of mixed volumes of convex bodies and give a derivation of the fundamental inequalities of Brunn–Minkowski, Alexandrov, and Fenchel. In Section 8 of Chapter 2 we recall well known facts about general, linear and quasilinear elliptic operators and equations. Also considered are Monge–Ampere operators and differential operators in global problems of differential geometry. We conclude Section 8 with the classical maximum principles and their applications to quasilinear equations and uniqueness theorems in global differential geometry.

Part II, encompassing Chapters 3–6, is devoted to the theory of elliptic convex solutions of n-dimensional Monge–Ampere equations. Chapter 3 contains the most important connections with partial differential equations where the geometric theory of the Dirichlet problem for Monge–Ampere equations,

$$\det(u_{ij}) = f(x, u, Du) \tag{1}$$

is presented. In this chapter we introduce the normal mapping of convex hypersurfaces. Normal mapping is the generalization of tangential mapping by support hyperplanes of the convex hypersurface. The second important construction which we introduce is the R-curvature of convex functions, which is a generalization of the integral Gaussian curvature. R-curvature is the operator set function $\omega(R, u, e) = \int_{\chi_u(e)} R(p)dp$ which we define on Borel subsets e of a convex domain B by a positive locally summable function $R(p)$, a convex function $u(x)$ defined in B and the normal image $\chi_u(e) \subset P$ defined by the function $u(x)$ on the set $e \subset B$. For convex functions $u(x)$ of class C^2 we have

$$\omega(R, u, e) = \int_e R(Du) \det(u_{ij})dx. \tag{2}$$

The main concepts used for defining generalized solutions of Monge–Ampere equations (1) is the normal mapping and R-curvature of convex functions. On the basis of these concepts we obtain a significant number of important properties of generalized solutions of the Dirichlet problem for equations (1) such as a priori estimates of C-norms of these solutions, criteria for weak and strong compactness, the conditions of pointwise and uniform convergence in the closed domain B, and others.

In Sections 11 and 12 of Chapter 3 we derive various existence and uniqueness theorems for the Dirichlet problem for generalized convex solutions of Monge–Ampere equations (1).

In Chapter 4 a conjecture of R. Courant and D. Hilbert that the minimizer of some naturally constructed variational problem is a generalized solution of the Dirichlet problem for the n-dimensional Monge–Ampere equation

$$\det(u_{ij}) = f(x)$$

is solved in detail. The solution is based on the study of special geometric functionals in the original class of general convex functions and certain profound geometric constructions and inequalities in the functional dual space. The later part of Chapter 4 deals with several other variational problems related to the basic variational problem considered above.

Chapter 5 is devoted to noncompact problems for Monge–Ampere equations (1) in the entire space R^n. We establish existence theorems for solutions of equations (1) with prescribed asymptotic convex cones. The solution is based on the application of topological principles of fixed points in non-trivial functional spaces.

Chapter 6, consisting of Sections 19 and 20, deals with the theory of convex elliptic solutions of class C^m, $m \geq 2$, for n-dimensional Monge–Ampere equations. In Section 19 we present the complete proof of the existence of strictly convex solutions of class C^{m+a}, $m \geq 4$, $0 < a < 1$, of the Minkowski problem (see Chapter 2, Section 8). It is possible to reduce this problem to the existence of a strictly convex solution of class C^{m+a} of an invariant Monge–Ampere equation on the unit hypersphere S in E^{n+1}. The proof that such a solution exists is based on a priori estimates of the derivatives of solutions of Monge–Ampere equations up to the third order.

In Section 20 we present a series of basic facts related to convex solutions of class C^m, $m \geq 2$, of n-dimensional Monge-Ampere equations. In Subsection 20.1 we establish uniqueness and comparison theorems for solutions of the Dirichlet problem for Monge–Ampere equations

$$\det(u_{ij}) = f(x, u, Du)$$

where 1) $f(x,u,p)$ and its first derivatives $f_u(x,u,p)$ and $f_{p_i}(x,u,p)$, $i = 1,2,\ldots,n$, are continuous in $\overline{B} \times R \times R^n$; 2) $f(x,u,p) > 0$ and $f_u(x,u,p) \geq 0$ for $(x,u,p) \in \overline{B} \times R \times R^n$; and 3) B is a bounded domain in E^n.

In Subsections 20.2 through 20.4 we establish C-norms estimates for solutions and their gradients for solutions of class C^m, $m \geq 2$, of the Dirichlet problem for $\det(u_{ij}) = f(x,u,Du)$ without the assumption that $f_u(x,u,p) \geq 0$.

In Subsections 20.5 through 20.10 we present the complete proof of the existence theorem for the Dirichlet problem

$$\begin{aligned} \det(u_{ij}) = \psi(x) \quad & \text{in} \quad B \\ u = \phi(x) \quad & \text{on} \quad \partial B \end{aligned}$$

where $\psi \in C^\infty(\overline{B})$, $\psi > 0$ in $\overline{B}$ and $\phi \in C^\infty(\partial B)$. This proof is based on the continuity method. The central difficulties are connected with a priori estimates of the second and third derivatives and the estimates of the logarithmic continuity for second derivatives in B.

In Part III, consisting of Chapters 7 and 8, we study geometric methods in general elliptic equations and their applications to calculus of variations, differential geometry and continuum mechanics. In Chapter 7 we study applications of the theory of convex bodies to the Dirichlet problem and nonlinear elliptic Euler–Lagrange equations. Presented here are recent results related to geometric maximum principles for generalized solutions of nonlinear Euler-Lagrange equations. As the first applications of these results, we present a significant development of the Bernstein estimates for two-dimensional Euler-Lagrange equations to n-dimensions. These can not be obtained by analytical methods based on classical maximum principles. Other applications are given to solutions of the mean curvature equation in Euclidean and Minkowski–Lorentz spaces, elasticity theory and hydrodynamics.

Chapter 8 is devoted to the systematic presentation of global geometric methods for the study of more profound maximum principles for solutions with first and second Sobolev generalized derivatives of linear and quasilinear elliptic equations. These principles can be used to obtain more profound results for variational problems and problems in differential geometry as well as continuum mechanics.

It is well known that the proof of existence theorems for the Dirichlet problem for general quasilinear equations can be given by means of the continuity method. The general scheme of this proof is presented in Section 29. The successful implementation of this scheme is based on obtaining appropriate a priori estimates for C^1-norms of the desired solutions. We clarify the significance of such estimates for proofs by the continuity method. Great attention in Section 29 is given to different methods of deriving the a priori estimates mentioned above.

Contents

Part I. Elements of Convex Analysis

In Part I we present those elements of convex analysis that play a significant role in nonlinear elliptic problems and especially in those that arise from global differential geometry.

Chapter 1. Convex Bodies and Hypersurfaces

§1. Convex Sets in Finite–Dimensional Euclidean Spaces

1.1 Main Definition

The concept of a convex set can be introduced in any linear space L. A set K in L is called *convex* if the line segment ab is contained in K for any elements $a, b \in K$, i.e. $x_t = (1-t)a + tb \in K$ for any $a, b \in K$ and any $t \in [0,1]$.

Clearly every subspace of L is convex. The whole space L and the zero-element θ are trivial convex subsets of L. The empty space $\emptyset$ is convex by definition. The following simple lemma is often used in various geometric constructions.

Lemma 1.1. *The intersection K of any family $\{K_\alpha\}$ of convex subsets K_α of L is convex.*

Proof. The lemma is trivial if $K = \emptyset$ or K consists of one element. Suppose K contains at least two different elements a and b. Then all convex sets K_α contain the segment ab. Therefore $ab \subset K$. □

1.2 Linear and Convex Operations with Convex Sets. Convex Hull

Let $a_1, a_2, \dots, a_m$ be arbitrary elements of L and $\lambda_1, \lambda_2, \dots, \lambda_m$, be any real numbers. The element

$$a = \lambda_1 a_1 + \lambda_2 a_2 + \cdots + \lambda_m a_m \tag{1.1}$$

is called a *linear combination* of $a_1, a_2, \dots, a_m$. If all numbers $\lambda_1, \lambda_2, \dots, \lambda_m$ are nonnegative and $\lambda_1 + \lambda_2 + \cdots + \lambda_m = 1$, then the element a is called a convex combination of $a_1, a_2, \dots, a_m$.

Now let $Q_1, Q_2, \dots, Q_m$ be subsets of L and $\lambda_1, \lambda_2, \dots, \lambda_m$ be real numbers. A set Q consisting of all linear combinations

$$a = \lambda_1 a_1 + \lambda_2 a_2 + \cdots + \lambda_m a_m, \qquad a_i \in Q_i, \tag{1.2}$$

is called a *linear combinations of the sets* $Q_1, Q_2, \dots, Q_m$, and is denoted by

$$Q = \sum_i \lambda_i Q_i = \lambda_1 Q_1 + \cdots + \lambda_m Q_m. \tag{1.3}$$

If all λ_i are nonnegative and $\lambda_1 + \lambda_2 + \cdots + \lambda_m = 1$ then the set (1.3) is called a *convex combination of sets* $Q_1, Q_2, \dots, Q_m$.

Lemma 1.2. *Let K be a convex subset of L. Then every convex combination of $a_1, a_2, \ldots, a_m \in K$ is also an element of K.*

Proof. The lemma holds for the case of two elements. Assume our assertion is correct for m elements $a_1, a_2, \ldots, a_m$ ($m \geq 2$) and we shall prove it for $m+1$ elements $a_1, a_2, \ldots, a_m$, a_{m+1}. Let $\lambda_i \geq 0$; $i = 1, 2, \ldots, m$, $m+1$; and $\sum_{i=1}^{m+1} \lambda_i = 1$ be any real numbers. Set

$$\mu = \lambda_2 + \lambda_3 + \cdots + \lambda_m + \lambda_{m+1}.$$

Since only the case $\lambda_{m+1} > 0$ is of interest we can assume $\mu > 0$. According to the inductive assumption

$$a = \frac{\lambda_2}{\mu} a_2 + \frac{\lambda_3}{\mu} a_3 + \cdots + \frac{\lambda_m}{\mu} a_m + \frac{\lambda_{m+1}}{\mu} a_{m+1} \in K$$

as a convex combination of m elements $a_2, \ldots, a_{m+1}$. Since $\lambda_1 + \mu = 1$, we have

$$\lambda_1 a_1 + \lambda_2 a_2 + \cdots + \lambda_m a_m + \lambda_{m+1} a_{m+1} = \lambda_1 a_1 + \mu a \in K. \qquad \square$$

Lemma 1.3. *Let K_1 and K_2 be convex subsets of L. Then the sets λK_1, $K_1 + K_2$ and $K_1 - K_2 = K_1 + (-1)K_2$ are also convex in L.*

Proof. Let a and b be any elements of λK_1. Then there exist some elements $a', b' \in K_1$ such that $a = \lambda a'$ and $b = \lambda b'$. Since K_1 is convex, $(1-t)a' + tb' \in K_1$ for any $t \in [0, 1]$. Hence

$$(1-t)a + tb = \lambda[(1-t)a' + tb' \in \lambda K_1$$

for any $t \in [0, 1]$. Therefore λK_1 is a convex set in L.

Now let elements $a, b \in K_1 \pm K_2$. Then there exist some elements $a' \in K_1$, $a'' \in K_2$, $b' \in K_1$ and $b'' \in K_2$ such that

$$a = a' \pm a'' \quad \text{and} \quad b = b' \pm b''.$$

Since K_1 and K_2 are convex sets in L, we have

$$ta + (1-t)b = [ta' + (1-t)b'] \pm [ta'' + (1-t)b''] \in K_1 \pm K_2$$

for any $t \in [0, 1]$. Thus $K_1 \pm K_2$ is a convex set in L. $\square$

Theorem 1.1. *Every convex combination of a finite system of convex sets in L is also convex.*

The proof follows directly from Lemma 1.3. Theorem 1.1 will have applications in Section 7, devoted to Minkowski's mixed volumes of convex bodies.

Let Q be a set in L. The intersection of all convex sets in L containing Q is called the *convex hull* of Q and denoted by CoQ. From Lemma 1.1 it follows that CoQ is a convex set. If W is any convex set containing Q, then $CoQ \subset W$. Thus CoQ is the smallest convex set containing Q. Therefore

$$CoQ = Q$$

for every convex set Q in L. Conversely if

$$CoQ = Q,$$

then Q is obviously convex.

Theorem 1.2. *Let Q be any subset of a linear space L. Then CoQ is the union of all convex combinations of finite systems $a_1, a_2, \ldots, a_m \in Q$.*

Proof. Let W be the set of all convex combinations of all various finite systems of elements of Q. If some element $a \in W$, then $a \in CoQ$. This fact follows from Lemma 1.2. Thus

$$W \subset CoQ. \tag{1.4}$$

On the other hand, if $b, c \in W$, then

$$b = \sum_{i=1}^{m} \lambda_i b_i, \qquad c = \sum_{j=1}^{k} \mu_j c_j,$$

where $b_1, \ldots, b_m \in Q$, $c_1, \ldots, c_k \in Q$, and $\lambda_i \geq 0$, $\mu_j \geq 0$, $\sum\limits_{i=1}^{m} \lambda_i = 1$, $\sum\limits_{j=1}^{k} \mu_j = 1$. Let $t \in [0,1]$. Then we consider the element

$$d_t = (1-t)b + tc.$$

Clearly

$$d_t = \sum_{i=1}^{m} \gamma_i b_i + \sum_{j=1}^{k} \gamma_{m+j} c_j,$$

where

$$\begin{aligned} \gamma_i &= (1-t)\lambda_i, \quad & i &= 1,2,\ldots,m, \\ \gamma_{m+j} &= t\mu_j, & j &= 1,2,\ldots,k. \end{aligned}$$

Clearly $\gamma_i \geq 0$, $i = 1,2,\ldots,m+k$ and $\sum\limits_{i=1}^{m+k} \gamma_i = 1$. Hence $d_t \in W$ for all $t \in [0,1]$. Thus W is a convex set and $W \supset CoQ$. Combining this inclusion with (1.4) we obtain $W = CoQ$. □

Theorem 1.3. *Let α be any real number and let Q, Q_1, Q_2 be any sets in L. Then the following relations*

a) $Co(\alpha Q) = \alpha(CoQ)$,
b) $Co(Q_1 + Q_2) = Co\, Q_1 + CoQ_2$,
c) $Co(Q_1 \cup Q_2) = Co(CoQ_1 \cup CoQ_2)$,
d) $Co(Q_1 \cap Q_2) \subset CoQ_1 \cap CoQ_2$

hold.

Proof. The equality a) holds for $\alpha = 0$. Let $\alpha \neq 0$. Then

$$\alpha Q \subset \alpha CoQ$$

because $Q \subset CoQ$. From Lemma 1.3 it follows that αCoQ is a convex set. Therefore

$$Co(\alpha Q) \subset \alpha CoQ. \tag{1.5}$$

If we set $T = \alpha Q$, then we obtain $Q = \frac{1}{\alpha}T$ and

$$Co\left(\frac{1}{\alpha}T\right) \subset \frac{1}{\alpha}CoT.$$

Therefore

$$\alpha CoQ \subset Co(\alpha Q). \tag{1.6}$$

From (1.5) and (1.6) it follows that the equality a) is correct.

Now

$$Q_1 + Q_2 \subset CoQ_1 + CoQ_2.$$

Since CoQ_1 and CoQ_2 are convex sets, we get from Lemma 1.3 that

$$CoQ_1 + CoQ_2 = Co(CoQ_1 + CoQ_2) \supset Co(Q_1 + Q_2). \tag{1.7}$$

If $b \in Q_2$ and $a = \lambda_1 a_1 + \cdots + \lambda_m a_m \in CoQ_1$, then, by Theorem 1.2,

$$Co(Q_1 + b) = CoQ_1 + b,$$

because $a + b = \lambda_1(a_1 + b) + \cdots + \lambda_m(a_m + b)$. Therefore

$$CoQ_1 + Q_2 \subset Co(Q_1 + Q_2) \tag{1.8}$$

and replacing Q_2 with CoQ_2 and interchanging Q_1 and Q_2 we obtain

$$CoQ_1 + CoQ_2 \subset Co(CoQ_1 + Q_2). \tag{1.9}$$

From (1.8) and (1.9) we obtain

$$\begin{aligned} CoQ_1 + CoQ_2 \subset Co(CoQ_1 + Q_2) \subset Co(Co(Q_1 + Q_2)) = \\ = Co(Q_1 + Q_2). \end{aligned} \tag{1.10}$$

Thus from (1.7) and (1.10) we obtain relation b). Since $Q \subset CoQ_1$ and $Q_2 \subset CoQ_2$, we have $Q_1 \cup Q_2 \subset CoQ_1 \cup CoQ_2$. Hence

$$Co(Q_1 \cup Q_2) \subset Co(CoQ_1 \cup CoQ_2). \tag{1.11}$$

On the other hand we have $Co(Q_1 \cup Q_2) \supset CoQ_1$ and $Co(Q_1 \cup Q_2) \supset CoQ_2$. Hence

$$Co(Q_1 \cup Q_2) = Co(Co(Q_1 \cup Q_2)) \supset Co(CoQ_1 \cup CoQ_2). \tag{1.12}$$

From (1.11) and (1.12) we obtain relation c). Relation d) follows from the facts:

1) the set $CoQ_1 \cap CoQ_2$ is convex,
2) $Q_1 \cap Q_2 \subset CoQ_1 \cap CoQ_2$.

□

1.3 The Properties of Convex Sets in Linear Topological Spaces

Theorem 1.4. *The closure of any convex set in a linear topological space is convex.*

Proof. Let K be any convex set in a linear topological space L. The set K is convex in L if and only if the mapping $f\colon L \times L \times [0,1] \to L$ maps $K \times K \times [0,1]$ into K, where

$$f(a,b,t) = (1-t)a + tb.$$

Since f is continuous and $\overline{K \times K \times [0,1]} = \overline{K} \times \overline{K} \times [0,1]$, we have

$$\begin{aligned} f(\overline{K} \times \overline{K} \times [0,1]) &= f(\overline{K \times K \times [0,1]}) \\ &\subset \overline{f(K \times K \times [0,1])}. \end{aligned} \tag{1.13}$$

Since K is convex,

$$\overline{f(K \times K \times [0,1])} \subset \overline{K}. \tag{1.14}$$

Thus from (1.13) and (1.14) it follows that

$$f(\overline{K} \times \overline{K} \times [0,1]) \subset \overline{K}.$$

Thus $\overline{K}$ is convex. □

Let L be a linear topological space. Then the closed convex hull can also be considered for any subset Q of this space. The *closed convex hull of* Q is defined as the intersection of all closed convex subsets of L containing Q and is denoted by $\overline{Co}Q$.

Theorem 1.5. *Let L be a linear topological space. Then the equalities*

a) $\overline{Co}Q = \overline{CoQ}$;
b) $\overline{Co}(\alpha Q) = \alpha\overline{Co}(Q)$

hold for any subset $Q \in L$ and any real number α.

Proof. The set $\overline{Co}(Q)$ is closed and contains CoQ. Therefore

$$\overline{Co}(Q) \supset \overline{Co(Q)}.$$

Since the closure of any convex set is convex (see Theorem 1.4), $\overline{Co(Q)}$ is convex and contains Q. Therefore

$$\overline{Co}(Q) \subset \overline{Co(Q)}.$$

Thus assertion a) holds.

Assertion b) follows immediately from assertion a) of Theorem 1.3 and Theorem 1.5. □

1.4 Euclidean Space E^n

Let $n \geq 1$ be an integer. A linear space L is called *n-dimensional* if there exists a system of n linearly independent elements of L and any system consisting of m elements, $m > n$, is linearly dependent. Let L be any linear space. A function $(\ ,\ )$: $L \times L \to R$ is called a *scalar product* in L if this function satisfies the following conditions:

1. $(a, b) = (b, a)$;
2. $(a_1 + a_2, b) = (a_1, b) + (a_2, b)$;
3. $(\lambda a, b) = \lambda(a, b)$;
4. If, for some $a \in L$, $(a, b) = 0$ for all $b \in L$, then $a = \theta$.

In conditions 1,2,3 elements a, b, a_1, b_1 are arbitrary elements of L and λ is any real number. A linear n-dimensional space V^n with a fixed scalar product is called a n-dimensional linear (or vector) Euclidean space. We assume that the theories of n-dimensional linear and Euclidean spaces are well known.

We now consider the concept of an n-dimensional Euclidean point space E^n. This concept is important in differential geometry and differential equations. E^n consists of two sets W and V^n and of certain relations between them. The elements of the first set are called *points* of E^n and are denoted by $A, B, C, \ldots$. The second space is an n-dimensional vector Euclidean space, its element are called *vectors* of E^n and are denoted by $a, b, c, \ldots$. We assume that there is a mapping g: $W \times W \to V^n$ transforming any ordered pair $(A, B) \in W \times W$ into a certain vector $g(A, B) = \overline{AB}$. The point A is called the *initial point* and the point B is called the *end* of the vector $\overline{AB}$. We also assume that the mapping g satisfies the following conditions:

a) for every point $A \in W$ and any vector $a \in V^n$ there exists a point $B \in W$ such that $\overline{AB} = g(A, B) = a$;

b) for any three points A, B, C the identity

$$g(A, B) + g(B, C) + g(C, A) = \overline{AB} + \overline{BC} + \overline{CA} = \theta \tag{1.15}$$

holds, where θ is the zero-element of V^n.

c) from the equation $g(A, B) = \theta$ it follows that points A and B coincide.

A pair of sets W and V^n together with a mapping g: $W \times W \to V^n$ satisfying conditions a), b), c) is called an *n-dimensional Euclidean point space* E^n. These axioms for E^n were given by H. Weyl.

From axioms a), b), c) of E^n and the axioms of V^n we obtain the following conclusions:

1. $\overline{AA} = \theta$ for all points $A \in E^n$;
2. $\overline{AB} = -\overline{BA}$ for all points $A, B \in E^n$;
3. $\overline{AB} + \overline{BC} + \overline{AC}$ for all points $A, B, C \in E^n$;
4. If $\overline{AB} = \overline{AC}$, then the points B and C coincide.

We fixed an arbitrary point 0 in E^n and associate the vector $r(X) = \overline{0X}$ with every point $X \in E^n$. This defines a one-to-one mapping $r\colon W \to V^n$. $r(X)$ is called the *position vector* of the point X with the initial point 0. Let $e_1, e_2, \dots, e_n$ be some orthonormal basis in V^n. The equation

$$\overline{0X} = x_1e_1 + x_2e_2 + \cdots + x_ne_n \tag{1.16}$$

defines a one-to-one correspondence between points X and ordered n-tuples $(x_1, \dots, x_n)$ of real numbers. We call $(x_1, \dots, x_n)$ the Cartesian coordinates of the point X in the system with origin 0 and with basis vectors $e_1, \dots, e_n$. Clearly, if X has coordinates $(x_1, \dots, x_n)$ and Y has coordinates $(y_1, \dots, y_n)$, then

$$\overline{XY} = \overline{0Y} - \overline{0X} = (y_1 - x_1)e_1 + \cdots + (y_n - x_n)e_n. \tag{1.17}$$

1.5 The Simple Figures in E^n

A point X is called a *convex combination of the points* $X_1, X_2, \dots, X_m$ if

$$\overline{0X} = \lambda_1\overline{0X_1} + \lambda_2\overline{0X_2} + \cdots + \lambda_m\overline{0X_m} \tag{1.18}$$

where $\lambda_1 + \lambda_2 + \cdots + \lambda_m = 1$ and $\lambda_i \geq 0$ $(i = 1, 2, \dots, m)$. This definition doesn't depend on the position of the origin 0. If the point $0'$ is a new origin, then

$$\overline{0X} = \overline{00'} + \overline{0'X} \tag{1.19}$$

and

$$\lambda_i\overline{0X_i} = \lambda_i\overline{00'} + \lambda_i\overline{0'X_i}. \tag{1.20}$$

Substituting (1.19) and (1.20) and (1.18) and using the condition $\lambda_1 + \lambda_2 + \cdots + \lambda_m = 1$ we obtain

$$\overline{0'X} = \lambda_1\overline{0'X_1} + \lambda_2\overline{0'X_2} + \cdots + \lambda_m\overline{0'X_m}.$$

Therefore we denote by

$$X = \lambda_1X_1 + \cdots + \lambda_mX_m$$

the convex combinations of the points $X_1, \dots, X_m$ with coefficients $\lambda_1, \lambda_2, \dots, \lambda_m$. The set of all convex combinations for two different given points X and Y is called the *segment* XY. The points X and Y are called the ends of the segment XY. Since all the relations for convex combinations of vectors hold for convex combinations of points, all of the definitions and facts about convex sets in linear spaces proved above (see Subsection 1.1) hold for convex sets in E^n also.

The points $X_0, X_1, \dots, X_k$ are in a *general location* if the vectors $\overline{X_0X_1}$, $\overline{X_0X_2}, \dots, \overline{X_0X_k}$ are linearly independent. Let $X_0, X_1, \dots, X_k$ be the points in a general location. The convex hull of this system of points is called the *k-dimensional simplex* with vertices $X_0, X_1, \dots, X_k$. The following theorem is an obvious consequence of the definition of the convex hull and Lemma 1.2.

Theorem 1.6. *If a convex set $Q \subset E^n$ contains $k+1$ points in a general location then Q contains the k-simplex with the vertices at these points.*

A convex closed set M in E^n is called a *k-convex body*, if M contains $k+1$ points in a general location, but does not contain $k+2$ points in a general location. If $k = n$ then we shall omit the term "n-dimensional". Clearly any k-dimensional simplex σ^k is a k-convex body ($k = 1, 2, \dots, n$).

Let A be a point in E^n and Q^k be a k-dimensional subspace of V^n. The set P^k consisting of all the points $X \in E^n$ and satisfying the condition $\overline{AX} \in Q^k$ is called the *k-plane*, determined by the point A and by subspace Q^k. Let the vectors $e_1, e_2, \dots, e_k$ form a basis in Q^k; then these vectors are called the direction vectors of P^k. All the points of P^k are equivalent because if A and B are in P^k then the pair A and Q^k determines the same k-plane in E^n as the pair B and Q^k.

Any 1-plane is called a *line* and any $(n-1)$-plane is called a *hyperplane*.

Any k-convex body is contained in a unique k-dimensional plane, because there exists only one k-plane passing through $k+1$ given points in a general location. Therefore from Theorem 1.6 it follows that any k-convex body has interior points with respect to this k-plane.

Let L be a line in E^n, A be some point of L, and e be a direction vector of L. The sets

$$L_A^+ = \{X \in L, \overline{AX} = \lambda e, \quad \lambda \geq 0\}$$

and

$$L_A^- = \{X \in L, \overline{AX} = \lambda e, \quad \lambda \leq 0\}$$

are called the rays of the line L with vertices at the point A.

Let the points 0_1 and 0_2 be the vertices of rays L_1 and L_2. If $\overline{0_1X_1} = \alpha\overline{0_2X_2}$ and $\alpha > 0$, then we say the rays L_1 and L_2 have the *same direction* (X_1 and X_2 are points of L_1 and L_2 respectively). If $\overline{0_1X_1} = \alpha\overline{0_2X_2}$ and $\alpha < 0$, then we say the rays L_1 and L_2 have the *opposite directions*. We use the notation $L_1 \uparrow\uparrow L_2$ in the first case and the notation $L_1 \uparrow\downarrow L_2$ in the second one.

Any convex set consisting of the union of rays with a common vertex is called a *convex cone*. The common vertex of all these rays is called the *vertex* of this cone. The simplest examples of convex cones are rays, k-planes, the whole space E^n and the halfspaces in E^n.

1.6 Spherical Convex Sets

Let S^{n-1} be an $(n-1)$-dimensional sphere in E^n. Let A and B be any points of S^{n-1}. The 2-plane P^2, passing through A, B, and the center of S^{n-1}, intersects S^{n-1} in some circle S^1, whose radius is equal to the radius of S^{n-1}. Such a circle is called the great circle of the sphere S^{n-1}, passing through the points A and B. If A and B are not the ends of a diameter of S^{n-1}, then there is only one great circle S passing through these points. If A and B are the ends of a diameter of S^{n-1}, then there are infinitely many great circles passing through A and B, and each great circle is divided by the diameter into two equal arcs.

Let G be a subset of S^{n-1}. G is called *spherically convex*, if the lesser arc AB of any great circle, passing through points A and B, lies in G for every pair $A, B \in G$. In the case that A and B are the ends of a diameter, we modify this condition in the following way: at least one-half of some great circle passing through A and B belongs to G.

There is a close connection between convex cones and convex subsets of S^n. Namely, let K be a convex cone with vertex 0 in E^n and S^{n-1} be the hypersphere with the center 0. Then

$$G = K \cap S^{n-1}$$

is a spherically convex set. Conversely let G be a spherically convex set in S^{n-1}, then the set

$$K = \bigcup_{X \in G} L_X$$

is a convex cone, where L_X is the ray with the origin at 0 containing the point X.

From this relationship it follows that: 1) if K is a closed convex cone, then G is a closed convex set on S^{n-1}; 2) if the cone K is a k-convex body, then G is a $(k-1)$-dimensional closed convex domain in S^{n-1} $(1 \leq k \leq n-1)$. The converse assertions are also true.

1.7 Starshapedness of Convex Bodies

Theorem 1.7. *Let M be a convex body in E^{n+1} and X be any interior point of M. Then every ray L_X starting from X is either contained in M or crosses ∂M at a unique point.*

Proof. Every ray is a connected set. Therefore if some ray does not lie in M, then this ray crosses ∂M.

If theorem is not correct, then there must exist a point $0 \in \operatorname{int} M$ and a ray L starting from 0 such that L crosses ∂M in at least two different points A and B. We suppose for convenience that A is an interior point of the segment $0B$. Since 0 is an interior point of the set M, there exists an open ball U with center 0 and radius r such that $U \subset M$. Let V be the set of all segments BX, where X belongs to U. Since A is an interior point of the segment $0B, V$ contains some open ball with center A and radius r' which is not more than the positive number $\frac{\overline{AB}}{\overline{0B}}r$. Since $V \subset M, A$ is an interior point of M. This is a contradiction. □

The set $M \subset E^n$ is called *starshaped with respect to the point* $A \in M$ if any ray L_A starting from A crosses ∂M in not more than one point. Thus Theorem 1.7 can be stated in the following way: "Any convex body is starshaped with respect to all of its interior points."

1.8 Asymptotic Cone

The concept of an asymptotic cone is very important to the study of infinite convex bodies and hypersurfaces in E^n. This concept is fundamental in the theory of non-compact boundary value problems for elliptic solutions of Monge–Ampere equations (see Parts I and II of this book).

We denote by $K_A(M)$ the set of points lying on the rays starting from the point $A \in M$ and contained in M. We set

$$K_A(M) = A$$

if there are no such rays.

Theorem 1.8. *If M is a convex set then the set $K_A(M)$ is convex; moreover, if $K_A(M) \neq A$, then $K_A(M)$ is a convex cone.*

Proof. Let B and C be points of $K_A(M)$.

To prove the theorem it suffices to show, for $0 \leq t \leq 1$ and $\lambda \geq 0$, that $\lambda(X_t - A)$ is in M, where $X_t = (1-t)B + tC$.

Let t in [0,1] and $\lambda \geq 0$ be fixed. Since B and C are in $K_A(M)$, $\lambda(B-A)$ and $\lambda(C-A)$ are in M. Thus, since M is convex and

$$\lambda(X_t - A) = (1-t)[\lambda(B-A)] + t[\lambda(C-A)],$$

we have $\lambda(X_t - A)$ is in M. □

Theorem 1.9. *Let M be a convex closed set in E^n and L_A and L_B be rays starting from different points A, B of the set M. If $L_A \subset M$ and $L_A \upuparrows L_B$, then $L_B \subset M$.*

Proof. Let C be any point of L_A. Then segment BC is contained in M, because M is a convex set. Every point $X \in L_B$ is the limit of the sequence of points lying on the segments BC_n, where $C_n \in L_A$. Since M is closed, $X \in M$. Therefore $L_B \subset M$. □

Corollary 1. *If M is a closed convex set in E^n, then the cone $K_M(B)$ can be obtained by parallel translation from the cone $K_M(A)$, where A and B are any points of M.*

Therefore any cone $K_X(M)$ is called the *asymptotic cone* of the closed convex set M, where X is any point of M.

Corollary 2. *If M is a closed convex set in E^n which contains a line L, then M is a cylinder formed by lines parallel to L.*

Theorem 1.10. *Let A be any point of a closed convex set M. If $K_A(M) = A$, then M is a bounded set.*

Proof. Suppose that this theorem does not hold. Then there exists a sequence of points $X_n \in M$ such that the lengths of the segments AX_n increase infinitely. Therefore we can choose a subsequence from AX_n convergent to some ray L.

Since all the segments $AX_n \subset M$ and since M is a closed set, we have $L \subset M$; i.e. $K_A(M)$ contains the ray L. This is a contradiction. □

It is clear that the convex cone $K_A(M)$ can be constructed as the topological limit of the convex sets M_k, which are the images of M by dilations g_k. Every dilation g_k has center at A and coefficient k. The limit mentioned above is taken under the conditions $k > 0$ and $k \to 0$.

1.9 Complete Convex Hypersurfaces in E^{n+1}

A *complete n-dimensional surface* is an entire component of the boundary of a convex body in E^{n+1}.

Theorem 1.11. *If a complete convex hypersurface F lies on the boundary of a convex body $M \subset E^{n+1}$, then there are only the following possibilities for the set $K_A(M)$, where A is an interior point of M:*

1. *$K_A(M) = A$ so that F is homeomorphic to an n-dimensional sphere;*
2. *$K_A(M)$ contains a line so that F is an n-dimensional cylinder;*
3. *$K_A(M)$ contains a ray, but does not contain a line. In this case F is homeomorphic to E^n.*

Proof. In the first case the convex body M is bounded. This follows from Theorem 1.10. Therefore M lies inside the sphere S^n_R with center A and radius R, where R is a sufficiently large positive number. Thus the desired homeomorphism can be established by the central projection from the point A.

In the second case Theorem 1.11 follows directly from Corollary 2 of Theorem 1.9.

We now consider the third case. *If $n = 1$, then $K_A(M)$ is either a ray or a convex angle.* Let $A \in \text{int } M$, and S^1 be a circle with center A. Then the set

$$L = S^1 \cap K_A(M)$$

is either a point or an arc of S^1, which is less than a semicircle of S^1. Thus the central projection with center A establishes the desired homeomorphism between F and the open arc $S^1 \backslash L$.

Let now $n > 1$. Let $A \in \text{int } M$. We denote by S^n the sphere with center A and radius 1. Finally let $p \subset K_A(M)$ be a ray with vertex A and q be the ray opposite to p. Let π be the central projection from the point A on the sphere S^n, and $F' = \pi(F)$. Recall that $F \subset \partial M$.

We now prove that F' is starshaped with respect to the point $B = q \cap S^n$, i.e. every great semicircle on S^n, which has initial point B, intersects $\partial F'$ only at one point. We take any point $Y' \in F'$ and consider the two-dimensional plane Q passing through points A, B, and Y'. From considerations made above it follows that $\pi(Q \cap F)$ is an arc of a great circle passing through the point B. Hence F' is starshaped with respect to the point B.

Let E^n be a tangent hyperplane of S^n at the point B and let F'' be the stereographic projection of F' onto E^n from the point $C = p \cap S^n$. Clearly F''

is starshaped with respect to B, because the set F' is starshaped. Hence F'' is homeomorphic to E^n. Since F is homeomorphic to F', F' is homeomorphic to F'', and F'' is homeomorphic to E^n, we have F is homeomorphic to E^n. □

Remark. From Theorem 1.11 it follows that in cases 1) and 3) the set ∂M has only one component. Thus ∂M is a complete convex hypersurface in these cases. In case 1) $F = \partial M$ is called an *n-dimensional convex closed hypersurface* and in case 3) $F = \partial M$ is call an *n-dimensional complete infinite convex surface*. If $n = 1$, then F is called either a *closed convex curve* or an *complete infinite convex curve*.

§2. Supporting Hyperplanes

2.1 Supporting Hyperplanes. The Separability Theorem

Let E^n be a n-dimensional Euclidean point space and let M be a set in E^n. A hyperplane α is called *supporting* to the set M if $\alpha \cap \overline{M} \neq \emptyset$ and the whole set M lies on one side of α, i.e. all points of M lie in one closed halfspace with the boundary α and $\alpha \cap \overline{M} \neq \emptyset$. If α is a supporting hyperplane of a set M, then α can not pass through interior points of M and hence

$$\alpha \cap \overline{M} \subset \partial M. \tag{2.1}$$

Theorem 2.1. *Let H be a closed convex set in E^n and A be a point of E^n. If $A \notin H$, then there exists a supporting hyperplane of H, which separates A from H.*

Proof. Since A and H are closed subsets of E^n and A is a bounded set, we have at least one point $B \in H$ such that

$$\operatorname{dist}(A, H) = |AB|,$$

where $|AB|$ is the length of the line segment AB. We denote by α the hyperplane, which passes through B and which is orthogonal to the line AB. We now prove that α is the desired supporting hyperplane. First of all $\alpha \cap H \neq \emptyset$, because $B \in \alpha \cap H$ and $\operatorname{dist}(A, \alpha) > 0$. We now prove that all points of H do not lie in the open half-space γ such that $A \in \gamma$ and $\partial\gamma = \alpha$. If our assertion is incorrect, then there exists a point $C \in H$ such that $C \in \gamma$ and C does not lie in segment AB. But this is impossible. Indeed we have two cases:

a) the angle ACB in triangle ACB is acute;
b) the angle ACB in triangle ACB is $\geq 90°$.

In case a) we consider the height $|AD|$ of triangle ACB. Clearly

$$|AD| < |AB| = \operatorname{dist}(A, H). \tag{2.2}$$

But D is an interior point of the line segment BC and $B, C \in H$. Since H is convex, we have $D \in H$. This contradicts inequality (2.2).

In case b) we have $|AC| < |AB|$ and we again have a contradiction with (2.2). Thus α is a supporting hyperplane of H. □

Corollary 1. *Let Q_α be the halfspace with boundary α which contains H. Then α separates the point A from Q_α.*

Corollary 2. *Clearly B is the unique nearest point of Q_α with respect to A, therefore B is also the unique nearest point of H with respect to A (because $H \subset Q_\alpha$).*

The point B is called the *projection* of $A \notin H$ on the closed convex set H. By Corollary 2, every point $A \notin H$ always has one and only one projection on the closed convex set H.

2.2 The Main Properties of Supporting Hyperplanes

In E^n every hyperplane α is determined by any point $A \in \alpha$ and by any unit vector ν orthogonal to α. Hence the hyperplanes α_m converge to the hyperplane α if we can choose points $A_m \in \alpha_m$ and unit normals ν_m of α_m which converge to a point $A \in \alpha$ and to an unit normal ν of α respectively. Let M be a set in E^n and let A be any point of a supporting hyperplane α of M. Then the unit normal $\nu = \overline{AB}$ of α is called *inward* (with respect to M) if the terminal point B of $\overline{AB}$ lies in the same open halfspace with the boundary α as the set M.

Let $\xi_1, \xi_2, \ldots, \xi_n$ be Cartesian coordinates in E^n and α be a supporting hyperplane of the set in E^n, passing through the point $A(\xi_1, \xi_2, \ldots, \xi_n)$ and having the inward unit normal $\nu = \overline{AB} = (p_1, p_2, \ldots, p_n)$. We denote by $X(x_1, x_2, \ldots, x_n)$ any point of α. Clearly

$$(\overline{AX}, \nu) \equiv \sum_{i=1}^{n} p_i(x_i - \xi_i) = 0 \tag{2.3}$$

is the equation of the hyperplane α, and

$$(\overline{AX}, \nu) \equiv \sum_{i=1}^{n} p_i(x_i - \xi_i) \geq 0 \tag{2.4}$$

is the equation of the closed halfspace Q_α with boundary α which contains M.

Theorem 2.2. *Let H be a bounded closed convex set in E^n and let $\alpha_1, \alpha_2, \ldots, \alpha_m, \ldots$ be any sequence of supporting hyperplanes of H convergent to a hyperplane α. Then α is also a supporting hyperplane of H.*

Proof. Choose points $X_m \in H \cap \alpha_m$. Since the sequence X_m is bounded, it has a convergent subsequence for which we use the same notation X_m. Since H is a closed set, $X_o = \lim_{m\to\infty} X_m \in H$. Now let ν_m be the unit inward normals of α_m. From the condition of Theorem 2.2 it follows that the normals ν_m converge to the unit inward normal ν_o of α. If we pass to the limit in inequalities (2.4) for supporting hyperplanes α_m, we obtain the proof that α is a supporting hyperplanes of H. □

Theorem 2.2′. *Let H be a closed, optionally unbounded convex set. Let all conditions of Theorem 2.2 hold, and in addition assume there exists a sequence of points $X_m \in H \cap \alpha_m$ which converges to a point $X_o \in H$. Then the conclusion of Theorem 2.2 holds and $X_o \in \alpha$.*

The proof of this theorem is similar to the proof of Theorem 2.2.

Theorem 2.3. *There exits at least one supporting hyperplane passing through any boundary point of any convex closed set in E^n.*

Proof. Let H be any closed convex set in E^n and X be any point of ∂H. Then there exists a sequence of the points $X_m \notin H$ such that X_m converge to X. We denote by B_m the projection of X_m on H for every m. From Theorem 2.1 it follows that $B_m \in \partial H$ and there exists a supporting hyperplane α_m of H passing through B_m and separating X_m from H. Clearly

$$\operatorname{dist}(B_m, X) \leq |B_m X_m| + |X_m X| \leq 2|X_m X|.$$

Therefore B_m converges to X. We denote by ν_m the inward unit normal to α_m and take a convergent subsequence of ν_m with limit ν_o. By Theorem 2.2′ there exists a supporting hyperplane α of H which passes through the point X whose unit inward normal is ν_o.

Theorem 2.4. *Every closed convex set H in E^n is the intersection of all halfspaces containing H, the boundaries of which are supporting hyperplanes of H.*

Proof. Let α be any supporting hyperplane of H. We denote by Q_α the halfspace of E^n containing H, whose boundary is α. Let $W = \bigcap_\alpha Q_\alpha$. Clearly

$$H \subset W. \tag{2.5}$$

We now assume that there is a point $X \in W \backslash H$. Then there exists a supporting hyperplane α_o separating X from the halfspace Q_{α_o}. Therefore X does not belong to W. But this contradicts our assumption. Thus

$$W \backslash H = \emptyset. \tag{2.6}$$

From (2.5) and (2.6) we obtain $H = W$. □

§3. Convex Hypersurfaces and Convex Functions

3.1 Convex Hypersurfaces and Convex Functions

A set F is called an *n-dimensional convex surface* (or a *convex hypersurface*) in E^{n+1} if F is a domain of the boundary of a $(n+1)$-dimensional convex body H in E^{n+1}, i.e. F is a connected and open subset of ∂H in the topology of ∂H induced by E^{n+1}. Analogously, any subdomain G of the boundary of a convex $k+1$-body H is called a *k-dimensional convex surface*. If $k = 1$, then G is called a *convex curve*.

If F is a convex hypersurface, satisfying two conditions: 1) ∂F lies in a hyperplane P in E^{n+1}, 2) F has a one-to-one orthogonal projection into P, then F is called an *n-dimensional convex cap*. Analogously a *k-dimensional convex* cap can be defined.

Let H be a convex body in E^{n+1} and $Q = \partial H$ be a complete convex hypersurface. We fix a line L and denote by U_L the set of lines parallel to L and passing through all interior points of H. From Subsections 1.7 and 1.9 it follows that the lines belonging to U_L satisfy one and only one of the following three possibilities:

1. all lines intersect Q in only one point,
2. all lines intersect Q in only two distinct points,
3. none of the lines intersect Q.

We denote by F the set of all points of intersection of ∂H with lines of the set U_L.

In the third case the set F is empty. We now consider the first case. Let L^+ be a ray of the line L, which intersects Q. We denote by U_L^+ the set of all rays, having the same direction as L^+ and starting from all interior points of H. Then F consists of all points of intersection of ∂H with the rays of the set U_L^+.

Let P be a hyperplane orthogonal to L. Then the projection of F onto P is an open convex domain G. Clearly G is the projection of the set *int* H onto the same hyperplane P. Let $x_1, \ldots, x_n, x_{n+1} = z$ be Cartesian coordinates in E^{n+1} such that the hyperplane P has the equation $z = 0$.

Clearly F is the graph of a function $z = f(x_1, \ldots, x_n)$ which is defined in G. We now prove that this function $f(x_1, \ldots, x_n)$ is continuous in G. Let X_o be any point of G and let the points $X_m \in G$ converge to X_o. We denote by Y_i $(i = 0, 1, 2, \ldots)$ the point of F which projects onto X_i. It suffices to show

$$Y_o = \lim_{m \to \infty} Y_m.$$

From Section 1.7 it follows that the sequence Y_m is bounded. Since $Y_m \in \partial H$, the limit of any convergent subsequence of the points Y_m belongs to ∂H. Moreover it must lie on a ray of the set U_L^+. Clearly Y_0 is a unique point satisfying these conditions. Thus

$$Y_o = \lim_{m \to \infty} Y_m$$

and the function $f(x_1, \ldots, x_n)$ is continuous in G. Since F is the graph of a continuous function $f(x_1, \ldots, x_n)$ defined in an open convex domain G and $F \subset \partial H$, F is a convex hypersurface.

We now consider the second case. Then F decomposes into two components. The first one consists of intersection points ∂H with the rays of the set U_L^+, and the second one consists of intersections points ∂H with the rays starting from the inner points of H and having the opposite direction from the ray L^+.

Thus every point $A \in \partial H$ has a neighborhood $U \subset \partial H$, which is projected one-to-one in a hyperplane.

Let $W(G)$ be the set of convex hypersurfaces in E^{n+1} which project orthogonally and one-to-one onto a convex open domain $G \subset P$. Let $x_1, \ldots, x_n, x_{n+1} = z$ be Cartesian coordinates in E^{n+1} and let the hyperplane have the equation $z = 0$ in these coordinates. Then any convex hypersurface $F \in W(G)$ can be determined by the equation

$$z = f(X).$$

where $X = (x_1, x_2, \ldots, x_n) \in G$. We denote by (X, z) a point $(x_1, x_2, \ldots, x_n, z) \in E^{n+1}$. Clearly X is the orthogonal projection of (X, z) onto the hyperplane P.

Let $(X, f(X))$ be any point of F. Then there exists at least one supporting hyperplane Q passing through this point. Here $X \in G$. If (X, z) is any point of Q and $X \in G$, then either

$$z \leq f(X) = f(x_1, \ldots, x_n), \tag{3.1}$$

or

$$z \geq f(X) = f(x_1, \ldots, x_n) \tag{3.2}$$

holds for all $X \in G$. In the first case the function $F(x_1, \ldots, x_n)$ is called *convex* and in the second one-*concave*.

Let X_0 and X_1 be any points of the convex domain G. Then the point

$$X_t = (1-t)X_0 + tX_1 \in G \tag{3.3}$$

for every $t \in [0,1]$. Since $(X_0, f(X_0)) \in F$ and $(X_1, f(X_1)) \in F$, these points also belong to the convex body H. Therefore the line segment L, with its ends at these points, belongs to H. The points of the segment L have the form (X_t, z_t), where X_t is defined by (3.3) and $z_t = (1-t)f(X_0) + tf(X_1)$, $t \in [0,1]$.

Thus we obtain the following conclusions:

a) If $f(X)$ is any convex function in G, then the inequality

$$\begin{aligned} f[(1-t)X_0 + tX_1] = f(X_t) \leq z_t = \\ = (1-t)f(X_0) + tf(X_1) \end{aligned} \tag{3.4}$$

holds for all $X_0, X_1 \in G$ and all $t \in [0,1]$.

b) If $f(X)$ is any concave function in G, then the inequality

$$\begin{aligned} f[(1-t)X_0 + tX_1] = f(X_t) \geq z_t = \\ = (1-t)f(X_0) + tf(X_1) \end{aligned} \tag{3.5}$$

holds for all $X_0, X_1 \in G$ and all $t \in [0,1]$.

Note that (3.4) and (3.5) can hold at the same time, if and only if $f(X)$ is a linear function of $x_1, x_2, \ldots, x_n$. Let $W^+(G)$ and $W^-(G)$ be the subsets of $W(G)$ associated with convex and concave functions defined in G. Then

$$W(G) = W^+(G) \cup W^-(G)$$

and the set $W^+(G) \cap W^-(G)$ consists only of convex domains lying in hyperplanes

$$z = a_1x_1 + a_2x_2 + \cdots + a_nx_n + b,$$

where $(x_1, x_2, \ldots, x_n) \in G$.

Conversely let G be a convex domain in the hyperplane $z = 0$ and let $f(X)$ be a continuous function satisfying either inequality (3.4) or inequality (3.5) for any $X_0, X_1 \in G$ and any $t \in [0,1]$. We assume for example that $f(X)$ satisfies inequality (3.4). Let H be the set of point $(X, z) \in E^{n+1}$ such that $X \in G$ and for every $X \in G$ the pair (X, z) satisfies the inequality

$$f(X) \leq z.$$

The set $\overline{H}$ is a closed convex set and $\partial H = F \cup M$, where M is the set projected onto $\partial G \subset P$. Therefore $F \cap M = \emptyset$. Since $\overline{H}$ contains interior points, $\overline{H}$ is a convex body. Clearly $F \subset \partial\overline{H}$. Hence $F \in W^+(G) \subset W(G)$.

Thus the mutual relations between convex and concave functions and convex hypersurfaces are described.

3.2 Test of Convexity of Smooth Functions

This test is useful for investigations of smooth convex functions.

Let G be a convex domain in E^n. We assume that Cartesian coordinates $x_1, x_2, \ldots, x_n$ are introduced in E^n.

Theorem 3.1. *A function $f(x_1, \ldots, x_n) \in C^k(G)$, $k \geq 2$, is convex (concave) if and only if the quadratic form*

$$A_f(X) = \sum_{i,k=1}^{n} \frac{\partial^2 f(x_1, \ldots, x_n)}{\partial x_i \partial x_k} \xi_i \xi_k$$

is non-negative (non-positive) for all $X = (x_1, x_2, \ldots, x_n) \in G$, where $\xi = (\xi_1, \xi_2, \ldots, \xi_n)$ is any unit vector from V^n (see Section 1.4).

Proof. The statement of convexity (concavity) of the function f is equivalent to convexity (concavity) of the restriction of this function $f|_\ell$ where ℓ is any segment contained in G. Since $f \in C^k(G)$, we have the statement of convexity of $f|_\ell$ is equivalent to the non-negativeness of the second derivative of $f|_\ell$ with respect to the length of an arc for any segment $\ell \subset G$. (Analogous for concave functions $f|_\ell$.)

Let f be a convex function. Take any point $X = (x_1, \ldots, x_n) \in G$ and any unit vector $\xi = (\xi_1, \xi_2, \ldots, \xi_n) \in V^n$. Let ℓ be the segment with the origin X and direction vector ξ. Then

$$Y = s(\xi_1 e_1 + \cdots + \xi_n e_n) + X,$$

where Y is an arbitrary point of ℓ, the vectors $e_1, \ldots, e_n$ are the basis corresponding to Cartesian coordinates $x_1, \ldots, x_n$ and s is the length of the segment XY. Also

$$0 \leq \left.\frac{d^2 f|_\ell}{ds^2}\right|_{s=0} = A_f(X) = \sum_{i,k=1}^{n} \frac{\partial^2 f(x_1, \ldots, x_n)}{\partial x_i \partial x_k} \xi_i \xi_k. \tag{3.6}$$

Since X is any point in G and ξ is any unit vector in V^n, the quadratic form $A_f(X)$ is nonnegative in G.

Conversely if the quadratic form $A_f(X)$ is nonnegative for all $X \in G$, then the function $f(X)$ is convex. This follows directly from (3.6). □

3.3 Convergence of Convex Functions

Let G be a bounded convex domain in E^n. A sequence of convex functions $f_n(X)$ defined in G is said to *converge* to a convex function $f(X)$, also defined in G, if

$$\lim_{n \to \infty} f_n(X) = f(X) \tag{3.7}$$

for each point $X \in G$. The pointwise convergence (3.7) is not necessarily uniform in the whole domain G. However we prove in this section that pointwise convergence (3.7) implies uniform convergence of convex functions $f_n(X)$ in any subset G' of the domain G which is a positive distance from ∂G.

The *slope* of a hyperplane α is defined as the absolute value of the tangent of the angle in the space E^{n+1} between the hyperplane α and the hyperplane E^n: $z = 0$.

Lemma 3.1. *Let $f(X)$ be a convex function defined in a convex bounded domain G and let F be the graph of $f(X)$ in E^{n+1} (clearly F is a convex hypersurface in E^{n+1}). If G' is any subset of G with a distance from ∂G not less than $2\delta > 0$, then the slopes of supporting hyperplanes of F on G' are uniformly bounded by the number M_1/δ, where*

$$M_1 = \sup_{G'} |f(X) - f(Y)|.$$

Proof. Let X_0 be any point of the set G'. Then $\operatorname{dist}(X_0, \partial G) \geq 2\delta$. We consider an arbitrary two-dimensional plane α passing through the points $(X_0, 0)$ and $A_0(X_0, f(X_0))$. Then α intersects F in a curve γ. Let γ_α be the arc of γ lying over the segment

$$\ell_\alpha = \alpha \cap \overline{U}(X_0, \delta),$$

where $U(X_0, \delta)$ is the n-ball with center X_0 and radius δ. Clearly the slope of any supporting line to γ_α at the point A_0 is not more than the maximum of the slopes of the segments A_0B_α and A_0C_α, where B_α and C_α are the ends of γ_α. The previous two slopes are estimated from above by M_1/δ. Since α is any hyperplane at the arbitrary point $X_0 \in G'$, our lemma is proved. □

Now let $f_n(X)$ be a sequence of convex functions which converges to a convex function $f(X)$ in G. Then $f_n(X)$ is uniformly bounded on every set $G' \subset G$ if $\operatorname{dist}(G', \partial G) = \delta > 0$. Therefore $f_n(X)$ converges uniformly in G'. We can use the same technique to prove the following:

Theorem 3.2. *Let a sequence of convex functions $f_n(X)$ be uniformly bounded in G. Then it is possible to take some subsequence $f_{n_k}(X)$ convergent to some function $f(X)$ defined in G.*

We shall also consider the convergence of convex functions $f_n(X)$ when every function $f_n(X)$ is defined in a domain G_n depending on the number n. Assume that G_n $(n = 1, 2, \ldots)$ satisfy the following conditions:

a) every G_n is an open convex domain;
b) $G_1 \subset G_2 \subset \cdots \subset G_n \subset \cdots$
c) the set

$$G = \bigcup_{n=1}^{\infty} G_n$$

is a bounded convex open domain in E^n.

Since any point $X \in G$ is in G_n for sufficiently large n, the convergence of $f_n(X) \in W(G_n)$ to the function $f(X) \in W(G)$ is defined by the equality

$$\lim_{n \to \infty} f_n(X) = f(X).$$

Theorem 3.3. *If a sequence of convex functions $f_n(X) \in W^+(G_n)$ is uniformly bounded and the convex domains G_n satisfy conditions a), b), c), then there exists a subsequence of $f_n(X)$ convergent to a convex function $f(X) \in W^+(G)$.*

3.4 Convergence in Topological Spaces

Let X be a topological space and $a_1, a_2, \ldots, a_n, \ldots$ be a sequence of points in X. A point $a \in X$ is called *a limit of the sequence* $a_1, a_2, \ldots, a_n, \ldots$ if, for each neighborhood U of a, there exists the number n_U such that $a_n \in U$ for every $n > n_U$.

If X is a Hausdorff topological space with a countable basis, then a point a is a limit point of a set $H \subset X$ if and only if there exists some sequence of different points $a_n \in H$ which converges to a. Therefore the subset H of X is closed if and only if the limits of all convergent sequences $\{a_n\} \subset H$ also belong to H.

Now we consider a sequence of subsets F_n in the topological space X. The set consisting of all limits of all convergent sequences $a_{p_1}, a_{p_2}, \ldots, a_{p_q}, \ldots$, where $a_{p_q} \in F_{p_q}$ and

$$p_1 < p_2 < \cdots < p_q < \cdots$$

is called *the superior (upper) topological limit of the sets* F_n and is denoted by $\overline{\lim}_{n \to \infty} F_n$. The set consisting of all limits of all convergent sequences $a_1, a_2, \ldots,$

$a_n, \ldots$, where $a_n \in F_n$, is called *the inferior (lower) topological limit of the sets* F_n and is denoted by $\underline{\lim}_{n\to\infty} F_n$. Clearly

$$\underline{\lim}_{n\to\infty} F_n \subset \overline{\lim}_{n\to\infty} F_n.$$

If

$$\underline{\lim}_{n\to\infty} F_n = \overline{\lim}_{n\to\infty} F_n$$

then we say that the sequence F_n has *a topological limit* and we write $\lim_{n\to\infty} F_n$.

Theorem 3.4. *The sets* $\underline{\lim}_{n\to\infty} F_n$ *and* $\overline{\lim}_{n\to\infty} F_n$ *are closed in any Hausdorff space with a countable basis.*

We leave the proof of this theorem as an exercise for the reader.

3.5 Convergence of Convex Bodies and Convex Hypersurfaces

A sequence of convex bodies H_m is called *convergent* if the topological limit of H_m exsits. Since E^{m+1} is a Hausdorff space, $\lim H_m$ is a closed convex set in E^{n+1} for every convergent sequence of convex bodies H_m. But $\lim H_m$ does not always contain interior points in the topology of E^{n+1}. Therefore $\lim H_m$ is not always a convex body in E^{n+1}. The convergence of complete convex hypersurfaces is defined analogously. Since complete convex hypersurfaces are boundaries of convex bodies or components of them, the statements of convergent convex bodies and convex complete hypersurfaces are equivalent.

Theorem 3.5. *(Blaschke) Let* $Q = \{F_\alpha\}$ *be a family of convex bodies and let* U_R *and* U_r *be two closed* $(n+1)$*-balls such that*

$$U_r \subset F_\alpha \subset U_R$$

for all convex bodies F_α *where* r *and* R *are the radii of these balls. Then we can take a convergent sequence* $F_{\alpha_1}, F_{\alpha_2}, \ldots, F_{\alpha_k}, \ldots$ *from the family* Q *such that*

$$F_0 = \lim_{k\to\infty} F_{\alpha_k}$$

is a convex body.

Proof. Suppose that the distances from all convex bodies F_α to ∂U_R are more than $\delta > 0$. If S_α is the boundary of F_α then S_α is a closed convex hypersurface. Let X be any point of S_α and P be some supporting hyperplane to S_α passing through the point X. Let $L_{X,P}$ be the ray with the vertex X orthogonal to the hyperplane P and which lies in the halfspace Q_P such that $Q_P = P$ and $Q_P \cap F_\alpha = \emptyset$. We denote by $Y(X,P)$ the point of intersection of the ray $L_{X,P}$ and the n-sphere $\sigma_R = \partial U_R$. Then every point $X \in S_\alpha$ will be associated with the set $\{Y(X,P)\}$ of points $Y(X,P)$, where P is any supporting hyperplane of F_α passing through X. Clearly the set $Y(X,P)$ lies on the sphere σ_R and the sets $\{Y(X,P)\}$ do not have common points for different X. It is also evident that every point z of the σ_R can belong to one and only one of the sets $\{Y(X,P)\}$. Therefore the mapping

$$g_\alpha\colon \sigma_R \to U_R$$

is defined for every convex body F_α. This mapping transforms any point $Y_0 \in \{Y(X_0,P)\}$ into $X_0 \in S_\alpha$.

Since $g_\alpha(\sigma_R) = S_\alpha$, the family of mapping g_α is uniformly bounded. From Lemma 3.2 (see the text above) it follows that all of the mappings g_a satisfy the inequality

$$\operatorname{dist}(g_\alpha(Y'), g_\alpha(Y'')) \leq \operatorname{dist}(Y', Y'') \tag{3.8}$$

for every $Y', Y'' \in \sigma_R$, if the angle between vectors $\overline{0Y'}$ and $\overline{0Y''}$ is less than $\frac{\pi}{4}$ (0 is the center of σ_R).

Therefore we can choose some sequence of mappings $g_{\alpha_1}, g_{\alpha_2}, \dots, g_{\alpha_k}, \dots$ uniformly converging to the continuous mapping $h\colon \sigma_R \to U_R$ and satisfying inequality (3.8). Let $\beta_h(Y)$ be the hyperplane passing through the point $h(Y)$ orthogonal to the segment with ends $h(Y)$ and Y. We denote by $Q_h(Y)$ the closed halfspace bounded by the hyperplane $\beta_h(Y)$, which does not contain Y. Let

$$H = \bigcap_{Y \in \sigma_R} Q_h(Y). \tag{3.9}$$

We shall prove that H is a convex body and

$$H = \lim_{k \to \infty} F_{\alpha_k}. \tag{3.10}$$

Analogously, if $\beta_{g_\alpha}(Y)$ is the hyperplane passing through the point $g_\alpha(Y)$ orthogonal to the segment with ends $g_\alpha(Y)$ and Y, and $Q_{g_\alpha}(Y)$ is the closed halfspace bounded by the hyperplane $\beta_{g_\alpha}(Y)$ and containing F_α, then

$$F_\alpha = \bigcap_{Y \in \sigma_R} Q_{g_\alpha}(Y). \tag{3.11}$$

Since the mappings $g_{\alpha_k}(Y)$ converge to $h(Y)$ uniformly on σ_R, from (3.9) and (3.11) it follows that the equality (3.10) is valid. Since the $(n+1)$-ball U_r is contained in all of the convex bodies F_α, we have

$$U_r \subset H.$$

Thus H is a $(n+1)$-convex body. □

Lemma 3.2. *Let P_1 and P_2 be two different half-hyperplanes with common boundary V (V is an $(n-1)$-plane), and let the union $P = P_1 \cup P_2$ not form a hyperplane. Then*

$$\mathit{dist}(A', B') \leq \mathit{dist}(A, B), \tag{3.12}$$

where the points A and B do not lie in the open convex set K bounded by P and A', B' are the projections of A and B onto $\overline{K}$.

Proof. Suppose that $A' \in P_1 - V, B' \in P_2 - V$. (Note that the proof of this lemma is trivial in other cases.) Denote by V' the $(n-1)$-plane passing through A' parallel to V. The two-dimensional plane Q passing through B and orthogonal to V and V' contains B', intersects V' at some point C', and contains the normal of the hyperplane P_1 at the point C'. Let C be the point lying outside K and belonging to the normal of P_1 at C' and let C satisfy the condition

$$\operatorname{dist}(C, C') = \operatorname{dist}(A, A').$$

Then

$$\begin{aligned}\operatorname{dist}^2(A, B) =&\operatorname{dist}^2(A, C) + \operatorname{dist}^2(C, B) = \\ &\operatorname{dist}^2(A', C') + \operatorname{dist}^2(C, B).\end{aligned}$$

Let L and L' be the lines parallel to line $B'C'$ and passing correspondingly through the points B and B'. Both cases are similar. We consider the first one and denote by D the point of intersection L and BB'. Since A and B lie outside K, the projection of B onto the line CC' must lie on the ray DC' with the vertex D. Therefore the angle BDC' is not obtuse and

$$\operatorname{dist}(B, C) \geq \operatorname{dist}(B', C').$$

Hence

$$\operatorname{dist}^2(B, C) \geq \operatorname{dist}^2(B, D) \geq \operatorname{dist}^2(B', C).$$

Thus

$$\operatorname{dist}^2(A, B) \geq \operatorname{dist}^2(A', C') + \operatorname{dist}^2(C', B') \geq \operatorname{dist}^2(A', C').$$

□

§4. Convex Polyhedra

Many fundamental facts in the theory of convex bodies, including various global existence theorems and various important inequalities and estimates, were established for convex polyhedra first and extended to general convex bodies by approximation with convex polyhedra. In this section we consider constructions of convex polyhedra by means of convex hulls of finite systems of points (closed convex polyhedra) and of finite systems of points and convex cones (infinite convex polyhedra). Finally we consider the approximation of convex bodies by convex polyhedra.

4.1 Definitions. Description of Convex Polyhedra by the Convex Hull of Their Vertices

A convex body F in E^{n+1} is called an $(n+1)$-*convex solid polyhedron*, if F is the intersection of a finite number of closed halfspaces. The boundary of the convex solid polyhedron is called a *complete convex polyhedron*, if the boundary consists of only one component. From Section 1.9 it follows that if F has more than one component, then there are exactly two components and they are hyperplanes.

If the solid polyhedron F is a bounded set in E^{n+1}, then its boundary ∂F is called a *closed convex polyhedron*. If F is a solid infinite convex polyhedron, then ∂F is called an *infinite convex polyhedron*. We will also call convex polyhedra in E^{n+1} n-convex polyhedra. Ever convex polyhedron S can be decomposed into the finite union of n-solid convex polyhedra lying in some hyperplanes. These hyperplanes are the boundaries of the halfspaces Q_i such that

$$S = \partial\left(\bigcap_j Q_j\right).$$

The above mentioned n-solid convex polyhedra are called *n-faces* of the convex polyhedron S. Applying this process of decreasing dimensions of the faces of S we finally obtain the zero-faces of S which are called the *vertices* of S or the *vertices* of the corresponding $(n+1)$-solid convex polyhedron F such that $S = \partial F$.

If S is a closed convex polyhedron, then the vertices of S define it completely.

Theorem 4.1. *Every solid bounded convex polyhedron is the convex hull of its vertices. Moreover such a polyhedron with prescribed vertices is unique.*

Proof. Every 1-dimensional solid bounded convex polyhedron is a straight line segment. Therefore the theorem obviously holds for all such 1-dimensional polyhedra. Now suppose that the theorem is true for any k-dimensional solid bounded convex polyhedron, where $k \leq n$. Then the theorem will be proved if it is established for the case $k = n+1$.

Clearly the convex hull of vertices for any solid bounded convex polyhedron is contained in this polyhedron. Therefore it is sufficient to prove the converse: that any $(n+1)$-dimensional solid bounded convex polyhedron is contained in the convex hull of its vertices. Let X be any point of such an $(n+1)$-dimensional polyhedron F. If $X \in \partial F$, then X belongs to the convex hull of its vertices, because induction can be used. Now let X be an interior point of F. Then from the boundedness and starshapedness of F it follows that any straight line passing through X intersects ∂F in two points A and B. Since X belongs to the segment AB and this segment AB is contained in the convex hull of the vertices F by induction, we have X belongs to the above mentioned convex hull. □

The next theorem is valid for infinite convex polyhedra.

Theorem 4.2. *Every solid infinite convex polyhedron is the convex hull of its vertices and its asymptotic convex polyhedral angle, which is placed at one of its vertices.*

The proof of this theorem is based on the same idea as the proof of Theorem 4.1. In addition, it makes use of the properties of asymptotic cones (see Section 1.8). Note that the asymptotic cone of any infinite solid convex polyhedron is a convex solid polyhedral angle. We call it an asymptotic angle (of an infinite convex polyhedron). Thus any infinite solid convex polyhedron is defined completely by its vertices and the asymptotic angle. The same statement is true for the boundary of this polyhedron.

4.2 Convex Hull of a Finite System of Points

Theorem 4.3. *Let M be a finite system of points $A_1, A_2, \dots, A_m$ in E^{n+1}. If there are $n+2$ points in a general location in M, then $P = CoM$ is a bounded solid $(n+1)$-convex polyhedron, and the vertices of P can only be the points $A_1, A_2, \dots, A_m$. The point A_i $(i = 1, 2, \dots, m)$ is a vertex of P if and only if A_i does not belong to the convex hull of all other points of M.*

Proof. According to the conditions of this theorem $m \geq n+2$, and there are $n+2$ points situated in a general location belonging to the set M. Without loss of generality, we denote these $n+2$ points by $A_1, A_2, \dots, A_{n+2}$.

Let Q be the convex hull of the points $A_1, A_2, \dots, A_{n+2}$. Then Q is the $(n+1)$-simplex in E^{n+1} with vertices $A_1, A_2, \dots, A_{n+1}$. We know (see Section 2.2) that Q is a bounded convex body in E^{n+1}.

Now we assume that the theorem is correct for $k-1$ points $A_1, A_2, \dots, A_{k-1}$, where $k-1 \geq n+2$, and prove it for k points $A_1, A_2, \dots, A_{k-1}, A_k$.

We denote by V the convex hull of the points $A_1, A_2, \dots, A_{k-1}$. From induction V is a bounded solid $(n+1)$-convex polyhedron. The vertices of V can be only the points $A_1, A_2, \dots, A_{k-1}$. We denote by $B_1, B_2, \dots, B_s$ the faces of V. If $A_k \in V$, then the convex hull of points $A_1, A_2, \dots, A_{k-1}, A_k$ coincides with V and A_k is not a vertex of V.

Therefore assume that A_k does not belong to V. Let W be the convex hull of the points $A_1, A_2, \dots, A_{k-1}, A_k$. Then

$$W = \bigcup_{\alpha=1}^{s} \pi_\alpha \tag{4.1}$$

where π_α is the pyramid with the vertex A and the base B_α. Actually

$$\bigcup_{\alpha=1}^{s} \pi_\alpha \subset W \tag{4.2}$$

because $V \subset W$ and every segment $A_m Z$ is contained in W, where Z is any point of V. The set $\bigcup_{\alpha=1}^{s} \pi_\alpha$ is therefore a solid $(n+1)$-polyhedron. Let B and C be arbitrary points of $\bigcup_{\alpha=1}^{s} \pi_\alpha$. Then B and C lie in segments $A_k B_o$ and $A_k C_o$ respectively, connecting A_k with points B_0 and C_0 belonging to the convex polyhedron V. Therefore the segment B_0C_0 is contained in V. Thus the triangle $A_kB_0C_0$ is contained in W together with the segment BC. Therefore $\bigcup_{\alpha=1}^{s} \pi_\alpha$ is a convex polyhedron containing the points $A_1, A_2, \dots, A_k$. Hence

$$W \subset \bigcup_{\alpha=1}^{s} \pi_\alpha. \tag{4.3}$$

From (4.2) and (4.3) it follows

$$W = \bigcup_{\alpha=1}^{s} \pi_\alpha.$$

Finally the point A_m does not belong to W. Therefore A_m can be separated by some hyperplane from the bounded convex polyhedron V (see Theorem 2.1, Section 2.1). Hence A_m can not lie inside W. □

The proof of the next theorem follows directly from Theorems 4.1 and 4.3.

Theorem 4.4. *Let $A_1, A_2, \dots, A_m$ be the given system of points in E^{n+1}. Then there exists a closed n-convex polyhedron in E^{n+1} with vertices $A_1, \dots, A_m$ if and only if the points $A_1, A_2, \dots, A_m$ are in a general location and any point A_k does not belong to the convex hull of other points $A_1, A_2, \dots, A_{k-1}, A_{k+1}, \dots, A_m$ $(k = 1, 2, \dots, m)$. Moreover, such a polyhedron with the given vertices is unique.*

The next theorem describes the structure of the convex hull for any finite system of points in E^{n+1}.

Theorem 4.5. *Let M be a finite system of points in E^{n+1}, satisfying the following conditions:*

a) there are $k+1$ points in a general location in M;
b) there are not $k+2$ points in a general location in M where $k = 0, 1, 2, \dots, n+1$.

Then $P = CoM$ is a bounded solid k-convex polyhedron P and the vertices of P can only be points of the set M. A point $A \in M$ is a vertex of P if and only if A does not belong to the convex hull of all other points of the set M.

I leave the proof of this theorem as a useful exercise. I also leave to the reader the proof of the following:

Theorem 4.6. *Let $A_1, A_2, \dots, A_m$ be a finite system of points, and V be an $(n+1)$-convex solid polyhedral angle in E^{n+1}, which has its vertex in one of the points $A_1, A_2, \dots, A_m$. Let*

$$M = A_1 \cup A_2 \cup \dots \cup A_m \cup V.$$

Then

$$P = CoM$$

is an $(n+1)$-convex infinite solid polyhedron in E^{n+1}, V is the asymptotic angle of P, and the vertices of P can only be the points $A_1, A_2, \ldots, A_m$. Moreover, the point A_k $(k = 1, 2, \ldots, m)$ is a vertex of P if and only if A_k does not belong to

$$Co(A_1 \cup \cdots \cup A_{k-1} \cup A_{k+1} \cup \cdots \cup A_m \cup V).$$

Note that the $(n+1)$-convex polyhedron P considered in Theorem 4.6 can be an $(n+1)$-convex polyhedral angle.

The last theorem has a natural generalization if V is any k-convex solid polyhedral angle in E^{n+1}. The proof of this generalization will be left as a useful exercise.

4.3 Approximation of Closed Convex Hypersurfaces by Closed Convex Polyhedra

Let S be any closed convex hypersurface in E^{n+1}. Denote by F the bounded solid convex body such that $S = \partial F$. Remember that closed n-convex polyhedra are considered as closed convex hypersurfaces in E^{n+1}. The closed n-convex polyhedron P is said to be inscribed in the closed hypersurface S if all its vertices belong to S.

Theorem 4.7. *There exists a sequence of closed n-convex polyhedra inscribed in any convex hypersurface S which converges to S.*

Proof. Let ε be a given positive number. Divide E^{n+1} into cubes with sides $\frac{\varepsilon}{\sqrt{n+1}}$ and take all of them having common points with S. Choose the point belonging to every such cube and denote these points by $A_1, A_2, \ldots, A_m$. If $A_1, A_2, \ldots, A_m$ lie in one hyperplane, then we can add other points of S so that these points and $A_1, \ldots, A_m$ do not lie together in any hyperplane. Therefore we can suppose that $A_1, A_2, \ldots, A_m$ do not lie in any hyperplane.

The cube with the side $\frac{\varepsilon}{\sqrt{n+1}}$, containing the point A_i is contained in the ball with the center A_i and the radius ε. Since the selected cubes cover the hypersurface S, the union of all the balls of radius ε and centers $A_1, A_2, \ldots, A_m$ cover S too; i.e. $A_1, A_2, \ldots, A_m$ form an ε-net of S.

Let $P = \partial Co(A_1 \cup A_2 \cup \cdots \cup A_m)$. From Theorem 4.3 it follows that P is a closed n-convex polyhedron inscribed in S and the vertices of P can only be the points $A_1, A_2, \ldots, A_m$.

Now take a sequence of positive numbers $\varepsilon_1, \varepsilon_2, \ldots, \varepsilon_k, \ldots$ convergent to zero and construct the points $A_1^k, \ldots, A_{m_k}^k$ for every ε_k. Let $P_k = \partial Co(A_1^k \cup \cdots \cup A_{m_k}^k)$. Our theorem will be proved if we establish that

$$\lim_{k \to \infty} P_k = S.$$

Let Z be any point of the hypersurface S. Then there exists a point $A_{s_k}^k$ such that $\mathrm{dist}(Z, A_{s_k}^k) < \varepsilon_k$ for every $k = 1, 2, 3, \ldots$. Since $\varepsilon_k \to 0$, the points $A_{s_k}^k$

converge to Z. Therefore each point of the hypersurface S is the limit of points belonging to the polyhedrons P_k, because $A^k_{s_k} \in P_k$.

Let Z_j be the sequence of points belonging to different polyhedrons P_j and let Z be the limit of this sequence. We denote by F the convex body in E^{n+1} such that $S = \partial F$.

Since

$$Z_j \in P_j = \partial(Co(A^j_{j_1} \cup A^j_{j_2} \cup \dots \cup A^j_{j_m})) \subset \subset Co(A^j_{j_1} \cup A^j_{j_2} \cup \dots \cup A^j_{j_m}) \subset F$$

the point $Z \in F$. Let Q_j be the supporting planes to the polyhedra P_j, passing through the points Z_j. We can choose the subsequence Q_j such that the unit outward normals to Q_{j_s} converge. Therefore the hyperplanes Q_{j_s} converge to some hyperplane Q passing through the point Z. From Section 3.3 it follows that Q is a supporting hyperplane to the set of all limits of the sequence of points lying on the polyhedrons P_{j_s}. But we proved that every point of S is such a limit. Therefore S lies on one side of the hyperplane Q. Since $Z \in F \cup Q$, we have $Z \in \partial F = S$. Thus $S = \lim_{k\to\infty} P_k$. □

By the same method the following theorem can be proved.

Theorem 4.8. *Every closed k-convex surface S is the limit of closed k-convex polyhedra inscribed in S, where* $k = 1, 2, \dots, n$.

§5. Integral Gaussian Curvature

5.1 Spherical Mapping and the Integral Gaussian Curvature

We denote by S^n the unit hypersphere in E^{n+1} with center 0. Let F be a convex hypersurface in E^{n+1} and M be a subset of F. We consider all supporting hyperplanes of F, which touch F at points of the set M. The terminal points of the unit exterior normals of these hyperplanes fill out some set of S^n, which we denote by $\psi_F(M)$. The set $\psi_F(M)$ is called the *spherical image* of M. The symbol ψ_F also denotes the mapping transforming each subset M of F onto the set $\psi_F(M)$. The mapping ψ_F is called the *spherical mapping* of F. This concept together with the concept of the *integral Gaussian curvature* (the area of $\psi_F(M)$) for general convex surfaces were introduced and studied by Alexandrov [3], see also Alexandrov [1] and Busemann [1].

Further, we will mainly consider closed convex hypersurfaces in E^{n+1}, because any nonclosed bounded convex hypersurface can be completed to the closed one. Infinite convex hypersurfaces will be considered in Subsection 5.3.

Lemma 5.1. *Let M be a closed set of a convex hypersurface F. Then the spherical image of M is a closed subset of* S^n.

Proof. Let a sequence of points $Y_k \in \psi_F(M)$ converges to the point $Y_0 \in S^n$. Let X_k be the point of F satisfying the condition: there exits a supporting hyperplane α_k of F, passing through X_k with unit exterior normal $\overline{0Y}_k$. Since

M is a closed subset of F, we can choose a convergent subsequence X_{k_j} from the sequence of the points X_k. Let X be the limit of X_{k_j}. From Theorem 2.3 it follows that there exists a supporting hyperplane α_0, passing through X with exterior unit normal $\overline{0Y}_0$. Hence $\psi_F(M)$ is a closed subset of S^n.

□

A supporting hyperplane α of a convex hypersurface F is called *singular* if there are at least two different points X and Y of F lying in α. If two points $X, Y \in F \cap \alpha$, then the whole segment XY belongs to F.

Let F be a convex hypersurface in E^{n+1}. All singular supporting hyperplanes of F define some subset N of S^n by the spherical mapping ψ_F.

Lemma 5.2. *The set N has measure zero on S^n.*

Proof. Let $L_1, L_2, \dots, L_{n+1}$ be an orthonormal frame with the origin 0 in E^{n+1}. We fixed this frame and decompose N into $n+1$ classes $N = N_1 \cup N_2 \cup \cdots \cup N_{n+1}$ such that for $i \leq n$ the set N_i consists of the terminal points of unit normals of all singular supporting hyperplanes α for which F contains a line segment q_α such that q_α is not perpendicular to L_i. The set N_{n+1} consists of the terminal points of unit normals to all singular supporting hyperplanes, whose intersections $F \cap \alpha$ don't contain any segment which is not perpendicular at least to one of axes $L_1, L_2, \dots, L_n$. Clearly the respective singular supporting hyperplane contains only one line segment parallel to L_{n+1}.

It is clear that N_{n+1} lies in one great $(n-1)$-dimensional geodesic sphere of S^n. Hence the measure of N_{n+1} is equal to zero. Using a symmetry consideration, it remains only to prove that N_1 has measure zero. First we prove that N_1 is measurable. Let $N_1^{k,m}$ be the set of terminal points of unit normals to all singular supporting hyperplanes which have common segments with F of length not less than $1/k$ and

$$\left|\frac{\pi}{2} - \gamma\right| \geq \frac{\pi}{m}$$

where γ is the angle between this segment and vector L_1.

It is clear that

$$N_1 = \bigcup_{k,m=1}^{\infty} N_1^{k,m}.$$

Since F is a closed convex hypersurface, from Lemma 5.1 it follows from the usual considerations that every set $N_1^{k,m}$ is closed. Then N_1 is a measurable subset of S^n as a countable union of closed sets.

We shall prove that N_1 has measure zero by induction on dimension. It is obviously true when $n = 1$. Now let L be any unit vector perpendicular to L_1. Project F along L to a hyperplane H. Then the image of F is a convex body G in H. Furthermore every supporting hyperplane in L_1 which is parallel to L projects onto a supporting hyperplane of G whose intersection with G is a non-trivial line segment.

Let C be the great $(n-1)$-dimensional geodesic sphere of S^n which is perpendicular to L. The observation considered above and the induction hypothesis then give the result that the $(n-1)$-dimensional Hausdorff measure of $C \cap L_1$ is zero. Letting L vary in a great circle which is perpendicular to L_1 we conclude by Fubini's theorem that L_1 has measure zero. Lemma 5.2 is proved. □

Let F be a closed convex hypersurface in E^{n+1}. Denote by $\sigma_F(M)$ the area (measure) of the spherical image $\psi_F(M)$ for every subset M of F and call it the *integral Gaussian curvature* of M.

Theorem 5.1. *For any convex hypersurface F the integral Gaussian curvature is a completely additive set function on the ring of Borel subsets of F.*

Proof. From Lemma 5.1 it follows that the spherical image $\psi_F(M)$ is measurable for any closed subset M of F.

Now let $\psi_F(M)$ be measurable for some set $M \subset F$. Then

$$\psi_F(F\backslash M) = (S^n\backslash\psi_F(M)) \cup (\psi_F(M) \cap \psi_F(F\backslash M)), \tag{5.1}$$

since $\psi_F(F) = S^n$ for every closed hypersurface. If

$$N \in \psi_F(M) \cap \psi_F(F\backslash M)$$

then the supporting hyperplane α with exterior unit normal $\overline{0N}$ contains at least one point belonging to $F\backslash M$. Hence α is a singular supporting hyperplane of F. Therefore

$$\operatorname{meas}(\psi_F(M) \cap \psi_F(F\backslash M)) = 0.$$

Hence $\psi_F(F\backslash M)$ is measurable, since $\psi_F(M)$ is measurable. From (5.1) it follows that

$$\sigma_F(F\backslash M) = \sigma_n - \sigma_F(M), \tag{5.2}$$

where σ_n is the area of S^n.

Now the complement of any closed set is an open set. Hence every open set $G \subset F$ has a measurable spherical image and

$$\sigma_F(G) = \sigma_n - \sigma_F(F\backslash G). \tag{5.3}$$

Now let the spherical images of sets $M_1, M_2, \ldots$ be measurable. It is clear that

$$\psi_F\left(\bigcup_{n=1}^{\infty} M_n\right) = \bigcup_{n=1}^{\infty} \psi_F(M_n).$$

Hence the set $\psi_F\left(\bigcup_{n=1}^{\infty} M_n\right)$ is measurable as a countable union of measurable sets, and the measurability of the spherical images for all Borel subsets of F

will be established if we proof the measurability of the set $\psi_F\left(\bigcap_{i=1}^{\infty} M_i\right)$ for any sets $M_1, M_2, \ldots$ which have measurable spherical images.

Note the obvious equality

$$\bigcap_{i=1}^{\infty} M_i = F\backslash \bigcup_{i=1}^{\infty}(F\backslash M_i) \tag{5.4}$$

where F is a closed convex hypersurface containing $M_1, M_2, \ldots$. We proved above that if the sets $F\backslash M_i$ have measurable spherical images, then the sets

$$\bigcup_{i=1}^{\infty}(F\backslash M_i) \quad \text{and} \quad F\backslash \bigcup_{i=1}^{\infty}(F\backslash M_i)$$

also have measurable spherical images (see the previous part of the proof of this theorem).

Thus all Borel subsets M of F have the definite integral Gaussian curvature $\sigma_F(M)$.

Now we shall prove that $\sigma_F(M)$ is a completely additive set function. Let M_1, M_2 be non-intersecting Borel subsets of F. Then the set $\psi_F(M_1) \cap \psi_F(M_2)$ consists of the terminal points of unit normals to singular supporting hyperplanes of F. From Lemma 5.2 it follows that

$$\text{meas}[(\psi_F(M_1) \cap \psi_F(M_2)] = 0.$$

Hence

$$\sigma_F(M_1 \cup M_2) = \sigma_F(M_1) + \sigma_F(M_2).$$

The sequence of sets $M_1, M_2, \ldots$ is called vanishing if

$$M_1 \supset M_2 \supset \cdots \supset M_k \supset \cdots$$

and

$$\bigcap_{k=1}^{\infty} M_k = \emptyset.$$

The finite additive set function $f(M)$ is called *continuous* if

$$\lim_{k\to\infty} f(M_k) = 0$$

for every vanishing sequence of sets $M_1, M_2, \ldots, M_k, \ldots$. This definition was introduced by Frechet. From measure theory it is well known that the finite additive set function $f(M)$ is completely additive if and only if $f(M)$ is continuous.

Thus let $M_1, M_2, \ldots, M_k, \ldots$ be a vanishing sequence of sets. Then

$$\psi_F(M_1) \supset \psi_F(M_2) \supset \cdots \supset \psi_F(M_k) \supset \cdots .$$

If

$$\bigcap_{k=1}^{\infty} \psi_F(M_k) = \emptyset$$

then from the properties of the Lebesgue measure of S^n, it follows that

$$\lim_{k\to\infty} \sigma_F(M_k) = 0.$$

Let

$$\bigcap_{k=1}^{\infty} \psi_F(M_k) \neq \emptyset.$$

Then

$$N \in \bigcap_{k=1}^{\infty} \psi_F(M_k)$$

is the terminal point of a unit outward normal, corresponding to some point X_k of each set M_k. All the points X_k can not coincide, because

$$\bigcap_{k=1}^{\infty} M_k = \emptyset.$$

Therefore N is the terminal point of a normal corresponding to different points of F. From Lemma 5.2 it follows that the measure of all such points on the hypersurface F is equal to zero. Therefore

$$\operatorname{meas}\left(\bigcap_{k=1}^{\infty} \psi_F(M_k)\right) = 0.$$

Since

$$\psi_F(M_1) \supset \psi_F(M_2) \supset \cdots \supset \psi_F(M_k) \supset \cdots,$$

we have

$$\lim_{k\to\infty} \sigma_F(M_k) = \operatorname{meas} \bigcap_{k=1}^{\infty} \psi_F(M_k) = 0. \qquad \square$$

5.2 The Convergence of Integral Gaussian Curvatures

Let $g(M)$ and $g_k(M)$, $k = 1, 2, \ldots$ be completely additive set functions defined on the class of Borel subsets of S^n. If

$$\lim_{k\to\infty} \int_{S^n} f(x)dg_k = \int_{S^n} f(x)dg$$

for any continuous function $f(x)$ on S^n, then the convergence of the set functions $g_k(M)$ to the set function $g(M)$ is called weak.

If the functions $g_k(M)$ and $g(M)$ are non-negative then necessary and sufficient conditions for weak convergence of g_k to g are

1. The inequality

$$g(M) \geq \overline{\lim_{k \to \infty}} g_k(M), \tag{5.5}$$

holds for any closed subset M of S^n.

2.

$$\lim_{k \to \infty} g_k(S^n) = g(S^n). \tag{5.6}$$

The next theorem establishes the weak convergence of the integral Gaussian curvatures of convergent convex hypersurfaces to the integral Gaussian curvature of the limiting convex hypersurface.

Theorem 5.2. *Let a sequence of closed convex hypersurfaces F_m converge to a closed convex hypersurface F and a sequence of closed subsets M_m of F_m converge to a closed subset M of F; then*

$$\sigma_F(M) \geq \overline{\lim_{m \to \infty}} \sigma_{F_m}(M_m). \tag{5.7}$$

Remarks: 1. Note that $\sigma_F(S^n) = \sigma_{F_m}(S^n) = \sigma_n$ for all closed convex hypersurfaces F, F_m $(m = 1, 2, \ldots)$, where σ_n is the area of S^n. Thus from Theorem 5.2 it follows that the integral Gaussian curvatures $\sigma_{F_m}(M)$ converge weakly to the integral Gaussian curvature $\sigma_F(M)$.

2. $\sigma_F(M)$ may not be equal to $\lim_{m \to \infty} \sigma_{F_m}(M_m)$. For example, if M is the vertex of a closed convex polyhedron F, then $\sigma_F(M) > 0$. Now let M_m be the points of smooth closed convex hypersurfaces F_m convergent to M, then $\sigma_{F_m}(M_m) = 0$.

Thus

$$\sigma_F(M) > 0 = \lim_{m \to \infty} \sigma_{F_m}(M_m).$$

Moreover

$$\lim_{m \to \infty} \sigma_{F_m}(M_m)$$

may not exist. An example is the sequence of the sets $M_1, M, M_2, M, \ldots$ which converges to M, but the limit of the integral Gaussian curvatures does not exist.

Proof of Theorem 5.2. If $\sigma_F(M) = \sigma_n$, then the inequality (5.7) is trivial. Now assume that

$$\sigma_F(M) < \sigma_n. \tag{5.8}$$

Then there exists an open subset G of S^n such that $\psi_F(M) \subset G$ and

$$\text{meas } G < \sigma_F(M) + \varepsilon, \tag{5.9}$$

where ε is a given positive number. We now prove that there exists the positive integer N_ε such that

$$\psi_{F_m}(M_m) \subset G$$

for $m > N_\varepsilon$. If it were not the case, then there would exist the subsequence of points $N_{m_1}, N_{m_2}, \ldots, N_{m_s}, \ldots$ such that:

1. $m_1 < m_2 \cdots < m_s < \cdots$ and $m_s \to \infty$;
2. $N_{m_s} \in \psi_{F_{m_s}}(M_{m_s})$, $s = 1, 2, 3, \ldots$;
3. $N_{m_s} \notin G$, $s = 1, 2, 3, \ldots$.

Without loss of generality we can suppose that N_{m_s} converge to the point N. It is clear, that N does not belong to G. Let X_{m_s} be the point of $M_{m_s} \subset F_{m_s}$ lying in the supporting hyperplane of F_{m_s} with the unit exterior normal $\overline{0N}_{m_s}$. Since M_{m_s} converge to M, without loss of generality we can assume that $\overline{0N}$ is the unit exterior normal to some supporting hyperplane of F passing through X. Therefore

$$N \in \psi_F(M) \subset G$$

and this contradicts our assumption.

Thus there exists N_ε such that

$$\psi_{F_m}(M_m) \subset G$$

for $m > N_\varepsilon$. But from (5.9) it follows that

$$\sigma_{F_m}(M_m) \leq \text{meas } G < \sigma_F(M) + \varepsilon,$$

if $m > N_\varepsilon$. Thus

$$\overline{\lim_{m \to \infty}} \, \sigma_{F_m}(M_m) \leq \sigma_F(M) + \varepsilon.$$

Since ε is an arbitrary positive number we have

$$\overline{\lim_{m \to \infty}} \, \sigma_{F_m}(M_m) \leq \sigma_F(M). \qquad \square$$

5.3 Infinite Convex Hypersurfaces

We shall only consider complete infinite convex hypersurfaces in E^{n+1}, since any infinite convex hypersurface can be completed. Let F be a complete infinite convex hypersurface and Q be the convex body such that $\partial Q = F$. Then Q contains at least one ray L. Clearly there exists a supporting hyperplane of the ray L parallel to every supporting hyperplane of the hypersurface F. Therefore

$$\psi_F(F) \subset \psi_L(L).$$

But $\psi_L(L)$ is a hemisphere of S^n. Hence

$$\sigma_F(F) \leq \frac{1}{2}\sigma_n$$

where σ_n is the area of S^n, and $\sigma_F(F)$ is the integral Gaussian curvature of the hypersurface F.

The next theorem is the natural generalization of Theorem 5.1 to infinite convex hypersurfaces.

Theorem 5.3. *The integral Gaussian curvature is a completely additive non-negative set function on the ring of Borel subsets of any infinite convex hypersurface F.*

The proof of this theorem can be obtained from Theorem 5.1 by means of an approximation with closed convex hypersurfaces.

§6. Supporting Function

The supporting function of a convex body was introduced by Minkowski [2]. By means of the supporting function we can describe the boundary of any convex body F as a generalized envelope of F. Below we present the definition and main properties of the supporting function for general convex bodies. We also present the differential geometry of the supporting function for any convex body with C^2-smooth boundary and various applications.

6.1 Definition and Main Properties

Let F be a bounded k-convex body in E^{n+1}, $k = 0, 1, 2, \dots, n+1$, and let

$$(m, r) = H(m) \tag{6.1}$$

be the equation of any supporting hyperplane α of F, where m is the exterior normal of α and r is the position vector of any point of α. Note that m is not necessarily a unit vector. If $m = \nu$ is a unit vector, then $H(\nu)$ is the distance from the origin of E^{n+1} to the supporting hyperplane α with the appropriate sign. If the origin is inside F, then $H(\nu)$ is positive and therefore it is exactly the distance from the origin to α. We denote by $h(\nu)$ the function $H(m)$ for unit vectors $m = \nu$. Let $m \neq 0$ be any vector. Then

$$\nu = \frac{m}{|m|}$$

is a unit vector and $\nu \uparrow\uparrow m$. Therefore

$$H(m) = |m| H\left(\frac{m}{|m|}\right) = |m| h(\nu). \tag{6.2}$$

We now set $H(0) = 0$ and define the function $H(m)$ for all vectors $m \in E^{n+1}$ by (6.2). The function $H(m)$ is called the supporting function of the convex body F.

Lemma 6.1. *The supporting function of any point $A \in E^{n+1}$ is a linear function.*

Proof. Clearly any point A of E^{n+1} can be considered as a 0-convex body. Let

$$r = \overline{0A} = a_1 e_1 + \cdots + a_{n+1} e_{n+1}$$

be the position vector of A, where 0 is the origin in E^{n+1}. We consider an arbitrary supporting hyperplane α of A with the normal $m \neq 0$. Let $H(m)$ be the supporting function of the point A and $\nu = \frac{m}{|m|}$ be a unit vector. Since α passes through the point A we have

$$H(m) = |m| h(\nu). \tag{6.3}$$

We know that $h(\nu)$ is the distance from the origin 0 to the hyperplane α with the appropriate sign. From (6.1) it follows that

$$h(\nu) = |r| \cos(r, m). \tag{6.4}$$

From (6.4) and (6.3) it follows that

$$H(m) = |m|\,|r| \cos(r, m) = a_1 m_1 + \cdots + a_{n+1} m_{n+1}, \tag{6.5}$$

where $m_1, m_2, \ldots, m_{n+1}$ are the components of the normal m. Clearly any hyperplane α passing through A is a supporting hyperplane of A. Hence (6.5) is correct for every vector $m \neq 0$. The equation

$$a_1 m_1 + \cdots + a_{n+1} m_{n+1} = 0$$

is evident for $m = 0$. Thus (6.5) is the supporting function of the point A. □

Lemma 6.2. *Let F_1 and F_2 be bounded k_1 and k_2-convex bodies respectively, and let*

$$0 \leq k_2 \leq k_1 \leq n+1 \quad \text{and} \quad F_2 \subset F_1.$$

Then

$$H_2(m) \leq H_1(m) \tag{6.6}$$

for every vector m, where $H_1(m)$ and $H_2(m)$ are the supporting functions of F_1 and F_2. Moreover if

$$H_2(m) = H_1(m)$$

for some $m \neq 0$, then F_1 and F_2 have a common supporting hyperplane with exterior normal m.

Proof of this lemma follows directly from the equation (6.4) and the definition of the supporting function.

Theorem 6.1. *If F is a bounded k-convex body, then its supporting function $H(m)$ satisfies the following properties:*

1. $H(\lambda m) = \lambda H(m)$ (6.7)
for all vectors m and all non-negative real numbers λ.

2. $H(m_1 + m_2) \leq H(m_1) + H(m_2)$ (6.8)
for all vectors m_1 and m_2.

3. If F is a bounded $(n+1)$-convex body, then

$$\sum_{i=1}^{n+1} H(m_i) > -H\left(-\sum_{i=1}^{n+1} m_i\right) \tag{6.9}$$

for any set of $n+1$ linearly independent vectors $m_1, m_2, \ldots, m_{n+1}$.

Conversely if the function $H(m)$ is defined for all vectors m of E^{n+1} and satisfies the conditions 1 and 2, then $H(m)$ is the supporting function for some bounded k-convex body F in E^{n+1}, $(0 \leq k \leq n+1)$.

If $H(m)$ satisfies the additional condition (6.9), then the convex body F mentioned above has dimension $n+1$.

Remark. From inequalities (6.7) and (6.8) it follows that $H(m)$ is a continuous function. Therefore the supporting function is continuous for every bounded k-convex body in E^{n+1}.

Proof. Let F be any bounded k-convex body in E^{n+1}, $0 \leq k \leq n+1$, and $H(m)$ be the supporting function of F. Clearly the equality (6.7) holds if either m is zero vector or $\lambda = 0$. Now let $m \neq 0$ be any vector and λ be any positive number. Then from (6.5) it follows, that

$$H(\lambda m) = (r, \lambda m) = \lambda(r, m) = \lambda H(m).$$

Thus relation (6.7) is proved. The inequality (6.8) is trivial if at least one of the vectors m_1 and m_2 is zero vector. Therefore we assume that $m_1 \neq 0$ and $m_2 \neq 0$ are two arbitrary vectors. We denote by A the point of ∂F such that there exists a supporting hyperplane α of F passing through A with exterior normal $m_1 + m_2$. Let $H'(m)$ be the supporting function of the point A and r_A be the position vector of A*). Then

$$H(m_1 + m_2) = (r_A, m_1 + m_2) = H'(m_1 + m_2). \tag{6.10}$$

But

$$H'(m) = (r_A, m).$$

Therefore

$$H'(m_1 + m_2) = (r_A, m_1) + (r_A, m_2) = H'(m_1) + H'(m_2). \tag{6.11}$$

From Lemma 6.2 we obtain the inequalities

$$H'(m_1) \leq H(m_1) \quad \text{and} \quad H'(m_2) \leq H(m_2), \tag{6.12}$$

since $A \in F$*). Now using (6.10–12) we obtain

$$H(m_1 + m_2) \leq H(m_1) + H(m_2).$$

---

*) In both cases we consider the point A as a convex body.

The inequality (6.8) is proved.

We now suppose that F is a bounded $(n+1)$-convex body. Let 0 be the origin of E^{n+1} and m be any non-zero vector in E^{n+1}. We denote by $\alpha(m)$ the supporting hyperplane of F with exterior normal m. Then F is contained in the half-space

$$(r, m) \leq H(m), \tag{6.13}$$

where $r = \overline{0X}$ is the positive vector of any point X of this half-space. Clearly the equality holds in (6.13) if and only if $X \in \alpha(m)$. Let $m_1, m_2, \ldots, m_{n+1}$ be linearly independent vectors in E^{n+1} and

$$m_0 = -(m_1 + m_2 + \cdots + m_{n+1}). \tag{6.14}$$

Let A be an interior point of the $(n+1)$-convex bounded body F. Then there exist the supporting hyperplanes $\alpha(m_0), \alpha(m_1), \ldots, \alpha(m_{n+1})$ of F with exterior normals $m_0, m_1, \ldots, m_{n+1}$ for which the following inequalities

$$(r_A, m_i) < H(m_i) \tag{6.15}$$

hold, $i = 0, 1, 2, \ldots, n+1$.

From (6.15) and (6.14) it follows that

$$\begin{aligned} H(m_0) > (r_A, m_0) &= \left(r_A, -\sum_{i=1}^{n+1} m_i\right) \\ &= -\sum_{i=1}^{n+1}(r_A, m_i) > -\sum_{i=1}^{n+1} H(m_i). \end{aligned} \tag{6.16}$$

Now from (6.14) and (6.16) we obtain

$$-H[(-(m_1 + m_2 + \cdots + m_{n+1})] < H(m_1) + H(m_2) + \cdots + H(m_{n+1}).$$

Thus the inequality (6.9) is proved. This completes the proof of the direct assertion of Theorem 6.1.

Let $H(m)$ be some function which satisfies conditions 1 and 2 of Theorem 6.1. Therefore $H(m)$ is a continuous function in E^{n+1} (see the remark mentioned above). Hence the function

$$h(\nu) = H\left(\frac{m}{|m|}\right), \quad \nu = \frac{m}{|m|},$$

is also continuous on the unit hypersphere $S^n \subset E^{n+1}$. Thus the function $h(\nu)$ is bounded on S^n. Let

$$F = \bigcap_{\nu \in S^n} Q_\nu$$

where Q_ν is the closed halfspace in E^{n+1} defined by

$$(r,\nu) \leq h(\nu) \tag{6.17}$$

for all $\nu \in S^n$, where r is the position vector of any point of Q_ν. Clearly F is a closed bounded convex set*) and the inequality

$$h_F(\nu) \leq h(\nu) \tag{6.18}$$

holds for any $\nu \in S^n$. In (6.18) $h_F(\nu)$ is the supporting function of the set F for all unit vectors $\nu \in S^n$.

Let G be the closed subset of S^n which consists of all unit vectors ν such that

$$h_F(\nu) = h(\nu). \tag{6.19}$$

Clearly

$$F = \bigcap_{\nu \in G} Q_\nu. \tag{6.20}$$

From (6.20) it follows that if there exists at least one supporting hyperplane with the exterior normal $v \in G$, which passes through any point of ∂F*'), then the spherical image of X is a closed convex subset $\psi_{\partial F}(X)$ of S^n **) and there exist not more than $n+2$ unit vectors $\nu_1, \nu_2, \ldots, \nu_s$ belonging to $G \cap \psi_{\partial F}(X)$ such that

$$\nu = \sum_{k=1}^{s} a_k \nu_k \tag{6.21}$$

for every vector $\nu \in \psi_{\partial F}(X)$, where $a_k \geq 0$ $(k = 1, 2, \ldots, s)$ and $s \leq n+2$ ***).

If $G = S^n$ then the present theorem is proved. Thus we can assume that there exists the unit vector $\nu_0 \in S^n \backslash G$. From the definition of the set G, it follows that

$$h_F(\nu_o) < h(\nu_o). \tag{6.22}$$

Since F is a bounded closed convex set in E^{n+1}, there is a point $X_0 \in \partial F$ such that the supporting hyperplane of ∂F passes through X_0.

The point X_o is a singular point of ∂F, because there is at least one supporting hyperplane of ∂F with the exterior normal $\nu \in G$, passing through X_o.

---

*) i.e. F is a bounded k-convex body $(0 \leq k \leq n+1)$.

*') i.e. there exist at least two different supporting hyperplanes of ∂F passing through X.

**) $\psi_{\partial F}(X)$ is either contained in some closed hemisphere of S^n or coincides with S^n. The last possibility is realized if F is a point.

***) The vectors $\nu_1, \nu_2, \ldots, \psi_s$ can depend on the choice of the vector $\nu \in \psi_{\partial F}(X)$; it suffices to use not more than $n+1$ vectors $\nu_1, \nu_2, \ldots, \nu_s$, if F is not a point.

Thus from the considerations mentioned above we obtain

$$\nu_o = \sum_{n+1}^{n} a_k \nu_k$$

where the unit vectors ν_k belong to $G \cap \psi_{\partial F}(X_o)$, $a_k \geq 0$ $(k = 1, 2, \ldots, s)$ and $s \leq n+2$.

Let $H_F(m)$ be the supporting function of F, $m \in E^{n+1}$. Then

$$H_F(\nu_k) = h_F(\nu_k) = h(\nu_k) = H(\nu_k)$$

for all unit vectors $\nu_k \in G \cap \psi_{\partial F}(X_o)$, $k = 1, 2, \ldots, s$.

From the properties of the supporting function $H_F(m)$, the prescribed function $H(m)$, and the positiveness of the numbers a_k $(k = 1, 2, \ldots, s)$ we obtain

$$\begin{aligned} h_F(\nu_o) = H_F(\nu_o) = (r_{X_o}, \nu_o) = \left(r_{X_o}, \sum_{k=1}^{s} a_k \nu_k \right) = \\ = \sum_{k=1}^{s} a_k (r_{X_o}, \nu_k) = \sum_{k=1}^{s} a_k H_F(\nu_k) = \sum_{k=1}^{s} a_k H(\nu_k) = \\ = \sum_{k=1}^{s} H(a_k \nu_k) \geq H(\nu_o) = h(\nu_o). \end{aligned}$$

Thus

$$h_F(\nu_o) \geq h(\nu_o). \tag{6.23}$$

Now the inequalities (6.22) and (6.23) are incompatible. Hence the set G coincides with S^n, and $H(m)$ is the supporting function of the closed convex set F.

If the function $H(m)$ satisfies the condition (6.9), then F is a bounded $(n+1)$-convex body. The proof of the last assertion would be useful to the readers. □

Let G be a closed convex domain of S^n contained in one closed hemisphere. Then the definition of the supporting function of convex infinite complete hypersurfaces is the same as for closed bounded convex hypersurfaces (or bounded convex bodies). Theorem 6.1 is also valid for the supporting functions of complete convex infinite hypersurfaces with the spherical image G.

6.2 Differential Geometry of Supporting Function

Let S^n be the unit sphere in E^{n+1}. We consider a smooth convex hypersurface $S \subset E^{n+1}$ whose spherical mapping

$$\psi_S \colon S \to S^n,$$

is one-to-one. We also suppose that S is a C^2-smooth hypersurface.

Clearly any tangential hyperplane α of S has only one common point with S. If ν is the unit normal of α, then we denote by $X(\nu)$ the common point of α and S. We denote by $r(\nu)$ the position vector $\overline{0X(\nu)}$.

Let $h(\nu) = (r(\nu), \nu)$. Then S is the envelope of the family of hyperplanes with the equations

$$(\nu, y) = h(\nu)$$

where y is any position vector of the tangential hyperplane.

Let $G = \psi_F(S) \subset S^n$ be the domain of the function $h(\nu)$. Since ν is completely defined by any intrinsic coordinates in S^n, the position vector $r(\nu)$ of S and the function $h(\nu)$ are functions of these coordinates. We put

$$H(m) = |m| h\left(\frac{m}{|m|}\right) \tag{6.24}$$

for every $m \neq 0$ satisfying the condition that $\frac{m}{|m|}$ belongs to G, and set $H(0) = 0$. As we know the function $H(m)$ is called the *supporting function* of the hypersurface S. If S is a closed convex smooth hypersurface, then

$$G = S^n$$

and $H(m)$ is the supporting function of S introduced in Section 6.1. Let the origin 0 of E^{n+1} be inside S. Then from the definition of $H(m)$ it follows that $H(m)$ is a positive homogeneous function of the first degree. If $m = (\alpha_1, \alpha_2, \ldots, \alpha_n, \alpha_{n+1})$, then $H(m)$ is a function $\alpha_1, \alpha_2, \ldots, \alpha_{n+1}$ and

$$H(\alpha_1, \alpha_2, \ldots, \alpha_{n+1}) = |\alpha| h\left(\frac{\alpha_1}{|\alpha|}, \frac{\alpha_2}{|\alpha|}, \ldots, \frac{\alpha_n}{|\alpha|}\right),$$

where $|\alpha|^2 = \sum_{i=1}^{n} \alpha_i^2$. Since H is a homogeneous function of the first degree,

$$\alpha_1 H_1 + \alpha_2 H_2 + \cdots + \alpha_{n+1} H_{n+1} = H, \tag{6.26}$$

where

$$H_i = \frac{\partial H}{\partial \alpha_i}.$$

The equation of the tangential hyperplane of S is

$$\alpha_1 x_1 + \alpha_2 x_2 + \cdots + \alpha_{n+1} x_{n+1} = H \tag{6.27}$$

where $x_1, x_2, \ldots, x_{n+1}$ are Cartesian coordinates in E^{n+1}. We now fix the variables $\alpha_2, \ldots, \alpha_{n+1}$ in (6.27) and change only α_1, i.e. we construct the cylinder around S with $(n-1)$-generators orthogonal to the x_1 axis. Differentiating (6.27) with respect to α_1 we obtain

$$x_1 = \frac{\partial H}{\partial \alpha_1}, \tag{6.28}$$

where x_1 is the coordinate of the unique common point of the hyperplane (6.27) with the hypersurface S. In a similar way we obtain

$$x_i = \frac{\partial H}{\partial \alpha_i}$$

$(i = 2, 3, \ldots, n+1)$. Since $H_i(\alpha_1, \alpha_2, \ldots, \alpha_{n+1})$ are homogeneous zero degree functions,

$$x_i = H_i(\lambda\alpha_1, \lambda\alpha_2, \ldots, \lambda\alpha_{n+1}),$$

where λ is any real positive number. We now consider the infinitesimal displacement from a given point of S along the principle direction on S. Then from Rodrigue's formula we obtain

$$dx_i - R d\nu_i = 0, \quad (i = 1, 2, \ldots, n+1), \tag{6.29}$$

where $\nu_1, \nu_2, \ldots, \nu_{n+1}$ are the components of the unit exterior normal of S, and R is the radius of the normal curvature in the direction of this displacement. If we use formula (6.28), then equalities (6.29) become

$$\sum_{k=1}^{n+1} H_{ik} d\nu_k - R d\nu_i = 0 \tag{6.30}$$

$(i = 1, 2, \ldots, n+1)$.

Since $d\nu \neq 0$ we obtain that

$$\det \begin{Vmatrix} H_{11} - R & H_{12} & \ldots & H_{1n} \\ H_{21} & H_{22} - R & \ldots & H_{2n} \\ \ldots & \ldots & \ldots & \ldots \\ \ldots & \ldots & \ldots & \ldots \\ H_{n1} & H_{2n} & \ldots & H_{nn} - R \end{Vmatrix} = 0. \tag{6.31}$$

Differentiating the identity (6.26) with respect to α_i we obtain

$$\sum_{k=1}^{n+1} H_{ik}\alpha_k = 0 \quad (i = 1, 2, \ldots, n+1). \tag{6.32}$$

From (6.32) it follows that

$$\det \|H_{ik}\| = 0.$$

Hence the set of the roots of equation (6.31) consists of the principal radii of normal curvature of S and of zero. From (6.31) we obtain

$$(R_1 + R_2 + \cdots + R_n) = H_{11} + H_{22} + \cdots + H_{n+1\ n+1} \tag{6.33}$$

and

$$R_1 R_2 \dots R_n = \det \begin{Vmatrix} H_{11} & \dots & H_{1n} \\ \dots & \dots & \dots \\ \dots & \dots & \dots \\ H_{n1} & \dots & H_{nn} \end{Vmatrix} + \cdots + \det \begin{Vmatrix} H_{22} & \dots & H_{2\ n+1} \\ \dots & \dots & \dots \\ \dots & \dots & \dots \\ H_{n+1\ 2} & \dots & H_{n+1\ n+1} \end{Vmatrix}. \tag{6.34}$$

In (6.33-34) we calculate the functions H_{ik} at the unit vectors $\nu = (\nu_1, \dots, \nu_{n+1})$, i.e. $\nu_1^2 + \nu_2^2 + \cdots + \nu_{n+1}^2 = 1$.

We now displace the convex body F along the vector $a = (a_1, a_2, \dots, a_{n+1})$. Let $H(m)$ and $\widetilde{H}(m)$ be the supporting functions of F before and after this displacement respectively. Then

$$\widetilde{H}(m) = H(m) + \sum_{k=1}^{n} a_k m_k. \tag{6.35}$$

Indeed $H(m) = (r, m)$, where r is the position vector of the point of ∂F, where the supporting hyperplane of F with the normal m touches ∂F. Thus

$$\widetilde{H}(m) = (r + a, m) = (r, m) + (a, m) = H(m) + \sum_{k=1}^{n} a_k m_k.$$

Let now G be a domain, which lies in one open hemisphere S_+^n of S^n. Without loss of generality we can assume that $G = S_+^n$. We also assume that the component of a unit vector $\nu \in S_+^n$, with respect to the axis $x_{n+1} = z$, is strictly positive. Below it is convenient to use the function

$$p(\nu) = \frac{h(\nu)}{(e, \nu)},$$

where e is the unit vector of the z axis. Since $(e, \nu) > 0$ for all $\nu \in G$ we have $p(\nu) \in C^2(G)$. For any vector $m \neq 0$ such that

$$\nu = \frac{m}{|m|} \in G$$

we have

$$p(\nu) = \frac{|m| h\left(\frac{m}{|m|}\right)}{|m|\left(e, \frac{m}{|m|}\right)} = \frac{H(m)}{(e, m)}$$

We now use the positive homogeneity of the supporting function $H(m)$. Then the last formula becomes

$$p = \frac{1}{m_{n+1}} H(m_1, m_2, \ldots, m_{n+1}) = \tag{6.36}$$
$$= H\left(\frac{m_1}{m_{n+1}}, \frac{m_2}{m_{n+1}}, \ldots, \frac{m_n}{m_{n+1}}, 1\right),$$

where $m_1, m_2, \ldots, m_{n+1}$ are the components of a vector m.

We set

$$v_i = \frac{m_i}{m_{n+1}}, \quad i = 1, 2, \ldots, n.$$

Then equation (6.36) becomes

$$p \equiv p(v_1, v_2, \ldots, v_n) = H(v_1, v_2, \ldots, v_n, 1). \tag{6.37}$$

From (6.36-37) and (6.28) it follows that

$$\begin{aligned} x_i &= H_i = p_i, \quad i = 1, 2, \ldots, n; \\ x_{n+1} &= H_{n+1} = p - \sum_{j=1}^{n} p_j v_j. \end{aligned} \tag{6.38}$$

Thus the following formulas

$$\begin{aligned} m_{n+1} H_{jk} &= p_{jk}, \quad (j, k = 1, 2, \ldots, n); \\ m_{n+1} H_{j\,n+1} &= -\sum_{s=1}^{n} p_{sj} v_s; \\ m_{n+1} H_{n+1\;n+1} &= \sum_{s,t=1}^{n} p_{st} v_s v_t \end{aligned} \tag{6.39}$$

hold for the second derivatives of the function H. Hence the second derivatives of H are only expressed by the second derivatives of the function p.

Finally from (6.33–34) and (6.38–39) we obtain the following important formulas:

$$R_1 + R_2 + \cdots + R_n = (1 + v_1^2 + \cdots + v_n^2)^{1/2} \left(\Delta p + \sum_{i,j=1}^{n} v_i v_j p_{ij}\right), \tag{6.40}$$

$$R_1 R_2 \ldots R_n = (1 + v_1^2 + \cdots + v_n^2)^{\frac{n}{2}+1} \det\left(\frac{\partial^2 p}{\partial v_i \partial v_j}\right). \tag{6.41}$$

where Δp is the Laplacian of the function $p(v)$.

Exercises and Problems

A. Convex Sets in V^n and E^n

1. Let $F_i, i = 1, 2, \dots, k$, be a finite system of closed subsets of V^n. Prove that the set $\mathrm{Co}\left(\bigcup_{i=1}^{k} F_i\right)$ is also a closed subset of V^n.
2. Let H be a subset of V^n. The *dimension* m of the set H in V^n is the minimal dimension of linear sets $x_0 + L$, which contain H, where x_0 is any point of V^n and L is any subspace of V^n. We will use the traditional notation $\dim H = m$. Prove the following assertion.

   If $\dim H = m$, then every point $x \in H$ is a convex combination of $m + 1$ points of the set H. (These $m + 1$ points can depend on the point x and it is not necessary that all $m + 1$ points should be different.)

3. Let a subset H of V^n contains at least $n+2$ different points. Prove that H can be decomposed into two nonempty sets H_1, H_2 such that $H = H_1 \cup H_2$, $H_1 \cap H_2 = \emptyset$, and $\mathrm{Co}\ H_1 \cap \mathrm{Co}\ H_2 \neq \emptyset$.
4. There are k compact convex bodies in V^n, where $k \geq n+2$. Prove that the intersection of all these bodies is non-empty, if every $n + 1$ bodies of this family have non-empty intersection. (This assertion is known as Kelly's theorem.)

   Hint: Use induction. This theorem is also valid for $k = +\infty$.

5. The sum, intersection and convex combination of a finite family of convex polyhedra are convex polyhedra.*)
6. The intersection of a finite family of closed halfspaces of V^n is a convex polyhedra if this intersection is a bounded set in V^n.

B. Wedges and Proper Convex Cones

It is convenient to develop the concept of a convex cone into the following two more specific concepts. A set K in a linear space L is called a *wedge* if the following conditions hold:

a) if $x_1, x_2 \in K$, then $x_1 + x_2 \in K$;
b) if $x \in K$, then $\alpha x \in K$ for all $\alpha \geq 0$.

A wedge K is called a *proper cone* if from $x \in K$ and $x \neq 0$ it follows that $-x \notin K$.

The dimension of the smallest halfspace which contains a wedge K is called the dimension of K and it is denoted by $\dim K$. Wedges of maximal dimension are called *solid*.

---

*) In exercises 5 and 6 convex polyhedra can be either the empty set or have dimensions $0, 1, 2, \dots, n$.

Let $K \subset V^n$ be a wedge. The largest subspace L of V^n, which is contained in K, is called the linear part of K.

Any subspace of V^n is a wedge. Let $K_x = \{\alpha x, \alpha \geq 0\}$, where $x \neq 0$. Clearly K_x is the simplest example of a wedge, which is not a subspace of V^n. Moreover K_x is a proper cone. This cone is called a *ray*.

Let $V^{*,n}$ be the *dual (conjugate) space* for V^n, i.e. $V^{*,n}$ is the space consisting of all linear functionals on V^n with natural linear operations, generated by linear operations in V^n. $V^{*,n}$ is an n-dimensional vector space. Let $f \in V^{*,n}$ and $f \neq 0$, then the set $K(f) \subset V^n$, consisting of all $x \in V^n$ such that $f(x) \geq 0$, is a wedge. Clearly $\dim K(f) = n - 1$ and $K(f)$ is a closed set in V^n. $K(f)$ is called a *closed halfspace* of V^n.

Let $L_1, L_2, \ldots, L_m$ be a finite family of subsets of the space V^n. The set $L \subset V^n$, which consists of all sums $\sum_{k=1}^{m} x_k$, $x_k \in L_k$, is called the sum of $L_1, L_2, \ldots, L_m$. We will use the notation

$$L = \sum_{k=1}^{m} L_k.$$

If F_α is an infinite family of subspaces then $L = \Sigma F_\alpha$ is the set of all finite sums Σx_{α_k}, where $x_{\alpha_k} \in F_{\alpha_k}$ and x_{α_k} are mutually different.

7. Let $\{K_\alpha\}$ be a family of wedges in the space V^n. Prove that the sets $K_\Sigma = \Sigma K_\alpha$ and $K_\cap = \cap K_\alpha$ are wedges in V^n.

   The wedge K_Σ is the smallest wedge, which contains all K_α. The wedge $K_\cap$ is the largest wedge which is contained in all K_α.

8. The sum of a family of rays is a wedge. Prove also the dual assertion: the intersection of a family of closed halfspaces is a wedge.
9. If the vectors, which define the rays in problem 8, are linearly independent, then the wedge K_Σ is a proper cone.
10. Every wedge in V^n can be represented as the sum of its linear part and some proper cone.
11. If a wedge $K \subset V^n$ is represented as sum of a subspace and a proper cone, then this subspace is the linear part of K.

    (Thus a wedge $K \subset V^n$ is a proper cone if and only if the linear part of K is the zero element of V^n.)

12. Let K be a wedge in V^n. We consider linear functionals $f \in V^{*,n}$ such that $f(x) \geq 0$ for all $x \in K$. Let K^* be the set of all such functionals. Prove that K^* is a wedge in $V^{*,n}$.

    The wedge $K^* \subset V^{*,n}$ is called the *dual (conjugate) cone* for K.

13. The dual wedge is always closed: $\overline{K^*} = K^*$.
14. $(\overline{K})^* = K^*$.

15. $(K_1 + K_2)^* = K_1^* \cap K_2^*$.
$(K_1 \cap K_1)^* \supset K_1^* + K_2^*$.

In addition: If K_1, K_2 are subspaces, then $(K_1 \cap K_2)^* = K_1^* + K_2^*$.

16. If $K_1 \subset K_2$; then $K_1^* \supset K_2^*$, i.e. the operation $K \to K^*$ is monotone.
17. The wedge K^* is solid in $V^{*,n}$ if and only if the initial wedge is a proper cone in V^n.
18. Let a proper cone $K \subset V^n$ be defined by the inequality

$$x_n \geq \left(\sum_{k=1}^{n-1} |x_k|^p \right)^{1/p}, \quad 1 < p < +\infty$$

in a basis $e_1, e_2, \ldots, e_n$. Then the dual cone $K^* \subset V^{*,n}$ is defined by

$$y_n \geq \left(\sum_{k=1}^{n-1} |y_k|^q \right)^{1/q}$$

in the dual basis $e_1^*, e_2^*, \ldots, e_n^*$, where

$$\frac{1}{p} + \frac{1}{q} = 1.$$

Finally we consider wedges in n-dimensional Euclidean vector spaces. In this case $V^{*,n} = V^n$. Therefore $K^* \subset V^n$ for any wedge $K \subset V^n$.

A wedge K of an Euclidean space V^n is called *acute* if $(x, y) \geq 0$ for all $x, y \in K$. A wedge $K \subset V^n$ is called *obtuse* if for any $y \notin K$ there exists $x \in K$ such that $(x, y) < 0$. Finally a wedge $K \subset V^n$ is called *right* if K is simultaneously acute and obtuse.

19. $K + K^* = V^n$;
20. An acute wedge is a proper cone.
21. An obtuse wedge is solid.
22. A wedge is acute if and only if $K^* \supset K$.
23. A wedge is obtuse if and only if $K^* \subset K$.
24. A wedge K is obtuse if and only if K^* is acute.
25. Let $e_1, e_2, \ldots, e_n$ be an orthonormal basis in V^n. We define by the inequality

$$x_n \geq \left(\sum_{k=1}^{n-1} |x_k|^p \right)^{1/p}, \quad 1 < p < +\infty$$

a convex cone K_p in V^n.
Prove that

a) $K_p^* = K_q$, where $\frac{1}{p} + \frac{1}{q} = 1$;
b) K_p is acute if $1 < p \leq 2$ and obtuse if $2 \leq p < +\infty$;

Remark. Thus K_2 is a right proper cone. This cone called the round Minkowski cone.

C. Spherical Mapping of Convex Polyhedra

The spherical mapping of general convex hypersurfaces in E^{n+1} was presented in Section 5. This mapping is already nontrivial in the simplest case of two-dimensional convex polyhedra in E^3. Therefore we will consider problems related to the spherical mappings of two-dimensional convex polyhedra.

Let F be any convex surface in E^3 and let S^2 be any unit sphere in E^3. We fix a set $M \subset F$ and consider all supporting planes of F passing through all points of M. The terminal points of all unit exterior normals of such planes fill out a set $M^* \subset S^2$. This set is called the *spherical image* of M.

We now apply this definition to two-dimensional convex polyhedra. If a polyhedron F has a boundary, then supporting planes α, satisfying the conditions $\alpha \cap \partial F \neq \emptyset$, $\alpha \cap$ int $F = \emptyset$, should be excluded from M^*, where M is any subset of F.

If $M \subset F$ is the set of all interior points lying on one face, then M^* consists of one point.

If M is an edge with excluded endpoints, then M^* is an arc L of the great circle of S^2. The endpoints of L are spherical images of the faces, which have M as a common part of their boundaries.

Indeed every intermediate plane passing through M and lying between the appropriate faces of F is a supporting hyperplane. Clearly these appropriate faces of F have M as the common part of their boundaries.

Finally if M consists of one vertex $A \in F$, then M^* is a part of S^2, which is cut by the solid angel Q of the unit exterior normals of all supporting planes at the vertex A.

We now formulate a few problems related to the polyhedral angle Q.

26. Q is a convex angle.
27. The faces of Q are plane convex angles, which are orthogonal to the edges of F with the initial point A. Let β_i be a dihedral angle of Q, which completes to π the plane angle α_i between two edges of F with common vertex A, i.e.

$$\alpha_i = \pi - \beta_i.$$

    Finally prove that A^* is a spherical convex polygon with angles β_i. The vertices of A^* are spherical images of the faces of F, passing through A and the sides of A^* are spherical images of the appropriate edges with excluded endpoints.

28. Let F be a closed convex polyhedron in E^3. Describe in detail the spherical mapping of F.

    Let V be a convex polyhedral angle with n faces, whose plane angles are $\alpha_1, \alpha_2, \ldots, \alpha_n$. Then $V^* \subset S^2$ is a convex spherical polygon with the angles $\beta_i = \pi - \alpha_i$.

The number $\theta = \sum_{i=1}^{n} \alpha_i$ is called the total angle at the vertex A of V.

29. Prove that

$$\sigma_F(V) = \sigma_F(A) = 2\pi - \theta,$$

where $\sigma_F(V)$ and $\sigma_F(A)$ are the integral Gaussian curvatures of V and A respectively.

Hint: First prove by induction that

$$\sigma_F(A) = \text{meas } A^* = \sum_{i=1}^{n} \beta_i - (n-2)\pi.$$

30. The spherical image of an infinite convex polyhedron coincides with the spherical image of its asymptotic cone.
31. Let P be a closed convex polyhedron in E^3 with vertices $A_1, A_2, \ldots, A_m$, $m \geq 4$. Prove that the sum of all plane angles (of all faces) is equal to $2\pi(m-2)$.

D. Finite Dimensional Normed Spaces and Supporting Function of Convex Bodies

Let L^n be a n-dimensional real vector space. The functional $N(x)$ in L^n is called a *norm* if it satisfies the following conditions:

1. $N(x) > 0$ for all non-zero vectors $x \in L^n$;
2. $N(\alpha x) = |\alpha| N(x)$ for all $x \in L^n$;
3. $N(x_1 + x_2) \leq N(x_1) + N(x_2)$ (the triangle inequality) for all $x_1; x_2 \in L^n$.

From the basic properties 1,2,3 it follows that

a) $N(0) = 0$;
b) $N(-x) = N(x)$ for all $x \in L^n$;
c)
$$|N(x_1) - N(x_2)| \leq N(x_1 \pm x_2) \leq N(x_1) + N(x_2)$$

for all $x_1, x_2 \in L^n$.

If some norm is fixed in L^n, then L^n is called a *normed space*. The chosen norm is denoted by $\|x\|$. A normed space is also called a *Minkowski space*.

A space L^n can be normed in many different ways.

Let $\sigma = \{e_k; k = 1, 2, \ldots, n\}$ be a basis in L^n, and $x = x_1 e_1 + \cdots + x_n e_n$ be any vector in L^n.

Solve the following problems.

32. Prove that the functional

$$N_c(x, \sigma) = \max_{1 \leq k \leq n} |x_k|$$

is a norm in L^n. This norm is called the c-norm (or the uniform norm).

33. Prove that the functional

$$N_\ell(x,\sigma) = \sum_{k=1}^{n} |x_k|$$

is a norm in L^n. This norm is called the ℓ-norm.

34. Let p be any number such that $p \geq 1$. The functional

$$N_{\ell^p}(x,\sigma) = \left(\sum_{k=1}^{n} |x_k|^p\right)^{\frac{1}{p}}$$

is a norm in L^n. This norm is called the ℓ^p-norm. If $p = 1$, then the ℓ^p-norm becomes the ℓ-norm in L^n. If $p = \infty$, then the ℓ^p-norm becomes the c-norm in the same space.

The arithmetic space $R^n = \{x = (x_1, x_2, \ldots, x_n)\}$ with the ℓ^p-norm introduced by any basis in R^n is called the space ℓ^p. If $p = 1$ or $p = \infty$, then we obtain the spaces ℓ and c respectively. The triangle inequality in ℓ^p is the well-known Minkowski's inequality

$$\left(\sum_{k=1}^{n} |x_k + y_k|^p\right)^{1/p} \leq \left(\sum_{k=1}^{n} |x_k|^p\right)^{\frac{1}{p}} + \left(\sum_{k=1}^{n} |y_k|^p\right)^{\frac{1}{p}}.$$

We now consider the space ℓ^2, $p = 2$. Let

$$(x,y) = \sum_{k=1}^{n} x_k y_k$$

be a scalar product in ℓ^2. Then

$$\|x\| = \left(\sum_{k=1}^{n} x_k^2\right)^{1/2} = \sqrt{(x,x)}$$

is the ℓ^2-norm with respect to any orthonormal basis. The ℓ^2-norm $\|x\| = \sqrt{(x,x)}$ and the ℓ^2-space are called *Euclidean*.

The triangle inequality in ℓ^2 is closely connected with the well-known Schwartz inequality

$$|(x,y)| \leq \|x\|\,\|y\|$$

35. A norm $\|x\|$ in L^n is Euclidean if and only if

$$\|x+y\|^2 + \|x-y\|^2 = 2(\|x\|^2 + \|y\|^2)$$

for all vectors $x, y \in L^n$.

36. Let

$$m(L^n) = \sup_{x,y \in L^n} \frac{\|x+y\|^2 + \|x-y\|^2}{2(\|x\|^2 + \|y\|^2)}.$$

Then $m(L^n)$ is a measure of how much L^n deviates from Euclidean space.

Prove that

$$1 \le m(L^n) \le 2.$$

A normed space L^n is Euclidean if and only if $m(L^n) = 1$.

A set M in L^n is called *absolutely convex* if for any $x, y \in M$ and for any α, β, satisfying the condition $|\alpha| + |\beta| = 1$, the inclusion

$$\alpha x + \beta y \in M$$

holds.

37. Prove the following facts:
 a) Any absolutely convex set M is symmetric with respect to $x = 0$, i.e. $-x \in M$ if $x \in M$;
 b) Any non-empty absolutely convex set contains the zero-element;
 c) If M is any absolutely convex set and $x \in M$, then $\alpha x \in M$ for all α such that $|\alpha| \le 1$.

A closed absolutely convex set M is called an *absolutely convex body*, if M contains some neighborhood of $x = 0$. A functional $p(x)$ in L^n is called a *seminorm* if it satisfies the following two conditions

$$p(x_1 + x_2) \le p(x_1) + p(x_2),$$
$$p(\alpha x) = |\alpha| p(x)$$

for all $x, x_1, x_2 \in L^n$ and for all real numbers α.

38. Prove the following properties of seminorms:
 a) $p(0) = 0$;
 b) $p(-x) = p(x)$ for all $x \in L^n$;
 c) $p(x) \ge 0$;
 d) a seminorm $p(x)$ is a norm if and only if $p(x) \ne 0$ for all $x \ne 0$.

39. Let $p(x)$ be a seminorm in L^n and Ker p be the set of all vectors $x \in L^n$ such that $p(x) = 0$. Prove that Ker p is a subspace of L^n.
40. Prove that the relation $x \equiv y(\text{mod Ker } p)$ implies $p(x) = p(y)$.
41. Let $[x]$ be an element of the factor-space $L^n/\text{Ker } p$. Prove that the functional $\tilde{p}[([x]) = p(x)$ is a norm in the factor-space $L^n/\text{Ker } p$.

A seminorm p_1 is *subordinate* to a seminorm p_2, if there exists a constant C such that

$$p_1(x) \le C p_2(x)$$

for all $x \in L^n$.

42. Prove that a seminorm $p(x)$ is *subordinate* to any norm.

Thus if $p(x)$ is a seminorm in a normed space, then there exists a constant C such that

$$p(x) \leq C\|x\|$$

for all $x \in L^n$. The smallest C in this inequality is called the norm of the functional $p(x)$. We denote it by $\|p\|$. Hence

$$\|p\| = \sup_x \frac{p(x)}{\|x\|} = \sup_{\|x\|=1} p(x).$$

A vector $x \in L^n$ is called normalized if $\|x\| = 1$. The set of all normalized vectors $x \in L^n$ is called the unit sphere in the normed space L^n.

Let p_1 and p_2 be two seminorms in $\mathcal{L}^n$. If p_1 is subordinate to p_2 and p_2 is subordinate to p_1, then p_1 and p_2 are called *equivalent*.

43. A seminorm p_1 is subordinate to p_2 if and only if $\operatorname{Ker} p_2 \subset \operatorname{Ker} p_1$.
44. Two seminorms p_1 and p_2 are equivalent if and only if $\operatorname{Ker} p_1 = \operatorname{Ker} p_2$.

From problems 43 and 44 it follows that the classes of equivalent seminorms correspond one-to-one to subspaces of L^n. In addition the zero-subspace corresponds to the class of norms, and L^n corresponds to the unique zero-seminorm.

45. If $p(x)$ is a seminorm, then the unit ball $p(x) \leq 1$ is a absolutely convex body.
46. Let $p_1(x)$ and $p_2(x)$ be two seminorms and let the unit balls corresponding to them coincide. Then $p_1(x) = p_2(x)$.

Let now Q be a absolutely convex body. Let A be the set of positive numbers α such that

$$\frac{1}{\alpha} x \in Q,$$

where x is a fixed element of L^n. We now introduce the functional

$$p_Q(x) = \inf_{\alpha \in A} \alpha.$$

47. The functional $p_Q(x)$ is a seminorm, and the unit ball corresponding to $p_Q(x)$ coincides with Q.

Thus if the origin of L^n coincides with the center of symmetry of Q, then $p_Q(x)$ is the supporting function of Q. The functional $p_Q(x)$ is also called the *Minkowski functional*.

48. A seminorm $p(x)$ is a norm in a normed space L^n if and only if the unit ball corresponding to $p(x)$ is bounded, i.e. this ball is contained in some ball $\|x\| < r = \text{const} < +\infty$.

Chapter 2. Mixed Volumes. Minkowski Problem. Selected Global Problems in Geometric Partial Differential Equations

§7. The Minkowski Mixed Volumes

7.1 Linear Combinations of Sets in E^{n+1}

Let V^{n+1} be a $(n+1)$-dimensional Euclidean vector space. We denote by E^{n+1} the $(n+1)$-dimensional Euclidean point space associated with V^{n+1} (see §1). Then every vector $a \in V^{n+1}$ generates a parallel translation

$$p_a: E^{n+1} \to E^{n+1} \tag{7.1}$$

in the following way: for any point $X \in E^{n+1}$

$$p_a(X) = X',$$

where X' is the terminal point of the vector $\overline{XX'} = a$.

Let 0 be any point of E^{n+1}. We fix 0 and associate V^{n+1} with the set of all position vectors $\overline{0X}$, $X \in E^{n+1}$. The point 0 is called the *origin* in E^{n+1}.

Let M be a given set of points in E^{n+1}. Then M can be identified with the set H of all position vectors $\overline{0X} \in E^{n+1}$, $X \in M$. Thus we have a one-to-one mapping

$$f_0:\ V^{n+1} \to E^{n+1} \tag{7.2}$$

such that

$$f_0(a) = X \tag{7.3}$$

where X is the terminal point of the vector $\overline{0X} = a$. Note that the mapping f_0 depends on the position of the origin 0. Thus

$$M = f_0(H) \tag{7.4}$$

for the sets $H \subset V^{n+1}$ and $M \subset E^{n+1}$.

We now can transfer concepts and constructions from vector spaces V^{n+1} to point spaces E^{n+1}. Let $H_1, H_2, \ldots, H_k$ be subset of V^{n+1} and let $\lambda_1, \lambda_2, \ldots,$ λ_k be any real numbers. Then the set

$$\lambda_1 H_1 + \lambda_2 H_2 + \cdots + \lambda_k H_k$$

is called a *linear combination of the sets* $H_1, H_2, \ldots, H_k$ (see the definition of a linear combination of sets $H_1, H_2, \ldots, H_k$ in §1). In the particular case when $H_1, H_2, \ldots, H_k$ are vectors $a_1, a_2, \ldots, a_k$ in V^{n+1} we obtain

$$X_1 = f_0(a_1), X_2 = f_0(a_2), \ldots, X_k = f_0(a_k)$$

are terminal points of the position vectors $\overline{0X_1}, \overline{0X_2}, \ldots, \overline{0X_k}$. Hence

$$a = \lambda_1 a_1 + \lambda_2 a_2 + \cdots + \lambda_k a_k = \lambda_1 \overline{0X_1} + \lambda_2 \overline{0X_2} + \cdots + \lambda_k \overline{0X_k} = \overline{0X}.$$

Thus

$$X = f_0(a).$$

and we can say that X is a *linear combination of points* $X_1, X_2, \ldots, X_k$ *with respect to the origin* 0. We also use the notation

$$X = \lambda_1 X_1 + \lambda_2 X_2 + \cdots + \lambda_k X_k. \tag{7.5}$$

Clearly this concept only depends on the choice of the origin 0.

Now let $H_1, H_2, \ldots, H_k$ be subsets of the space V^{n+1} and let

$$H = \lambda_1 H_1 + \lambda_2 H_2 + \cdots + \lambda_k H_k. \tag{7.6}$$

Then

$$\begin{aligned} M = f_0(H) &= f_0(\lambda_1 H_1 + \cdots + \lambda_k H_k) = \\ &= \lambda_1 M_1 + \cdots + \lambda_k M_k, \end{aligned}$$

where $M_i = f_0(H_i)$, $i = 1, 2, \ldots, k$. Thus the set M is the set of all linear combinations of all points

$$X = \lambda_1 X_1 + \cdots + \lambda_n X_k, \quad X_i \in M_i, \quad i = 1, 2, \ldots, k.$$

The set M is called a *linear combination of the sets* $M_1, M_2, \ldots, M_k$ *with coefficients* $\lambda_1, \lambda_2, \ldots, \lambda_k$. The numbers $\lambda_1, \lambda_2, \ldots, \lambda_k$ are fixed in our previous considerations.

Theorem 7.1. *Let*

$$M = \lambda_1 M_1 + \lambda_2 M_2 + \cdots + \lambda_k M_k$$

be a linear combination of the sets $M_1, M_2, \ldots, M_k$ with respect to the given origin 0. Let 0′ be the new origin. We consider the linear combination of the sets $M_1, M_2, \ldots, M_k$ with the same coefficients $\lambda_1, \lambda_2, \ldots, \lambda_k$ with respect to the new origin 0′. Then the last linear combination can be obtained from the initial one by some parallel translation.

Proof. Let $X_1 \in M_1, X_2 \in M_2, \ldots, X_k \in M_k$ be an arbitrary system of points. Then

$$\overline{0'X_i} = \overline{0'0} + \overline{0X_i}, \quad i = 1, 2, \ldots, k. \tag{7.7}$$

Hence

$$\sum_{i=1}^{k} \lambda_i \overline{0'X_i} = (\lambda_1 + \lambda_2 + \cdots + \lambda_k)\overline{0'0} + \sum_{i=1}^{k} \lambda_i \overline{0X_i}. \tag{7.8}$$

Clearly the linear combination of the points $X_1, X_2, \ldots, X_k$ with respect to the origin $0'$ is the terminal point X' of the vector

$$\overline{0'X'} = \sum_{i=1}^{k} \lambda_i \overline{0'X_i}. \tag{7.9}$$

Since

$$\overline{0X'} = \overline{00'} + \overline{0'X'}$$

we obtain from (7.8) and (7.9) that

$$\begin{aligned} \overline{0X'} &= \sum_{i=1}^{k} \lambda_i \overline{0X_i} + (1 - \lambda_1 - \cdots - \lambda_k)\overline{00'} = \\ &= \overline{0X} + (1 - \lambda_1 - \cdots - \lambda_k)\overline{00'}. \end{aligned}$$

Hence

$$X' = X + (1 - \lambda_1 - \lambda_2 - \cdots - \lambda_k)0'. \tag{7.10}$$

If $a = (1-\lambda_1-\lambda_2-\cdots-\lambda_k)\overline{00'}$, then the parallel translation $p_a\colon E^{n+1} \to E^{n+1}$ maps the point X into the point X'. From (7.10) it follows that

$$M' = p_a(M),$$

where M' is the linear combination of the sets $M_1, M_2, \ldots, M_k$ with respect to the new origin $0'$. □

Remark. If $\lambda_1 + \lambda_2 + \cdots + \lambda_k = 1$, then the vector a is equal to zero. Hence p_a is the identity transformation of E^{n+1}.

For a convex linear combination of the sets $M_1, M_2, \ldots, M_k$ we have $\lambda_1 \geq 0$, $\lambda_2 \geq 0, \ldots, \lambda_k \geq 0$ and $\lambda_1 + \lambda_2 + \cdots + \lambda_k = 1$. Hence a convex combination of any sets $M_1, M_2, \ldots, M_k$ does not depend on the position of the origin 0 in E^{n+1}.

Theorem 7.2. *Let M be a linear combination of sets $M_1, M_2, \ldots, M_k$ with coefficients $\lambda_1, \lambda_2, \ldots, \lambda_k$. We consider this linear combination with respect to the origin 0 of E^{n+1}.*

Let $p_{a_1}, p_{a_2}, \ldots, p_{a_k}$ be parallel translations of E^{n+1} and let

$$M_1' = p_{a_1}(M_1), \ldots, M_k' = p_{a_k}(M_k). \tag{7.11}$$

Finally let

$$M' = \sum_{i=1}^{k} \lambda_i M_i'$$

be a linear combination of $M_i^!$ with respect to the same origin 0. Then

$$M' = p_{\lambda_1 a_1 + \lambda_2 a_2 + \cdots + \lambda_k a_k}(M),$$

i.e. M' can be obtained from M by the parallel translation $P_{\lambda_1 a_1 + \lambda_2 a_2 + \cdots + \lambda_k a_k}$.

Proof. The set M consists of all terminal points of vectors $\overline{0X}$, where

$$X = \lambda_1 X_1 + \cdots + \lambda_k X_k$$

and $X_1 \in M_1, \ldots, X_k \in M_k$. If $Y_1 = p_{a_1}(X_1)$, $Y_2 = p_{a_2}(X_2), \ldots, Y_k = p_{a_k}(X_k)$, then the set M' consists of all terminal points of vectors $\overline{0Y}$, where

$$\begin{aligned} \overline{0Y} &= \sum_{i=1}^{k} \lambda_i \overline{0Y_i} = \sum_{i=1}^{k} \lambda_i (\overline{0X_i} + \overline{X_i Y_i}) = \\ &= \overline{0X} + \lambda_1 \overline{0A_1} + \cdots + \lambda_k \overline{0A_k}. \end{aligned} \tag{7.12}$$

In (7.12) we define the points $A_1, A_2, \ldots, A_k$ of E^{n+1} by the formulas

$$\overline{0A_i} = \overline{X_i Y_i} = a_i, \quad i = 1, 2, \ldots, k. \tag{7.13}$$

According to the axioms of E^{n+1} there is only one point A_i for equation (7.13).
Thus

$$Y = p_a(X),$$

where $a = \lambda_1 a_1 + \cdots + \lambda_k a_k$. Hence

$$M' = p_a(M). \qquad \square$$

Finally from Theorem 7.1 and 7.2 it follows that a linear combination of sets $M_1, M_2, \ldots, M_k$ with fixed coefficients $\lambda_1, \lambda_2, \ldots, \lambda_k$ does not change to within parallel translations of the origin and sets $M_1, M_2, \ldots, M_k$.

Let now $\lambda_1, \lambda_2, \ldots, \lambda_k$ be non-negative numbers. We consider dilations of the sets $M_1, M_2, \ldots, M_k$ with coefficients $\lambda_1, \lambda_2, \ldots, \lambda_k$ and obtain the sets $\lambda_1 M_1, \lambda_2 M_2, \ldots, \lambda_k M_k$. We fix an arbitrary point X in the set $\lambda_2 X_2$ and translate $\lambda_2 X_2$ such that X is any point of $\lambda_1 X_1$. The set obtained as the union of all such admissible translations of $\lambda_2 M_2$ is clearly the set $\lambda_1 M_1 + \lambda_2 M_2$. Starting from the set $\lambda_3 M_3$ and using the same construction with respect to the set $\lambda_1 M_1 + \lambda_2 M_2$ we obtain the set $\lambda_1 M_1 + \lambda_2 M_2 + \lambda_3 M_3$. Finally we obtain the set $\lambda_1 M_1 + \lambda_2 M_2 + \cdots + \lambda_k M_k$ by these constructions.

We now consider the general case, where $\lambda_1, \lambda_2, \ldots, \lambda_k$ *are any real numbers*. Clearly $\lambda_1 M_1 + \lambda_2 M_2 + \cdots + \lambda_k M_k = |\lambda_1| M_1' + |\lambda_2| M_2' + \cdots + |\lambda_k| M_k'$, where M_s' coincides with M_s if $\lambda_s \geq 0$ and M_s' is symmetric to M_s with respect to the origin, if $\lambda_s < 0$.

Clearly

$$\sum_{s=1}^{k} \lambda_{i_s} M_{i_s} = \lambda_1 M_1 + \lambda_2 M_2 + \cdots + \lambda_k M_k$$

since the addition of vectors is a commutative operation. In the last equation $i_1, i_2, \ldots, i_k$ is a rearrangement of $1, 2, \ldots, k$.

7.2 Exercises and Problems to Subsection 7.1

In all problems of this subsection $\lambda_1, \lambda_2, \ldots, \lambda_k$ are non-negative numbers.
Problem 1 (Theorem 7.3). A linear combination of convex sets in E^{n+1} is again a convex set.

Definition. *A subset H of a closed set M is called* a face *of M, if*

$$H = M \cap \alpha,$$

where α is a supporting hyperplane of M.

Example. If M is a 3-dimensional solid closed convex polyhedron, then a *face* of M can be an ordinary face, edge or vertex.
Problem 2. Formulate and prove the generalization of this fact for 1-dimensional convex polyhedra.
Problem 3 (Theorem 7.4). Let α and β be supporting hyperplanes of the sets M_1 and M_2 and let α and β have parallel exterior normals. We denote by H_1 and H_2 the faces of M_1 and M_2 with exterior normals parallel to the exterior normals of α and β. Then

$$\gamma = \lambda_1 \alpha + \lambda_2 \beta$$

is the supporting hyperplane of the set

$$M = \lambda_1 M_1 + \lambda_2 M_2,$$

and

$$H = \lambda_1 H_1 + \lambda_2 H_2$$

is the face of M with the same exterior normal as H_1 and H_2.
Problem 4 (Theorem 7.5). Let $P_1, P_2, \ldots, P_s$ be a system of k-convex polyhedra, $k \leq n+1$, lying in parallel k-planes of E^{n+1}. Then

$$P = \lambda_1 P_1 + \lambda_2 P_2 + \cdots + \lambda_s P_s,$$
$$\lambda_1 \geq 0, \lambda_2 \geq 0, \ldots, \lambda_s \geq 0,$$

is a k-convex polyhedron lying in a k-plane, which is parallel to the planes of $P_1, \ldots, P_s$.
Problem 5 (Theorem 7.6). Linear combinations of any $(n+1)$-convex solid polyhedra with non-negative coefficients are again $(n+1)$-convex polyhedra.
Problem 6 (Theorem 7.7). A linear combination

$$P = \lambda_1 P_1 + \lambda_2 P_2, \quad \lambda_1 \geq 0, \lambda_2 \geq 0$$

of $(n+1)$-convex solid polyhedra P_1, P_2 is again a $(n+1)$-convex solid polyhedron.

If Q is a *face* of P, then $Q = \lambda_1 Q_1 + \lambda_2 Q_2$ are *faces* of P_1 and P_2; Q_1 and Q_2 lie in supporting hyperplanes of P_1 and P_2 respectively and have the parallel exterior normals.

Hint. Use Theorems 7.3–7.6 (Problems 1–5) in the proof of Theorem 7.7 (Problem 6).

7.3 Minkowski Mixed Volumes for Convex Polyhedra

Let $\lambda_1, \lambda_2, \ldots, \lambda_k$ be real non-negative numbers and $P_1, P_2, \ldots, P_k$ be convex bounded solid polyhedra in E^{n+1}. From Subsection 7.2 it follows that

$$P = \lambda_1 P_1 + \lambda_2 P_2 + \cdots + \lambda_k P_k \tag{7.14}$$

is also a convex bounded solid polyhedron. According to problem 6 (Theorem 7.7) each face Q of P is a linear combination

$$Q = \lambda_1 Q_1 + \lambda_2 Q_2 + \cdots + \lambda_k Q_k, \tag{7.15}$$

where Q_s are faces of P_s, $s = 1, 2, \ldots, k$, which lie in planes parallel to Q. Not that the parallelism of faces is understood in terms of exterior normals. Finally, the faces Q_i forming the n-face Q can have the dimensions k_i, where $0 \leq k_i \leq n$.

Let Q be a n-face of P, ν be the exterior unit normal of Q and $h(\nu)$ be the supporting function of P. The number

$$h = h(\nu) \tag{7.16}$$

is called the *supporting number* of the n-face Q. Let $h_1, h_2, \ldots, h_k$ be the supporting numbers of the faces $Q_1, Q_2, \ldots, Q_k$ of the polyhedra $P_1, P_2, \ldots, P_k$, then

$$h = \lambda_1 h_1 + \lambda_2 h_2 + \cdots + \lambda_k h_k \tag{7.17}$$

is the supporting number of the face Q of $P = \lambda_1 P_1 + \lambda_2 P_2 + \cdots + \lambda_k P_k$.

Indeed, if $\overline{0X_1}, \overline{0X_2}, \ldots, \overline{0X_k}$ are the position vectors of points $X_i \in Q_i$, $i = 1, 2, \ldots, k$, and

$$\overline{0X} = \lambda_1 \overline{0X_1} + \lambda_2 \overline{0X_2} + \cdots + \lambda_k \overline{0X_k}, \tag{7.18}$$

then X is a point of $Q = \lambda_1 Q_1 + \lambda_2 Q_2 + \cdots + \lambda_k Q_k$. Above we denote by ν the common exterior unit normal of the faces $Q, Q_1, \ldots, Q_k$. Therefore according to the properties of the supporting function we obtain

$$h = (\nu, \overline{0X}), h_i = (\nu, \overline{0X_i}), \quad i = 1, 2, \ldots, k. \tag{7.19}$$

From (7.18) and (7.19) it follows that

$$h = (\nu, \overline{0X}) = \sum_{i=1}^{k} \lambda_i (\nu, \overline{0X_i}) = \sum_{i=1}^{k} \lambda_i h_i.$$

Thus formula (7.17) is proved. □

Theorem 7.8. *Let $\lambda_1, \lambda_2, \ldots, \lambda_k$ be arbitrary non-negative numbers and let $P_1, P_2, \ldots, P_k$ be given convex bounded solid polyhedra in E^{n+1}. Finally let*

$$P = \lambda_1 P_1 + \lambda_2 P_2 + \cdots + \lambda_k P_k.$$

Then the volume $V(P)$ of a convex bounded solid polyhedron P is a homogeneous polynomial of degree $(n+1)$ with respect to the real variables $\lambda_1, \lambda_2, \ldots, \lambda_k$.

Proof. We use induction. If $n = 0$, then our assertion is obvious since P_i are segments and the length of

$$P = \sum_{i=1}^{k} \lambda_i P_i$$

is equal to a linear combination of the lengths of P_i with the same non-negative coefficients $\lambda_1, \lambda_2, \ldots, \lambda_k$.

We now assume that the theorem is valid in E^n. Let convex bounded solid polyhedra $P_1, P_2, \ldots, P_k$ lie in E^{n+1}. Then the volume of the polyhedron

$$P = \lambda_1 P_1 + \lambda_2 P_2 + \cdots + \lambda_k P_k \tag{7.20}$$

can be expressed by the formula

$$V(P) = \frac{1}{n+1} \sum_{i=1}^{k} h_i F_i, \tag{7.21}$$

where $h_i, i = 1, 2, \ldots, k$, are the supporting numbers of P and F_i are the areas of n-dimensional faces of P, which correspond to the numbers h_i.

These n-faces of P are the same linear combinations of the appropriate parallel faces of convex polyhedra $P_1, P_2, \ldots, P_k$. Therefore from Theorems 7.1 and 7.2 it follows that every face of P is a linear combination of convex polyhedra in E^n. Hence by induction the n-volumes of these faces, i.e. the areas F_i, are homogeneous polynomials of degree n with respect to $\lambda_1, \lambda_2, \ldots, \lambda_k$.

According to (7.17) the supporting numbers h_i of the convex polyhedron P are linear functions of $\lambda_1, \lambda_2, \ldots, \lambda_k$. Therefore the right side of (7.21) is a homogeneous polynomial of degree $n+1$ with respect to $\lambda_1, \lambda_2, \ldots, \lambda_k$. □

It is useful to write the volume $V(P)$ of a convex bounded solid polyhedron

$$P = \lambda_1 P_1 + \lambda_2 P_2 + \cdots + \lambda_k P_k$$

in the form

$$V(P) = \sum_{i_1, \ldots, i_{n+1}}^{k} \lambda_{i_1} \lambda_{i_2} \ldots \lambda_{i_{n+1}} V_{i_1 i_2 \ldots i_{n+1}}, \tag{7.22}$$

where every index i_j runs independently from all others from 1 to k. Hence the product $\lambda_{i_1} \lambda_{i_2} \ldots \lambda_{i_{n+1}}$ is met as many times as the number of rearrangements of the positive integers $i_1, i_2, \ldots, i_{n+1}$. The coefficients $V_{i_1 i_2 \ldots i_{n+1}}$ are defined as numbers independent of the order of $i_1, i_2, \ldots, i_{n+1}$.

Let $\lambda_{i_1}\lambda_{i_2}\ldots\lambda_{i_{n+1}}$ be a product of the numbers taken from $\lambda_1, \lambda_2, \ldots, \lambda_k$. Set the values of λ_j equal to zero except the chosen numbers $\lambda_{i_1}, \lambda_{i_2}, \ldots, \lambda_{i_{n+1}}$ in the formula

$$P = \sum_j \lambda_j P_j. \tag{7.23}$$

Then the corresponding polyhedra P_j are not included in the linear combination (7.23). Thus the coefficient $V_{i_1 i_2 \ldots i_{n+1}}$ for the product $\lambda_{i_1}\lambda_{i_2}\ldots\lambda_{i_{n+1}}$ only depends on the polyhedra $P_{i_1}, P_{i_2}, \ldots, P_{i_{n+1}}$. The polyhedra $P_{i_1}, P_{i_2}, \ldots, P_{i_{n+1}}$, are not necessarily distinct because the positive integers $i_1, i_2, \ldots, i_{n+1}$ are not necessarily distinct.

We consider the special case $\lambda_1 = 1$, $\lambda_2 = \cdots = \lambda_k = 0$ as a useful example. Then $P = P_1$. From (7.22) it follows that

$$V_{11\ldots 1} = V(P_1).$$

The coefficients $V_{i_1 i_2 \ldots i_{n+1}}$ are called the *Minkowski mixed volumes* (or briefly *mixed volumes*) of the polyhedra P_{i_1},$P_{i_2}, \ldots$,$P_{i_{n+1}}$. We denote these volumes by $V(P_{i_1}, P_{i_2}, \ldots, P_{i_{n+1}})$. From the definition it follows that $V(P_{i_1}, P_{i_2}, \ldots, P_{i_{n+1}})$ do not depend on order of indices $i_1, i_2, \ldots, i_{n+1}$, i.e. they are symmetric functions of $P_{i_1}, P_{i_2}, \ldots, P_{i_{n+1}}$.

For the case of two convex bounded solid polyhedra P_1 and P_2 the formula (7.22) takes the form

$$V(\lambda_1 P_1 + \lambda_2 P_2) = \tag{7.24}$$
$$= \sum_{j=1}^{n+1} \lambda_1^{n-j+1}\lambda_2^j C_{n+1}^j V(\underbrace{P_1, \ldots, P_1}_{n-j+1}; \underbrace{P_2, \ldots, P_2}_{j}).$$

Theorem 7.9. *Let $V(P_0, P_1, \ldots, P_1)$ be a mixed volume, $F_i(P_1)$ be the areas of faces of the polyhedron P_1, and h_i^0 be the corresponding supporting numbers of the polyhedron P_0, then*

$$V(P_0, P_1, \ldots, P_1) = \frac{1}{n+1}\sum_i h_i^0 F_i(P_1). \tag{7.25}$$

Proof. Let $P = P_1 + \lambda P_0$, where P_1 and P_0 are convex bounded solid polyhedra in E^{n+1}. Since

$$V(P) = \sum_i h_i F_i(P),$$

we have

$$\frac{\partial V(P)}{\partial h_i} = F_i(P).$$

From (7.17) it follows that

$$h_i = h_i^1 + \lambda h_i^0,$$

where h_i^1 are the corresponding supporting numbers of the polyhedron P_1. Therefore

$$\frac{\partial V(P)}{\partial \lambda} = \sum_i \frac{\partial V}{\partial h_i}\frac{dh_i}{d\lambda} = \sum_i h_i^0 F_i(P).$$

Since

$$\lim_{\lambda\to 0} P = \lim_{\lambda\to 0}[P_1 + \lambda P_0] = P_1$$

we have

$$\lim_{\lambda\to 0} F_i(P) = F_i(P_1)$$

for all i. Thus

$$\left.\frac{\partial V(P)}{\partial \lambda}\right|_{\lambda=0} = \sum_i h_i^0 F_i(P_1). \tag{7.26}$$

On the other hand from (7.22) it follows that

$$\begin{aligned} V(P) &= V(P_1 + \lambda P_0) = \\ &= V(P_1, \ldots, P_1) + (n+1)\lambda V(P_0, P_1, \ldots, P_1) + \cdots \end{aligned}$$

where terms with $\lambda^2, \lambda^3, \ldots, \lambda^n, \lambda^{n+1}$ are denoted by points. From the last equation we obtain

$$\left.\frac{\partial V(P)}{\partial \lambda}\right|_{\lambda=0} = (n+1)V(P_0, P_1, \ldots, P_1). \tag{7.27}$$

Now (7.25) follows from (7.26) and (7.27). □

Remark. The formula (7.25) can be extended to mixed volumes of several convex polyhedra.

Theorem 7.10. *Let $P_1, P_2, \ldots, P_{n+1}$ be convex bounded solid polyhedra in E^{n+1} and let*

$$P = \lambda_1 P_1 + \lambda_2 P_2 + \cdots + \lambda_{n+1} P_{n+1},$$

where $\lambda_1 \geq 0$, $\lambda_2 \geq 0, \ldots, \lambda_{n+1} \geq 0$. Let Q^i be a n-face of P. Then there exist faces $Q_1^i, Q_2^i, \ldots, Q_{n+1}^i$ of $P_1, P_2, \ldots, P_{n+1}$ respectively such that the equations

$$Q^i = \lambda_1 Q_1^i + \lambda_2 Q_2^i + \cdots + \lambda_{n+1} Q_{n+1}^i;$$

and

$$V(P_1, P_2, \ldots, P_{n+1}) = \frac{1}{n+1}\sum_i h_1^i F(Q_2^i, \ldots, Q_{n+1}^i), \tag{7.28}$$

hold, where $F(Q_2^i, \ldots, Q_{n+1}^i)$ is the mixed n-volume (area) of $Q_2^i, \ldots, Q_{n+1}^i$.

We offer the proof of Theorem 7.10 as a useful exercise.

Hint: Use the definition of mixed volumes and Theorem 7.9.

7.4 The Minkowski Mixed Volumes for General Bounded Convex Bodies

Let $H_1, H_2, \ldots, H_{n+1}$ be given bounded convex bodies in E^{n+1}, and

$$H = \lambda_1 H_1 + \lambda_2 H_2 + \cdots + \lambda_{n+1} H_{n+1},$$

where $\lambda_1 \geq 0, \lambda_2 \geq 0, \ldots, \lambda_{n+1} \geq 0$. If bounded solid convex polyhedra $P_1^{(i)}, P_2^{(i)}, \ldots, P_{n+1}^{(i)}$, $i = 1, 2, \ldots$ converge to $H_1, H_2, \ldots, H_{n+1}$ respectively, then the bounded solid convex polyhedra

$$P^{(i)} = \lambda_1 P_1^{(i)} + \lambda_2 P_2^{(i)} + \cdots + \lambda_{n+1} P_{n+1}^{(i)}$$

converges to H.

Let $F = \partial H$. Then F is a closed convex hypersurface in E^{n+1}. If ψ_F is the spherical mapping of F, then $\psi_F(F) = S^n$, where S^n is the unit hypersphere in E^{n+1}.

We denote by $M \subset F$ the preimage of any set $N \subset S^n$ with respect to the mapping ψ_F. Clearly M is a subset of F and for any point $X \in M$ there is at least one supporting hyperplane α of F such that $X \in \alpha \cap F$ and $\psi_F(X) \subset N$.

The set function

$$\mu_F(N) = \text{area } M$$

is called *the surface function* of F.

Theorem 7.11. *Let H be a bounded convex body in E^{n+1} and $F = \partial H$ be a closed convex hypersurface in E^{n+1}. Then the surface function $\mu_F(N)$, $N \subset S^n$, is a non-negative, completely additive set function on the ring of Borel subsets of S^n. Moreover if bounded convex bodies $H^{(i)}$ converge to a bounded convex body H as $i \to +\infty$, then $\mu_{F^{(i)}}(N)$ converge weakly to $\mu_F(N)$, where*

$$F^{(i)} = \partial H^{(i)}, \qquad F = \partial H.$$

The proof of this theorem was given independently by Alexandrov [3], and Fenchel–Jessen [1]; see also Busemann [1]. We refer the readers to these papers.

We now study a few formulas for the Minkowski mixed volumes.

Let H be a bounded convex body in E^{n+1} and $F = \partial H$. Clearly the formula

$$V(H) = \int_{S^n} h(\nu)\mu_F(de_\nu) \tag{7.29}$$

holds, where $h(\nu)$ and $\mu_F(e)$ are the supporting and surface functions of H respectively.

Indeed

$$V(P) = \sum_i [h(\nu_i) \cdot F_i] \tag{7.30}$$

for any bounded solid convex polyhedron P, where $\nu_1, \nu_2, \ldots, \nu_m$ are the unit exterior normals to faces of P and $F_i = \mu_{\partial P}(\nu_i)$ are the areas of the faces of P with these normals.

If bounded solid convex polyhedra $P^{(i)}$ converge to a bounded convex body H in E^{n+1}, then from (7.30) and Theorem 7.11 it follows that

$$V(H) = \lim_{i \to +\infty} V(P^{(i)}). \tag{7.31}$$

From (7.31) and Theorem 7.11 we obtain the validity of (7.29).

Thus we derive the following.

Theorem 7.12. *The volume of a linear combination of bounded convex bodies in E^{n+1} is a homogeneous polynomial of degree $n+1$ with respect to its coefficients.*

Note that Theorems 7.9 and 7.10 can be generalized to bounded convex bodies in E^{n+1}. For example the equation (7.25) takes the form

$$V(H_0, H_1, \ldots, H_1) = \int_{\partial H_1} h^0(\nu) dS^{(1)}, \tag{7.32}$$

where $h^0(\nu)$ is the supporting function of H_0 computed for unit vector ν, and $dS^{(1)}$ is the element of the area ∂H_1.

Now we briefly describe a few simple properties of mixed volumes:

1) Let $H_1, H_2, \ldots, H_{n+1}$ be bounded convex bodies in E^{n+1}. They are necessarily distinct. The important case is that two of them are distinct. In this case we use the special notation

$$V_m(H_1, H_2) = V(\underbrace{H_1, \ldots, H_1}_{n+1-m}, \underbrace{H_2, \ldots, H_2}_{m}) = \tag{7.33}$$
$$= V_{n+1-m}(H_1, H_2).$$

Therefore

$$V(\lambda_1 H_1 + \lambda_2 H_2) = \sum_{m=0}^{n+1} C_m^{n+1} \lambda_1^{n+1-m} \lambda_2^m V_m(H_1, H_2),$$

and

$$nV_1(H_1, H_2) = \lim_{\lambda \to 0} \frac{V(H_1 + \lambda H_2) - V(H_1)}{\lambda}, \tag{7.34}$$

where $\lambda \geq 0$.

2) A translation of every H_i in an arbitrary direction induces only a translation of $H = \sum_i \lambda_i H_i$. Therefore the mixed volumes are invariants of such translations. This is not correct for other types of motions of bodies H_i.

3) The equation

$$V(H_1, H_2, \ldots, H_n, X) = 0 \tag{7.35}$$

holds for any point $X \in E^{n+1}$ and any convex bounded bodies $H_1, H_2, \ldots, H_n$ in E^{n+1}. The equation (7.35) follows directly from the identity

$$V(\lambda_1 H_1 + \cdots + \lambda_n H_n) = V(\lambda_1 H_1 + \cdots + \lambda_n H_n + \lambda_{n+1} X).$$

If convex bodies $H_1, H_2, \ldots, H_{n+1}$ coincide with the same convex body H, then

$$V\left(\sum_i \lambda_i H\right) = \left(\sum_i \lambda_i\right)^{n+1} V(H).$$

Thus

$$V(H, H, \ldots, H) = V(H)$$

or

$$V_0(H_1, H_2) = V(H_1), \quad V_{n+1}(H_1, H_2) = V(H_2).$$

4) The monotonicity property is given by the inequality

$$V(H_1, H_2, \ldots, H_{n+1}) \leq V(H_1', H_2', \ldots, H_{n+1}'), \tag{7.36}$$

if $H_i \subset H_i'; i = 1, 2, \ldots, n+1$. We offer the proof of this assertion as an useful exercise.

5) If we replace H_{n+1} with a point $X \in E^{n+1}$, then from (7.35) and (7.36) it follows that mixed volumes take only non-negative values.

Obviously $V(H_1, H_2, \ldots, H_{n+1}) > 0$ if and only if we can find non-degenerate segments $\ell_i \subset H_i$, $i = 1, 2, \ldots, n+1$, such that $Co(\ell_1 \cup \ell_2 \cup \cdots \cup \ell_{n+1})$ does not lie in any hyperplane of E^{n+1}.

6) We now assume that bounded convex bodies $H_1, H_2, \ldots, H_{n+1}$ are strictly convex and their supporting functions are of class C^2 for all non-zero vectors . From (7.29), (7.30) and (7.32) it follows that

$$V_1(H_1, H_2) = \frac{1}{n+1} \int_{S^n} h_2(\nu) D_n(H_1, \nu) d\sigma, \tag{7.37}$$

where ν is an unit vector, $d\sigma$ is the element of the area of S^n, $h_1(\nu)$, $h_2(\nu)$ are the supporting functions of H_1 and H_2, and $D_m(H_1, \nu)$ denotes the sum of all principal minors of order m of the Hessian of $h_1(\nu)$.

The extension of (7.37) is as follows

$$V(H_1, H_2, \ldots, H_{n+1}) = \tag{7.38}$$
$$= \frac{1}{n+1} \int_{S^n} h_{n+1}(\nu) D_n(H_1, H_2, \ldots, H_n, \nu) d\sigma,$$

where ν is a unit vector, $H_1, H_2, \ldots, H_{n+1}$ are bounded strictly convex bodies with supporting functions of class C^2, and $D_n(H_1, H_2, \ldots, H_n, \nu)$ is the multiplier at $\lambda_1 \lambda_2 \ldots \lambda_n$ in $D_n(H_1, \ldots, H_n, \nu)$ divided by $n!$

We obtain from (7.35) that (7.38) vanishes if H_{n+1} degenerates to a point X. Since

$$h_{n+1} = (\overline{0X}, \nu), \tag{7.39}$$

there results

$$\int_{S^n} u_i D_n(H_1, \ldots, H_n, \nu) d\sigma = 0,$$

$i = 1, 2, \ldots, n+1$; $\nu = (u_1, u_2, \ldots, u_{n+1})$, $|\nu| = 1$.

7.5 The Brunn–Minkowski Theorem. The Minkowski Inequalities

Let $V(Q)$ be the $(n+1)$-dimensional volume of a k-convex body $Q \subset E^{n+1}$, $0 \le k \le n+1$. Note that

$$V(Q) \begin{cases} = 0 & \text{if} \quad k < n+1; \\ > 0 & \text{if} \quad k = n+1. \end{cases} \tag{7.40}$$

Theorem 7.13 (The Brunn–Minkowski Theorem). *Let Q_0 and Q_1 be convex bounded bodies in E^{n+1}, which have dimensions k and ℓ respectively, $0 \le k \le n+1$, $0 \le \ell \le n+1$. Then*

$$g(t) = V^{\frac{1}{n+1}}((1-t)Q_0 + tQ_1) \tag{7.41}$$

is a concave function. This function is linear if and only if Q_0 and Q_1 are either homothetic or lie in parallel hyperplanes.)*

Remark 1 (The Second Statement of the Brunn–Minkowski Theorem). If Q_0 and Q_1 are the convex bounded bodies mentioned in Theorem 7, then the inequality

$$\begin{aligned} g(t) = V^{\frac{1}{n+1}}((1-t)Q_0 + tQ_1) \ge \\ \ge (1-t)V^{\frac{1}{n+1}}(Q_0) + tV^{\frac{1}{n+1}}(Q_1) \end{aligned} \tag{7.42}$$

holds for all $t \in [0,1]$.

Clearly inequality (7.42) follows from the concavity of the function $g(t)$, if the convex bodies Q_0 and Q_1 are fixed. Thus the second statement of the Brunn–Minkowski Theorem is a special case of the first one. We now prove the converse statement.

Let Q_t be the convex body $(1-t)Q_0 + tQ_1$ for all $t \in [0,1]$. Then

$$Q_t = \frac{t_2 - t}{t_2 - t_1} Q_{t_1} + \frac{t - t_1}{t_2 - t_1} Q_{t_2},$$

$$\frac{t_2 - t}{t_2 - t_1} > 0, \quad \frac{t - t_1}{t_2 - t_1} > 0, \quad \frac{t_2 - t}{t_2 - t_1} + \frac{t - t_1}{t_2 - t_1} = 1$$

*)Note that the coincidence of hyperplanes is a particular case of their parallelism.

for all t_1, t, t_2 such that $0 \le t_1 < t < t_2 \le 1$.

We now apply inequality (7.42) to Q_t for various systems t_1, t, t_2 such that $0 \le t_1 < t < t_2 \le 1$ and obtain the concavity of the function $g(t)$.

Remark 2. If $V(Q_0) = V(Q_1) = 0$, then the dimensions of Q_0 and Q_1 are not greater than n. There are only two possibilities for Q_0 and Q_1:

a) There exist two parallel hyperplanes T_0 and T_1 such that $Q_0 \subset T_0$ and $Q_1 \subset T_1$. Then $Q_t \subset T_t$, where $Q_t = (1-t)Q_0 + tQ_1$, $0 < t < 1$, and T_t is a hyperplane parallel to T_0 and T_1. Therefore $V(Q_t) = 0$ and $g(t)$ is a linear function vanishing for all $t \in [0, 1]$.

b) It is impossible to find any parallel hyperplanes T_0 and T_1 containing Q_0 and Q_1 respectively. In this case

$$V(Q_t) > 0$$

for all $t \in (0, 1)$. Hence $g(t)$ is not a linear function.

Remark 3. Suppose $V(Q_0) = 0$ and $V(Q_1) > 0$. Clearly the function $g(t)$ will be linear if and only if the body Q_0 is a point.

Then

$$V(Q_t) = t^{n+1} V(Q_1)$$

and all bodies Q_t are homothetic to Q_1 for $t \in (0, 1)$. Clearly

$$V(Q_t) > t^{n+1} V(Q_1)$$

if Q_0 is a convex bounded k-body, $1 \le k \le n$, $0 < t < 1$.

Thus without loss of generality we assume Q_0 and Q_1 have non-zero volumes. The proof is based on the induction method and follows the idea used in Bonnesen and Fenchel [1].

If Q_0 and Q_1 are convex bodies in E^1, then inequality (7.42) is trivial, because we deal with segments on one line.

Lemma 7.1. *If $V(Q_0) = V(Q_1) = 1$ for convex $n+1$-bodies Q_0 and Q_1, then*

$$V(Q_t) \ge 1 \tag{7.43}$$

for $0 < t < 1$, and the equality sign in (7.43) holds if and only if Q_0 is obtained from Q_1 by some parallel translation.

First we show how the general Brunn–Minkowski Theorem follows from Lemma 7.1.

Let Q_0 and Q_1 be any convex bounded (n_1)-bodies, then the $(n+1)$-bodies

$$H_0 = (V(Q_0))^{-\frac{1}{n+1}} Q_0$$

and

$$H_1 = (V(Q_1))^{-\frac{1}{n+1}} Q_1$$

both have volume one. Therefore from Lemma 7.1 it follows that

$$V(H_s) \geq 1 \tag{7.44}$$

where $H_s = (1-s)H_0 + sH_1$, $0 < s < 1$.

$$s = \frac{tV^{\frac{1}{n+1}}(Q_1)}{(1-t)V^{\frac{1}{n+1}}(Q_0) + tV^{\frac{1}{n+1}}(Q_1)},$$

then

$$H_s = \frac{(1-t)Q_0 + tQ_1}{(1-t)V^{\frac{1}{n+1}}(Q_0) + tV^{\frac{1}{n+1}}(Q_1)}.$$

Thus (7.44) is equivalent to the inequality

$$V(Q_t) \geq \left[(1-t)V^{\frac{1}{n+1}}(Q_0) + tV^{\frac{1}{n+1}}(Q_1)\right]. \tag{7.45}$$

(7.45) is exactly the Brunn–Minkowski inequality (see Remark 1). The equality sign holds in (7.45) if and only if it also holds in (7.44), i.e. when H_0 and H_1 are congruent and parallel. Hence Q_0 and Q_1 are homothetic.

Proof of Lemma 7.1: Let Q_0 and Q_1 be convex $(n+1)$-bodies and

$$V(Q_0) = V(Q_1) = 1.$$

Let T be some hyperplane in E^{n+1}. We transfer the bodies Q_0 and Q_1 in a parallel way so that T becomes a common supporting hyperplane to Q_0 and Q_1, and both bodies lie on one side of T. Clearly this does not change the volume of the body $Q_t = (1-t)Q_0 + tQ_1$. We now introduce the variable $v \in [0,1]$ and associate hyperplanes parallel to T cutting from Q_0 and Q_1 pieces with the same volume v in the direction of the exterior normal of T. Let the corresponding cross-sections be $G_0(v)$ and $G_1(v)$. Let $x_0(v)$ and $x_1(v)$ be the distances from $G_0(v)$ and $G_1(v)$ to the hyperplane T, and let $s_0(v)$ and $s_1(v)$ be the areas of $G_0(v)$ and $G_1(v)$, then

$$s_0(v)dx_0(v) = dv \quad \text{and} \quad s_1(v)dx_1(v) = dv.$$

Hence

$$\frac{dx_0(v)}{dv} = \frac{1}{s_0(v)} \quad \text{and} \quad \frac{dx_1(v)}{dv} = \frac{1}{s_1(v)}. \tag{7.46}$$

The linear combination of $G_0(v)$ and $G_1(v)$

$$G_t(v) = (1-t)G_0(v) + tG_1(v)$$

is clearly contained in the body Q_t. Moreover $G_t(v)$ lies in the hyperplane parallel to hyperplane T. The distance $x_t(v)$ from $G_t(v)$ to T is

$$x_t(v) = (1-t)x_0(v) + tx_1(v). \tag{7.47}$$

If $s_t(v)$ is the area of $G_t(v)$, then

$$V(Q_t) \geq \int_0^1 s_t(v)dx_t(v) = \int_0^1 s_t(v)x_t'(v)dv, \tag{7.48}$$

where $V(Q_t)$ is the $(n+1)$-volume of Q_t. By induction, it follows that the Brunn–Minkowski Theorem (Theorem 7.13) is applicable to the cross-sections $G_0(v)$ and $G_1(v)$. Therefore

$$s_t(v) \geq [(1-t)s_0^{\frac{1}{n}}(v) + ts_1^{\frac{1}{n}}(v)]^n. \tag{7.49}$$

Moreover, from (7.46) and (7.47) we obtain

$$\frac{dx_t(v)}{dv} = \frac{1-t}{s_0(v)} + \frac{t}{s_1(v)}. \tag{7.50}$$

Thus from (7.49), (7.50) and (7.48) it follows that

$$V(Q_t) \geq \int_0^1 ((1-t)s_0^{\frac{1}{n}}(v) + ts_1^{\frac{1}{n}}(v))^n \left[\frac{1-t}{s_0(v)} + \frac{t}{s_1(v)}\right] dv. \tag{7.51}$$

Now we show that the integrand in (7.51) is always greater than 1, i.e.

$$\left((1-t)s_0^{\frac{1}{n}} + ts_1^{\frac{1}{n}}(v)\right)^n \left(\frac{1-t}{s_0(v)} + \frac{t}{s_1(v)}\right) \geq 1 \tag{7.52}$$

for every $s_0(v), s_1(v) > 0$ and $0 < t < 1$ only if

$$s_0(v) = s_1(v).$$

Let

$$q = \frac{s_1(v)}{s_0(v)},$$

then the left part of (7.52) can be expressed by the function

$$f(q) = \left[1 - t + tq^{\frac{1}{n}}\right]^n \left(1 - t + \frac{t}{q}\right).$$

Clearly

$$f(0) = f(+\infty) = +\infty.$$

Therefore the function $f(q)$ has at least one local minimum. Since $0 < t < 1$ and

$$f'(q) = \frac{t(1-t)}{q^2}(1 - t + tq^{\frac{1}{n}})^{n-1}(q^{\frac{n+1}{n}} - 1)$$

we have $f'(q) = 0$ only at the point $q = 1$. Thus

$$\inf_{q>0} f(q) = f(1) = 1$$

and

$$f(q) > 1$$

if $q \neq 1$. Hence the inequality (7.52) is proved under the accompanying condition. Thus we proved that

$$V(Q_t) \geq 1.$$

Now we show that $V(Q_t) = 1$ if and only if the bodies Q_0 and Q_1 are congruent and parallel. If $V(Q_t) = 1$, then

$$q = \frac{s_1(v)}{s_2(v)} = 1$$

for every $v \in [0,1]$. Thus

$$s_0(v) = s_1(v)$$

for every $v \in [0,1]$. Since $x_0(0) = x_1(0) = 0$ (i.e. the bodies Q_0 and Q_1 have one and the same supporting hyperplane T), from (7.46) it follows that

$$x_0(v) = x_1(v)$$

for all $v \in [0,1]$. Thus hyperplanes parallel to T cutting equal volumes from Q_0 and Q_1 coincide. Hence the centroids of Q_0 and Q_1 lie on one side of the hyperplane T and are equidistant from T. Now we transfer Q_0 in a parallel way so that the centroids of Q_0 and Q_1 coincide. Then T will again be the common supporting hyperplane for both bodies Q_0 and Q_1. Therefore Q_0 and Q_1 lie on one side of T. But the hyperplane T is arbitrarily chosen. Therefore all supporting hyperplanes of Q_0 and Q_1 coincide if the centroids of these bodies coincide. Hence the convex $(n+1)$-bodies Q_0 and Q_1 coincide after the parallel translation considered in the beginning of the proof. Thus Q_0 and Q_1 are congruent and parallel under this translation. □

Theorem 7.14 (The Minkowski Inequalities). *Let Q_0 and Q_1 be two convex bodies in E^{n+1}. Then the Minkowski inequalities*

$$V_n^{n+1}(Q_1, Q_0) \geq V(Q_1)V^n(Q_0) \tag{7.53}$$

and

$$V_1^{n+1}(Q_1, Q_0) \geq V^n(Q_1)V(Q_0) \tag{7.54}$$

hold.

Proof. First of all remember that $V(Q_0)$ and $V(Q_1)$ are $(n+1)$-volumes of the convex bodies Q_0 and Q_1, and that $V_n(Q_1, Q_0)$ and $V_1(Q_1, Q_0)$ are their mixed volumes which are defined by the formulas

$$V_n(Q_1, Q_0) = V(Q_1, \underbrace{Q_0, Q_0, \ldots, Q_0})$$

and

$$V_1(Q_1, Q_0) = V(\underbrace{Q_1, Q_1, \ldots, Q_1}, Q_0).$$

It suffices to prove the inequality (7.53) since (7.54) can be obtained from (7.53) by interchanging Q_0 and Q_1.

From the Brunn–Minkowski Theorem it follows that the function

$$h(t) = V^{\frac{1}{n+1}}(Q_t) - (1-t)V^{\frac{1}{n+1}}(Q_0) - tV^{\frac{1}{n+1}}(Q_1) \tag{7.55}$$

is non-negative and concave for all $t \in [0,1]$. Since $h(0) = h(1) = 0$, $h'(0) \geq 0$. Moreover $h'(0) = 0$ if and only if $h(t) = 0$ for all $t \in [0,1]$, i.e. the convex bodies Q_0 and Q_1 are either homothetic or lie in parallel hyperplanes, because $h(t)$ is a concave function, vanishing at $t = 0$ and $t = 1$. From formula (7.24) (see Section 7.3) it follows that

$$V(Q_t) = \sum_{k=0}^{n+1} C_{n+1}^k (1-t)^{n+k+1} t^k V_k(Q_0, Q_1)$$

where

$$V_k(Q_0, Q_1) = V(\underbrace{Q_0, \ldots, Q_0}_{n-k+1}, \underbrace{Q_1, \ldots, Q_1}_{k}).$$

Therefore

$$\begin{aligned} h'(0) = {} & [V(Q_0)]^{-\frac{n}{n+1}}[V_1(Q_0, Q_1) - V_0(Q_0, Q_1)] + \\ & + V^{\frac{1}{n+1}}(Q_0) - V^{\frac{1}{n+1}}(Q_1). \end{aligned} \tag{7.56}$$

Since

$$V_1(Q_0, Q_1) = V_n(Q_1, Q_0)$$

and

$$V_0(Q_0, Q_1) = V(Q_0)$$

(see Section 7.4), the equality (7.56) takes form

$$h'(0) = [V(Q_0)]^{-\frac{n}{n+1}} V_n(Q_1, Q_0) - V^{\frac{1}{n+1}}(Q_1).$$

Thus the inequality $h'(0) \geq 0$ is equivalent to the Minkowski inequality

$$V_n^{n+1}(Q_1, Q_0) \geq V(Q_1) V^n(Q_0).$$

The equality sign holds if and only if $h'(t) = 0$, i.e. when Q_0 and Q_1 are homothetic or lie in parallel hyperplanes. □

We also consider so-called quadratic Minkowski inequalities which follow from the inequality $h''(0) \leq 0$, where $h(t)$ is the concave function introduced in the proof of Lemma 7.1.

Theorem 7.15 (The Quadratic Minkowski Inequalities). *Let Q_0 and Q_1 be two convex bodies, then*

$$V_1^2(Q_0,Q_1) \geq V(Q_0)V_2(Q_0,Q_1) \tag{7.57}$$

and

$$V_n^2(Q_0,Q_1) \geq V(Q_1)V_2(Q_0,Q_1). \tag{7.58}$$

Proof. It suffices to prove only the inequality (7.57) because (7.58) can be obtained from (7.57) by interchanging Q_0 and Q_1. From (7.55), (7.24) and (7.35) it follows that

$$h''(0) = -\frac{1}{V(Q_0)}[-V(Q_0)+V_1(Q_0,Q_1)]^2 + [V(Q_0)+V_2(Q_0,Q_1)]. \tag{7.59}$$

Since

$$0 \geq h''(0) = 2V_1(Q_1,Q_1) - \frac{V_1^2(Q_0,Q_1)}{V(Q_0)} + V_2(Q_0,Q_1)$$

and

$$V_1(Q_0,Q_1) \geq 0$$

we obtain

$$V_1^2(Q_0,Q_1) \geq V(Q_0)V_2(Q_0,Q_1).$$

The proof of Theorem 7.15 is completed. □

Remark 4. Assume that Q_0 and Q_1 do not lie in parallel hyperplanes. Then the equality holds in (7.57) not only when Q_0 and Q_1 are homothetic but also when Q_0 is homothetic to a Kappenkörper of Q_1 (see Bonnesen and Fenchel [1], pp. 17, 92). Minkowski conjectured and Bol [1] proved, that this is the only case in which the equality takes place. The same statement holds for (9.38) if we replace Q_0 for Q_1 and Q_1 for Q_0.

7.6 Alexandrov's and Fenchel's Inequalities

In this section we consider a brief review of the Alexandrov and Fenchel inequalities between mixed volumes of convex bodies. The basic inequality proved by Fenchel [1] and Alexandrov [3] independently is the natural generalization of (7.57) and (7.58).

Theorem 7.16. *Let $C_1, C_2, \ldots, C_{n-1}, Q_0, Q_1$ be convex bodies in E^{n+1}, then*

$$\begin{aligned} &V^2(C_1,C_2,\ldots,C_{n-1},Q_0,Q_1) \geq \\ &V(C_1,C_2,\ldots,C_{n-1},Q_0,Q_0)V(C_1,C_2,\ldots,C_{n-1},Q_1,Q_1). \end{aligned} \tag{7.60}$$

There are three proofs of Theorem 7.16, one by Fenchel [1] and two by Alexandrov [3]. Alexandrov's second proof seems to be the simplest. It generalizes

Hilbert's proof of the inequality (7.57) (see Bonnesen and Fenchel [1], pp. 102–104). For an outline of Alexandrov's proof containing all the essential steps, see Busemann [1], pp. 51–60.

We now show that the inequality (7.60) implies generalizations of Theorems 7.13 and 7.14. We set

$$V_{m,k}(C,Q_0,Q_1) = \\ = V(C_1,C_2,\dots,C_{n+1-m},\underbrace{Q_0,\dots,Q_0}_{m-k},\underbrace{Q_1,\dots,Q_1}_{k}) \tag{7.61}$$

$0 \le m \le n+1$, $0 \le k \le m$. Then (7.60) becomes

$$V_{2,1}^2(C,Q_0,Q_1) \ge V_{2,0}(C,Q_0,Q_1)V_{2,2}(C,Q_0,Q_1). \tag{7.62}$$

Theorem 7.17 (The General Brunn–Minkowski Theorem). *If*

$$Q_t = (1-t)Q_0 + tQ_1,$$

then

$$g(t) = V_{m,0}(C,Q_t,Q_1) = V(C_1,C_2,\dots,C_{n+1-m},\underbrace{Q_t,\dots,Q_t}_{m}),$$

$m \ge 2$ is a concave function of $t \in [0,1]$.

Proof. The equality

$$Q_t = \widetilde{Q}_{t'} = (1-t')Q_{t_1} + t'Q_1$$

holds for all $t \in [t_1,1]$, where $0 < t_1 < 1$ is a fixed number and

$$t' = (t-t_1)(1-t_1)^{-1}.$$

We introduce the function

$$\tilde{g}(t') = V_{m,0}^{\frac{1}{m}}(C,Q_{t'},Q_1).$$

Then

$$\tilde{g}''(0) = (1-t_1)^2 g''(t_1). \tag{7.63}$$

From (7.63) it follows that the function $g(t)$ will be concave if $g''(0) \le 0$. We use the formula:

$$V_{m,0}(C,Q_t,Q_1) = \sum_{k=0}^{m} C_m^k (1-t)^{m-k} t^k V_{m,k}(C,Q_0,Q_1)$$

(see Bonnesen and Fenchel [1], p. 40) and obtain

$$g''(0) = (m-1)V_{m,0}^{\frac{1}{m-2}}(C,Q_0,Q_1) \\ [V_{m,0}(C,Q_0,Q_1)V_{m,2}(C,Q_0,Q_1) - V_{m,1}^2(C,Q_0,Q_1)]. \tag{7.64}$$

Now insert $C_{n+1-m+1} = C_{n+1-m+2} = \cdots = C_{n-1} = Q_0$ into (7.62). Then from (7.64) and (7.62) it follows that $g''(0) \leq 0$. (The tacit assumption $V_{m,0} > 0$ is easily removed by a limiting process.)

Theorem 7.17 is proved. □

Remark 5. The inequality

$$g'(0) \geq g(1) - g(0)$$

which follows from the concavity of $g(t)$, gives the inequality

$$V^m_{m,1}(C, Q_0, Q_1) \geq V^{m-1}_{m,0}(C, Q_0, Q_1) V_{m,m}(C, Q_0, Q_1), \tag{7.65}$$

with equality if and only if $g(t)$ is a linear function.

Remark 6. From (7.60) it directly follows that

$$\begin{aligned} &V^2(C_1, C_2, \ldots, C_{n+1-m}, \underbrace{Q_0, \ldots, Q_0}_{m}, \underbrace{Q_1, \ldots, Q_1}_{k}) \geq \\ &V(C_1, C_2, \ldots, C_{n+1-m}, \underbrace{Q_0, \ldots, Q_0}_{m-k+1}, \underbrace{Q_1, \ldots, Q_1}_{k-1}) \\ &V(C_1, C_2, \ldots, C_{n+1-m}, \underbrace{Q_0, \ldots, Q_0}_{m-k-1}, \underbrace{Q_1, \ldots, Q_1}_{k+1}). \end{aligned} \tag{7.66}$$

Inequality (7.66) can be briefly written in the form

$$V^2_{m,k}(C, Q_0, Q_1) \geq V_{m,k-1}(C, Q_0, Q_1) V^k_{m,k+1}(C, Q_0, Q_1); \tag{7.67}$$

Now (7.67) yields the following generalization of the preceding inequalities

$$V^m_{m,k}(C, Q_0, Q_1) \geq V^{m-k}_{m,0}(C, Q_0, Q_1) V^k_{m,m}(C, Q_0, Q_1); \tag{7.68}$$

which becomes an equality when the function $g(t)$ is linear.

Alexandrov [3] noticed that the inequalities (7.68) lead to the more general inequality

$$\begin{aligned} &V^m(C_1, C_2, \ldots, C_{n+1}) \geq \\ &\geq \prod_{k=0}^{m-1} V(C_1, \ldots, C_{n+1-m}, C_{n+1-k}, \ldots, C_{n+1-k}), \end{aligned}$$

which becomes

$$V^{n+1}(C_1, C_2, \ldots, C_{n+1}) \geq V(C_1) V(C_2) \ldots V(C_{n+1}) \tag{7.69}$$

for $m = n+1$.

Inequality (7.68) becomes an equality for any $m > 2$, if the function $g(t)$ is linear. Finally we formulate a theorem due to Alexandrov [3].

Theorem 7.18. *If Q_0 and Q_1 are convex bodies of dimensions of at least m and if $C_1, C_2, \ldots, C_{n+1-m}$ are regular, then the function*

$$g(t) = V^{\frac{1}{m}}(C_1, C_2, \ldots, C_{n+1-m}, Q_t, \ldots, Q_t)$$

is linear if and only if Q_0 and Q_1 are homothetic, where

$$Q_t = (1-t)Q_0 + tQ_1, \quad 0 \le t \le 1.$$

Remember that a convex body C or its boundary ∂C is called regular if C contains interior points, and ∂C is a C^2 hypersurface with positive Gaussian curvature.

§8. Selected Global Problems in Geometric Partial Differential Equations

8.1 Minkowski's Problem for Convex Polyhedra in E^{n+1}

We consider closed convex bounded solid polyhedra in Euclidean space E^{n+1}. If P is such a polyhedron, then we denote by $\eta_1, \eta_2, \ldots, \eta_m$ the unit exterior normals of its n-dimensional faces and by $F_1, F_2, \ldots, F_m$ the areas of these faces.*) For every closed convex polyhedron P the normals $\eta_1, \eta_2, \ldots, \eta_m$ and the numbers $F_1, F_2, \ldots, F_m$ satisfy the following necessary conditions:

1) The vectors $\eta_1, \eta_2, \ldots, \eta_m$ are not coplanar; i.e. $\eta_1, \eta_2, \ldots, \eta_m$ do not lie in any hyperplane in E^{n+1};
2) all numbers F_i are positive:

$$F_i > 0, \quad i = 1, 2, \ldots, m. \tag{8.1}$$

3)
$$\sum_{i=1}^{n} \eta_i F_i = 0. \tag{8.2}$$

The necessity of the first two conditions is obvious. The third condition is equivalent to the statement that the vector area of P is equal to zero. The simple proof of this statement is as follows.

Let Q be any hyperplane in E^{n+1} and η be a unit normal of Q. Then the scalar product (η, η_i) is the cosine between η and η_i. Hence $(\eta, \eta_i)F_i$ is the area of projection of the n-face with the exterior normal η_i. This area is taken with its sign depending on the angle between η and η_i. The projection of P on the hyperplane Q is some bounded convex polyhedron V. Clearly V is covered twice by this projection. One covering has a positive area, and the area of the second one is negative. Thus

$$\sum_{i=1}^{n} (\eta, \eta_i) F_i = 0 \tag{8.3}$$

*) We use the term area for n-dimensional volumes of n-faces of P.

or

$$\left(\eta, \sum_{i=1}^{n} \eta_i F_i\right) = 0. \tag{8.4}$$

Since the hyperplane Q is arbitrarily chosen, (8.4) holds for any unit vector η. Therefore equality (8.2) is proved.

Theorem 8.1 (Minkowski's Theorem for Polyhedra). *The three conditions formulated above are not only necessary but also sufficient for the existence of a closed convex polyhedron with prescribed unit exterior normals $\eta_1, \eta_2, \ldots, \eta_m$ of its n-faces and prescribed areas $F_1, F_2, \ldots, F_m$ of these faces.*

Proof. (Minkowski's method.) For a closed convex polyhedron P its supporting function $h(\eta)$ can be replaced by the system of *supporting numbers* $h_1, h_2, \ldots, h_m$, where $h_i = h(\eta_i)$, $i = 1, 2, \ldots, m$; η is an unit vector; and $\eta_1, \eta_2, \ldots, \eta_m$ are unit exterior normals of all n-faces of P.

From the condition

$$\sum_{i=1}^{m} \eta_i F_i = 0$$

it follows that vectors $\eta_1, \eta_2, \ldots, \eta_m$ can not be directed to any closed halfspace of E^{n+1}.* Therefore we assume that the unit vectors $\eta_1, \eta_2, \ldots, \eta_m$ are pairwise distinct and not directed to any closed halfspace of E^{n+1}.

Let 0 be the origin in E^{n+1} and let $\ell_1, \ell_2, \ldots, \ell_m$ be the rays directed along the vectors $\eta_1, \eta_2, \ldots, \eta_m$ with vertices at the origin 0, such that

$$\operatorname{dist}(0, Q_i) = h_i > 0, \quad i = 1, 2, \ldots, m.$$

Clearly these hyperplanes bound a finite convex solid polyhedron P. Since the rays $\ell_1, \ell_2, \ldots, \ell_m$ are prescribed, the polyhedron P is completely defined by the distances $h_1, h_2, \ldots, h_m$. Generally speaking, it is possible to have for arbitrary numbers $h_i > 0$ some hyperplanes Q_i which do not touch the convex polyhedron P. Hence P may not have all n-faces with the prescribed exterior normals $\eta_1, \eta_2, \ldots, \eta_m$.

We consider the set U, consisting of all such polyhedra P, with n-faces having exterior normals $\eta_1, \eta_2, \ldots, \eta_m$. We allow some n-faces to have zero-area.

Clearly the volume of the polyhedron $P \in U$ is a differentiable function of its supporting numbers $h_1, h_2, \ldots, h_m$. Moreover either $\frac{\partial V(P)}{\partial h_i} = F_i$, where F_i is the area of the corresponding n-face of P, or $\frac{\partial V(P)}{\partial h_i} = 0$, if the hyperplane Q_i does not give any n-face of P.

If Q_i does not give any n-face of P, then it is also convenient to assume that such an n-face exists but its area is equal to zero. Thus the formula

$$\frac{\partial V(P)}{\partial h_i} = F_i \tag{8.5}$$

*In the opposite case $\sum_{i=1}^{m} \eta_i F_i \neq 0$, because all $F_i > 0$.

will be valid in this way without any stipulation.

According to the statement of Minkowski's Theorem for closed convex polyhedra, these are the following prescribed data:

1) the unit vectors $\eta_1, \eta_2, \ldots, \eta_m$, which arenot coplanar and not directed into any closed halfspace of E^{n+1};
2) the system of positive numbers

$$F_i^0 > 0, \quad i = 1, 2, \ldots, m \tag{8.6}$$

satisfying the condition

$$\sum_{i=1}^{m} \eta_i F_i^0 = 0. \tag{8.7}$$

Above, we introduced convex solid polyhedra bounded by hyperplanes Q_i with the unit normals η_i and

$$\operatorname{dist}(0, Q_i) = h_i > 0, \tag{8.8}$$

where 0 is the origin of E^{n+1}. Below, we consider only these polyhedra. We denote this set of polyhedra by U. Let T be a subset of U consisting of all P, for which the condition

$$\sum_{i=1}^{m} h_i F_i^0 = 1 \tag{8.9}$$

holds.

Now consider the properties of the set T.

1) *T is not empty.* Actually, if all numbers $h_i > 0$, $i = 1, 2, \ldots, m$, are sufficiently small, then

$$\sum_{i=1}^{m} h_i F_i^0 < 1.$$

But if all these numbers $h_i > 0$ are sufficiently large then

$$\sum_{i=1}^{m} h_i F_i^0 > 1,$$

because $F_i^0 > 0$ for $i = 1, 2, \ldots, m$. Since the function $\sum\limits_{i=1}^{m} h_i F_i^0$ is continuous with respect to $h_1, h_2, \ldots, h_m$, there exist positive numbers $h_1^0, h_2^0, \ldots, h_m^0$ such that equation (8.9) is satisfied for $h_i = h_i^0$, $i = 1, 2, \ldots, m$. Thus the set T is not empty.

Let $f_0 = \max\limits_{1 \leq i \leq m} \left\{\frac{1}{F_i}\right\}$. Clearly $0 < f_0 < +\infty$. We denote by $G_0 \in U$ the convex solid polyhedron whose supporting numbers are

$$h_1^0 = h_2^0 = \cdots = h_m^0 = f_0. \tag{8.10}$$

A second property of the set T is as follows.

2) *Every polyhedron $P \in T$ lies inside the convex solid polyhedron G_0*, i.e

$$P \subset G_0. \tag{8.11}$$

It is sufficient to prove the inequalities

$$h_i \leq f_0$$

for all $i = 1, 2, \dots, m$, where $h_1, h_2, \dots, h_m$ are supporting numbers of P. Since

$$h_1 F_1^0 + h_2 F_2^0 + \dots + h_m F_m^0 = 1$$

and

$$h_1 > 0,\ h_2 > 0, \dots, h_m > 0,$$

we have

$$h_i < \frac{1}{F_i^0} \leq f_0.$$

Thus property 2 is proved.

Hence all convex polyhedra $P \in T$ are uniformly bounded.

3)
$$V_0 = \sup_{P \in T} V(P) \leq V(G_0) < +\infty, \tag{8.12}$$

where $V(P)$ and $V(G_0)$ are the volumes of convex solid polyhedra P and G_0.

This assertion follows directly from property 2).

4) *There exists a convex solid polyhedron $P_1 \in T$ such that $V_0 = V(P_1)$.*

The proof of this property requires a few preliminary remarks. First of all it is clear that convex solid polyhedra $P^k \in U$ converge to a convex solid polyhedron $P^0 \in U$ (see §4.5, Chapter 1) if and only if

$$\lim_{k \to \infty} h_i^k = h_i^0, \quad i = 1, 2, \dots, m, \tag{8.13}$$

where h_i^k and h_i^0 are supporting numbers of polyhedra P^k and P^0.

Both sets U and T are open with respect to this convergence. If we add convex polyhedra whose supporting numbers can also take zero values, then we extend the sets U and T to the closed sets $\overline{U}$ and $\overline{T}$ with respect to the same convergence. Clearly $\overline{U}$ and $\overline{T}$ are closures of U and T with respect to topology induced by this convergence. From the property 2 it follows that every $P \in \overline{T}$ lies inside G_0, i.e. $P \subset G_0$. Clearly the supporting numbers of P satisfy equality (8.9) and

$$V_0 = \sup_{P \in T} V(P) = \sup_{P \in \overline{T}} V(P).$$

Since $\overline{T}$ is a compact set of convex solid polyhedra and $V(P)$ is continuous in $\overline{T}$, there exists a convex solid polyhedron P_0 such that $V_0 = V(P_0)$.

The proof of property 4 will be completed, if we establish that the convex polyhedron of the maximal volume V_0 can be chosen in the open set T. Thus only the case $P_0 \in \overline{T} \backslash T$ is interesting for further consideration.

If $h_1^0, h_2^0, \dots, h_m^0$ are supporting numbers of P_0, then

$$h_1^0 \geq 0,\ h_2^0 \geq 0, \dots, h_m^0 \geq 0 \tag{8.14}$$

and

$$h_1^0 F_1^0 + h_2^0 F_2^0 + \dots + h_m^0 F_m^0 = 1. \tag{8.15}$$

If we displace all hyperplanes Q_i^0, bounding the polyhedron P_0, by one and the same vector α, then P_0 is also displaced by the vector α and its volume does not change. Therefore we choose the displacement so that the origin 0 of E^{n+1} is inside of the displaced polyhedron. We denote by P_1 the displaced polyhedron. Clearly all its supporting numbers h_i^1 are positive and

$$V(P_1) = V(P_0) = \sup_{P \in T} V(P). \tag{8.16}$$

The proof of property 4 will be complete, if we establish the equality

$$\sum_{i=1}^{m} h_i^1 F_i^0 = 1. \tag{8.17}$$

If the hyperplane Q_i^0 with a normal η_i is displaced by the vector α to the hyperplane Q_i^1, then

$$h_i^1 = \operatorname{dist}(0, Q_i^1) = h_i^0 + (\eta_i, \alpha).$$

Therefore the sum on the left side of (8.17) should be replaced by

$$\sum_{i=1}^{m} (h_i^0 + (\eta_i, \alpha)) F_i^0 = \sum_{i=1}^{m} h_i^0 F_i^0 + \left(\alpha, \sum_{i=1}^{m} \eta_i F_i^0 \right).$$

According to conditions (8.14) and (8.15) imposed on the numbers h_i^0 and F_i^0 we obtain $\sum_{i=1}^{m} h_i^0 F_i^0 = 1$ and $\sum_{i=1}^{m} \eta_i F_i^0 = 0$. Hence

$$\sum_{i=1}^{m} h_i^1 F_i^0 = 1.$$

Thus the convex solid polyhedron P_1 of the maximal volume belongs to T. Property 4 is proved.

Thus the differentiable function $V(P) = V(h_1, h_2, \dots, h_m)$ achieves an absolute maximum at the polyhedron P_1, which is an inner point of the open set T. According to the Lagrangian rule

$$\frac{\partial}{\partial h_i}\left[V(h_1, h_2, \dots, h_m) + \lambda \sum_{j=1}^{m} F_j^0 h_j\right] = 0$$

where λ is some real multiplier. As we mentioned above (see formula (8.5))

$$\frac{\partial}{\partial h_i}[V(h_1, h_2, \dots, h_m)] = F_i^1,$$

where F_i^1 are the areas of the n-faces of the polyhedron P_1. Therefore

$$\mu F_i^1 = F_i^0, \quad i = 1, 2, \dots, m \tag{8.18}$$

where $\mu = -\frac{1}{\lambda}$. Equation (8.18) mean that the areas of n-faces of the polyhedron P_1 are proportional to prescribed numbers F_i^0. Therefore all these areas are positive. Thus the desired polyhedron is

$$P_1^* = (\mu)^{\frac{1}{n-1}} P_1. \tag{8.19}$$

The proof of Minkowski's Theorem for closed convex polyhedra is complete.□

8.2 The Classical Minkowski Theorem

This theorem is related to existence and uniqueness of a convex closed hypersurface with prescribed Gaussian curvature. One assumes that the prescribed data is given by a positive and continuous function $K(\eta)$ defined on the hypersphere S^n. More explicitly, Minkowski's problem consists of finding a convex closed hypersurface F, for which $K(\eta)$ is its Gaussian curvature at the point with the outward unit normal η.

The Gaussian curvature at point X of a convex hypersurface F is defined as

$$\lim_{G \to X} \frac{\sigma(G)}{s(G)}, \tag{8.20}$$

where G is a domain of the hypersurface F, which shrinks to X, and $\sigma(G)$ and $s(G)$ are the integral Gaussian curvature*) and area of G respectively. This definition of the Gaussian curvature does not assume the regularity **) of a convex hypersurface F, because both set functions $\sigma(G)$ and $s(G)$ are

*) The set function $\sigma(G)$ is often called the area of the spherical image of G (see §6).

**) A hypersurface F is called regular if F has a local C^m-representation, $m \geq 2$ in some neighborhood of any point of F by a vector-function $r = r(u_1, \dots, u_n)$, and $(dr)^2$ is a positive form for admissible values of local coordinates $u_1, \dots, u_n$.

defined for all Borel subsets of any general convex hypersurface F. Therefore the solutions of Minkowski's problem considered above are called *generalized*. These solutions do not contain any information concerning their regularity even if we assume the analyticity of $K(\eta)$.

Simple examples were constructed in this way by Alexandrov [3]. They show that existence and continuity of the Gaussian curvature of a convex surface in E^3 do not provide the existence of second derivatives at all points of such a surface. Here are the examples:

1) The surface

$$2z = ax^2 + \frac{1}{a}y^2$$

is convex in some neighborhood of the point $x = 0$, $y = 0$. The Gaussian curvature of this surface is continuous at all points of this neighborhood and equal to one at the point $x = 0$, $y = 0$. But any normal section at the point $x = 0$, $y = 0$ of this surface does not have a definite value of curvature.

2) The Gaussian curvature of the convex surface

$$z^2 = |x|^7 + |y|^3, \quad z \geq 0$$

is continuous in some neighborhood of the point $x = 0$, $y = 0$ and its value is zero at the point $x = 0$, $y = 0$. But at the same point $x = 0$, $y = 0$ one of the principal normal curvatures is equal to zero and the value of the second one is $+\infty$.

If the Gaussian curvature of a convex hypersurface is prescribed as a positive continuous function on the unit sphere S^n, then the surface function of this hypersurface is given by the formula

$$\mu(H) = \int_H \frac{d\sigma}{K(\eta)}, \tag{8.21}$$

where H is any Borel subset of S^n.

We now consider the classical Minkowski Theorem.

Theorem 8.2 (Minkowski). *Let $K(\eta)$ be a positive continuous function on S^n, satisfying the condition*

$$\int_{S^n} \frac{\eta d\sigma}{K(\eta)} = 0. \tag{8.22}$$

Then there exists a convex closed hypersurface F for which $K(\eta)$ is its Gaussian curvature at a point with the exterior unit normal η.

Proof. We decompose S^n into the small domains $g_1, g_2, \ldots, g_m$ and define the numbers $F_1, F_2, \ldots, F_m$ and unit vectors $\eta_1, \eta_2, \ldots, \eta_m$ by the conditions

$$F_k\eta_k = \int_{g_k} \frac{\eta d\sigma}{K(\eta)}, \tag{8.23}$$

$k = 1, 2, \dots, m$. Clearly

$$\sum_{k=1}^{m} \eta_k F_k = 0. \tag{8.24}$$

According to the Minkowski Theorem for polyhedra (see Subsection 8.1) there exists a convex polyhedron P with n-faces, whose areas are $F_1, F_2, \dots, F_m$ and whose exterior normals are $\eta_1, \eta_2, \dots, \eta_m$. We choose the origin of E^{n+1} at an interior point of the closed bounded convex solid polyhedron P.

We now consider a sequence of decompositions V_s of the unit hypersphere S^n, $s = 1, 2, \dots$ such that

$$\delta_s = \max_{1 \le k \le m_s} \{\mathrm{diam}(g_k^s)\}^{*)}$$

converges to zero, when $s \to +\infty$. Without loss of generality we can assume that

$$\delta_1 > \delta_2 > \cdots > \delta_s > \cdots . \tag{8.25}$$

Let P^s be the convex polyhedron which is constructed above for the decomposition V_s. Below we introduce a few useful geometric figures and notations. Let A_s, B_s be two points of P^s such that $d_s = \mathrm{diam}\ P^s = \mathrm{dist}(A_s, B_s)$. We denote by γ_s the hyperplane passing through the middle of the segment A_sB_s orthogonal to this segment and by Q^s the orthogonal projection of P^s onto the hyperplane γ_s. Let $S(Q^s)$ be the area of Q^s. Finally let σ_n be the area of S^n, $\sigma(M)$ be the area of a set $M \subset S^n$ and $S(P^s)$ be the area of polyhedron P^s. If e_s is the unit vector collinear to the vector $\overrightarrow{A_sB_s}$, then

$$|(\eta, e_s)| = |\cos(\eta, e_s)| \le 1. \tag{8.26}$$

For any $0 \le \varepsilon < 1$ we introduce the subset M_ε of S^n such that for all points $\eta \in M_\varepsilon$ the inequality

$$|(\eta, e_s)| > \varepsilon \tag{8.27}$$

holds.

M_ε is the union of two symmetric spherical segments with parallel boundaries and

$$\lim_{\varepsilon \to 0} \sigma(M_\varepsilon) = \sigma_n. \tag{8.28}$$

Thus there exists $\varepsilon_0 \in (0, 1)$ such that

$$\sigma(M_{\varepsilon_0}) = \frac{3}{4}\sigma_n. \tag{8.29}$$

We denote by $\widetilde{M_\varepsilon^s}$ the union of all domains g_k^s such that for all points $\eta \in g_k^s$ the inequality

$$|(\eta, e_s)| \ge \varepsilon > 0 \tag{8.30}$$

holds. From (8.28), (8.29) and (8.30) it follows that Lemma 8.1 is valid.

---

*) m_s is the number of the decomposition V_s, which consists of small domains $g_1^s, g_2^s, \dots, g_{m_s}^s$.

Lemma 8.1. *There exists a positive integer* s_{ε_0} *such that*

$$\sigma(\widetilde{M}^s_{\varepsilon_0}) > \frac{1}{2}\sigma_n, \tag{8.31}$$

if $s > s_{\varepsilon_0}$.

Lemma 8.2. *If* $s > s_0$, *then the inequality*

$$S(Q^s) > \frac{\varepsilon_0 \sigma_n}{4a} \tag{8.32}$$

holds, where $0 < a = \sup_{S^n} K(\eta) < +\infty$.

Proof.

$$S(Q^s) = \frac{1}{2}\sum_{k=1}^{m} F^s_k |(\eta_k, e_s)| =$$
$$= \frac{1}{2}\sum_{k=1}^{m_s} \left| \int_{g^s_k} \frac{(\eta, e_s) d\sigma}{K(\eta)} \right|, \tag{8.33}$$

where e_s is the unit vector collinear to vector $\overrightarrow{A_s B_s}$.

Since $K(\eta)$ is a positive continuous function on S^n,

$$0 < b = \inf_{S^n} K(\eta) \le \sup_{S^n} K(\eta) = a < +\infty. \tag{8.34}$$

Therefore

$$S(Q^s) \ge \frac{1}{2}\sum_{k=1}^{m_s} \left| \int_{g^s_k} (\eta, e_s) d\sigma \right|. \tag{8.35}$$

Let s_{ε_0} be the positive integer taken from Lemma 8.2. We consider domains $g^s_k \subset \widetilde{M}^s_{\varepsilon_0}$ for every fixed $s > s_{\varepsilon_0}$. Then the estimate

$$\left| \int_{g^s_k} (\eta, e_s) d\sigma \right| = \int_{g^s_k} |(\eta, e_s)| d\sigma > \varepsilon_0 \int_{g^s_k} d\sigma \tag{8.36}$$

holds for any such domain. Now from Lemma 8.1 and (8.35–36) it follows that

$$S(Q^s) > \frac{\varepsilon_0 \cdot \sigma_n}{4a}$$

for any $s > s_{\varepsilon_0}$. □

Lemma 8.2 is very important in the proof of Minkowski's Theorem. First of all this lemma will be applied to establish a uniform estimate from above for the diameters d_s of convex polyhedra P^s. Since the origin 0 lies inside all

P^s, the existence of such an estimate leads immediately to the compactness of the sequence of convex polyhedra P^s (see Theorem 3.5, Chapter 1, §3). The existence of a generalized solution P^0 of Minkowski's problem can be derived by the weak convergence of the surface functions for a convergent subsequence of convex polyhedra P^{s_j}. Lemma 8.2 will also be applied to establish the non-degeneracy of the generalized solution P^0 in the final part of the proof of Theorem 8.2.

The uniform estimate of d_s mentioned above follows directly from Lemma 8.3 and 8.4.

Lemma 8.3. *The inequality*

$$V(P^s) \geq \frac{d_s \varepsilon_0 \sigma_n}{4a(n+1)} \tag{8.37}$$

holds for all $s > s_{\varepsilon_0}$.

Lemma 8.4. *The inequality*

$$V(P^s) \leq V^0 = \text{const} < +\infty$$

holds, where V^0 depends only on the number $b = \inf_{S^n} K(\eta)$ and the constants of the isoperimetric inequality in E^{n+1}.

Proof of Lemma 8.3. We apply Steiner's symmetrization to the polyhedron P^s with respect to the hyperplane γ_s (see Bonnesen–Fenchel [1, p. 69]). This symmetrization does not change the volume $V(P^s)$ of P^s and preserves the set $Q^s = P^s \cap \gamma_s$. Hence the common volume of two convex cones with vertices A_s, B_s and bases Q^s is not more than the volume $V(P^s)$. Thus

$$V(P^s) \geq \frac{1}{n+1} \cdot S(Q^s) d_s \geq \frac{d_s \varepsilon_0 \sigma_n}{4a(n+1)} \tag{8.38}$$

for $s > s_{\varepsilon_0}$. □

Proof of Lemma 8.4. Clearly there is the following estimate for the areas of all convex polyhedra P^s, $s = 1, 2, \dots, s, \dots$:

$$S(P^s) = \sum_{k=1}^{m_s} F_k^s \leq \sum_{k=1}^{m_s} \int_{g_k^s} \frac{d\sigma}{K(\eta)}. \tag{8.39}$$

From (8.34) and (8.39) we obtain

$$S(P^s) \leq \frac{1}{b} \sigma_n.$$

This estimate together with the isoperimetric inequality provides the uniform upper estimate for the volume $V(P^s)$:

$$V(P^s) \leq V^0 = \text{const} < +\infty. \tag{8.40}$$

Clearly the constant V^0 depends only on the number b and the constants in the isoperimetric inequality in E^{n+1}. From (8.37) and (8.40) the uniform estimate

$$d_s \leq \frac{4(n+1)V^0 a}{\varepsilon_0 \cdot \sigma_n} \tag{8.41}$$

can be obtained. □

Since the origin of E^{n+1} is an interior point of all polyhedra P^s (see the proof of Minkowski's Theorem for convex polyhedra in Subsection 8.1), from (8.41) it follows that the family of closed convex polyhedra P^s is uniformly bounded. Now we use Theorem 3.5 (see Chapter 1, §3) and extract from P^s a convergent subsequence of closed convex polyhedra for which we keep the same notation P^s. Now we prove that the limit convex hypersurface P^0 is not degenerate. If this is not the case then there exists a hyperplane α in E^{n+1} such that $P^0 \subset \alpha$. Let $\delta > 0$ be any arbitrary number. We denote by $\sqcap_\delta$ the open layer between two parallel hyperplanes α' and α'', for which α' and α'' are parallel to α. For any $\delta > 0$ there exists a positive integer s_δ such that any polyhedron P^s is contained in $\sqcap_\delta$, if $s > s_\delta$. Let β be any hyperplane orthogonal to the hyperplane α. We denote by $S(Q^s_\beta)$ the area of the projection Q^s_β of the polyhedron P^s on the hyperplane β in the direction of the normal of this hyperplane. Clearly

$$S(Q^s_\beta) \leq d_s^{n-1}\delta,$$

where d_s is the diameter of P^s. Thus from the estimate (8.41) it follows that the number $S(Q^s)$ is arbitrarily small together with the number $\delta > 0$. But on the other hand

$$S(Q^s_\beta) \geq \frac{\varepsilon_0 \sigma_n}{4a}$$

for all $s > s_{\varepsilon_0}$ according to Lemma 8.2. Thus we obtain a contradiction. Hence P^0 is a non-degenerated closed convex hypersurface.

Since the surface functions of convex polyhedra P^s converge weakly to the surface function of the hypersurface P^0, for the surface function of P^0 the expression

$$\mu_{P^0}(H) = \int_H \frac{d\sigma}{K(\eta)}$$

holds for all Borel subsets of the unit hypersphere S^n. Thus the existence of at least one generalized solution for the Minkowski's problem is established.

We now consider the uniqueness of a generalized solution. Let two closed convex hypersurfaces F_1 and F_2 have the same surface function, and let H_1 and H_2 be two bounded convex bodies such that $\partial H_1 = F$, and $\partial H_2 = F_2$. If h_1 and h_2 denote the supporting functions of H_1 and H_2 respectively, and σ_1, σ_2 denote their surface area functions, then

$$V(H_1, H_2, \ldots, H_2) = \frac{1}{n+1}\int_\Omega h_2 d\sigma_1.$$

Since $\sigma_1 \equiv \sigma_2$, the right-hand side of this equation is the volume of H_2. Thus

$$V(H_1, H_2, \ldots, H_2) = V(H_2). \tag{8.42}$$

By the Minkowski inequality

$$V^{n+1}(H_1, H_2, \ldots, H_2) \geq V^n(H_2)V(H_1). \tag{8.43}$$

From (8.42) and (8.43) it follows that

$$V(H_2) \geq V(H_1). \tag{8.44}$$

Since H_1 and H_2 are symmetric in the proof of inequality (8.44) we can interchange H_1 and H_2. Thus, we obtain

$$V(H_1) \leq V(H_2).$$

Therefore $V(H_1) = V(H_2)$. Thus equality holds in Minkowski's inequality (8.43). This is possible if and only if H_1 and H_2 differ by translation. The uniqueness of generalized solutions of the Minkowski problem is proved. □

8.3 General Elliptic Operators and Equations

Let B be a bounded domain in the Euclidean space E^n and let $x_1, x_2, \ldots, x_n$ be Cartesian coordinates in E^n. Let

$$F(x, z, p, r) = F(x_1, \ldots, x_n, z, p_1, p_2, \ldots, p_n, r_{11}, \ldots, r_{nn}) \tag{8.45}$$

be a C^1-function satisfying the condition

$$\frac{\partial F}{\partial r_{ik}} = \frac{\partial F}{\partial r_{ki}}, \quad i, k = 1, 2, \ldots, n, \tag{8.46}$$

where $x \in B$, $z \in R$, $p = (p_1, p_2, \ldots, p_n) \in R^n$ and $r = (r_{11}, r_{12}, \ldots, r_{nn}) \in R^{n^2}$. It is convenient to consider p and r as vectors in the Euclidean spaces R^n and R^{n^2}. Every function $F(x, z, p, q)$ generates the operator

$$\phi(z) = F(x, z(x), Dz(x), D^2z(x)) \tag{8.47}$$

on the set of functions $z(x) \in C^2(B) \cap C(\overline{B})$.

The operator $\phi(z)$ is called *elliptic* on a function $z_0(x) \in C^2(B) \cap C(\overline{B})$ if the quadratic form

$$T(\phi, z_0) = \sum_{i,k=1}^{n} \frac{\partial F^0}{\partial r_{ik}} \xi_i \xi_k \tag{8.48}$$

is defined for all $x \in B$ and $\xi = (\xi_1, \xi_2, \ldots, \xi_n) \in R^n \backslash \{\theta\}$, where

$$\frac{\partial F^0}{\partial r_{ik}} = \frac{\partial F(x, z_0(x), Dz_0(x), D^2z_0(x))}{\partial r_{ik}}, \tag{8.49}$$

$i, k = 1, 2, \ldots, n$, and all the eigenvalues of the matrix $\|\frac{\partial F^0}{\partial r_{ik}}\|$ do not vanish in B and they have one and the same sign at any point of B. The coefficients of the quadratic form $T(\phi, z_0)$ are continuous in B. Hence this form keeps a definite sign in B. Thus we can introduce the following definition:

The operator $\phi(z)$ is *positive (negative) elliptic* on $z_0(x)$ if $T(\phi, z_0)$ is a positive (negative) definite form at any point of B.

A differential equation

$$F(x, z, Dz, D^2 z) = 0 \tag{8.50}$$

is called *elliptic* if the operator $\phi(z)$ is elliptic on all solutions of equation (8.50).

In all considerations and definitions mentioned above it is sufficient to require that $F(x, z, p, r)$ is only a C^1-function with respect to components r_{ik} of vectors $r \in R^{n^2}$. Below we use only this assumption when considering a few important classes of partial differential equations.

8.4 Linear Elliptic Operators and Equations

We consider linear operators

$$Lu = \sum_{i,j=1}^{n} a_{ij}(x) u_{ij}(x) + \sum_{i=1}^{n} b_i(x) u_i(x) + c(x) u(x), \tag{8.51}$$

where $a_{ij}(x) = a_{ji}(x)$ for all $x \in B$ and $u(x) \in C^2(B) \cap C(\overline{B})$.

For linear equations the equality

$$T(L, z_0) = \sum_{i,j=1}^{n} a_{ij}(x) \xi_i \xi_j \tag{8.52}$$

holds. Hence $T(L, z_0)$ is independent of any function $z_0(x) \in C^2(B) \cap C(\overline{B})$. Thus the operator L is elliptic if the matrix $\|a_{ij}(x)\|$ is positive or negative everywhere in B. Without loss of generality we can assume that $\|a_{ij}(x)\|$ is positive in B, that is, if $\lambda(x)$ and $\Lambda(x)$ denote respectively the minimum and maximum eigenvalues of $\|a_{ij}(x)\|$, then

$$0 < \lambda(x) |\xi|^2 \le \sum_{i,j=1}^{n} a_{ij}(x) \xi_i \xi_j \le \Lambda(x) |\xi|^2 \tag{8.53}$$

for all vectors $\xi = (\xi_1, \xi_2, \ldots, \xi_n) \in R^n \backslash \{\theta\}$. In the opposite case we replace the operator Lu by $-Lu$.

The operator L is called *strictly elliptic* in B if $\lambda(x) \ge \lambda_0 = \text{const} > 0$ for all $x \in B$.

If

$$\frac{\Lambda(x)}{\lambda(x)} \le \Lambda_0 = \text{const} < +\infty \tag{8.54}$$

for all $x \in B$, then L is called *uniformly elliptic*.

The different versions of the maximum principle for linear elliptic operators (8.51) require additional assumptions for the lower terms

$$\sum_{i=1}^{n} b_i(x)u_i \quad \text{and} \quad c(x)u.$$

It is convenient to use the following assumptions for their statements.

Assumption 8.1. The inequalities

$$\frac{|b_i(x)|}{\lambda(x)} \le \beta = \text{const} \ < +\infty; \quad i = 1, 2, \ldots, n \tag{8.55}$$

hold for all $x \in B$.

Assumption 8.2. The inequality

$$\frac{|c(x)|}{\lambda(x)} \le \gamma = \text{const} \ < +\infty \tag{8.56}$$

holds for all $x \in B$.

Assumption 8.3. The inequalities

$$\frac{a_{ii}(x)}{\lambda(x)} \le \alpha = \text{const} \ < +\infty; \quad i = 1, 2, \ldots, n \tag{8.57}$$

hold for all $x \in B$.

Clearly Assumption 8.1,2,3 are fulfilled if B is a bounded domain in E^n and all functions $a_{ij}(x)$, $b_i(x)$, $c(x)$ are continuous in $\overline{B}$.

8.5 Quasilinear Elliptic Operators and Equations

The operator

$$Mu = \sum_{i,k=1}^{n} A_{ik}(x, u, Du)u_{ik} + B(x, u, Du) \tag{8.58}$$

is called *quasilinear* and the equation

$$Mu = 0 \tag{8.59}$$

is called *quasilinear*. In this chapter we assume that $A_{ik}(x, u, p)$ and $B(x, u, p)$ are functions defined in $\overline{B} \times R \times R^n$. Clearly the positiveness of the symmetric matrix $(A_{ik}(x, z, p))$ on $\overline{B} \times R \times R^n$ is a sufficient condition for ellipticity of operator (8.58) and equation (8.59). It is easy to prove that this condition is also necessary.

8.6 The Classical Monge–Ampere Equations

These equations have the following form:

$$\phi(z) = z_{11}z_{22} - z_{12}^2 + \\ + \sum_{i,k=1}^{2} A_{ik}(x, z, Dz)z_{ik} + B(x, z, Dz) = 0. \tag{8.60}$$

We assume as above that $A_{ik}, B \in C^1(B \times R \times R^n)$ and $A_{12}(x, z, p) = A_{21}(x, z, p)$. Then the function $F(x, z, p, r) = r_{11}r_{22} - r_{12}r_{21} + A_{11}r_{11} + A_{12}r_{12} + A_{21}r_{21} + A_{22}r_{22} + B$ has the following derivatives

$$\frac{\partial F}{\partial r_{11}} = r_{22} + A_{11}, \quad \frac{\partial F}{\partial r_{12}} = -r_{21} + A_{12},$$
$$\frac{\partial F}{\partial r_{21}} = -r_{12} + A_{21}, \quad \frac{\partial F}{\partial r_{22}} = r_{11} + A_{22}.$$

Let $z \in C^2(G) \cap C(\overline{G})$ be a solution of equation (8.60). Then the form $T(\phi, z)$ becomes

$$T(\phi, z) = (z_{22} + A_{11})\xi_1^2 - 2(z_{12} - A_{12})\xi_1\xi_2 + (z_{11} + A_{22})\xi_2^2.$$

Therefore the ellipticity of equation (8.60) is equivalent, the inequality

$$(z_{22} + A_{11})(z_{11} + A_{22}) - (z_{12} - A_{12})^2 > 0$$

which should be fulfilled for all solutions of equation (8.60).

Since

$$(z_{22} + A_{11})(z_{11} + A_{22}) - (z_{12} - A_{12})^2 = -B + A_{11}A_{22} - A_{12}^2,$$

a necessary and sufficient condition for ellipticity of equation (8.60) can be expressed by the inequality

$$A_{11}A_{12} - A_{12}^2 - B > 0 \tag{8.61}$$

where $x \in B, z \in R$ and $p \in R^2$.

Now we consider the simplest Monge–Ampere equation

$$z_{11}z_{22} - z_{12}^2 = \varphi(x, z, p). \tag{8.62}$$

From (8.61) it follows that the ellipticity of equation (8.62) is equivalent to the positiveness of the function $\varphi(x, z, p)$ in $G \times R \times R^2$. Thus the ellipticity of equation (8.62) is equivalent to the statement that all solutions of (8.62) are convex or concave functions.

The equation

$$\det(z_{ij}) = \varphi(x, z, Dz) \tag{8.63}$$

for functions of n variables is an analog of equation (8.62). This equation also can have only two classes of elliptic solutions. The first one consists of convex solutions and the second one consists of concave solutions.

The theory of generalized elliptic solutions for equations (8.62) and (8.63) will be presented in Part II of this book. The theory of classical solutions and the investigations of smoothness for generalized elliptic solutions will also be presented in the Part II of this book.

8.7 Differential Equations in Global Problems of Differential Geometry

In this subsection we consider a few global problems in Differential Geometry whose investigation can be reduced to nonlinear elliptic differential equations.

We consider two classes of hypersurfaces in E^{n+1}. The first one consists of hypersurfaces with one-to-one projection on some fixed hyperplane and the second one consists of hypersurfaces with one-to-one spherical mapping.

A) *Hypersurfaces With One-To-One Projection on a Fixed Hyperplane.* Let some hyperplane E^n be fixed in E^{n+1} and let $B \subset E^n$ be some open bounded domain. We introduce the Cartesian coordinates $x_1, x_2, \ldots, x_n, x_{n+1} = z$ such that the hyperplane E^n has equation $z = 0$. Let S be a hypersurface in E^{n+1} and

$$z = z(x_1, x_2, \ldots, x_n) = z(x), \quad x \in B \tag{8.64}$$

be the equation of S. We assume that $z(x) \in C^2(B) \cap C(\overline{B})$. Let $k_1 \geq k_2 \geq \ldots \geq k_n$ be the principal normal curvatures of S. We choose the normal of S in the way that its angle with the positive direction of the z-axis is not greater than $\frac{\pi}{2}$. Then the signs of the principal normal curvatures k_i, $i = 1, 2, \ldots, n$, are fixed uniquely.

There are many investigations concerning existence and uniqueness for hypersurfaces with prescribed values for a given function $\phi(k_1, k_2, \ldots, k_n)$ and even more general function $\phi(k_1, k_2, \ldots, k_n, p, z, x)$. In the last case, ϕ depends also on a point $(x, z) \in S$ and on the tangential hyperplane of S at the point (x, z). The case when ϕ is some elementary symmetric function of $k_1, k_2, \ldots, k_n$ is the most important. If $n = 2$ these functions are $\phi_1 = k_1 k_2$, $\phi_2 = k_1 + k_2$. ϕ_1 is the Gaussian curvature of S and ϕ_2 is the doubled mean curvature of S.

If the existence and uniqueness problems are considered in a bounded domain B in E^n, then the border of the desired solution must be also prescribed. For example these problems concerning the functions $\phi_1 = k_1 k_2$ and $\phi_2 = k_1 + k_2$ are the Dirichlet problems

$$k_1 k_2 = \frac{z_{11} z_{22} - z_{12}^2}{(1 + z_1^2 + z_2^2)^2} = \varphi(x), \quad z|_{\partial B} = h(x) \tag{8.65}$$

and

$$k_1 + k_2 = \frac{(1+z_2^2)z_{11} - 2z_1 z_2 z_{12} + (1+z_1^2)z_{22}}{(1+z_1^2+z_2^2)^{3/2}} \tag{8.66}$$

$$= 2\varphi(x),$$

$$z|_{\partial B} = h(x). \tag{8.66-a}$$

From (8.65) and (8.66) it follows that the functions $k_1 k_2$ and $k_1 + k_2$ are differential operators of the second order on the functions $z(x)$, which define the desired surfaces by the explicit equations $z = z(x)$, $x \in B$.

Now we consider the conditions providing the ellipticity of the differential operator of the second order, which corresponds to a general function

$$\phi(k_1, k_2, \ldots, k_n, p, z, x). \tag{8.67}$$

Theorem 8.3. *Let $\phi(k_1, k_2, \ldots, k_n, p, z, x)$ be C^1-function with respect to $x \in \overline{B}$, $z \in R$, $p \in R^n$, $-\infty < k_n \leq k_{n-1} \leq \ldots \leq k_1 < +\infty$ and let $\frac{\partial \phi}{\partial k_i} = \frac{\partial \phi}{\partial k_j}$ if $k_i = k_j$. Then the function*

$$F(D^2 z, Dz, z, x) = \phi(k_1, \ldots, k_n, Dz, z, x)$$

is C^1-function with respect to z_{ij}, z_i, z. Moreover if $\frac{\partial \phi}{\partial k_i} > 0$, $i = 1, 2, \ldots, n$, then the form $T(F, z)$ is positive definite in $\overline{B}$, i.e. the operator F is elliptic on the function $z(x)$.

This theorem was proved by Alexandrov [4]. We consider the scheme of the proof and the statements of the main lemmas.

Let $g = g_{jk}\xi^j\xi^k$, $b = b_{jk}\xi^j\xi^k$ be two quadratic forms. We assume that the form g is positive definite. Then the eigenvalues k_i of the form b with respect to form g are the extremums of b/g. If for example $g = \sum_{i=1}^{n}(\xi^i)^2$, then k_i are the eigenvalues of the form b.

If b, g are correspondingly the second and the first fundamental forms of S, where S is the graph of the function $z(x)$, then k_i are the principal normal curvatures of S. Since $k_1 \geq k_2 \geq \ldots \geq k_n$, it is easy to prove that every eigenvalue k_i is a definite continuous function of b_{jk} and g_{jk}. Moreover k_i are the roots of an algebraic equation. Therefore if k_i is not a multiple root, then k_i is a smooth function of b_{jk} and g_{jk}. In the case of the multiple roots, i.e. on the corresponding algebraic manifolds in the space of coefficients b_{jk} and g_{jk}, these functions become non-smooth. We can obviously see this fact for eigenvalues of forms with two independent variables. But the condition

$$\frac{\partial \phi}{\partial k_i} = \frac{\partial \phi}{\partial k_j} \tag{8.68}$$

if $k_i = k_j$, permits us to prove the following lemmas.

Lemma 8.4. *Let $k_1 \geq k_2 \geq \ldots \geq k_n$ be the eigenvalues of the form $b = b_{jk}\xi^j\xi^k$ with respect to the positive definite form $g = g_{jk}\xi^j\xi^k$. Then the function $\phi(k_1, \ldots, k_n)$ is also a function of b_{jk} and g_{jk}:*

$$\psi(b_{jk}, g_{jk}) = \phi(k_1, \ldots, k_n).$$

Moreover if ϕ is a C^1-function of $k_1, \ldots, k_n$ and satisfies condition (8.68), then ψ is also a C^1-function of b_{jk} and g_{jk}.

Lemma 8.5. *Let the conditions of Lemma 8.4 be satisfied and let the forms b and g be reduced to the canonical form*

$$b = \sum_{i=1}^{n} b_{ii}\xi_i^2, \quad g = \sum_{i=1}^{n} \xi_i^2,$$

where $b_{11} \geq b_{22} \geq \ldots \geq b_{nn}$ such that $b_{ii} = k_i$. Then the form

$$\frac{\partial \psi}{\partial b_{11}}\xi_1^2 + \frac{\partial \psi}{\partial b_{12}}\xi_1\xi_2 + \cdots$$

becomes

$$\frac{\partial \phi}{\partial k_1}\xi_1^2 + \frac{\partial \phi}{\partial k_2}\xi_2^2 + \cdots .$$

Now we consider the proof of Theorem 8.3. If we express k_i by the coefficients of the first and second fundamental forms of the hypersurface S, then we obtain

$$\phi(k_i, Dz, z, x) = \psi(b_{jk}, g_{jk}, Dz, z, x).$$

According to Lemma 1 ψ is a C^1-function with respect to b_{jk} and g_{jk}. Clearly ψ is also a C^1-function with respect to Dz, z and x. Since b_{jk}, g_{jk} are C^1-functions with respect to z_{ik}, z_i, z and x, the first part of Theorem 8.3 is proved. □

Now we prove that the operator $F(D^2z, Dz, z, x)$ is elliptic on the function $z(x)$. First of all we have the equality

$$F(D^2z, Dz, z, x) = \psi(b_{jk}, g_{jk}, Dz, z, x),$$

where only the coefficients b_{jk} contain the second derivatives of the function $z(x)$ and $b_{jk} = az_{jk}$, $a = \cos(n, z) > 0$, is the unit normal of S: $z = z(x)$. Hence

$$\frac{\partial F}{\partial z_{jk}} = a\frac{\partial \psi}{\partial b_{jk}}.$$

Therefore

$$\frac{\partial F}{\partial z_{11}}\xi_1^2 + \frac{\partial F}{\partial z_{12}}\xi_1\xi_2 + \cdots = a\left(\frac{\partial \psi}{\partial b_{11}}\xi_1^2 + \frac{\partial \psi}{\partial b_{12}}\xi_1\xi_2 + \cdots\right).$$

Now we choose the coordinates $x_1, \dots, x_n$ such that the forms b and g have the canonical forms

$$b = \sum_{i=1}^{n} b_{ii}(dx_i)^2, \quad g = \sum_{i=1}^{n}(dx_i)^2$$

at any given point $x \in B$. Without loss of generality we can assume that

$$b_{11} \geq b_{22} \geq \dots \geq b_{nn}.$$

Then $k_i = b_{ii}$, $i = 1, 2, \dots, n$. From Lemma 8.5 it follows that

$$\frac{\partial F}{\partial z_{11}}\xi_1^2 + \frac{\partial F}{\partial z_{12}}\xi_1\xi_2 + \cdots = a\left(\frac{\partial \phi}{\partial k_1}\xi_1^2 + \frac{\partial \phi}{\partial k_2}\xi_2^2 + \cdots\right). \tag{8.69}$$

Thus the form in the left side of (8.69) is positive definite if and only if the inequalities

$$\frac{\partial \phi}{\partial k_1} > 0, \frac{\partial \phi}{\partial k_2} > 0, \dots, \frac{\partial \phi}{\partial k_n} > 0$$

hold. The proof of Theorem 8.3 is complete. □

B) *Hypersurfaces With One-To-One Spherical Mappings.*

Let $S \subset E^{n+1}$ be a hypersurface with one-to-one spherical mapping ψ_s: $S \to S^n$. Here S^n is the unit sphere $|x| = 1$ in E^{n+1}. Then there is only one point of tangency in every tangent hyperplane Q of S with the unit normal ν. Let $x(\nu)$ be the position vector of this point, whose initial point is the origin 0 of E^{n+1} and let

$$h(\nu) = (x(\nu), \nu).$$

Then S is the envelope of the family of hyperplanes with equations

$$(\nu, y) = h(\nu), \tag{8.70}$$

where y is a moving point of a tangent hyperplane. Let $G \subset S^n$ be the domain of the function $h(\nu)$. Since ν is completely defined by any intrinsic coordinates in S^n, both $x(\nu)$ and $h(\nu)$ are also functions of these coordinates.

The function $h(\nu)$ is called the *supporting function* of S. The supporting function of S can be extended for any vectors $m \neq 0$ by

$$H(m) = |m| h\left(\frac{m}{|m|}\right) \tag{8.71}$$

for any $m \neq 0$ such that $\nu = \frac{m}{|m|} \in G$.

A few of the important properties of the supporting function of a convex hypersurface S were obtained in Subsection 6.2 (see §6, Chapter 1). It turns out that they are independent of the convexity of S and can be transferred to the hypersurfaces with the spherical one-to-one mapping. We present briefly these facts.

Let $m = \alpha_1 e_1 + \alpha_2 e_2 + \cdots + \alpha_{n+1} e_{n+1}$ be any non-zero vector in E^{n+1}, where $e_1, e_2, \ldots, e_{n+1}$ is any orthonormal frame in E^{n+1}. From Euler's Theorem it follows that

$$H = \sum_{i=1}^{n+1} \alpha_i H_i, \tag{8.72}$$

where

$$H_i = \frac{\partial H}{\partial \alpha_i}, \quad i = 1, 2, \ldots, n+1. \tag{8.73}$$

The equation of the tangent hyperplane of S is

$$\alpha_1 x_1 + \alpha_2 x_2 + \cdots + \alpha_{n+1} x_{n+1} = H. \tag{8.74}$$

Now we fix $\alpha_2, \alpha_3, \ldots, \alpha_{n+1}$ and change only α_1 in (8.72). If we differentiate (8.72) with respect to α_1, then we obtain

$$x_1 = \frac{\partial H}{\partial \alpha_1}.$$

The formulas

$$x_i = \frac{\partial H}{\partial \alpha_i} \tag{8.75}$$

can be obtained in the same way. Since H_i are homogeneous functions of degree 0,

$$x_i = H_i(\lambda\alpha_1, \lambda\alpha_2, \ldots, \lambda\alpha_{n+1}),$$

where λ is any real number.

Let $R_i = \frac{1}{k_i}$ be the principal radii of curvature of a hypersurface S. According to Rodrigue's formula we obtain

$$dx_i - R_i d\xi_i = 0, \quad i = 1, 2, \ldots, n+1 \tag{8.76}$$

in the principal directions at any given point of S, where $\xi_1, \ldots, \xi_{n+1}$ are components of a unit normal ν. Using (8.75) and the condition $d\nu \neq 0$ we deduce from (8.76) the following equation

$$\left\| \begin{matrix} H_{11} - R & H_{12} & H_{1\ n+1} \\ H_{21} & H_{22} - R & H_{2\ n+1} \\ \cdots\cdots & \cdots\cdots & \cdots\cdots \\ \cdots\cdots & \cdots\cdots & \cdots\cdots \\ H_{n+1\ 1} & H_{n+1\ 2} & H_{n+1\ n+1} - R \end{matrix} \right\| = 0. \tag{8.77}$$

(8.72) becomes

$$\sum_{k=1}^{n+1} H_{ik} \alpha_k = 0. \tag{8.78}$$

after differentiation with respect to α_i. From (8.78) it follows that

$$\det(H_{ik}) = 0.$$

Hence one of the roots of equation (8.77) is always equal to zero and other n roots are the principal normal curvatures of S. From (8.77) we obtain

$$(R_1 + R_2 + \cdots + R_n) = \Delta H(\alpha_1, \alpha_2, \ldots, \alpha_{n+1}) \tag{8.79}$$

and

$$R_1 R_2 \ldots R_n = \begin{vmatrix} H_{11} & \ldots & H_{1n} \\ \cdots & \cdots & \cdots \\ \cdots & \cdots & \cdots \\ H_{n1} & \ldots & H_{nn} \end{vmatrix} + \cdots + \begin{vmatrix} H_{22} & \ldots & H_{2\ n+1} \\ \cdots & \cdots & \cdots \\ \cdots & \cdots & \cdots \\ H_{n+1\ 2} & \ldots & H_{n+2\ n+2} \end{vmatrix},$$

where we set the components $\xi_1, \xi_2, \ldots, \xi_{n+1}$ of the unit vector ν after the formal differentiation with respect to $\alpha_1, \alpha_2, \ldots, \alpha_{n+1}$.

Theorem 8.4. *Let $\phi(R_1, R_2, \ldots, R_n, p, h, \nu)$ be a C^1-function for $\nu \in \overline{G}$, $h \in R$, $p = (p_1, p_2, \ldots, p_n) \in R^n$, $-\infty < R_1 \le R_2 \le \ldots \le R_n < +\infty$ and let moreover*

$$\frac{\partial \phi}{\partial R_i} = \frac{\partial \phi}{\partial R_j}, \tag{8.80}$$

if $R_i = R_j$. Then

$$F(D^2 h, Dh, h, \nu) = \phi(R_1, R_2, \ldots, R_n, Dh, h, \nu) \tag{8.81}$$

is a C^1-function with respect h_{ij}, h_i, h, ν, where $h(\nu)$ is the supporting function of the hypersurface S. If moreover $\frac{\partial \phi}{\partial R_i} > 0$ (< 0), $i = 1, 2, \ldots, n$, then the operator F is elliptic on the function $h(\nu)$.

This theorem was proved by Alexandrov [4] in the same way as Theorem 8.3. In the proof of this theorem b is the second fundamental form of S and $g = (d\nu)^2$.

8.8 The Classical Maximum Principles for General Elliptic Equations

Let

$$Lu = \sum_{i,k=1}^{n} a_{ik}(x) u_{ik} + \sum_{i=1}^{n} b_i(x) u_i + c(x) u \tag{8.82}$$

be a linear elliptic operator on functions $C^2(B) \cap C(\overline{B})$, where B is a bounded domain in E^n.

A function $u(x) \in C^2(B) \cap C(\overline{B})$ satisfying the equation

$$Lu = 0 \tag{8.83}$$

in B is called a *solution of* $Lu = 0$ *in* B. A function $u(x) \in C^2(B) \cap C(\overline{B})$ satisfying the inequality

$$Lu \geq 0 \tag{8.84}$$

$$(Lu \leq 0) \tag{8.85}$$

in B is called a *subsolution (supersolution) of equation (8.83)*. If Lu is Laplacian these concepts are generalizations of harmonic, subharmonic and superharmonic functions respectively. We also consider two particular types of general linear elliptic operators (8.82):

$$L_1 u = \sum_{i,k=1}^{n} a_{ik}(x) u_{ik} + \sum_{i=1}^{n} b_i(x) u_i \tag{8.82$'$}$$

and

$$L_2 u = \sum_{i,k=1}^{n} a_{ik}(x) u_{ik}. \tag{8.82$''$}$$

Theorem 8.5. *Let B be a bounded domain in E^n and let L_1 be a linear elliptic operator (8.82′) satisfying Assumption 8.3. Then for any subsolution (supersolution) $u(x) \in C^2(B) \cap C(\overline{B})$ of equation (8.83) the maximum (minimum) of $u(x)$ in $\overline{B}$ is achieved on ∂B, that is,*

$$\sup_{B} u(x) = \sup_{\partial B} u(x) \quad (\inf_{B} u(x) = \inf_{\partial B} u(x)). \tag{8.86}$$

Proof. The proof consists of two parts a) and b).

a) Here we prove that $u(x)$ can not achieve an interior maximum if

$$L_1 u > 0 \quad \text{in} \quad B. \tag{8.87}$$

If this statement is incorrect, then $u(x)$ achieves its maximum at some interior point x_0 of B. Clearly

$$u_1(x_0) = u_2(x_0) = \cdots = u_n(x_0) = 0$$

and the matrix $D^2 u(x_0) = (u_{ij}(x_0))$ is nonpositive. But the matrix $(a_{ij}(x_0))$ is positive since L_1 is elliptic. Hence we obtain the inequality

$$L_1(u(x_0)) = \sum_{i,k=1}^{n} a_{ik}(x_0) u_{ik}(x_0) \leq 0$$

contradicting (8.87). This proves part a).

b) From Assumption 8.2 it follows that

$$\frac{|b_i(x)|}{\lambda(x)} \leq \beta, \quad i = 1, 2, \ldots, n \quad \text{for all} \quad x \in B.$$

Since $a_{ii}(x) \geq \lambda(x) > 0$, $i = 1, 2, \ldots, n$, for all $x \in B$, there is a sufficiently large constant k for which

$$L_1(e^{kx_1}) = (k^2 a_{11} + kb_1)e^{kx_1} \geq \lambda(k^2 - k\beta)e^{k_1 x} > 0.$$

Hence for any $\eta > 0$

$$L_1(u + \eta e^{kx_1}) > 0$$

in B, so that

$$\sup_B (u + \eta e^{kx_1}) = \sup_{\partial B} (u + \eta e^{kx_1})$$

by part a).

If we take η to zero, then we obtain

$$\sup_B u = \sup_{\partial B} u. \qquad \square$$

Now we introduce two functions

$$u^+(x) = \max\{u(x), 0\} \tag{8.88}$$

and

$$u^-(x) = \min\{u(x), 0\}. \tag{8.89}$$

It is convenient to use the functions $u^+(x)$ and $u^-(x)$ in the statement of the maximum principle for general linear operators (8.82).

Theorem 8.6. *Let B be a bounded domain in E^n and let L be a linear general elliptic operator (8.82), satisfying Assumption 8.1 and the condition $c(x) \leq 0$ in B. Then for any subsolution (supersolution $u(x) \in C^2(B) \cap C(\overline{B})$ of equation (8.83) the positive maximum (negative minimum) of $u(x)$ in $\overline{B}$ is achieved on ∂B, that is,*

$$\sup_B u(x) \leq \sup_{\partial B} u^+(x) \tag{8.90}$$

$$(\inf_B u(x) \geq \inf_{\partial B} u^-(x)). \tag{8.90'}$$

If $Lu = 0$ in B, then

$$\sup_B |u(x)| = \sup_{\partial B} |u(x)|.$$

Proof. Let B^+ be the subset of B, where $u > 0$. Clearly if $Lu \geq 0$ in B, then

$$L_1 u = \sum_{i,j=1}^n a_{ij}(x)u_{ij} + \sum_{i=1}^n b_i(x)u_i = -c(x)u \geq 0$$

in B^+. Therefore the maximum of u on B^+ must be achieved on ∂B^+ and hence also on ∂B (see Theorem 8.5). Other cases can be proved in the same way. $\square$

Remark. If $c(x) > 0$, then Theorem 8.3 is not true, because there exist positive eigenfunctions for the Dirichlet problem $\Delta u + ku = 0$, $u|_{\partial B} = 0$, $k = \text{const} > 0$.

The following uniqueness and comparison theorem is valid.

Theorem 8.7. *Let L be a linear elliptic operator (8.82) in B satisfying Assumption 8.1 and the condition $c(x) \le 0$. Let $u(x), v(x) \in C^2(B) \cap C(\overline{B})$ satisfy the conditions*

$$Lu = Lv \quad \text{in} \quad B \quad \text{and} \quad u = v \quad \text{on} \quad \partial B.$$

Then $u(x) = v(x)$ in B. Moreover if $Lu \ge Lv$ in B and $u \le v$ on ∂B, then $u(x) \le v(x)$ in B.

The proof of this theorem follows directly from Theorem 8.6. □

8.9 Hopf's Maximum Principle for Uniformly Elliptic Linear Equations

Although the classical maximum principles (Theorems 8.5 and 8.6) suffice for many important applications, it is often necessary to exclude the possibility of existence a non-trivial interior maximum (minimum) for subsolutions (supersolutions) of elliptic equations (8.83). This problem can be solved by more strong form of the maximum principle in the class of uniformly elliptic operators, which is called Hopf's maximum principle.

Theorem 8.8. *Let Lu and L_1u be linear uniformly elliptic operators defined respectively by (8.82) and (8.82′) in a domain $B \subset E^n$ (B is not necessarily bounded). Let Lu and L_1u satisfy Assumptions 8.1–3. Finally let $z(x) \in C^2(B) \cap C(\overline{B})$ be any function.*

Then the following assertions hold:

I. If $L_1 z \ge 0$ in B and $z(x)$ achieves its maximum at an interior point of B, then $z(x)$ is a constant.

II. If $Lz \ge 0$, $c(x) \le 0$ in B, and $z(x)$ achieves its maximum at an interior point of B, then $z(x)$ is a constant.

III. Let $L_1 z \ge 0$ and let $z(x)$ achieves the greatest value at a point $\tilde{x} \in \partial B$. We assume that there exists a closed ball $\Omega \subset \overline{B}$ with $\tilde{x} \in \partial\Omega$. Then either $z(x)$ is a constant or the exterior normal derivative of $z(x)$ at $\tilde{x}$ satisfies the strict inequality

$$\frac{\partial z(\tilde{x})}{\partial n} > 0. \tag{8.91}$$

IV. Let $L(z) \ge 0$ and $c(x) \le 0$ in B, and $z(x)$ achieves its maximum at a point $\tilde{x} \in \partial B$. We assume that the point $\tilde{x}$ satisfies the condition of Statement III. If the maximum of $z(x)$ is positive, then either $z(x)$ is a constant or the exterior normal derivative of z at $\tilde{x}$ is positive.

Remark. In Statements III and IV we suppose in addition that $\frac{\partial z}{\partial n}(\tilde{x})$ exists. Below we will see that both assertions are true for the lower exterior derivative of $z(\tilde{x})$ if $\frac{\partial z(\tilde{x})}{\partial n}$ does not exist. Moreover if $L_1 z \ge 0$ and $Lz \ge 0$ are replaced in Statements I–IV by inequalities $L_1 z \le 0$, $Lz \le 0$, and $z(x)$ achieves its minimum at all corresponding points, then all Statements I–IV will be correct.

Proof of Theorem 8.8. a) First of all we prove the Statement III. We assume in addition that $z < z(\tilde{x})$ in B. Let S_0 and S_1 be two concentric spheres such that the larger sphere S_0 passes through the point $\tilde{x}$ and all points in the closure B_0 of the annulus between S_0 and S_1, except $\tilde{x}$, belong to B. Now we suppose that there exists a function $g(x) \in C^2(\overline{B})$, such that $L_1(g) > 0$ in B_0, $g = 0$ on the larger sphere S_0 and $\frac{\partial g}{\partial n} < 0$ at the point $\tilde{x}$. Let $v = u + \varepsilon g$, where $\varepsilon > 0$ is so small that $v < \tilde{u}(x)$ on the smaller sphere S_1. Then $L_1(v) > 0$ in B_0 and $v < u(\tilde{x})$ on S_1. If $v(x)$ achieves the maximum at a point $x_0 \in B_0$, then $v_1(x_0) = v_2(x_0) = \cdots = v_n(x_0) = 0$ and $d^2 v \le 0$. Therefore $L_1(v(x_0)) \le 0$, but this is impossible because according to the definition of the function $v(x)$ we have the opposite inequality $L_1(v(x_0)) > 0$. Hence $v(x)$ achieves its maximum in B_0 only on S_0. Since $v = 0$ on S_0, the maximum of $v(x)$ is achieved only at the point $\tilde{x}$. Therefore

$$\frac{\partial v}{\partial n} = \frac{\partial z}{\partial n} + \varepsilon \frac{\partial g}{\partial n} \ge 0.$$

Since $\frac{\partial g}{\partial n} < 0$, $\frac{\partial z}{\partial n} > 0$ at the point $\tilde{x}$.

Below we construct the function $g(x)$ by the Hopf's method. This function is

$$g = e^{-\tau r^2} - e^{-\tau r_0^2} \tag{8.92}$$

where $\tau > 0$ is a constant, r is the distance from the common center of spheres S_0 and S_1 and $r_0 > 0$ is the radius of the exterior sphere S_0. We choose the origin at the center of the concentric spheres S_0 and S_1 and denote by r_1 the radius of S_1, then

$$L_1(g) = e^{-\tau r^2}(4\tau^2 m r_1^2 - 2n\tau K r_0 - 2n\tau K),$$

where $\sum_{i,k=1}^{n} a_{ik}\xi_i\xi_k \ge m \sum_{i=1}^{n} \xi_i^2$, $|a_{ii}| \le K$, $|b_i| \le K$. According to Assumptions 23.1, 2, 3 and the condition of uniformal ellipticity we can obtain constants m and K such that $L_1(g) > 0$ in B_0 if τ is sufficiently large.

Now we consider the proof of the Assertion 1. Let us assume that the maximum μ of the function $z(x)$ is achieved at an interior point $x_0 \in B$ and let $\delta > 0$ is so small that the ball $(U_{x_0}, 2\delta) \subset B$. Let us prove that $z(x) = \mu$ if $x \in U(x_0, \delta)$. Let $z(x_1) < \mu$ for some $x_1 \in U(x_0, \delta)$. Then there exists $\rho \in (0, \delta)$ for which $z(x) < \mu$ in the ball $U(x_1, \rho)$ and $z = \mu$ at some boundary point $\tilde{x}$ of this ball. From the facts established above the derivative of $z(x)$ at the point $\tilde{x}$ in some direction should be positive. But this is impossible, because at the maximum point all first derivatives are equal to zero.

Let Ω be the set of all points $x \in B$, where $z(x) = \mu$. We prove above that Ω is an open set. Since $z(x)$ is continuous, Ω is also a closed set. Clearly Ω is a connected set. Hence $\Omega = B$ and $z(x) = \mu =$ constant.

Now all additional assumptions made in the proof of Assertion III can be omitted.

Let us consider the proof of the Assertion II. Let $z(x)$ achieve a positive maximum at a point $x_0 \in B$. Then $cz \leq 0$ in a small neighborhood of the point x_0. Hence $L_1 z = Lz - az \geq 0$ in the same neighborhood. According to Assertion I we obtain $z = z(x_0)$ in some neighborhood of the point x_0. Therefore the set of points where $z = z(x_0)$ is open. But the same set is closed, because the function $z(x)$ is continuous, and B is a connected domain. Hence $z(x) = \text{const.}$ in B. The Assertion II is proved.

Finally we consider the proof of Assertion IV. Let $\tilde{x}$ be a point of ∂B such that $\mu = \max z = z(\tilde{x}) > 0$. Then $L_1(z) = L(z) - cz \geq 0$ near $\tilde{x}$. From Assertion III it follows that either $\frac{\partial z}{\partial n} > 0$ or $z \equiv \mu$ at some interior points of B close to $\tilde{x}$. In the last case $z(x) = \text{const.}$ according to Assertion II. The proof of Theorem 8.8 is complete. □

8.10 Uniqueness Theorem for General Nonlinear Elliptic Equations

Let $F(x, z, p, r) \in C^1(\overline{B} \times R \times R^n \times R^{n^2})$. this function generates the operator

$$\phi(z) = F(x, z, Dz, D^2 z)$$

on the set of functions $C^2(\overline{B})$. The operator $\phi(z)$ is called *elliptically convex* if the quadratic form

$$T(\phi, z) = \sum_{i,k=1}^{n} \frac{\partial F(x, z, Dz, D^2 z)}{\partial z_{ik}} \xi_i \xi_k$$

is positive (negative) definite for all functions $z_\tau(x) = \tau z_1(x) + (1 - \tau) z_0(x)$, $0 < \tau < 1$ whenever $T(\phi, z_0)$ and $T(\phi, z_1)$ are positive (negative) definite forms.

Theorem 8.9. *Let the function* $F(x, z, p, r) \in C^1(\overline{B} \times R \times R^n \times R^{n^2})$ *generate the elliptically convex operator* $\phi(z)$. *Let* B *be a bounded domain and* $z_1(x) = z_2(x)$ *on* ∂B. *Finally let*

$$F_z(x, z_\tau(x), Dz_\tau(x), D^2 z_\tau(x)) \leq 0, \qquad 0 \leq \tau \leq 1$$

or

$$(F_z(x, z_\tau(x), Dz_\tau x), D^2 z_\tau(x))) \geq 0, \qquad 0 \leq \tau \leq 1.$$

Then $z_0(x) = z_1(x)$ *in* ∂B, *where* $z_0(x), z_1(x) \in C^2(B)$ *are solutions of the equation*

$$F(x, z, Dz, D^2 z) = 0$$

on which the operator $\Phi(z)$ *is positive (negative) elliptic.*

Proof. First of all we introduce the following notations. Let $H(x, z, p, r)$ be a continuous function in $\overline{B} \times R \times R^n \times R^{n^2}$. Then we set

$$h(x, z) = H(x, z(x), Dz(x), D^2 z(x))$$

for any function $z(x) \in C^2(\overline{B})$. Clearly $h(x,z) \in C(\overline{B})$. We also set

$$\delta(x) = z_1(x) - z_0(x)$$

and

$$z_\tau(x) = (1-\tau)z_0(x) + \tau z_1(x), \quad 0 \le \tau \le 1.$$

Then

$$0 = f(x, z_1) - f(x, z_0) = \int_0^1 \frac{df(x, z_\tau)}{d\tau} d\tau =$$
$$= \sum_{i,k=1}^n a_{ik}(x)\delta_{ik} + \sum_{i=1}^n b_i(x)\delta_i + c(x)\delta,$$

where

$$a_{ik}(x) = \int_0^1 \frac{\partial f(x, z_\tau)}{\partial r_{ik}} d\tau, \quad b_i(x) = \int_0^1 \frac{\partial f(x, z_\tau)}{\partial p_i} d\tau,$$
$$c(x) = \int_0^1 \frac{\partial f(x, z_\tau)}{\partial z} d\tau.$$

Let $I = [0,1]$. Then the set $\overline{B} \times I$ is compact in $E^n \times R$. Clearly the functions $a_{ik}(x)$, $b_i(x)$, $c(x)$ are continuous in $\overline{B}$ and the inequality

$$\sum_{i,k=1}^n a_{ik}(x)\xi_i\xi_k > 0, \quad |\xi| = 1$$

holds everywhere in $\overline{B}$. From these facts it follows that there exist the constants $m > 0$ and $K < +\infty$ such that the inequalities

$$\sum_{i,k=1}^n a_{ik}(x)\xi_i\xi_k \ge m \sum_{i=1}^n \xi_i^2, \quad \xi \in R^n \backslash \{0\}; \tag{8.93}$$
$$|b_i(x)| \le K \quad 0 < a_{ii}(x) \le K, \quad c(x) \le 0$$

hold in $\overline{B}$. Thus all conditions of Hopf's maximum principle (see Theorem 8.8) are fulfilled for the equation

$$\sum_{i,k=1}^n a_{ik}(x)\delta_{ik} + \sum_{i=1}^n b_i(x)\delta_i + c(x)\delta = 0. \tag{8.94}$$

Since $\delta(x)$ vanishes on ∂B, from Theorem 8.8 we obtain the $\delta(x) = 0$ in B. Hence $z_1(x) = z_0(x)$ everywhere in $\overline{B}$. □

The following uniqueness theorems are direct corollaries of Theorem 8.9.

Theorem 8.10. *The Dirichlet problem for quasilinear equations*

$$\sum_{i,k=1}^{n} a_{ik}(x, Dz)z_{ik} + b(x, z, Dz) = 0 \tag{8.95}$$

has not more than one solution $z(x) \in C^2(\overline{B})$ in a bounded domain B, if $a_{ik}(x,p)$ and $b(x,z,p)$ are C^1-functions in $\overline{B} \times R \times R^n$ and the inequalities

a) $\sum_{i,k=1}^{n} a_{ik}(x,p)\xi_i\xi_k > 0,\ \xi \in R^n \backslash \{\theta\}$

b) $b_z(x,z,p) \leq 0$

hold in $\overline{B} \times R \times R^n$.

Theorem 8.11. *The Dirichlet problem for n-dimensional Monge–Ampere equations*

$$\det(z_{ij}) = f(x, z, Dz)$$

has not more than one strictly convex solution $z(x) \in C^2(\overline{B})$ if $f(x,z,p)$ is C^1-function in $\overline{B} \times R \times R^n$ and the inequalities

a) $f(x,z,p) > 0$

b) $f_z(x,z,p) \geq 0$

hold in $\overline{B} \times R \times R^n$.

A similar theorem can be obtained for strictly concave solutions.

The domain B is bounded in both Theorems 8.10 and 8.11.

Theorem 8.12. *Suppose $\phi(k_1, k_2, \ldots, k_n, p, z, x)$ is a C^1-function for $x \in \overline{B}$, $z \in R$, $p \in R^n$ and $-\infty < k_n \leq k_{n-1} \leq \ldots \leq k_1 < +\infty$, where B is a bounded domain, and the following conditions are satisfied*

a) $\frac{\partial \phi}{\partial k_i} = \frac{\partial \phi}{\partial k_j}$ if $k_i = k_j$;

b) ϕ is monotonically convex*);

c) $\frac{\partial \phi}{\partial z} \leq 0$ for all values of $x \in \overline{B}$ and any values of $z, p, k_1, \ldots, k_n$.

Now let $z_1(x), x_2(x) \in C_2(\overline{B})$ satisfy the conditions

d) $z_1(x)|_{\partial B} = z_2(x)|_{\partial B}$

e) the identity

$$\phi(k_i^{(1)}, Dz_1, z_1, x) = \phi(k_i^{(2)}, Dz_2, z_2, x)$$

holds everywhere in B.

f)

$$\left.\frac{\partial \phi}{\partial k_i}\right|_{z_1(x)} > 0, \quad \left.\frac{\partial \phi}{\partial k_i}\right|_{z_2(x)} > 0.$$

---

*)If the inequality $\partial\phi/\partial k_i > 0$ is fulfilled for all functions $z_\tau = (1-\tau)z_1 + \tau z_2$, then the function ϕ is called *monotonically convex*.

Then $z_1(x) = z_2(x)$ in B.

A similar theorem can be proved for C^2-hypersurfaces with one-to-one spherical mapping and with prescribed monotonically convex function $\phi(R_1, \dots, R_n, Dh, h, n)$ of the principal radii of normal curvature. We omit the statement of such a theorem, because it is similar to the statement and proof of Theorem 8.12.

8.11 The Maximum Principle for Divergent Quasilinear Elliptic Equations

In this subsection we present the variant of the maximum principle obtained by Serrin [4, § 6] for divergent quasilinear elliptic equations. This principle has interesting applications to first order variational problems for multiple integrals. In these problems the integrand $F(x, u, Du)$ is jointly convex in u and Du for each fixed value of x.

We write the basic equation in the form

$$\sum_{i=1}^{n} \frac{\partial[A_i(x, u, Du)]}{\partial x_i} = B(x, u, Du)$$

where A_i and B are of classes C^2 and C^1 respectively. We suppose further that the $(n+1) \times (n+1)$ matrix

$$\left\| \begin{matrix} \partial A_1/\partial p_1 & \dots & \partial A_1/\partial p_n & \partial A_1/\partial u \\ \dots\dots & \dots & \dots\dots & \dots\dots \\ \dots\dots & \dots & \dots\dots & \dots\dots \\ \partial A_n/\partial p_1 & \dots & \partial A_n/\partial p_n & \partial A_n/\partial u \\ \partial B/\partial p_1 & \dots & \partial B/\partial p_n & \partial B/\partial u \end{matrix} \right\| \tag{8.96}$$

is non-negative definite (here $(p_1, p_2, \dots, p_n)$ is the positive variable for Du, as usual).

Theorem 8.13. *Suppose that the equation*

$$A(x, u, Du)D^2u \equiv \sum_{i,j=1}^{n} A_{ij}u_{ij} = B(x, u, Du) \tag{8.97}$$

has divergence structure and that the matrix (8.96) is non-negative definite. Assume that $u \in C^2(\overline{B})$ and $v \in C^2(\overline{B})$ are solutions of (8.97) in B, such that $u - v \le 0$ on the boundary. Then $u \le v$ in B.

Proof. Let $k = \sup(u - v)$ in B, and suppose for contradiction that $k > 0$. For $\varepsilon < k$ put

$$w = \max\{u - v - \varepsilon, 0\}.$$

Clearly w is Lipschitz continuous in B, and vanishes in some neighborhood of the boundary. Consequently a straightforward process of integration by parts yields

$$\int \{Dw \cdot A(x,u,Du) + wB(x,u,Du)\}dx = 0$$

and

$$\int \{Dw \cdot A(x,v,Dv) + wB(x,v,Dv)\}dx = 0,$$

and integration taking place over the set where $w > 0$. If we subtract these equations, there results

$$\begin{aligned}\int [(Du - Dv)\cdot(A(x,u,Du) - A(x,v,Dv))\\ + (u-v)(B(x,u,Du) - B(x,v,Dv))]dx-\\ -\varepsilon \int [B(x,u,Du) - B(x,v,Dv)]dx = 0.\end{aligned}$$

Letting ε tend to zero then yields

$$\begin{aligned}\int [(Du - Dv)\cdot(A(x,u,Du) - A(x,v,Dv))\\ + (u-v)(B(x,u,Du) - B(x,v,Dv))]dx = 0,\end{aligned}$$

the integration now extending over the set where $u > v$.

The integrand is non-negative by the definiteness of the matrix (8.96), and consequently vanishes on the set where $u > v$. Now by a well-known application of the mean value theorem, the integrand becomes

$$\begin{aligned}(Du - Dv)\frac{\widetilde{\partial A}}{\partial p}(Du - Dv) + (Du - Dv)\\ \cdot\left(\frac{\widetilde{\partial A}}{\partial u} + \frac{\widetilde{\partial B}}{\partial p}\right)(u-v) + \frac{\widetilde{\partial B}}{\partial u}(u-v)^2,\end{aligned}$$

where the notation has an obvious meaning and the tilde denotes evaluation at some intermediate value of the variables. The matrix $\frac{\partial A}{\partial p}$ is positive definite due to ellipticity, and moreover the intermediate value in question lies in a compact set of the u, p space when $x \in \overline{B}$. Thus there exists a positive constant λ such that

$$(Du - Dv)\left(\frac{\widetilde{\partial A}}{\partial p}\right)(Du - Dv) \geq \lambda|Du - Dv|^2$$

for x in $\overline{B}$. Furthermore, by the Cauchy–Schwartz inequality

$$\begin{aligned}(Du - Dv)\cdot\left[\frac{\widetilde{\partial A}}{\partial u} + \frac{\widetilde{\partial B}}{\partial p}\right](u-v) \geq -\text{const.}|Du - Dv|\cdot|u-v| \geq\\ \geq -\frac{\lambda}{2}|Du - Dv|^2 - \text{const.}(u-v)^2.\end{aligned}$$

Consequently we find

$$|Du - Dv| \leq \text{const. } (u - v)$$

throughout the set where $u > v$.

This, however, is impossible. For let P be any point where $u > v$ and let Q be some nearest point to P where $u = v$. (Such a point exists since $u \leq v$ at the boundary of B.) By integrating the preceding differential inequality along the straight line from Q to P we then get

$$0 < u(P) - v(P) \leq [u(Q) - v(Q)] \cdot \exp(\text{const.}|P - Q| = 0,$$

and the theorem is thereby proved. □

We complete §8 with an important application of the strong maximum principle to isometric embeddings of two-dimensional Riemannian manifolds with positive Gaussian curvature in E^3.

8.12 Uniqueness Theorem for Isometric Embeddings of Two-dimensional Riemannian Metrics in E^3

Let V^2 be a C^2-Riemannian manifold and let (U, φ) be a map of C^2-atlas on V^2. Let the Riemannian metric is prescribed by the quadratic form

$$ds^2 = Edu^2 + 2Fdudv + Gdv^2. \tag{8.98}$$

Let E^3 be Euclidean space with Cartesian coordinates x, y, z. Let U be a C^r-isometrically $(r \geq 3)$ immersed in E^3, then

$$ds^2 = dx^2 + dy^2 + dz^2 \tag{8.99}$$

and the functions $x(u,v), y(u,v), z(u,v)$ describe this immersion. Clearly

$$(E - z_u^2)du^2 + 2(F - z_u z_v)dudv + (G - z_v^2)dv^2 = dx^2 + dy^2. \tag{8.100}$$

The metric $dx^2 + dy^2$ is Euclidean. Therefore its Gaussian curvature K_1 is equal everywhere to zero. Hence from the intrinsic formula for the Gaussian curvature

$$K(u,v) =$$

$$= \frac{1}{EG - F^2} \left\{ \begin{vmatrix} -\frac{1}{2}G_{uu} + F_{uv} - \frac{1}{2}E_{vv} & \frac{1}{2}E_u & F_u - \frac{1}{2}D_v \\ F_v - \frac{1}{2}G_u & E & F \\ \frac{1}{2}G_v & F & G \end{vmatrix} \right.$$

$$\left. - \begin{vmatrix} 0 & \frac{1}{2}E_v & \frac{1}{2}G_u \\ \frac{1}{2}E_v & E & F \\ \frac{1}{2}G_u & F & G \end{vmatrix} \right\}$$

it follows that the function $z(u,v)$ satisfies the Darboux equation

$$z_{11}z_{22} - z_{12}^2 = K(EG - F^2)(1 - \Delta_1 z) \tag{8.101}$$

where z_{11}, z_{12}, z_{22} are the second covariant derivatives with respect to the metric (8.98), K is the Gaussian curvature of the same metric (8.98) and

$$\Delta_1 z = \frac{Ez_v^2 - 2Fz_u z_v + Gz_u^2}{EG - F^2} \tag{8.102}$$

is the first Beltrami operator with respect to metric (8.98). Since both quadratic forms (8.98) and (8.100) are positive definite,

$$1 - \Delta_1 z = \frac{(E - z_u^2)(G - z_v^2) - (F - z_u z_v)^2}{EG - F^2} > 0. \tag{8.103}$$

Clearly (8.103) is a necessary condition for the function $z(u,v)$ to be a solution of (8.101).

We now suppose that $z(u,v)$ is a solution of equation (8.101) which satisfies condition (8.103). We will prove that there exists an immersion of U in E^3 and moreover the functions $x(u,v)$ and $y(u,v)$ can be found by means of $z(u,v)$ by quadratures. Indeed the Darboux metric

$$(E - z_u^2)du^2 + 2(F - z_u z_v)dudv + (G - z_v^2)dv^2 \tag{8.104}$$

is positive definite (see the identity (8.100)). The metric (8.104) has zero Gaussian curvature. Therefore this metric defines locally the metric of two-dimensional Euclidean plane. Such form can be reduced to the sum of squares of differentials of functions $\bar{x}(u,v)$ and $\bar{y}(u,v)$. Let $x = \bar{x}(u,v)$, $y = \bar{y}(u,v)$. Then (8.100) will be fulfilled. Hence the functions $x = \bar{x}(u,v)$, $y = \bar{y}(u,v)$, $z = z(u,v)$ describe the desired immersion.

The Darboux equation is a particular case of the Monge–Ampere equation. The type of equation (8.101) is completely defined by the sign of the Gaussian curvature of the original Riemannian metric (8.98).

Thus we reduce the problem of C^r-immersions and embeddings for the Riemannian metric (8.98) to the description of the set of all possible solutions $z(u,v)$ for the Darboux equation (8.101).

Below we establish the uniqueness theorem for C^2-solutions of elliptic Darboux equations with prescribed Dirichlet boundary data.

Theorem 8.14. *Let $z(u,v)$ and $Z(u,v)$ be any C^2-solutions of equation (8.101) in $\overline{U}$, where U is a bounded domain in Euclidean space E^2 with Cartesian coordinates u,v. We assume that 1) the Riemannian metric (8.98) has strictly positive Gaussian curvature K in $\overline{U}$; 2) both quadratic forms $T(\phi,z)$ and $T(\phi,Z)$ are positive definite in $\overline{U}$; 3) $z|_{\partial U} = Z|_{\partial U}$.*

Then $z(u,v) = Z(u,v)$ in U.

Proof. We consider any function $h(u,v) \in C^2(\overline{U})$. Then the expression

$$\phi(h) = h_{11}h_{22} - h_{12}^2 = K(EG - F^2)(1 - \Delta_1 h) \tag{8.105}$$

is called the Darboux operator with respect to the metric (8.98).

We should prove that $z(u,v) = Z(u,v)$ in U if both functions $z(u,v)$, $Z(u,v) \in C^2(\overline{U})$ are positive elliptic solutions of equations

$$\phi(z) = 0 \tag{8.106}$$

$$\phi(Z) = 0 \tag{8.107}$$

and $z(u,v) = Z(u,v)$ on ∂U. Clearly the quadratic forms $T(\phi, z)$ and $T(\phi, Z)$ are positive definite in $\overline{U}$.

The Darboux equation (8.101) contains only the first and second derivatives of the desired solution but does not contain the desired solution itself. Hence the proof of Theorem 8.11 follows directly from Theorem 8.6 if we establish the elliptic convexity of the operator $\phi(h)$ on the family of functions

$$z_\tau(u,v) = (1-\tau)z(u,v) + \tau Z(u,v)$$

for $0 < \tau < 1$.

More precisely we should prove that the quadratic form

$$T(\phi, h) = \sum_{i,k=1}^{2} \frac{\partial \phi}{\partial h_{u_i v_k}} \xi_i \xi_k$$

is positive definite for all functions $h = z_\tau(u,v) = (1-\tau)z(u,v) + \tau Z(u,v)$, $0 < \tau < 1$, if the forms $T(\phi, z)$ and $T(\phi, Z)$ are positive definite. From the formulas of second covariant derivatives of the function $h(u,v)$:

$$\begin{aligned} h_{11} &= h_{uu} - \Gamma_{11}^1(u,v)h_u - \Gamma_{11}^2(u,v)h_v, \\ h_{12} &= h_{uv} - \Gamma_{12}^1(u,v)h_u - \Gamma_{12}^2(u,v)h_v, \\ h_{22} &= h_{vv} - \Gamma_{22}^1(u,v)h_u - \Gamma_{22}^2(u,v)h_v, \end{aligned}$$

with respect to the Riemannian matrix (8.98) it follows that

$$T(\phi, z_\tau) = (1-\tau)T(\phi, z) + \tau T(\phi, Z)$$

for all $\tau \in [0,1]$.

Thus $T(\phi, z)$ and $T(\phi, Z)$ are positive definite forms in $\overline{U}$. The proof of Theorem 8.11 is complete. □

Part II. Geometric Theory of Elliptic Solutions of Monge–Ampere Equations

Introduction

Let $x_1, x_2, \ldots, x_n, x_{n+1}$ be Cartesian coordinates in the $(n+1)$-dimensional Euclidean space E^{n+1}. We denote by E^n the hyperplane $x_{n+1} = 0$. Below we use the notation

$$x_{n+1} = z$$

and call

$$x = (x_1, x_2, \ldots, x_n) \quad \text{and} \quad (x, z) = (x_1, x_2, \ldots, x_n, z)$$

points of E^n and E^{n+1}.

Let G be a bounded domain in E^n and $C^2(G)$ be the set of C^2-functions defined in G. Let

$$H(u) = \det(u_{ij})^{*)} \tag{II.1}$$

for all $u(x) \in C^2(G)$. Then the operator

$$H\colon\ C^2(G) \to C(G)$$

is called the *simplest n-dimensional Monge–Ampere operator* or more briefly the *Monge–Ampere operator*. Partial differential equations containing $H(u)$ as the principal term are called the *Monge–Ampere equations*. More precisely, these equations can be described in the following way:

(a) Classical Monge–Ampere Equations $(n = 2)$. Classical Monge–Ampere equations are related to functions with two independent variables x_1 and x_2. These equations have the form

$$u_{11}u_{22} - u_{12}^2 = Au_{11} + 2Bu_{12} + Cu_{22} + D, \tag{II.2}$$

*) We use the notations: $u_i = u_{x_i}$, $u_{ij} = u_{x_i x_j}$, $i, j = 1, 2, \ldots, n$ and $\det(u_{ij}) = \det\begin{pmatrix} u_{11} & u_{12} & \cdots & u_{1n} \\ \cdots & \cdots & \cdots & \cdots \\ \cdots & \cdots & \cdots & \cdots \\ \cdots & \cdots & \cdots & \cdots \\ u_{n1} & u_{n2} & \cdots & u_{nn} \end{pmatrix}$.

where A, B, C, D are given functions of x_1, x_2, u, u_1, u_2. The expression

$$\Delta = D + AC - B^2 \tag{II.3}$$

is called the discriminant of equation (II.2). For any solution $u(x_1, x_2) \in C^2(G)$ of equation (II.2) the identity

$$(u_{11} - C)(u_{22} - A) - (u_{12} + B)^2 = D + AC - B^2 \tag{II.4}$$

holds. This identity yields the ellipticity (hyperbolicity) of equation (II.2) if and only if $\Delta > 0$ ($\Delta < 0$) for all $(x_1, x_2) \in G$, $u \in R$, $(u_1, u_2) \in R^2$.

Even the theory of the simplest classical Monge–Ampere equations

$$u_{11}u_{22} - u_{12}^2 = D, \qquad A = B = C = 0 \tag{II.5}$$

is rich and has various deep applications to global differential geometry, linear and quasilinear PDE, calculus of variations, applied mathematics and others.

All solutions of elliptic equations (II.5) are necessarily convex or concave functions. All solutions of hyperbolic equations (II.5) necessarily have saddle graphs.

(b) The n-Dimensional Simplest Monge–Ampere Equations. These equations have the form

$$\det(u_{ij}) = D(x, u, \text{grad } u). \tag{II.6}$$

Here the relationship between the graphs of solutions and the type of equation (II.6) holds only for elliptic solutions of equation (II.6). The elliptic solutions of this equation are necessarily convex or concave functions. It is sufficient to consider only convex solutions of equation (II.6). If equation (II.6) has convex solutions, then the function D takes only positive values.

In Part II we are concerned with investigations of the boundary value problems for weak and generalized solutions of the Monge–Ampere equations (II.6). These solutions are general convex and concave functions. *The geometric theory of Monge–Ampere equations* is just the union of the concepts, techniques and results of these investigations. Just as in other branches of modern mathematics, the study of non-smooth objects, which are in the present case weak and generalized solutions of equation (II.6), is not the main goal. We present them just for deeper understanding of smooth subjects. The theory of weak and generalized solutions of various boundary value problems for equation (II.6) is an excellent illustration of this statement, because the proofs of all existence theorems for weak and generalized solutions of such boundary value problems are based on simple principles of compactness in appropriate function spaces. Moreover these theorems are proved under more general and natural conditions than the corresponding theorems for classical solutions.

Since weak and generalized solutions of equation (II.6) are in the set of all general convex and concave functions, the question of C^m-smoothness, $m \geq 2$, of such solutions arises. This question should be considered in two ways. Let G be the domain of equation (II.6). First of all we want to find conditions on the function $D(x,u,p)$, which provide the C^m-smoothness of weak and generalized solutions of equation (II.6) in G. The second problem is to find conditions on the function $D(x,u,p)$ and the boundary data, which provide the C^m-smoothness of the same solutions in $\overline{G} = G \cup \partial G$. The positive solution of these problems opens the way for the application of the techniques and results of the geometric theory of Monge–Ampere operators and equations to classical problems in PDE and Differential Geometry. We consider the C^m-smoothness of solutions for equation (II.6) in Chapter 6.

Chapter 3. Generalized Solutions of N-Dimensional Monge–Ampere Equations

First of all we introduce the concepts of the normal mapping of convex functions and the R-curvature of these functions. The R-curvature of convex functions is the extension of Monge–Ampere operators to the class of all general convex functions. We study in detail the properties of the normal mapping and R-curvature of convex functions and then investigate the solvability of the Dirichlet problem for weak and generalized elliptic solutions together with uniqueness and non-uniqueness theorems for these solutions.

§9. Normal Mapping and R-curvature of Convex Functions

In this section we present the extension of the special type of n-dimensional Monge–Ampere operator $R(\text{grad } u)\det(u_{ij})$ to the class of general convex (concave) functions. This extension will be given by a non-negative completely additive set function $\omega(R, u, e)$, where $R(p) > 0$ is a locally summable function in R^n and e is a subset of the domain of a convex function $u(x)$. The set function $\omega(R, u, e)$ is called the *R-curvature* of a convex function $u(x)$.

9.1 Some Notation

Let $x_1, x_2, \dots, x_n, x_{n+1}$ be Cartesian coordinates in $(n+1)$-dimensional Euclidean space E^{n+1} and let E^n be the hyperplane $x_{n+1} = 0$. As stated in the introduction to Part II, we use the notation $x_{n+1} = z$ and call

$$x = (x_1, x_2, \dots, x_n) \quad \text{and} \quad (x, z) = (x_1, x_2, \dots, x_n, z)$$

points of E^n and E^{n+1}. Let G be a bounded open convex domain in E^n. We denote by S_z the graph of a function z: $G \to R$. Finally $W^+(G)$ and $W^-(G)$ are the respective classes of all convex and all concave functions defined in G. If $z(x) \in W^+(G)$ or $z(x) \in W^-(G)$, then S_z is called a *convex* or *concave hypersurface*.

9.2 Normal Mapping

First of all we introduce a new n-dimensional space $R^n = \{p = (p_1, p_2, \dots, p_n)\}$ with the canonical scalar product

$$(p, q) = \sum_{i=1}^{n} p_i q_i$$

and denote by $|p| = (p,p)^{1/2}$ the length of any vector $p \in R^n$. We call R^n the *gradient space*.

Now let $z(x)$ be any convex function defined in G. Let α be an arbitrary supporting hyperplane of S_z. If

$$Z - z_0 = (p^0, X - x^0) = p_1^0(X_1 - x_1^0) + \cdots + p_n^0(X_n - x_n^0) \tag{9.1}$$

is the equation of α, then the point $(x^0, z^0) \in S_z \cap \alpha$. The point $p^0 = (p_1^0, p_2^0, \ldots, p_n^0) \in R^n$ is called the *normal image of the supporting hyperplane* α and is denoted by

$$p^0 = \chi_z(\alpha). \tag{9.2}$$

The set

$$\chi_z(x_0) = \bigcup_{\alpha} \chi_z(\alpha) \tag{9.3}$$

is called the *normal image of the point* x_0 (more precisely: $\chi_z(x_0)$ is the normal image of the point x_0 with respect to the function $z(x)$), where x_0 is any point of G, and α is any hyperplane of S_z at the point $(x_0, z(x_0)) \in S_z$. Clearly the set $\chi_z(x_0)$ is a closed convex subset of R^n. If $\chi_z(x_0)$ consists of one point, then the convex hypersurface S_z has a tangent hyperplane at the point $(x_0, z(x_0))$. The point $(x_0, z(x_0))$ is called *smooth* for S_z. If for example $z(x)$ is a piecewise linear convex function, then $\chi_z(x_0)$ is a convex closed polyhedron in the gradient space R^n, whose dimension can be $0, 1, 2, \ldots, n$.

Now let e by any subset of G. The set

$$\chi_z(e) = \bigcup_{x_0 \in e} \chi_z(x_0) \tag{9.4}$$

is called the *normal image of* e. Notice that $\chi_z(e)$ is a subset of the gradient space R^n. The mapping χ_z, which maps the sets $e \subset G$ in the sets $\chi_z(e) \subset R^n$, is called *normal*.

It will be proved that the set $\chi_z(e)$ is Lebesgue measurable in R^n if e is a Borel subset of G. Therefore we can introduce the set function

$$\omega(R, z, e) = \int_{\chi_z(e)} R(p)dp \tag{9.5}$$

on the ring of Borel subsets of G, where $R(p) > 0$ is a locally summable function in R^n. The set function (9.5) is called the *R-curvature of convex function* $z(x)$. The main properties of the normal mapping and R-curvature of convex functions will be considered in Subsections 9.4–9.7 and Section 10.

9.3 Convergence Lemma of Supporting Hyperplanes

The following well known lemma will be very useful in the proofs of many fundamental facts.

Lemma 9.1. *Let the sequence of convex hypersurfaces S_{z_k} converge to some convex hypersurface S_z, where $z_k(x)$ and $z(x)$ are convex functions defined in G. Let the points*

$$(x^k, z_k(x^k)) \in S_{z_k}$$

converge to the point $(x^0, z(x^0)) \in S_z$. Then the limit of any convergent sequence of supporting hyperplanes α_k of S_{z_k} at the points $(x^k, z_k(x^k))$ is some supporting hyperplane of S_z at the point $(x^0, z(x^0))$.

Proof. Every convex hypersurface S_{z_k} lies over its supporting hyperplane α_k for $k = 1, 2, \ldots$. Therefore S_z lies over the hyperplane $\alpha = \lim_{k\to\infty} \alpha_k$. Since α passes through the point $(x^0, z(x^0)) \in S_z$, α is some supporting hyperplane of S_z at the point $(x^0, z(x^0))$. □

Remark. If all convex hypersurfaces S_{z_k} coincide with the convex hypersurface S_z then the statement of Lemma 9.1 can be considered as the continuity lemma for supporting hyperplanes of a convex hypersurface S_z.

9.4 Main Properties of the Normal Mapping of a Convex Hypersurface

We recall that G is an open convex bounded domain of E^n.

A) Let $z_1(x)$ and $z_2(x)$ be convex functions defined in G such that $z_1|_{\partial G} = z_2|_{\partial G}$ and $z_1(x) \le z_2(x)$ for all $x \in G$. Then

$$\chi_{z_2}(G) \subset \chi_{z_1}(G). \tag{9.6}$$

B) Let $z(x)$ be any convex function defined in G. Then $\chi_z(F)$ is a bounded closed set in the gradient space R^n for any closed subset F of G.

If we let δ_F be the distance from F to ∂G, and let $M(z, \delta_F)$ be the maximum of $|z(x)|$ for all $x \in G$ such that $\operatorname{dist}(x, \partial G) \ge \delta_F$, then the inequality

$$\operatorname{diam} \chi_z(F) \le 8\delta_F^{-1} M(z, \delta_F) \tag{9.7}$$

holds, and $\chi_z(F)$ is contained in the ball $|p| \le 4\delta_F^{-1} M(z, \delta_F)$.

C) A supporting hyperplane α of a convex hypersurface S_z is called ***singular****) if the set $\alpha \cap S_z$ contains at least two different points. Clearly any singular supporting hyperplane α of S_z contains at least some line segment $\ell \subset \alpha \cap S_z$.

Let Q_z be the set of all singular hyperplanes of S_z. Then

$$\operatorname{mes}_{R^n} \left\{ \bigcup_{\alpha \in Q_z} \chi_z(\alpha) \right\} = 0. \tag{9.8}$$

If the set Q_z is empty, then S_z is called a ***strictly convex hypersurface***.

---

*) We also say that such supporting hyperplanes are strongly touching S_z. See the definition before Lemma 5.2 and the statement and proof of this lemma in § 5.

D) If e is any Borel subset of G, then the set $\chi_z(G)$ is Lebesgue measurable in the gradient space R^n.

E) If $z(x) \in W^+(G) \cap C^1(G)$, then the normal mapping can be reduced to a mapping of points, which coincides with the tangential mapping.

9.5 Proofs

In this subsection we consider the proofs of statements A)–E).

1) Proof of the statement A). Clearly any supporting hyperplane α_2 of the convex hypersurface S_{z_2} has a corresponding supporting hyperplane α_1 of the convex hypersurface S_{z_1}, which is parallel to α_2. Since

$$\chi_{z_1}(\alpha_1) = \chi_{z_2}(\alpha_2),$$

we have

$$\chi_{z_1}(\alpha_1) \supset \chi_{z_2}(\alpha_2).$$ □

2) Proof of the statement B). Since G is an open bounded convex domain, F is a compact subset of G and

$$\operatorname{dist}(F, \partial G) = \delta_F > 0.$$

The convex function $z(x)$ is continuous in G. Therefore

$$M(z, \delta_F) = \sup |z(x)| < +\infty, \quad x \in G, \quad \operatorname{dist}(x, \partial G) \geq \delta_F.$$

Let $\chi_z(F)$ be the normal image of F and let the points $p_i \in \chi_z(F)$ converge to the point p_0. From the compactness of the set F and Lemma 9.1 in the special case $S_{z_k} = S_z$ $(k = 1, 2, \ldots)$, it follows that $p_0 \in \chi_z(F)$.

The proof is realized by standard techniques and is therefore omitted. Thus $\chi_z(F)$ is a closed subset of the gradient space R^n.

Now we prove the estimate (9.7). We can assume that F is a closed convex subdomain of G, because for every closed subset F of G we can find such a convex domain containing F. Let

$$C = \sup_F z(x) < +\infty.$$

Since any convex function achieves its least upper bound on ∂F, there exists a point $c \in \partial F$ such that $C = z(c)$, and the hyperplane

$$\gamma\colon \; z = C$$

lies over the graph of $z(X)$ on the set F and passes through the point (c, C).

Since

$$\operatorname{dist}(F, \partial G) = \delta_F > 0,$$

$\gamma \cap S_z$ is some $(n-1)$-dimensional closed convex surface ℓ. Now consider the convex cone with the vertex $(x_0, z(x_0))$ and base ℓ, where x_0 is any point of the set F. Then

$$\chi_z(x_0) \subset \chi_k(x_0) = \chi_k(F)$$

(remember that F is a closed convex subdomain of G), since every supporting hyperplane of S_z at the point $(x_0, z(x_0))$ is also a supporting hyperplane to the cone K.

But $\chi_k(F)$ is contained in the ball

$$|p| \leq 4\delta_F^{-1} M(z, \delta_F).$$

Hence

$$\operatorname{diam} \chi_z(F) \leq 8\delta_F^{-1} M(z, \delta_F). \qquad \square$$

3) Proof of statement C). Statement C can be proved by a straightforward modification of Alexandrov's proof of Lemma 5.2 (see § 5). Therefore we omit the proof of the statement C).

4) Proof of statement D). Below, for simplicity, we shall refer to Lebesgue measurable sets in the gradient space R^n as measurable sets. From property B) it follows that the set $\chi_z(F)$ is a measurable set in R^n for every closed subset F of an open bounded convex domain G.

Since

$$G = \bigcup_{k=1}^{\infty} F_k \tag{9.9}$$

where $F_1 \subset F_2 \subset \cdots \subset F_k \subset \cdots$ are some closed subsets of G, the set

$$\chi_2(G) = \chi_2\left(\bigcup_{k=1}^{\infty} F_k\right) = \bigcup_{k=1}^{\infty} \chi_2(F_k) \tag{9.10}$$

is measurable in R^n.

Now assume that the set $\chi_2(e)$ is measurable in R^n, where e is some subset of G. Clearly

$$\chi_2(G\backslash e) = [\chi_2(G)\backslash\chi_2(e)] \cup [\chi_2(e) \cap \chi_2(G\backslash e)]. \tag{9.11}$$

The set $\chi_2(e) \cap \chi_2(G\backslash e)$ consists only of the normal images of the strongly touching supporting hyperplanes of S_z. These hyperplanes contain at least two different points belonging to the sets $\chi_z(e)$ and $\chi_z(G\backslash e)$ respectively. Hence

$$\operatorname{mes}_{R^n}[\chi_2(e) \cap \chi_2(G\backslash e)] = 0 \tag{9.12}$$

and the set $\chi_2(e) \cap \chi_2(G\backslash e)$ is measurable. The set $[\chi_z(G)\backslash\chi_z(e)]$ is measurable as the difference of two measurable sets. Now from (9.11) it follows that the set $\chi_z(G\backslash e)$ is measurable in R^n.

Borel subsets of G are generated by taking unions and intersections of countable sequences of closed subsets of G. Clearly $\chi_z\left(\bigcup_{k=1}^{\infty} F_k\right)$ is measurable in R^n for every sequence of closed subsets F_k of G.

From the identity

$$\bigcap_{k=1}^{\infty} F_k = G \Big\backslash \bigcup_{k=1}^{\infty} (G \backslash F_k)$$

(where $F_i, \ldots, F_k, \ldots$ are again closed subsets of G) it follows that the set

$$\chi_2\left(\bigcap_{k=1}^{\infty} F_k\right)$$

is measurable in R^n. Here we use all the previous properties of measurable sets established in this subsection. Statement D is proved. □

Statement E) is obvious.

9.6 R-curvature of convex functions

Let $R(p) > 0$ be a locally summable function in the gradient space R^n. Let $z(x)$ be any convex function defined in G. We introduce the set function

$$\omega(R, z, e) = \int_{\chi_z(e)} R(p)dp, \tag{9.13}$$

where e is any Borel subset of G. (Recall that G is a bounded open convex domain in E^n.)

The set function $\omega(R, z, e)$ is called the *R-curvature* of the convex function $z(x)$. R-curvature takes only non-negative values. Let

$$B(R) = \int_{R^n} R(p)dp. \tag{9.14}$$

Clearly $B(R) > 0$. The case $B(R) = +\infty$ is not excluded. The equality

$$\text{mes}_{R^n}[\chi_z(e_1) \cap \chi_z(e_2)] = 0$$

holds for every two disjunct Borel subsets e_1 and e_2 of G (see property C) of the normal mapping). Therefore from integral theory, it follows that the R-curvature $\omega(R, z, e)$ is a non-negative completely additive set function on the ring of Borel subsets of G.

9.7 Weak convergence of R-curvatures

We assume again that the domain G is open, bounded and convex although for the most part this is not necessary for further considerations.

Let the set functions $\phi_k(e)$, $k = 1, 2, \ldots$ and $\phi(e)$ be completely additive on the subsets e of G. Then $\varphi_k(e)$ are called weakly convergent inside G to $\varphi(e)$ if

$$\lim_{k\to\infty} \int_G h(x)\varphi_k(de) = \int_G h(x)\varphi(de)$$

for every continuous function $h(x)$ differing from zero only on some set H such that

$$H \subset \overline{H} \subset G,$$

where $\overline{H}$ is the closure of H.

The following main theorem holds.

Theorem 9.1. *If the convex functions $z_k(x) \in W^+(G)$ converge to the convex function $z(x) \in W^+(G)$ for all $x \in G$, then their R-curvatures $\omega(R, z_k, e)$ weakly converge inside G to $\omega(R, z, e)$.*

The proof of this theorem is based on the following lemmas:

Lemma 9.2. *If the convex functions $z_k(x) \in W^+(G)$ converge to the convex function $z(x) \in W^+(G)$ for all $x \in G$, then*

$$\overline{\lim_{k\to\infty}}\, \omega(R, z_k, F) \leq \omega(R, z, F) \tag{9.15}$$

for every closed subset F of G.

Proof. Since F is a closed subset of G, $\chi_z(F)$ is a closed and bounded subset of the gradient space R^n. The same statement is correct for all the sets $\chi_{z_k}(F)$. (See property B, Subsection 9.4). Now let Q be some open bounded set in R^n such that

$$\chi_z(F) \subset Q$$

and

$$\int_Q R(p)dp \leq \int_{\chi_2(F)} R(p)dp + \varepsilon,$$

where $\varepsilon > 0$ is a given arbitrary number. We shall prove that the sets $\chi_{z_k}(e)$ are contained in Q for all $k > K(Q)$. Otherwise there exists the sequences of points $p_{k_j} \in \chi_{z_{k_j}}(F)$ such that

$$k_1 < k_2 < \cdots < k_j < \cdots, \qquad \lim_{j\to\infty} k_j = +\infty$$

and

$$p_{z_{k_j}} \notin G.$$

Without loss of generality, we can assume that P_{z_k} converge to some point $p_0 \in R^n$. Here we use the uniform boundedness of all the closed sets $\chi_{z_k}(F)$ (which follows from property B, Subsection 9.4). Since the points $P_{z_{k_j}}$ lie outside of G, the point p_0 does not belong to the set $\chi_z(F)$. Further $p_{z_{k_j}} =$

$\chi_{z_{k_j}}(x_{k_j})$, where $x_{k_j} \in F$. Without loss of generality, we can assume that x_{k_j} converge to point $x_0 \in F$, because F is a closed and bounded subset of E^n.

Now from Lemma 9.1 it follows that $p_0 \in \chi_z(x_0) \subset \chi_2(F)$. Therefore our assumption is not correct. Hence there exists the number $K(Q)$ such that

$$\chi_{z_k}(F) \subset Q$$

for all $k > K(Q)$. Thus

$$\begin{aligned}\omega(R, z_k, F) = \int_{\chi_{z_k}(F)} R(p)dp \le \int_Q R(p)dp < \int_{\chi_2(F)} R(p)dp + \varepsilon \\ = \omega(R, z, F) + \varepsilon\end{aligned}$$

for all $k \geq K(Q)$. The last inequality yields

$$\overline{\lim_{k\to\infty}}\, \omega(R, z, F) \le \omega(R, z, F).$$ □

Lemma 9.3. *Let H be an open subset of G and $\overline{H} \subset G$, where $\overline{H}$ is the closure of H. Let the convex functions $z_k(x) \in W^+(G)$ converge to the convex function $z(x) \in W^+(G)$ for all $x \in G$ and $\omega(R, z, \partial H) = 0$. Then*

$$\lim_{k+\infty} \omega(R, z_k, H) = \omega(R, z, H), \tag{9.16}$$

$$\lim_{k\to\infty} \omega(R, z_k, \partial H) = \omega(R, z, \partial H), \tag{9.17}$$

$$\lim_{k\to\infty} \omega(R, z_k, \overline{H}) = \omega(R, z, \overline{H}), \tag{9.18}$$

if the equality

$$\lim_{k\to\infty} \omega(R, z_k, G) = \omega(R, z, G) < +\infty \tag{9.19}$$

holds.

Proof. The set $F = G\backslash H$ is a closed subset of G. Therefore from Lemma 9.2 it follows that

$$\omega(R, z, G\backslash H) \le \overline{\lim_{k\to\infty}}\, \omega(R, z_k, G\backslash H). \tag{9.20}$$

Since the R-curvature of convex functions is an additive set function,

$$\begin{aligned}\omega(R, z, G\backslash H) = \omega(R, z, G) - \omega(R, z, H)\\ \omega(R, z_k, G\backslash H) = \omega(R, z_k, G) - \omega(R, z_k, H).\end{aligned}$$

From two last equalities, (9.19) and (9.20) it follows that

$$\underline{\lim_{k\to\infty}}\, \omega(R, z_k, H) \ge \omega(R, z, H). \tag{9.21}$$

Since ∂H is a closed subset of G,

$$\overline{\lim_{k\to\infty}}\, \omega(R, z_k, \partial H) \le \omega(R, z, \partial H). \tag{9.22}$$

Since ∂H is a closed subset of G,

$$\underline{\lim_{k\to\infty}}\, \omega(R, z_k, H) \ge \omega(R, z, H). \tag{9.23}$$

Since $\omega(R, z_k, H) \ge 0$ (R-curvature takes only non-negative values) and since $\omega(R, z, \partial H) = 0$ (see the conditions of the present theorem), from (9.22) it follows that

$$\lim_{k\to\infty} \omega(R, z_k, H) = \omega(R, z, \partial H) = 0. \tag{9.24}$$

Thus (9.17) is proved. Now using the additivity of the R-curvature and (9.24) we obtain

$$\omega(R, z, H) = \omega(R, z, \overline{H}) - \omega(R, z, \partial H) = \omega(R, z, \overline{H})$$

and

$$\omega(R, z_k, H) = \omega(R, z_k, \overline{H}) - \omega(R, z_k, \partial H).$$

From two last equalities, Lemma 9.2 and (9.24) it follows that

$$\omega(R, z, H) \ge \overline{\lim_{k\to\infty}}\, \omega(R, z_k, H). \tag{9.25}$$

Now (9.19) and (9.25) yield to the equality

$$\omega(R, z, H) = \lim_{k\to\infty} \omega(R, z_k, H). \tag{9.26}$$

Finally (9.26) and (9.24) yield

$$\omega(R, z, \overline{H}) = \lim_{k\to\infty} \omega(R, z_k, \overline{H}).$$ □

Proof of Theorem 9.1. Let the convex functions $z_k(x)$ converge pointwise to the convex function $z(x)$ in G. Now let $h(x)$ be any continuous function in G different from zero only on some set H such that

$$H \subset \overline{H} \subset G,$$

where $\overline{H}$ is the closure of H. Since $\delta_H = \delta_{\overline{H}} = \operatorname{dist}(\overline{H}, \partial G) > 0$ and $\omega(R, z, e)$ is a completely additive set function, without loss of generality, we can assume that H is a convex subdomain of G and

$$\omega(R, z, \partial H) = 0.$$

From property B it follows that all closed sets $\chi_z(\overline{H})$ and $\chi_{z_k}(\overline{H})$ are contained in the ball

$$|p| \leq \delta_H^{-1}(4M(z,\delta_H)+1)$$

of the gradient space R^n, because $z_k(x)$ uniformly converge to $z(x)$ in $\overline{H}$. Thus the values of the weight function $R(P)$ will not be involved in our consideration if

$$|p| > \delta_H^{-1}(4M(z,\delta_H)+1).$$

Therefore we assume without losing generality that

$$B(R) = \int_{R^n} R(p)dp < +\infty. \tag{9.27}$$

Since $h(x) \equiv 0$ in $G \backslash H$, we extend $h(x)$ as a continuous function in $\overline{G}$ assuming $h|_{\partial G} = 0$. Let $\varepsilon > 0$ be any positive number. Then there exists a positive number δ such that

$$|h(x) - h(x')| < \varepsilon$$

if $\operatorname{dist}(x,x') < \delta$, where $x, x' \in \overline{H}$. Now we decompose the closed convex domain $\overline{H}$ in the union of Borel subsets $V_1, V_2, \ldots, V_m$ such that

$$\begin{aligned} &\text{a) } \operatorname{diam} V_k \leq \delta \quad k = 1,2,\ldots,m;\\ &\text{b) } V_i \text{ and } V_k \text{ are disjoint for } i \neq k, \quad i,k = 1,2,\ldots,m;\\ &\text{c) } \omega(R,z,V_i) = 0, \quad i = 1,2,\ldots,m. \end{aligned} \tag{9.28}$$

Let x_i be some point in V_i $(i = 1,2,\ldots,m)$. Then

$$\begin{aligned} \int_G h(x)\omega(R,z,de) &= \sum_{i=1}^m h(x_i)\omega(R,z,V_i)\\ &\quad + \sum_{i=1}^m \int_{V_i} [h(x) - h(x_i)]\omega(R,z,de) \end{aligned} \tag{9.29}$$

and

$$\begin{aligned} \int_G h(x)\omega(R,z_k,de) &= \sum_{i=1}^m h(x_i)\omega(R,z_k,V_i)\\ &\quad + \sum_{i=1}^m \int_{V_i} [h(x) - h(x_i)]\omega(R,z_k,de). \end{aligned} \tag{9.30}$$

From the assumptions of the present theorem, the condition (9.17) and Lemma 9.3, it follows that

$$\lim_{k\to\infty} \sum_{i=1}^m h(x_i)\omega(R,z_k,V_i) = \sum_{i=1}^m h(x_i)\omega(R,z,V_i). \tag{9.31}$$

Now from (9.16) and the properties of the sets V_i we obtain

$$\left|\sum_{i=1}^{m}\int_{V_i}[h(x)-h_i(x)]\omega(R,z,de)\right| \leq \tag{9.32a}$$
$$\leq \varepsilon \sum_{i=1}^{m}\omega(R,z,V_i) < \varepsilon\cdot B(R)$$

and

$$\left|\sum_{i=1}^{m}\int_{V_i}[h(x)-h_i(x)]\omega(R,z,de)\right| \leq \tag{9.32b}$$
$$\leq \varepsilon\cdot B(R), \qquad k=1,2,3,\dots\ .$$

Equality (9.31) and the inequalities (9.32a,b) give

$$\lim_{k\to\infty}\int_G h(x)\omega(R,z_k,de) = \int_G h(x)\omega(R,z,de).$$

Theorem 9.1 is proved. □

§10. The Properties of Convex Functions Connected With Their R-Curvature

10.1 The Comparison and Uniqueness Theorems

Theorem 10.1. *Let $z_1(x), z_2(x) \in W^+(G)$ and let $z_1(x) \geq z_2(x)$ on G, where G is a bounded open convex domain in E^n. Assume that*

$$\omega(R,z_1,e) \leq \omega(R,z_2,e)^{*)} \tag{10.1}$$

for every Borel subset e of G. *) *Then*

$$z_1(x) \geq z_2(x)$$

for all $x \in G$.

The proof of this theorem is based on the following lemmas:

Lemma 10.1. *Let $z(x) \in W^+(G)$ and C by any constant, then*

$$\omega(R,z+C,e) = \omega(R,z,e) \tag{10.2}$$

for all Borel subsets e of G.

The proof of this theorem follows directly from the obvious equality

$$\chi_{z+C}(e) = \chi_e(e)$$

*) We assume that the function $R(p)$ generating $\omega(R,z,e)$ is positive and locally summable in the gradient space R^n (see Section 9.6).

for all Borel subsets e of G. □

Lemma 10.2. *Let $z_1(x), z_2(x) \in W^+(G)$ and let Q be an open subdomain of G such that:*

a) $\overline{Q} \subset G$,
b) $z_1(x) < z_2(x)$ *for all* $x \in Q$,
c) $z_1(x) = z_2(x)$ *for all* $x \in \partial Q$.

If

$$\chi_{z_2}(x_0) \backslash \chi_{z_1}(x_0) \neq \emptyset \tag{10.3}$$

for at least in one point $x_0 \in \partial G$, then

$$\omega(R, z_1, Q) > \omega(R, z_2, Q).$$

Proof. Let S_{z_1} and S_{z_2} be the graphs of the functions $z_1(x)$ and $z_2(x)$. Let H_1 and H_2 be the domains of S_{z_1} and S_{z_2} projecting onto Q. Clearly the proof of Lemma 10.2 follows from the two statements.

a) $\chi_{z_1}(Q) \supset \chi_{z_2}(Q)$
b) int $\chi_{z_1}(Q) \neq$ int $\chi_{z_2}(Q)$,

where int $\chi_{z_i}(Q)$ is the interior of the set $\chi_{z_i}(Q)$, $i = 1, 2$.

The statement a) is clearly true since for any supporting hyperplane P of H_2 there is some supporting hyperplane P' of H_1 parallel to P. Now there exists such a supporting hyperplane T of S_{z_2} which will not be supporting for S_{z_1}. This fact follows directly from the conditions of the present lemma. The hyperplane T cuts the cap from H_1. Therefore the normal image $\chi(T)$ of T will be an interior point of $\chi_{z_1}(Q)$. Since $\chi(T) \in \partial\chi_{z_2}(Q)$,

$$\text{int } \chi_{z_1}(Q) \neq \text{int } \chi_{z_2}(Q).$$

Statement b) is proved. The proof is complete. □

Now consider the proof of our theorem. Let $z_1(x) < z_2(x)$ on some subset M of G. Clearly M is open. Let Q be one of the components of M. We can assume without loss of generality, that $\overline{Q} \subset G$.

We consider the set of points such that $z_1(x) < z_2(x) - \frac{\varepsilon_0}{2}$, where $\varepsilon_0 = \sup|z_2(x)z_1(x)|$. Then all the conditions of Lemma 10.2 will be fulfilled for the function $z_1(x)$ and $\tilde{z}_2(x) = z_2(x) - \frac{1}{2}\varepsilon_0$. This is provided by Lemma 10.1.

If the statement (10.3) is fulfilled at least at one point $x_0 \in G$, then Lemma 10.2 leads to a contradiction with the assumption (10.1). Therefore

$$\chi_{z_2}(x) \subset \chi_{z_1}(x)$$

for every point $x \in \partial Q$. Moreover

$$\chi_{z_2}(x) \subset \chi_{z_1}(x) \tag{10.4}$$

for all $x \in Q$. Otherwise we transfer the hypersurface S_{z_2} in a parallel way to the z-axis so that the points $(x, z_2(x))$ and $(x_1, z_1(x))$ coincide. Then the conditions of Lemma 10.2 will be fulfilled. Using Lemma 10.1 and 10.2 we shall obtain the contradiction.

The convex functions $z_1(x)$ and $z_2(x)$ have their first differential almost everywhere in Q. Therefore from (10.4) it follows that $dz_1 = dz_2$ almost everywhere in Q. Since $z_1(x)$ and $z_2(x)$ satisfy the Lipschitz condition in $\overline{Q}$, and $z_1(x) = z_2(x)$ for all $x \in \partial Q$, we have $z_1(x) \equiv z_2(x)$ in Q. This contradicts our assumption and proves Theorem 10.1. □

Theorem 10.2. *Let $z_1(x), z_2(x) \in W^+(G)$ and $z_1(x) = z_2(x)$ for all $x \in \partial G$. Let*

$$\omega(R, z_1, e) = \omega(R, z_2, e)$$

for all Borel subsets e of G. Then $z_1(x) = z_2(x)$ for all $x \in G$.

This theorem follows directly from Theorem 10.1. □

10.2 Geometric Lemmas and Estimates

Let G be a convex bounded open domain in E^n and let $u(x) \in C(\overline{G})$ be an arbitrary convex function vanishing on ∂G. Consider the convex cone K with the vertex $(x_0, u(x_0))$ and base ∂G, where x_0 is an interior point of G.

Lemma 10.3. *Let $R(p) > 0$ be any locally summable function in $R^n = \{p = (p_1, p_2, \ldots, p_n)\}$. Then*

$$\omega(R, u, G) \geq \omega(R, K, G) \geq \int_{|p| \leq \rho} R(p)dp, \tag{10.5}$$

where $\rho = \frac{|u(x_0)|}{d(G)}$ and $d(G)$ is the diameter of G.

Proof. Since $\chi_u(G) \supset \chi_K(G)$ (see the property A, Section 9.4),

$$\begin{aligned}\omega(R, K, G) &= \int_{\chi_K(G)} R(p)dp \\ &\leq \int_{\chi_u(G)} R(p)dp = \omega(R, u, G).\end{aligned} \tag{10.6}$$

Now consider the n-ball $S \subset E^n$ with the center x_0 and the radius $d(G)$. Let K_0 be the convex cone of revolution with the vertex $(x_0, u(x_0))$ and base ∂S. The set $\chi_{K_0}(S)$ is the n-ball in the gradient space R^n with the center $(0, 0, \ldots, 0)$ and the radius $\rho = \frac{u(x_0)}{d(G)}$. Clearly, $\chi_{K_0}(S) \subset \chi_K(G)$. Therefore

$$\omega(R, k_0, S) \leq \omega(R, K, G). \tag{10.7}$$

From (10.5), (10.6) and (10.7) it follows that

$$\omega(R, u, G) \geq \omega(R, K, G) \geq \int_{|p| \leq \rho} R(p)dp. \qquad \square$$

Remark. If we consider the condition

$$u|_{\partial G} = h = \text{const} \tag{10.8}$$

instead of the condition $u|_{\partial G} = 0$ then the inequalities (10.5) take the form

$$\omega(R, u, G) \geq \omega(R, K, G) \geq \int_{|\rho| \leq \rho_h} R(p)dp$$

where

$$\rho_h = |h - u(x_0)| \cdot (\text{diam } G)^{-1}. \tag{10.9}$$

Let $R(p) > 0$ be a locally summable function in the gradient space R^n. Now we introduce the function

$$g_R(\rho) = \int_{|p| < \rho} R(p)dp \tag{10.10}$$

for $\rho \in [0, +\infty)$. Evidently $g_R(\rho)$ is strictly increasing and continuous, and $g_R(0) = 0$, $g_R(\infty) = B(R)$.

We denote by T_R: $[0, (R)) \to [0, +\infty)$ the inverse of the function $g_R(\rho)$. Clearly $T_R(\tau)$ is also strictly increasing and continuous.

Theorem 10.3. *Let $u(x)$ be a convex function in G which satisfies two conditions*

a) $u|_{\partial G} = h = const$
b) $\omega(R, u, G) < B(R)$.

Then

$$h - T_R(\omega_u)d(G) \leq u(x) \leq h \tag{10.11}$$

where $\omega_u = \omega(R, u, G)$.

Remark. If $U(x)$ is a concave function in G satisfying the same conditions a) and b), then the inequalities (10.11) take the form

$$h \leq U(x) \leq h + T_R(\omega_u)d(G) \tag{10.12}$$

everywhere in G.

Proof. Let K_0 be the cone of revolution considered in Lemma 10.3. Then

$$g_R(\rho_h) = \int_{|p| \leq \rho_h} R(p)dp = \omega(R, K_0, S),$$

where

$$\rho_h = \frac{|h - u(x_0)|}{d(G)}.$$

We can take $x_0 \in G$ so that the equality

$$\sup_G |h - u(x)| = |h - u(x_0)|$$

holds.

Therefore

$$\rho_h = T_R(\omega(R, K_0, S)) \leq T_R(\omega(R, U, G)) = T_R(\omega_u),$$

because T_R is an increasing function. Thus

$$|h - u(x_0)| \leq T_R(\omega_u)d(G). \tag{10.13}$$

From (10.13) it follows that

$$h - T_R(\omega_u)d(G) \leq u(x) \leq h \tag{10.14}$$

for all $x \in G$, if $u(x)$ is convex in G. □

The proof of the inequalities (10.12) also follows from (10.13) if the function $u(x)$ is concave in G.

Theorem 10.4. *Let G be a convex bounded domain in E^n and let $V(\omega_0) = \{z(x)\}$ be the set of all convex and concave functions belonging to $W(G)$ and satisfying the following conditions:*

$$1) \qquad -\infty < m \leq z|_{\partial G} \leq M < +\infty; \tag{10.15}$$
$$2) \qquad \omega(R, z, G) \leq \omega_0 < B(R). \tag{10.16}$$

Then the inequalities

$$m - T_R(\omega_0)d(G) \leq z(x) \leq M \tag{10.17}$$

hold if $z(x)$ is convex and the similar inequalities

$$M \leq z(x) \leq M + T_R(\omega_0)d(G) \tag{10.18}$$

hold if $z(x)$ is concave.

The proof of this theorem follows directly from Lemmas 10.2 and 10.3.

10.3 The Border of a Convex Function

In this subsection we are concerned with properties of the limiting values of convex functions $u(x) \in W^+(G)$, where G is an open bounded convex domain. Let $u(x)$ be any function belonging to $W^+(G)$. Then $u(x)$ is continuous in G, and takes only finite values in G.

Let H_u be the union of points (x, z), where $x \in G$ and $z \leq u(x)$. The H_u is a convex set and $\overline{H}_u = H_u \cup \partial H_u$ is a convex body in E^{n+1} (see Subsection 3.1, Chapter 1). Let

$$Z = \partial G \times R$$

by the cylinder $\{(x,z); x \in \partial G, z \in (-\infty,+\infty)\}$. In Subsection 3.1 it was established that

$$\partial \overline{H}_u = S_u \cup M_u$$

where S_u is the graph of $u(x)$ in G and M_u is a closed subset of Z such that $M_u \cap S_u = \emptyset$, and the projection of M_u on E^n coincides with ∂G (see Subsection 3.1).

Now we prove that

$$u_0 = \inf_G u(x) > -\infty \tag{10.19}$$

for any function $u(x) \in W^+(G)$. Assume that

$$u_0 = \inf_G u(x) = -\infty \tag{10.20}$$

for some function $u(x) \in W^+(G)$. Then there exists the sequence of points $x_k \in G$ such that

$$\lim_{k\to\infty} u(x_k) = -\infty. \tag{10.21}$$

Since $\overline{G} = G \cup \partial G$ is a compact set in E^n, without loss of generality assume that the points x_k converge to some point $x_0 \in \overline{G}$. Clearly $x_0 \in \partial G$. Let

$$\ell_{x_0} = \{(x_0,z), z(-\infty,+\infty)\} \tag{10.22}$$

be the straight line orthogonal to the hyperplane E^n. From the equality (10.20) it follows that

$$\ell_{x_0} \subset M_u. \tag{10.23}$$

Since $\overline{H}$ is a convex body in E^{n+1}, $\overline{H} = \overline{G} \times R$. Hence $\partial\overline{H} = Z$, and S_u can not be a part of $\partial\overline{H}$. This contradicts assumption (10.20). Thus the inequality (10.19) is proved.

We denote by $L_u(x_0)$ the set of all limit points of the convex hypersurface S_u lying on the straight line ℓ_{x_0}. From (10.19) and the convexity of S_u it follows that $L_u(x_0)$ is either some point (x_0, z_0), some closed segment consisting of points $\{(x_0,z);\ z_0 \le z \le z_1\}$, or some closed ray consisting of points $\{(x_0,z); z_0 \le z < +\infty\}$.

Now we introduce the function

$$h_u\colon \partial G \to R \tag{10.24}$$

by means of the formula

$$h_u(x_0) = z_0$$

for any $x_0 \in \partial G$. We call the function h_u *the border of the function* $u(x) \in W^+(G)$.

Remark. The function h_u can be discontinuous for some functions $u(x) \in W^+(G)$. Here is the simplest example. Consider the convex cone K with the base ∂G and the vertex (x_0, z_0). Let $u(x)$ be the function setting K over the open domain G. Then the border of $u(x)$ is the discontinuous function

$$h_u(x) = \begin{cases} 0 & \text{if} \quad x \neq x_0 \\ z_0 & \text{if} \quad x = x_0. \end{cases}$$

10.4 Convergence of Convex Functions in a Closed Convex Domain. Compactness Theorems

Theorem 10.5 (The First Compactness Theorem). *Let $V^+(\omega_0)$ be the subset of all convex functions $z(x) \in W^+(G)$ satisfying the following conditions:*

a) G is a bounded open convex domain in E^n;
b) $\omega(R, z, G) \leq \omega_0 = const \; < B(R)$ (10.25)

where $R(p) > 0$ is a locally summable function in the gradient space R^n;
c) The border of convex functions $z(x) \in V^+(\omega_0)$ satisfy the inequalities

$$-\infty < m = const \; \leq h_z|_{\partial G} \leq M = const \; < +\infty. \tag{10.26}$$

Then the set $V^+(\omega_0)$ is compact with respect to pointwise convergence in G.

The proof of this theorem follows directly from Theorem 10.4 (see Section 10.2) and the Blaschke Theorem (see Section 3.5).

Here are some remarks concerning the convergence considered in Theorem 10.5. We assume that all the conditions of Theorem 10.5 are fulfilled.

Remark 1. Let G_δ be an open subdomain of G and $\delta = \text{dist}(G_\delta, \partial G) > 0$. Then every pointwise convergent sequence of convex functions converges uniformly in $\overline{G}_\delta$.

Proof. All convex functions $z(x) \in V^+(\omega_0)$ have the uniform estimate

$$m - T_R(\omega_0)d(G) \leq z(x) \leq M$$

(see Theorem 10.4; Section 10.2). Therefore all these functions have uniform estimates of norms in the space $C^{0,1}(\overline{G})^{*)}$ depending only on δ, $m - T_R(\omega_0)d(G)$ and M. The desired statement follows directly from these inequalities.

Remark 2. There exist pointwise convergent sequences of convex functions non-uniformly convergent in G.

One of the simplest examples is as follows: Let ∂G be an open bounded convex domain in E^n, 0 be some inner point of G and x be some point of ∂G. Consider the sequence of points $a_m \in G$ lying on the segment $0a_0$. Assume that $a_0 = \lim\limits_{m \to \infty} a_m$.

---

*) Here we consider the restrictions of functions $z(x)$ to the set $\overline{G}$.

Let K_m be the convex cone with base G and the vertex $(a_m, -1)$. Let $u_m(x) \in W^+(G)$ be a convex function defining the interior part of K_m, i.e. the graph of $u_m(x)$ coincides with $K_m \backslash \partial K_m$, where $K_m = \partial G$. Clearly $h_{u_m}(x) = 0$ for all $x \in \partial G$.

If $m \to \infty$ then the convex cones K_m converge to the convex cone K_0 with the vertex $(a_0, -1)$ and base ∂G. Let $u_0(x)$ be the function determining the inner part of the cone K_0. Then

$$u_m(x) \to u_0(x)$$

for all $x \in G$, but this convergence is non-uniform. Note that

$$h_{u_0}(x) = \begin{cases} 0 & \text{if } \quad x \in G \quad \text{ and } \quad x \neq a_0; \\ -1 & \text{if } \quad x = a_0. \end{cases}$$

Thus

$$h_{u_0}(x) = \lim_{m \to \infty} h_{u_m}(x) = 0$$

for all $x \in \partial G$ except the point a_0. Clearly, $h_{u_0}(a_0) = -1$, but

$$\lim_{m \to \infty} h_{u_m}(a_0) = 0.$$

Hence the convex functions $u_m(x)$ do not converge uniformly in G.

Now we consider the problem of the uniform convergence of convex functions in $\overline{G}$, where $\overline{G}$ is the closure of an open bounded convex domain G. Let a_0 be any point of ∂G. Then there exists a supporting $(n-1)$-plane α of ∂G passing through a_0, an open n-ball $U_\rho(a_0)$ with the center a_0, and the radius $\rho > 0$ such that the convex $(n-1)$-surface

$$\Gamma_\rho(a_0) = \partial G \cap U_\rho(a_0) \tag{10.27}$$

has the one-to-one orthogonal projection π_α: $\Gamma_\rho(a_0) \to \alpha$.

Moreover, the unit normal of α in the direction of the halfspace of E^n, where $\overline{G}$ lies, passes through interior points of G.

All considerations made above are also valid for every n-ball $U_{\rho'}(a_0)$ where $0 < \rho' \leq \rho$. Denote by $\sqcap_\rho(a_0)$ the set $\pi_\alpha(\Gamma_\rho(a_0))$ (see pic.). Let $x_1, x_2, \ldots, x_{n-1}, x_n, z$ be the Cartesian coordinates in E^{n+1} with the following properties: a_0 is the origin, the axes $x_1, x_2, \ldots, x_{n-1}$ lie in the plane α, the axis x_n is directed along the interior normal of ∂G at the point a_0, and the axis z is orthogonal to the hyperplane E^n. Clearly the convex $(n-1)$-surface $\Gamma_\rho(a_0)$ is the graph of some convex function $g(x_1, x_2, \ldots, x_{n-1}) \in W^+(\sqcap(a_0))$.

Obviously

$$g(0, 0, \ldots, 0) = 0 \tag{10.28}$$

and

$$g(x_1, x_2, \ldots, x_{n-1}) \geq 0 \tag{10.29}$$

for all points of the set $\sqcap_\rho(a_0)$. The function $g(x_1, x_2, \dots, x_{n-1})$ is called the *local explicit representation of the convex surface* ∂G near the marked point a_0. Note that a_0 is any point of ∂G.

We shall say that ∂G has a *local parabolic support of order* $\tau \geq 0$ at the point a_0 if there exist positive numbers $\rho_0 \leq \rho$*) and $b(x_0)$ such that

$$g(x_1, x_2, \dots, x_{n-1}) \geq b(x_0) \left(\sum_{i=1}^{n-1} x_i^2 \right)^{\frac{\tau+2}{2}} \tag{10.30}$$

for all $(x_1, x_2, \dots, x_{n-1}) \in \sqcap_\rho(a_0)$.

The equivalent statement of the last concept is as follows: the convex $(n-1)$-surface $\Gamma_{\rho_0}(a_0)$ can be touched from outside by the $(n-1)$-paraboloid

$$x_n = b(x_0) \left(\sum_{i=1}^{n-1} x_i^2 \right)^{\frac{\tau+2}{2}}$$

at the point a_0 for all $\rho_0 \leq \rho$.

We shall say that ∂G has a *parabolic support of order not greater than* $\tau =$ const ≥ 0, if the local parabolic support of ∂G has order not greater than τ at all points $a_0 \in \partial G$. Now we state a few assumptions providing for the uniform convergence of a sequence of convex functions in a closed convex domain $\overline{G}$.

Assumption 10.1. Let $R(p)$ be a locally summable positive function in the space R^n, and let the inequality

$$R(p) \geq C_0 |p|^{-2k}, \quad p \neq (0, 0, \dots, 0) \tag{10.31}$$

hold for all $p \in R^n$, where $k \geq 0$ and $C_0 > 0$ are some constants.

Assumption 10.2. Let G be an open bounded convex domain in E^n and let $G_1 \subset G_2 \subset G_3 \subset \dots \subset G_m \subset \dots \subset G$ be the sequence of open convex domains such that $G = \bigcup_{i=1}^{\infty} G_i$. Let $u_m(x)$ be a convex function defined in G_m.

We assume that the following conditions are fulfilled:

a) the sequence of convex functions $u_m(x)$ converge pointwise to $u(x) \in W^+(G)$;
b) there exists an n-ball $U_\rho(x_0)$ for any point $x_0 \in \partial G$ such that

$$\underline{\lim}_{m \to \infty} \omega(R, u_m, e \cap G_m) \leq a[\sup_e(\text{dist}(x, \partial G))]^\lambda \text{ mes } e, \tag{10.32}$$

*) We recall that $\rho > 0$ is the radius of the n-ball $U_\rho(a_0)$ used in the definition of the convex $(n-1)$-surface $\Gamma_\rho(a_0)$.

where $e \subset U_\rho(x_0) \cap G$ is a Borel subset, and $a > 0$, $\lambda \geq 0$ are constants.

Assumption 10.3. Let $h_m(x)$ be the borders of the convex functions $u_m(x)$, and $h(x)$ be the border of the limiting function $u(x)$. Let S_{h_m} and S_h be the graphs of $u_m(x)$ and $u(x)$ respectively. Then we assume that S_{h_m} are continuous, and the $(n-1)$-surfaces S_{h_m} converge to some $(n-1)$-surface S_{h_0}, which is the graph of some continuous function $h_0(x)$: $\partial G \to R$. (Clearly both surfaces S_{h_m} and S_{h_0} are homeomorphic to the $n-1$-unit sphere.)

Theorem 10.6. *Let G be an open bounded convex domain and let ∂G have a parabolic support of order not greater than $\tau =$ const ≥ 0. Let Assumptions 10.1–3 hold for the sequence of convex functions $u_m(x)$ and their R-curvatures respectively (see (10.32)). Now let the numbers k, λ and τ satisfy the additional inequalities*

$$k \leq K \quad \text{if} \quad 0 \leq k < 1 \quad \text{or} \quad k \geq \frac{n}{2}; \tag{10.33a}$$

$$k < K \quad \text{if} \quad 1 \leq k < \frac{n}{2}, \tag{10.33b}$$

where $K = \frac{n+\tau+1}{\tau+2} + \frac{\lambda}{2}$. Then the function $h_0(x)$ is the border of the function $u(x)$, i.e. $h_0(x) = h_u(x)$ for all $x \in \partial G$.

Proof. The proof has three parts.

Part 1. Suppose that the values of the border $h_u(x)$ of $u(x)$ do not coincide with the values of the function $h_0(x)$ for all $x \in \partial G$. Since the inequality

$$h_u(x) \leq h_0(x)$$

holds for all $x \in \partial G$, there exist at least one point $x_0 \in \partial G$ such that

$$h_u(x_0) < h_0(x_0).$$

Now we introduce special Cartesian coordinates in E^n and E^{n+1}. We put the axes $x_1, \ldots, x_{n-1}$ in the supporting $(n-1)$-plane α of ∂G. Further, the axis x_n is orthogonal to α, and has points inside G. Finally axis z is orthogonal to the hyperplane E^n. Thus the point $x_0 \in E^n$ can be denoted by $0(0, 0, \ldots, 0)$. Let

$$Q = (0, h_0(0))$$

and

$$\overline{Q} = (0, h_u(0))$$

be points of the z-axis and let $0 < \delta < 1$ be an arbitrary number.

We introduce two new points

$$Q'(0, h_0(0) - \delta\Delta h) \quad \text{and} \quad Q''(0, h_u(0) - \delta\Delta h)$$

lying on the z-axis, where

$$\Delta h = h_0(0) - h_u(0) > 0. \tag{10.34}$$

Clearly Δh is the length of the segment $Q'Q''$ and Q' lies inside the segment $Q\overline{Q}$, and Q'' lies under the point $\overline{Q}$. Now consider two hyperplanes β' and β'' given in E^{n+1} by equations:

$$\beta' \colon \ z = h_0(0) - \delta\Delta h - (1/\gamma)x_n; \tag{10.35}$$

$$\beta'' \colon \ z = h_u(0) - \delta\Delta h, \tag{10.36}$$

where δ is a sufficient small positive number.

Let $Z = \partial G \times R$ be a cylinder in E^{n+1} with basis ∂G whose generators are parallel to the z-axis. Then Z bounds some convex body K together with the hyperplanes β' and β''. Clearly the convex hypersurfaces S_{u_m} (graphs of the convex functions $u_m(x)$) have non-empty intersections with the convex body K. We denote by Q_m the nearest point of S_{u_m} to the point $\overline{Q}$. Now we introduce two sets

$$S_m(K) = S_{u_m} \cap K \tag{10.37}$$

and

$$\beta'(K) = \beta' \cap K. \tag{10.38}$$

If $\gamma > 0$ is sufficiently small and m is sufficiently large then

$$S_m(K) \cap \beta'' \neq \emptyset$$

and

$$S_m(K) \cap Z \neq \emptyset.$$

Therefore $Q_m \in S_m(K)$. Let $H_m(K)$ be the projection of $S_m(K)$ on E^n and let V_m be the convex cone with the vertex Q_m and the base $\beta'(K)$. Then the normal image of the set $S_m(K)$ covers the normal image of the cone V_m.

Thus

$$\omega(R, V_m) \leq \omega(R, u_m, S_m(K)). \tag{10.39}$$

Let $H(K)$ be the projection of the convex body K on E^n. Since

$$H_m(K) \subset H(K) \cap G_m,$$

then

$$\omega(R, u_m, H_m(K)) \geq \omega(R, u_m, H(K) \cap G_m).$$

Thus

$$\omega(R, V_m) \leq \omega(R, u_m, H(K) \cap G_m). \tag{10.40}$$

If $m \to \infty$, then the points Q_m converge to the point $\overline{Q}$. Hence

$$\lim_{m \to \infty} \omega(R, V_m) = \omega(R, V), \tag{10.41}$$

where V is a convex cone, the vertex of which is $\overline{Q}$ and the basis of which is $\beta'(K)$.

Now we employ Assumption 10.1, which is one of conditions of the present theorem, i.e. there is some n-ball $U_\rho(0)$ for which inequality (10.32) holds. If γ is sufficiently small, then the Borel set

$$H_m(K) \subset U_\rho(0) \cap G.$$

Hence

$$\varliminf_{m\to\infty} \omega(R, u_m, H(K) \cap G_m) \leq a[\sup_{H(K)} (\operatorname{dist}(x, \partial G)]^\lambda \cdot \operatorname{mes} H(K).$$

The last inequality together with (10.40) implies the inequality

$$\omega(R, V) \leq a[\sup_{H(K)} (\operatorname{dist}(x, \partial G)]^\lambda \operatorname{mes} H(K). \tag{10.42}$$

Clearly

$$\sup_{H(K)} [\operatorname{dist}(x, \partial G)] = \gamma(\Delta h).$$

Let T be the set of points $x \in E^n$ satisfying the conditions

$$b(0)\left(\sum_{i=1}^{n-1} x_i^2\right)^{\frac{\tau+2}{2}} \leq x_n \leq \gamma(\Delta h),$$

where $b(0) =$ const > 0. Hence

$$H(K) \subset T$$

and

$$\operatorname{mes} T = \int_T dx = \mu_{n-1} \int_0^{\gamma(\Delta h)} \left(\frac{h}{b(0)}\right)^{\frac{n-1}{\tau+2}} dh = \tag{10.43}$$
$$= \frac{(\tau+2)}{n+1} \mu_{n-1} \left(\frac{h}{b(0)}\right)^{\frac{n+\tau+1}{\tau+2}} \gamma^{\frac{n+\tau+1}{\tau+2}} = d_1 \gamma^{\frac{n+\tau+1}{\tau+2}},$$

where $d_1 =$ const > 0 depending only on given constants $\tau \geqq 0$, $b(0) > 0$. Δh and μ_{n-1}; we remind that Δh is the length of the segment $Q\overline{Q}$ and μ_{n-1} is the volume of the unit $(n-1)$-ball.

Now from (10.43) and (10.42) it follows that

$$\omega(R, V) \leq d_2 \gamma^{\lambda + \frac{n+\tau+1}{\tau+2}}, \tag{10.44}$$

where $d_2 = a d_1 (\Delta h)^\lambda =$ const > 0.

Estimate (10.44) completes the first part of the proof of the present theorem.

Part 2. We estimate $\omega(R, V)$ from below.

Let $\widetilde{Z} = S_0 \times R$ be the cylinder with the base S_0 and generators parallel to the z-axis. We recall that S_0 is the graph of the equation

$$x_n = b(x_0)\left[\sum_{i=1}^{n-1} x_i^2\right]^{\frac{r+2}{2}}$$

in the hyperplane E^n: $z = 0$.

Now we replace the cylinder Z by the cylinder $\widetilde{Z}$ in all of the constructions mentioned above. Then we obtain the convex cone $\widetilde{V}$ instead of the convex cone V. Clearly

$$\omega(R, V) \geq \omega(R, \widetilde{V}). \tag{10.45}$$

The estimate for $\omega(R, \widetilde{V})$ from below can be derived from the following lemma.

Lemma 10.4. *Let H^n be the cone of revolution with axis p_n, vertex $(0, 0, \ldots, 0, -\frac{\delta}{\gamma})$ and base H^{n-1}, where H^{n-1} is the $(n-1)$-dimensional ball*

$$p_1^2 + p_2^2 + \cdots + p_{n-1}^2 \leq (C')^2 \gamma^{-\frac{2}{r+2}}, \quad p_n = -C''\gamma^{-1}$$

where the values C' and C'' do not depend on γ and have positive limits as $\delta > 0$ approaches zero. Then the normal image of $\widetilde{V}$ contains the cone H^n.

Proof. We shall continue to use the Cartesian coordinates $x_1, x_2, \ldots, x_n$, z introduced above (see Theorem 10.6). We recall that the point $x_0 \in \partial G$ is the origin, the axes $x_1, x_2, \ldots, x_{n-1}$ lie in the supporting hyperplane α of ∂G at the point x_0, and the axis x_n is orthogonal to α and directed inside G; finally the axis z is orthogonal to the hyperplane E^n: $z = 0$.

Now we introduce Cartesian coordinates $p_1, p_2, \ldots, p_n$ in the gradient space R^n, which are associated with $x_1, x_2, \ldots, x_n$ in E^n in the following way:

$$p_1 = \frac{\partial z}{\partial x_1}, \ldots, p_n = \frac{\partial z}{\partial x_n}$$

for any C^1-function $z(x)$ in E^n.

Clearly the normal image of the convex cone $\widetilde{V}$ is some convex closed set in R^n. Therefore it is sufficient to prove the two following statements:

a) the point $(0, 0, \ldots, 0 - \frac{\delta}{\gamma})$ belongs to the normal image of $\widetilde{V}$;

b) the normal image of $\widetilde{V}$ contains the $(n-1)$-dimensional ball H^{n-1} mentioned above in the statement of Lemma 10.4.

Now consider the hyperplane

$$\beta''': \; z = h_u(0) - \frac{\delta}{\gamma} x_n$$

in E^{n+1}. Clearly β''' passes through the point $\overline{Q}$ and the $(n-1)$-plane $\alpha^* = \beta' \cap \beta''$. Thus β''' is a supporting hyperplane of the convex cone $\widetilde{V}$ at its vertex $\overline{Q}$.

Hence the normal image of β'''

$$\chi(\beta''') = \left(0, 0, \ldots, 0, -\frac{\delta}{\gamma}\right)$$

is contained in the normal image of the convex cone $\widetilde{V}$. The statement a) is proved. Now we consider the $(n-1)$-plane

$$\alpha: \begin{cases} x_n = 0, \\ z = 0, \end{cases}$$

introduced above (see Theorem 10.6).

Denote by $\rho, \theta_1, \theta_2, \ldots, \theta_{n-2}$ the spherical coordinates in the plane α. Let S^{n-2} be the unit $(n-2)$-sphere in the plane α with the center x_0 and let $\vec{q}(\theta_1, \theta_2, \ldots, \theta_{n-2})$ be the positive vector-function of S^{n-2} with the origin at the same point x_0. Clearly, for every point $y_0 \in S^{n-2}$ there exists a neighborhood $U(y_0)$ of a fixed size such that the spherical parameters $\theta_1, \theta_2, \ldots, \theta_{n-2}$ can be chosen under the condition that the vectors

$$\vec{q}, \frac{\partial \vec{q}}{\partial \theta_1}, \frac{\partial \vec{q}}{\partial \theta_2}, \ldots, \frac{\partial \vec{q}}{\partial \theta_{n-2}}$$

form an orthogonal basis in the $(n-1)$-plane α at every point $z \in U(y_0)$. Thus the convex cone $\widetilde{V}$ is described locally by the vector function

$$\vec{V}(\rho, \theta_1, \theta_2, \ldots, \theta_{n-2}, s) = \overrightarrow{0Q} + s\,\vec{r}(\rho, \theta_1, \theta_2, \ldots, \theta_{n-2})$$

where

$$\begin{aligned} \vec{r}(\rho, \theta_1, \theta_2, \ldots, \theta_{n-2}) &= \rho\,\vec{q}(\theta_1, \theta_2, \ldots, \theta_{n-2}) + b(0)\rho^{\tau+2}\,\vec{e}_n \\ &+ [h_0(0) - h_u(0) - \delta\Delta h - b(0) - \gamma^{-1}\rho^{\tau+2}]\,\vec{e}_{n+1}{}^{*)} \end{aligned}$$

and $0 \leq s \leq S_M$. Here we denote by M any point of the base of the convex cone $\widetilde{V}$. Clearly $S_M \leq S_0$, where the number S_0 depends only on δ and γ.

Since

$$\Delta h = h_0(0) - h_u(0),$$

we have

$$\begin{aligned} \vec{V}(\rho, \theta_1, \theta_2, \ldots, \theta_{n-2}, s) &= s[\rho\,\vec{q}(\theta_1, \theta_2, \ldots, \theta_{n-2}) \\ &+ b(0)\rho^{\tau+2}\vec{e}_n + ((1-\delta)\Delta h - b(0)\gamma^{-1}\rho^{\tau+2})\vec{e}_{n+1}] + \overrightarrow{0Q}. \end{aligned} \tag{10.46}$$

---

*) Since x_0 has coordinates $(0, 0, \ldots, 0)$ we write $b(0)$ instead of $b(x_0)$.

From (10.46) it follows that

$$\frac{\partial \vec{V}}{\delta \rho} = s\vec{q} + sb(0)(\tau+2)\rho^{\tau+1}\vec{e}_n - sb(0)\gamma^{-1}(\tau+2)\rho^{\tau+1}\vec{e}_{n+1}, \tag{10.47}$$

$$\frac{\partial \vec{V}}{\partial \theta_1} = s\rho\frac{\partial \vec{q}}{\partial \theta_1}, \dots, \frac{\partial \vec{V}}{\partial \theta_{n-2}} = s\frac{\partial \vec{q}}{\partial \theta_{n-2}},$$

$$\frac{\partial \vec{V}}{\partial s} = \vec{r} = \rho\vec{q} + b(0)\rho^{\tau+2}\vec{e}_n + [(1-\delta)\Delta h - b(0)\gamma^{-1}\rho^{\tau+2}]\vec{e}_{n+1}.$$

The vectors $\vec{q}, \frac{\partial \vec{q}}{\partial \theta_1}, \dots, \frac{\partial \vec{q}}{\partial \theta_{n-2}}, \vec{e}_n, \vec{e}_{n+1}$ form an orthogonal frame in E^{n+1}. Now we find the normal $\vec{m}$ of the convex cone $\widetilde{V}$ at the points of the set $\alpha^* = \beta' \cap \beta''$. First of all, this normal can be found by the external product of the derivatives of the vector-function $\vec{V}(\rho, \theta_1, \theta_2, \dots, \theta_{n-2}, s)$ with respect to all variables ρ, $\theta_1, \theta_2, \dots, \theta_{n-2}, s$. Secondly, normals used in the construction of the normal mapping have the component -1 at the unit vector e_{n+1}, i.e.

$$\vec{m} = p_0\vec{q} + p_1\frac{\partial \vec{q}}{\partial \theta_1} + p_2\frac{\partial \vec{q}}{\partial \theta_2} + \dots + p_{n-2}\frac{\partial \vec{q}}{\partial \theta_{n-2}} + p_{n-1}\vec{e}_n - \vec{e}_{n+1}.$$

Since

$$m = \det \begin{Vmatrix} \vec{q} & \vec{q}_{\theta_1} & \vec{q}_{\theta_2} & \dots & \vec{q}_{\theta_{n-2}} & \vec{e}_n & \vec{e}_{n+1} \\ s & 0 & 0 & \dots & 0 & A & B \\ 0 & s\rho & 0 & \dots & 0 & 0 & 0 \\ 0 & 0 & s\rho & \dots & 0 & 0 & 0 \\ \cdots & \cdots & \cdots & \cdots & \cdots & \cdots & \cdots \\ \cdots & \cdots & \cdots & \cdots & \cdots & \cdots & \cdots \\ 0 & 0 & 0 & \dots & s\rho & 0 & 0 \\ \rho & 0 & 0 & \dots & 0 & C & D \end{Vmatrix}$$

where

$$\begin{aligned} A &= sb(0)\rho^{\tau+1}(\tau+2), \\ B &= -sb(0)\gamma^{-1}\rho^{\tau+1}(\tau+2), \\ C &= b(0)\rho^{\tau+2} \\ D &= [(1-\delta)\Delta h - b(0)\gamma^{-1}\rho^{\tau+2}], \end{aligned}$$

we have

$$|p_0| = (s\rho)^{n-1}b(0)(\tau+2)\rho^{\tau}(1-\delta)\Delta h, \tag{10.48}$$

$$p_1 = p_2 = \cdots = p_{n-2} = 0,$$
$$|p_{n-1}| = (s\rho)^{n-1}[b(0)\rho^{\tau+1}(\tau+1)\gamma^{-1} + \frac{(1-\delta)\Delta h}{\rho}],$$
$$1 = |p_n| = (s\rho)^{n-1}b(0)\rho^{\tau+1}(\tau+1)$$

for the components of the vector $\vec{m}$ at all points of the convex cone $\widetilde{V}$ except at the vertex of this cone. Since $|\vec{q}| = |\vec{e}_n| = 1$, the normal image of $\widetilde{V}$ contains the $(n-1)$-dimensional ball Q^{n-1} with the center $(0, 0, \ldots, -|p_{n-1}|)$ and the radius $|p_0|$.

From the formulas (10.48) it follows that

$$|p_0| = \frac{\tau+2}{\tau+1}\frac{(1-\delta)\Delta h}{\rho} \tag{10.49}$$

and

$$|p_{n-1}| = \frac{1}{\rho^{\tau+2}b(0)(\tau+1)}(1-\delta)\Delta h + \frac{b(0)}{\gamma}(\tau+1)\rho^{\tau+2}. \tag{10.50}$$

The value of ρ in (10.49–50) can be found if we substitute the coordinates of the point $M \in \widetilde{V} \cap \alpha^*$ in (10.35) and (10.36). First of all we obtain the equality

$$b(0)\rho^{\tau+2} = \gamma\Delta h. \tag{10.51}$$

Thus the radius of the ball Q^{n-1} is

$$|p_0| = \frac{\tau+2}{\tau+1}(1-\delta)\left(\frac{b(0)}{\gamma}\right)^{\frac{1}{\tau+2}}(\Delta h)^{\frac{\tau+1}{\tau+2}} \tag{10.52}$$

and the center of this ball is the point $(0, 0, \ldots, 0, -|p_{n-1}|)$, where

$$|p_{n-1}| = \frac{\tau+2-\delta}{\tau+1}\gamma^{-1}. \tag{10.53}$$

Note that the constants C' and C'' mentioned in the statement of Lemma 10.4 are

$$C' = \frac{\tau+2}{\tau+1}(1-\delta)(\Delta h)^{\frac{\tau+1}{\tau+2}}[b(0)]^{\frac{1}{\tau+2}}, \tag{10.54}$$

$$C'' = \frac{\tau+2-\delta}{\tau+1}. \tag{10.55}$$

Lemma 10.4 is proved. □

Now we return to the proof of Theorem 10.6. We need only to establish the estimate $\omega(R, \widetilde{V})$ from below. It is clear that

$$\omega(R, \widetilde{V}) \geq \int_{H^n} R(p)dp \geq C_0 \int_{H^n} |p|^{-2k}dp,$$

where H^n is the convex cone described in Lemma 10.4 and C_0 is the positive constant taken from inequality (10.31). The convex cone H^n lies between two parallel hyperplanes

$$p_n = -\delta\gamma^{-1} \quad \text{and} \quad p_n = -\frac{\tau+2-\delta}{(\tau+1)\delta},$$

where δ and γ are two independent sufficiently small positive numbers. If we choose δ such that $0 < \delta < 1/2$, then

$$0 > -\frac{\delta}{\gamma} > -\frac{1}{2\gamma} > -\left(1+\frac{1}{2(\tau+1)}\right)\frac{1}{\gamma} > -\frac{\tau+2-\delta}{(\tau+1)-\delta}.$$

Hence the cone H^n lies inside the halfspace $p_n < -\delta\gamma^{-1}$. Let $\widetilde{H}^n$ be the convex cone symmetric to H^n with respect to the hyperplane $p_n = 0$. $\widetilde{H}_n$ lies between two parallel hyperplanes $p_n = \delta\gamma^{-1}$ and $p_n = \frac{\tau+2-\delta}{(\tau+1)\gamma}$, where $0 < \delta < 1/2$ and $0 < \delta\gamma^{-1} < \frac{\tau+2-\delta}{(\tau+1)\gamma} = C''\gamma^{-1}$. Here $C'' = \frac{\tau+2-\delta}{(\tau+1)}$ according to (10.55). Since

$$\int_{\widetilde{H}^n} |p|^{-2k} dp = \int_{H^n} |p|^{-2k} dp,$$

we have

$$\omega(R, \widetilde{V}) \geq C_0\sigma_{n-2} \int_{\delta\gamma^{-1}}^{C''\gamma^{-1}} dp_n \int_0^{\psi(p_n,\gamma,\delta,\tau)} \frac{h^2 dh}{(h^2+p^2)^k}. \tag{10.56}$$

We use the following notation in (10.56): σ_{n-2} is the area of the $(n-2)$-dimensional unit sphere;

$$h^2 = p_1^2 + \cdots + p_{n-1}^2; \tag{10.56a}$$

$$\psi(p_n, \gamma, \delta, \tau) = (p_n - \delta\gamma^{-1})\frac{C'\gamma^{1-\frac{1}{\tau+2}}}{C''-\delta}, \tag{10.56b}$$

where C' is the constant introduced in (10.54).

Now we use the inequality

$$h^2 + p_n^2 \leq (h+p_n)^2,$$

because $h \geq 0$ and $p_n > \delta\gamma^{-1} > 0$. Thus (10.56) can be reduced to the inequality

$$\omega(R, \widetilde{V}) \geq C_0\sigma_{n-2} \int_{\delta\gamma^{-1}}^{C''\gamma^{-1}} dp_n \int_0^{\psi(p_n,\gamma,\delta,\tau)} \frac{h^{n-2} dh}{(h+p_n)^{2k}}. \tag{10.57}$$

Now we introduce the new variables

$$p = p_n\gamma \quad \text{and} \quad q = h\gamma^{\frac{1}{\tau+2}}.$$

Then inequality (10.57) becomes

$$\omega(R,\widetilde{V}) \geq C_0\sigma_{n-2}\gamma^{-1+2k-\frac{n-1}{\tau+2}} \int_\delta^{C''} dp \int_0^{\frac{C'(p-\delta)}{C''-\delta}} \frac{q^{n-2}dq}{(p+\gamma^{\frac{\tau+1}{\tau+2}}q)^{2k}}. \tag{10.58}$$

Since

$$0 \leq q \leq \frac{C'(p-\delta)}{C''-\delta},$$

we have

$$\omega(R,\widetilde{V}) \geq \frac{1}{n-1}\left(\frac{C'}{C''-\delta}\right)^{n-1} B_k(\delta,\gamma), \tag{10.59}$$

where

$$B_k(\delta,\gamma) = \int_\delta^{C''} \frac{(p-\delta)^{n-1}dp}{[p+\gamma^{\frac{\tau+1}{\tau+2}}\cdot\frac{C'(p-\delta)}{C''-\delta}]^{2k}}.$$

Thus from (10.44) and (10.59) it follows that

$$d_2\gamma^{\lambda+\frac{n+\tau+1}{\tau+2}} \geq \omega(R,\widetilde{V}) \geq d_3\gamma^{2k-\frac{n+\tau+1}{\tau+2}} B_k(\delta,\gamma), \tag{10.60}$$

where d_2 and d_3 are some positive constants independent of δ and γ. Inequality (10.60) leads to the following inequality

$$1 \geq d_4\gamma^{2k-\lambda-\frac{2(n+\tau+1)}{\tau+2}} B_k(\delta,\gamma), \tag{10.61}$$

where $d_4 = d_3/d_2 =$ const > 0. First of all consider the estimate for $B_k(\delta,\gamma)$ from below. Let $t = p-\delta$ be the new variable of integration. Then the expression for $B_k(\delta,\gamma)$ becomes

$$B_k(\delta,\gamma) = \int_0^{C''-\delta} \frac{t^{n-1}dt}{(\mu_t,\delta)^{2k}}, \tag{10.62}$$

where

$$\mu = 1 + \frac{C'\gamma^{\frac{\tau+1}{\tau+2}}}{C''-\delta}. \tag{10.63}$$

If $k \leq 0$ then

$$B_k(\delta,\gamma) > \int_0^{C''-\delta} t^{n-1-2k}dt = \frac{(C''-\delta)^{n-2k}}{n-2k}.$$

Let $\delta > 0$ be a fixed positive small number. It is sufficient to take $\delta < \frac{1}{2}$. Then $C''-\delta > \frac{1}{2}$. Thus

$$B_k(\delta,\gamma) > \frac{1}{2^{n+2|k|}(n+2|k|)}$$

for all $k \le 0$.

Now let $k > 0$. Then from the Hölder inequality it follows that

$$(\mu_t + \delta)^{2k} \le (t^n + \delta^n)^{\frac{2k}{n}} (\mu^s + 1)^{\frac{2k}{s}}$$

where

$$\frac{1}{n} + \frac{1}{s} = 1.$$

Thus we obtain the estimate for $B_k(\delta, \gamma)$ from below

$$B_k(\delta,\gamma) \ge \tag{10.64}$$
$$\begin{cases} \dfrac{[\{(C''-\delta)^n+\delta^n\}^{1-\frac{2k}{n}} - \delta^{n(1-\frac{2k}{n})}]}{n(\mu^s+1)^{\frac{2k}{s}}(1-\frac{2k}{n})} & \text{if } \; 0 < \frac{2k}{n} < 1; \\ \dfrac{\ln[1+(C''-\delta)^n\delta^{-n}]}{n(\mu^s+1)^{2k/s}} & \text{if } \; \frac{2k}{n} = 1; \\ \dfrac{[\delta^{-n\left(\frac{2k}{n}-1\right)} - \{(C''-\delta)^n+\delta^n\}^{-\left(\frac{2k}{n}-1\right)}]}{n(\mu^s+1)^{\frac{2k}{s}}} & \text{if } \; \frac{2k}{n} > 1. \end{cases}$$

Let $0 < \theta < \frac{1}{6}$ be some small number. Below we shall choose θ more precisely. Now we set

$$\delta = \theta \frac{\tau + 2}{\tau + 1 + \theta}. \tag{10.65}$$

Then

$$\theta < \delta < 2\theta, \tag{10.66}$$

since the inequalities

$$1 \le \frac{\tau + 2}{\tau + 1 + \theta} \le 2$$

hold for all $\tau \ge 0$ and $0 \le \theta \le \frac{1}{6}$. Then from (10.55) it follows that

$$\delta = \theta \cdot C''. \tag{10.67}$$

Now we substitute the expression for δ into (10.64) and obtain

$$B_k(\theta,\gamma) > \begin{cases} \dfrac{(C'')^{n-2k}\{[(1-\theta)^n+\theta^n]^{1-\frac{2k}{n}} - \theta^{n(1-\frac{2k}{n})}\}}{n(\mu^s+1)^{\frac{2k}{s}}(1-\frac{2k}{n})} & \text{if } 0 < \frac{2k}{n} < 1; \\ \dfrac{\ln\left[1+\left(\frac{1-\theta}{\theta}\right)^n\right]}{n(\mu^2+1)^{\frac{2k}{s}}} & \text{if } \frac{2k}{n} = 1; \\ \dfrac{(C'')^{n-2k}}{n(\mu^s+1)^{\frac{2k}{s}}}\left[\theta^{\frac{1}{2k-n}} - \dfrac{1}{[(1-\theta)^n+\theta^n]^{\frac{2k-n}{n}}}\right] & \text{if } \frac{2k}{n} > 1. \end{cases}$$

Part 3. Theorem 10.6. will be proved if we show that the inequality (10.61) contradicts the conditions of this theorem. We established it above for non-positive values of k. Now we consider the cases: a) $0 < k < 1$; b) $1 \le k < \frac{n}{2}$; c) $k = \frac{n}{2}$; d) $k > \frac{n}{2}$.

a) Assume that $0 < k < 1$. Then

$$2k - \lambda - \frac{2(n+\tau+1)}{\tau+2} < 0$$

for all $\lambda \geq 0$, $\tau \geq 0$ and $n \geq 2$. The inequality (10.66) and the equalities (10.67), (10.63), (10.54), (10.55) provide

$$\lim_{\theta\to 0} B_k(\theta,\gamma) \geq \frac{\left(\frac{\tau+2}{\tau+1}\right)}{(n-2k)\cdot[1+(1+\psi\gamma^{\frac{\tau+1}{\tau+2}})^s]^{\frac{2k}{s}}},$$

where

$$0 < \psi = (\Delta h)^{\frac{\tau+2}{\tau+1}}(b(0))^{\frac{1}{\tau+2}}$$

(the definitions of Δh and $b(0)$ were given by (10.34) and (10.30)).

The small parameter $\gamma > 0$ was not defined up to now. Let $\gamma_0 > 0$

$$\psi\gamma_0^{\frac{\tau+1}{\tau+2}} = 1.$$

Then

$$1 + (1+\psi\gamma^{\frac{\tau+1}{\tau+2}})^s < 1 + 2^s$$

for all $0 < \gamma \leq \gamma_0$. Recall that $s = \frac{n}{n-1}$. Thus

$$\lim_{\theta\to 0} B_k(\theta,\gamma) \geq \frac{1}{(n-2k)(1+2^s)^{\frac{2k}{s}}}\left(\frac{\tau+2}{\tau+1}\right)^{n-2k}$$

for all $0 < \gamma \leq \gamma_0$. Now let $\theta_0 \in (0, 1/6)$ be a number such that the inequality

$$B_k(\theta_0,\gamma) \geq \frac{1}{2(n-2k)(1+2^s)^{\frac{2k}{s}}}\left(\frac{\tau+2}{\tau+1}\right)^{n-2k} \tag{10.68}$$

holds for all $0 < \theta \leq \theta_0$ and $0 < \gamma \leq \gamma_0$. Thus the inequality (10.61) yields the inequality

$$1 \geq d_4 \frac{1}{2(n-2k)(1+2^s)^{\frac{2k}{s}}}\left(\frac{\tau+2}{\tau+1}\right)^{n-2k} \cdot \gamma^{2k-\lambda-\frac{2(n+\tau+1)}{\tau+2}} \tag{10.69}$$

for all $\theta < (0, \theta_0]$ and $\gamma \in (0, \gamma_0]$.

But the inequality (10.69) does not hold for sufficiently small $\gamma \in (0, \gamma_0]$, since $n - 2k > 0$ and $k - \frac{\lambda}{2} - \frac{n+\tau+1}{\tau+2} < 0$. Thus Theorem 10.6 is proved for case a).

Now we introduce the number

$$K = \frac{n+\tau+1}{\tau+2} + \frac{\lambda}{2} \tag{10.70}$$

which is very important in cases b), c), d).

b) Assume that $1 \le k < \frac{n}{2}$. Then Theorem 10.6 is valid if

$$k < K. \tag{10.71}$$

Clearly $n > 2$ in this case, where n is the dimension of the domain G.

Case b) can be reduced to inequality (10.70) by the same procedure as in case a). From the condition

$$k < K$$

follows the inequality

$$k - \frac{\lambda}{2} - \frac{n+\tau+1}{\tau+2} < 0.$$

Hence the inequality (10.69) does not hold for very small $\gamma \in (0, \gamma_0]$ and $\theta \in (0, \theta_0]$. This contradiction leads to the validity of Theorem 10.6. Thus Theorem 10.6 is also proved for case b).

c) Assume that $k = \frac{n}{2}$. Then Theorem 10.6 is valid if

$$k \le K. \tag{10.72}$$

First we consider the estimate from $B_k(\theta, \gamma)$ from below for sufficiently small positive numbers θ and γ. Since $k = \frac{n}{2}$ and $\frac{1}{n} + \frac{1}{s} = 1$ then

$$\begin{aligned}
&B_k(\theta,\gamma)\\
&= B_{n/2}(\theta,\gamma)\frac{\ln\left[1+\frac{(1-\theta)^n}{\theta^n}\right]}{n\left\{\left[1+\gamma^{\frac{\tau+1}{\tau+2}}\frac{C'}{C''(1-\theta)}\right]^{\frac{n}{n-1}}+1\right\}^{n-1}} =\\
&= \frac{\ln(\theta^n+(1-\theta)^n)-n\ln\theta}{n\left[\left[1+\gamma^{\frac{\tau+1}{\tau+2}}\frac{(\tau-\delta+2)(1-\delta)(\Delta h)^{\tau+1/\tau+2}(b(0))^{1/\tau+1}}{(\tau+2)(1-\theta)}\right]^{n/n-1}+1\right]^{n-1}}.
\end{aligned}$$

Since $0 \le \theta \le \frac{1}{6}$ and $\theta \le \delta \le 2\theta$, we have

$$\begin{aligned}
1+\delta^{\frac{\tau+1}{\tau+2}}\frac{(\tau+2-\delta)(1-\delta)(\Delta h)^{\frac{\tau+1}{\tau+2}}[b(0)]^{\frac{n}{n-1}}}{(\tau+2)(1-\theta)} &\le\\
\le 1+\gamma^{\frac{\tau+1}{\tau+2}}(\Delta h)^{\frac{\tau+1}{\tau+2}}[b(0)]^{\frac{1}{\tau+1}}&.
\end{aligned}$$

Now let $\gamma_0 > 0$ be a number such that

$$\gamma_0^{\frac{\tau+1}{\tau+2}}(\Delta h)^{\frac{\tau+1}{\tau+2}}[b(0)]^{\frac{1}{\tau+1}} = 1.$$

Then the inequality

$$B_{n/2}(\theta,\gamma) \ge \frac{\ln(\theta^n+(1-\theta)^n)-n\ln n}{n[1+2^{n/n-1}]^{n-1}} \tag{10.73}$$

holds for all θ and γ satisfying the inequalities:

$$0 < \theta \le \frac{1}{6},$$
$$0 < \gamma \le \gamma_0.$$

Thus the basic inequality (10.34) leads to the inequality

$$1 \ge d_4 \gamma^{2k-\lambda-\frac{2(n+\tau+1)}{\tau+2}} \frac{\ln[\theta^n + (1-\theta)^n] - n\ln\theta}{n[1+2^{n/n-1}]^{n-1}} \tag{10.74}$$

for $2k = n$, $0 < \theta \le \frac{1}{6}$, $0 < \gamma \le \gamma_0$. Remember that the constant $d_4 > 0$ does not depend on θ and γ.

Since

$$k = \frac{n}{2} \le K = \frac{\lambda}{2} + \frac{n+\tau+1}{\tau+2}$$

(see the condition of case c)), we have

$$2k - \lambda - \frac{2(n+\tau+1)}{\tau+2} \le 0. \tag{10.75}$$

Thus we should consider two cases according to the inequality (10.75):

a) The first one is

$$2k - \lambda - \frac{2(n+\tau+1)}{\tau+2} < 0. \tag{10.76}$$

We set $\theta = \frac{1}{6}$. Then the inequality (10.74) becomes

$$1 \ge d_4 \gamma^{2k-\lambda-\frac{2(n+\tau+1)}{\tau+2}} \frac{\ln(1+5^n)}{n[1+2^{n/n-1}]^{n-1}} \tag{10.77}$$

for all $\gamma \in (0, \gamma_0]$. But from (10.76) it follows that (10.77) does not hold if $\gamma > 0$ is sufficiently small. Hence statement c) is proved.

b) The second case is

$$2k - \lambda - \frac{2(n+\tau+1)}{\tau+2} = 0.$$

Then (10.74) becomes

$$1 \ge d_4 \frac{\ln[\theta^n + (1-\theta)^n] - n\ln\theta}{n[1+2^{n/n-1}]^{n-1}} \tag{10.78}$$

if $0 < \theta \le \frac{1}{6}$ and $0 < \gamma \le \gamma_0$.

But

$$\lim_{\theta\to 0}[\ln\theta^n + (1-\theta)^n] - n\ln\theta = +\infty.$$

Therefore the inequality (10.78) is not valid if $\theta > 0$ is sufficiently small. Therefore statement c) is proved in the second case. Thus the proof of the statement c) is completed.
d) Finally we assume that $k > \frac{n}{2}$. Then Theorem 10.6 is valid if $k \leq K$. This statement can be proved similarly to case c). We offer it as a useful exercise.

The proof of Theorem 10.6 is complete. □

Theorem 10.7. *Let $u_m(x)$, $m = 1, 2, \ldots$ be a sequence of convex functions in a convex bounded domain G and let the borders $h_{u_m}(x)$ of $u_m(x)$ be continuous functions on ∂G. Let $u_0(x)$ be a convex function in G. We assume that:*

1) *the convex functions $u_m(x)$ converge pointwise to the convex function $u_0(x)$;*
2) *the borders $h_{u_m}(x)$ of $u_m(x)$ converge uniformly to some continuous function $v_0(x)$ on ∂G;*
3) *if x_0 is an arbitrary point of ∂G, then there exists some n-ball $U_\rho(x_0) \subset E^n$ such that the inequality*

$$\varliminf_{m \to \infty} \omega(R, u_m, e \cap G_m) \leq a[\sup_e (dist(x, \partial G))^\lambda] \text{ mes } e$$

holds, where $a =$ const > 0 and $\lambda =$ const ≥ 0;

4) *G is an open bounded convex domain and ∂G has a parabolic support of order not greater than $\tau =$ const ≥ 0. (See also the definition and Theorem 10.6);*
5) *$R(p)$ satisfies Assumption 10.1;*
6) *if $K = \frac{\lambda}{2} + \frac{n+\tau+1}{\tau+2}$ then*

$$k \leq K \quad \text{if} \quad k < 1 \quad \text{or} \quad k \geq \tfrac{n}{2},$$

$$k < K \quad \text{if} \quad 1 \leq k < \tfrac{n}{2}.$$

Then the function $v_0(x)$ is the border of the convex function $u_0(x)$, and $u_m(x)$ converge uniformly to $u_0(x)$ in $\overline{G}$.

Proof. If our statement is not true, then there exist the points $x_k \in G$ and positive integers

$$m_1 < m_2 < \cdots < m_k < \cdots$$

such that

$$|u_0(x_k) - u_{m_k}(x_k)| \geq \varepsilon_0 > 0, \tag{10.79}$$

where $\varepsilon_0 > 0$ is some number. Clearly the sequence of points $(x_k, u(x_k))$ is bounded in E^{n+1}.

Without loss of generality, we assume that the sequence of points $(x_k, u_{m_k}(x_k))$ converges to some point (x_0, u_0). The point x_0 belongs either to G or to ∂G. If x_0 is an interior point of G, then we obtain

$$|u_0(x_0) - u_0| \geq \varepsilon_0. \tag{10.80}$$

This inequality follows from inequalities (10.79) if $k \to +\infty$. But (10.80) can not be valid because the point (x_0, u_0) has to lie on the graph of $u_0(x)$. Hence $u_0 = u_0(x_0)$. This contradiction shows that $x_0 \in \partial G$. The conditions of Theorem 10.7 allows us to use Theorem 10.6 and establish the following statements:

a) $v_0(x)$ is the border of the convex function $u_0(x)$;

b) $u_m(x)$ and $u_0(x)$ can be extended as continuous convex functions in the closed convex bounded domain $\overline{G} = G \cup \partial G$, and

$$u_m(x)|_{\partial G} = h_{u_m}(x), \qquad u_0(x)|_{\partial G} = v_0(x)$$

and the continuous functions $h_{u_m}(x)$ converge uniformly to the continuous function $v_0(x)$ on ∂G;

c) the convex functions $u_m(x)$ converge to the convex function $u_0(x)$ for all $x \in G \cup \partial G$. From the properties of the convergence of convex hypersurfaces and functions (see § 3) it follows that

$$u_0 = \lim_{k\to\infty} u_{m_k}(x_k) = u_0(x_0) = v_0(x_0) \tag{10.81}$$

since $x_0 \in \partial G$. On the other hand if we take limit in (10.79), then

$$|v_0(x_0) - u_0| \geq \varepsilon_0 > 0. \tag{10.82}$$

The inequalities (10.81) and (10.82) are incompatible. This contradiction leads to the proof of Theorem 10.7.

§11. Geometric Theory of the Monge–Ampere Equations $\det(u_{ij}) = \varphi(x)/R(Du)$.

11.1 Introduction. Obstructions and Necessary Conditions of Solvability for the Dirichlet Problem

Let G be an open bounded convex domain in E^n and let R^n be the gradient space. Let $z(x) \in C^2(G) \cap C(\overline{G})$ be a strictly convex solution of the equation

$$\det(u_{ij}) = \frac{\varphi(x)}{R(Du)} \tag{11.1}$$

where $\varphi(x)$ and $R(p)$ are positive and continuous functions in G and R^n respectively. Then the tangential mapping

$$\chi_z \colon G \to R^n$$

is a C^1-diffeomorphism. Therefore

$$\int_G \varphi(x)dx = \int_G R(Dz)\det(z_{ij})dx = \int_{\chi_z(G)} R(p)dp$$
$$\leq \int_{R^n} R(p)dp = B(R).$$

The inequality

$$\int_G \varphi(x)dx \leq B(R) \tag{11.2}$$

is a necessary condition for the solvability of the Dirichlet problem for equation (11.1). If

$$B(R) = \int_{R^n} R(p)dp = +\infty$$

then the inequality (11.2) is not a restriction on the function $\varphi(x)$. Now we consider a second non-trivial necessary condition for the solvability of the Dirichlet problem

$$\det(z_{ij}) = \varphi(x)(1 + |Du|^2)^{\frac{n+2}{2}}, \tag{11.3}$$

$$z|_{\partial G} = Kx_1, \tag{11.4}$$

where G is the n-ball $\sum_{i=1}^{n} x_i^2 \leq r^2$, $K = \text{const} > 0$ and the function $\varphi(x)$ satisfies the inequalities

$$0 < a = \text{const} \leq \varphi(x) \leq b = \text{const} < +\infty.$$

Note that (11.3–4) is the Dirichlet problem for hypersurfaces with a prescribed Gaussian curvature. Let

$$\nu_z\colon G \to S^n$$

be the Gaussian (spherical) mapping of the hypersurface S_z: $z = z(x)$ where S^n is the unit sphere in E^{n+1}. Then $\nu_z(G)$ lies only in one hemisphere of S^n and

$$\frac{1}{2}\sigma_n \geq \sigma(\nu_z(G)) = \int_G \varphi(x)\sqrt{1 + |\text{grad } z|^2}\, dx \geq$$

$$\geq a\sigma(z), \tag{11.5}$$

where σ_n is the area of S^n, $\sigma(\nu_z(G))$ is the area of the spherical image of G and $\sigma(z)$ is the area of the hypersurface S_z.

From inequality (11.5), it follows that

$$\sigma(z) \leq \frac{1}{2a}\sigma_n. \tag{11.6}$$

The inequality (11.6) is an *a priori* estimate for the areas of all solutions of the equation (11.3). Let

$$r < b^{-n}.$$

Then

$$\int_G \varphi(x)dx \leq \int_G bdx = \mu_n br^n < \mu_n, \tag{11.7}$$

where μ_n is the volume of the n-unit ball in E^n. Since

$$\int_{R^n} (1+|p|^2)^{-\frac{n+2}{2}} dp = \mu_n,$$

inequality (11.7) shows that the necessary condition (11.2) is fulfilled for the Dirichlet problem (11.3–4). Let F_K be the n-dimensional ellipsoid defined by the graph of the function $z = Kx_1$, where $(x_1, x_2, \dots, x_n) \in \overline{G}$. Assume that

$$K > \frac{\sigma_n}{a\mu_n r^n} \tag{11.8}$$

and that $z(x) \in C^2(G) \cap C(\overline{G})$ is the solution of the Dirichlet problem (11.3–4). Then

$$z|_{\partial G} = Kx_1 \quad \text{and} \quad z(x) < Kx_1$$

everywhere in G. Hence

$$\sigma(z) > \sigma(F_k) = K\mu_n r^n.$$

From (11.8) it follows that

$$\sigma(z) > \frac{\sigma_n}{a}. \tag{11.9}$$

Since inequalities (11.6) and (11.9) are incompatible, the Dirichlet problem (11.3–4) does not have solution $z(x) \in C^2(G) \cap C(\overline{G})$ satisfying the condition (11.8) in the classical sense.

Thus there are at least two different kinds of obstructions to the solvability of the Dirichlet problem for equation (11.1).

11.2 Generalized and Weak Solutions for Equation (11.1)

Let G be a bounded open convex domain in E^n. We consider the Monge–Ampere equation

$$\det(u_{ij}) = \frac{\varphi(x)}{R(Du)}, \tag{11.10}$$

where $\varphi(x)$ is a non-negative summable function in G, and $R(p)$ is a locally summable positive function in the gradient space R^n. The equation

$$\omega(R, u, e) = \int_e \varphi(x)dx \tag{11.11}$$

is the extension of the Monge–Ampere equation (11.10) to the class of convex functions $W^+(G)$, where e is any Borel subset of G, i.e. the integration of Monge–Ampere equations (11.10) can be considered as the reconstruction of a convex function $z(x) \in W^+(G)$ by the prescribed values of its R-curvature.

Note that the values of the R-curvature are given by any absolutely continuous non-negative set function

$$\mu(e) = \int_e \varphi(x)dx. \tag{11.12}$$

We call the convex function $z(x) \in W^+(G)$ a *generalized solution* of Monge–Ampere equation (11.10) if the R-curvature $\omega(R, z, e)$ satisfies equation (11.11). Since every function $z(x) \in W^+(G)$ has first and the second differentials almost everywhere, we can give another definition of generalized solutions. The function $z(x) \in W^+(G)$ is called a *generalized solution* of the equation (11.10) if $z(x)$ satisfies (11.10) almost everywhere and $\omega(R, z, e)$ is an absolutely continuous set function. Obviously, both definitions of generalized solutions are equivalent.

If the R-curvature of a convex function $z(x) \in W^+(G)$ is not absolutely continuous, then $z(x)$ is not a generalized solution of the equation (11.10). The functions whose graphs are convex cones or convex polyhedra provide examples of this kind. We now enlarge the class of allowable solutions of equation (11.10) to include all convex functions. Let $\mu(e)$ be some non-negative completely additive set function on the ring of Borel subsets of G. Then we associate the equation

$$\omega(R, z, e) = \mu(e) \tag{11.13}$$

with equation (11.10). Clearly every convex function defined in G can be considered as solution of such an equation for a suitable set function $\mu(e)$. Equation (11.13) is the widest extension of the Monge–Ampere equation (11.10). The convex functions satisfying equation (11.13) are called *weak solutions* of equation (11.10). Clearly, the functions, whose graphs are convex cones or convex polyhedra are only weak solutions of the equation (11.10). Now we explain the structure of the right side of equation (11.13) when its solutions are convex cones or polyhedra.

Let K be any convex cone which is the graph of a convex function $k(x) \in W^+(G)$. We suppose that the vertex of K is projected on some interior point x_0 of G. Since

$$\omega(R, K, e) = \begin{cases} \int_{\chi_K(e)} R(p)dp & \text{if } \; x_0 \in e \\ 0 & \text{if } \; x_0 \notin e, \end{cases} \tag{11.14}$$

the function $k(x)$ is a solution of the equation (11.13) if and only if the set function $\mu(e)$ has the following form

$$\mu(e) = \begin{cases} \mu(x_0) > 0 & \text{if } \; x_0 \in e \\ 0 & \text{if } \; x_0 \notin e. \end{cases} \tag{11.15}$$

Thus the set function $\mu(e)$ is completely defined by its value at the point x_0 if we seek solutions $k(x)$ of the equation

$$\omega(R, z, e) = \mu(e),$$

whose graphs are convex cones with vertices projected on the point x_0.

Now let G be a bounded convex solid n-dimensional polyhedron in E^n. Let $A_1, A_2, \dots, A_k$ be a given system of interior points of G. Let $Q(A_1, A_2, \dots, A_k)$ be of convex functions satisfying the following conditions:

1. Every function $z(x) \in Q(A_1, A_2, \dots, A_k)$ is a piecewise linear convex function defined in G.
2. The vertices of every convex polyhedron S_z are projected only at the given system of points $A_1, A_2, \dots, A_k$, if $z(x) \in Q(A_1, A_2, \dots, A_k)$.

Since

$$\omega(R, z, e) = \sum_{A_i \in e} \int_{\chi_z(A_i)} R(p)dp \tag{11.16}$$

for every convex function $z(x) \in Q(A_1, A_2, \dots, A_k)$, the solutions $z(x)$ of equation (11.13) belong to $Q(A_1, A_2, \dots, A_k)$ if and only if $\mu(e)$ has the form

$$\mu(e) = \sum_{A_i \in e} \mu(A_i). \tag{11.17}$$

Thus the set function $\mu(e)$ is completely defined by means of its values at the points $A_1, A_2, \dots, A_k$ if we seek solutions of equation (11.13) in the set $Q(A_1, A_2, \dots, A_k)$.

11.3 The Dirichlet Problem in the Set of Convex Functions $Q(A_1, A_2, \dots, A_k)$

Let G be a bounded solid convex n-dimensional polyhedron in the Euclidean space E^n and let $B_1, B_2, \dots, B_m$ be the vertices of the closed convex $(n-1)$-dimensional polyhedron G.*) Let $Q(A_1, A_2, \dots, A_k)$ be the set of convex functions introduced in Subsection 11.2. Remember that $A_1, A_2, \dots, A_k$ are fixed interior points of G.

Clearly, we can reformulate the Dirichlet problem for the equation

$$\omega(R, z, e) = \mu(e) \tag{11.18}$$

in the set $Q(A_1, A_2, \dots, A_k)$ as the following boundary value problem:

Let $\mu_1 \geq 0$, $\mu_2 \geq 0, \dots, \mu_k \geq 0$ and $h_1, h_2, \dots, h_m$ be prescribed numbers. It is required to prove the existence of at least one convex function $z(x) \in Q(A_1, A_2, \dots, A_k)$ such that

$$\omega(R, z, A_i) = \mu_i \qquad (i = 1, 2, \dots, k), \tag{11.19}$$

$$z(B_s) = h_s \qquad (s = 1, 2, \dots, m). \tag{11.20}$$

*) We suppose that G is a closed subset of E^n and every point B_j $(j = 1, 2, \dots, m)$ is a real vertex of the polyhedron P. The last statement means that every vertex B_j $(j = 1, 2, \dots, m)$ does not lie in the convex hull of all other vertices of P.

Theorem 11.1. *If the numbers $\mu_1, \mu_2, \ldots, \mu_k$ satisfy the inequality*

$$\sum_{i=1}^{k} \mu_i < B(R) = \int_{R^n} R(p)dp, \tag{11.21}$$

then the Dirichlet problem (11.19–20) has exactly one solution

$$z(x) \in Q(A_1, A_2, \ldots, A_k).$$

Proof. Denote by H the subset of $Q(A_1, A_2, \ldots, A_k)$ which consists of function $u(x)$ such that:

$$\begin{aligned} &1)\ \omega(R, u, A_i) \le \mu_i \qquad (i = 1, 2, \ldots, k); \\ &2)\ u(B_j) = h_j \qquad (j = 1, 2, \ldots, m). \end{aligned} \tag{11.22}$$

The set H is not empty. We consider the convex hull M of the points (B_1, h_1), $(B_2, h_2), \ldots, (B_m, h_m)$ in the Euclidean space E^{n+1}. Then M is a bounded solid convex polyhedron in E^{n+1} with vertices at the points (B_1, h_1), $(B_2, h_2), \ldots, (B_m, h_m)$. The lower part of M is the graph of some piecewise linear function $z = u_0(x)$ satisfying the following conditions:

a) $u_0(x)$ is continuous in G (remember that G is a closed set in E^n);
b) the graph of $u_0(x)$ is a convex polyhedron S_{u_0} and the vertices of S_{u_0} lie only at the points (B_1, h_1), $(B_2, h_2), \ldots, (B_m, h_m)$;
c)

$$\omega(R, u_0, A_i) = 0 \qquad (i = 1, 2, \ldots, k).$$

Clearly, $u_0(x) \in H$. Thus the set H is not empty. Now let $u(x) \in H$ be an arbitrary convex function. Let

$$\mu_0 = \sum_{i=1}^{k} \mu_i.$$

Then

$$\omega(R, u, \text{int } G) = \sum_{i=1}^{k} \omega(R, u, A_i) \le \sum_{i=1}^{k} \mu_i = \mu_0 < B(R),$$
where $\text{int } G = G \backslash \partial G$.

Therefore from Theorem 11.4 it follows that the inequalities

$$\min_{1 \le i \le m} \{h_i\} - T_R(\mu_0) \text{diam } G \le u(x) \le \max_{1 \le i \le m} \{h_i\} \tag{11.23}$$

hold for all $x \in G$. Denote by $\delta > 0$ the least of the distances from the points A_i to ∂G $(i = 1, 2, \ldots, k)$. Then all the functions $u(x) \in H$ satisfy the

Lipschitz condition of degree 1 and the common constant $M = (\beta_2 - \beta_1 + T_R(\mu_0)\text{diam } G)\delta^{-1}$, where

$$\beta_1 = \inf_{1 \le i \le m} \{h_i\}, \quad \beta_2 = \sup_{1 \le i \le m} \{h_i\}.$$

Therefore H is a compact set in $C(G)$. Now let $V(u)$ be the volume of the solid convex $(n+1)$-polyhedron $\overline{Co}(S_u)$. Clearly $V(u)$ is a continuous function on the compact set H in the space $C(G)$. Hence

$$V_0 = \sup_H V(u) < +\infty$$

and there exists a function $\tilde{u}(x) \in H$ such that $V_0 = V(\tilde{u})$. We claim that $\tilde{u}(x)$ is the desired solution. In the opposite case we shall have

$$\omega(R, \tilde{u}, A_i) \le \mu_i, \quad i = 1, 2, \ldots, k \tag{11.24}$$

and there exists at least one point A_{i_0} such that

$$\omega(R, \tilde{u}, A_i) < \mu_{i_0}. \tag{11.25}$$

Now consider the point $(A_{i_0}, \tilde{u}(A_{i_0}) - \varepsilon)$ where $\varepsilon > 0$ is any number. Let S_v be the lower part of

$$\partial[\overline{Co}(S_{\tilde{u}} \cup \{(A_{i_0}, \tilde{u}(A_{i_0}) - \varepsilon)\}],$$

where $v(x)$ is the convex function whose graph is the convex polyhedron S_v. If $\varepsilon > 0$ is sufficiently small, then $v(x) \in Q(A_1, A_2, \ldots, A_k)$ and

$$\omega(R, S_v, A_{i_0}) < \mu_{i_0},$$
$$\omega(R, S_v, A_i) \le \mu_i$$

for all $i = 1, 2, \ldots, k$ except $i = i_0$. Thus $S_v \in H$. From the construction of the convex polyhedron S_v it follows that the volume $V(v)$ is strictly more than $V(u) = \sup_{u \in H} V(u)$. This contradiction proves that $u(x)$ is a solution of the Dirichlet problem (11.19-20). From Theorem 10.3 it follows that the considered Dirichlet problem has only one solution. □

11.4 Existence and Uniqueness of Weak Solutions of the Dirichlet Problem for Monge–Ampere Equations $\det(u_{ij}) = \varphi(x)/R(Du)$

We will present here existence and uniqueness theorems for the Dirichlet problem

$$\omega(R, z, e) = \mu(e), \tag{11.26}$$
$$u|_{\partial G} = h(x), \tag{11.27}$$

where G is a bounded open convex domain in E^n, $\mu(e)$ is a prescribed non-negative completely additive function of Borel subsets of G and $h(x)$ is a prescribed continuous function on ∂G. Solutions of equation (11.26) are called *weak solutions* of the Monge–Ampere equation (11.10). We denote by G_ε the open subdomain of G such that

$$\operatorname{dist}(x, \partial G) > \varepsilon$$

for all $x \in G_\varepsilon$, where $\varepsilon > 0$ is a sufficiently small number.

Theorem 11.2. *Let ∂G have a parabolic support of order not greater than τ, where $\tau = \text{const} \geq 0$ is any fixed number. Let a non-negative completely additive set function $\mu(e)$ satisfy two conditions:*

$$\mu(G \backslash G_\varepsilon) = 0 \tag{11.28}$$

and

$$\mu(G) = \mu(G_\varepsilon) < \int_{R^n} R(p)dp, \tag{11.29}$$

then the Dirichlet problem (11.26–27) has unique solution $u(x) \in W^+(G)$.

Proof. We denote by $P_\varepsilon^{(k)}$ an n-dimensional open bounded polyhedral*) domain such that

$$G \supset \overline{P}_\varepsilon^{(k)} \supset P_\varepsilon^{(k)} \supset \overline{G}_\varepsilon$$

$k = 1, 2, 3, \ldots$, where $\overline{G}_\varepsilon$ is the closure of the set B_ε, and $\overline{P}_\varepsilon^{(k)}$ is the closure of $P_\varepsilon^{(k)}$. Let $A_1^{(k)}$, $A_2^{(k)} \ldots, A_{m_k}^{(k)}$ be the system of arbitrary interior points of G_ε. Now consider the Dirichlet problem

$$\omega(R, u, A_i^{(k)}) = \mu_i^{(i)} \tag{11.30}$$
$$u(B_j^{(k)}) = b_j^{(k)} \tag{11.31}$$

*) i.e. the boundary of $P_\varepsilon^{(k)}$ is a closed $(n-1)$-dimensional convex polyhedron $\partial P_\varepsilon^{(k)}$. Thus the closure of $P_\varepsilon^{(k)}$ is a n-dimensional closed bounded solid convex polyhedron.

where $\mu_i^{(k)}$ $(i = 1, 2, \ldots, m_k)$ are arbitrary non-negative numbers, $B_j^{(k)}$ $(j = 1, 2, \ldots, s_k)$ are the vertices of the polyhedron $\overline{P}_\varepsilon^{(k)}$ and $b_j^{(k)}$ are arbitrary real numbers. We suppose that

$$\sum_{i=1}^{m_k} \mu_i^{(k)} \leq \mu(G_\varepsilon) \tag{11.32}$$

for all positive integers k. Since

$$\mu(G_\varepsilon) = \mu(G) < \int_{R^n} R(p)dp \tag{11.33}$$

(see the condition (11.29)), from Theorem 11.1 it follows that the Dirichlet problem (11.30–31) has only one solution $u_k(x)$ for all positive integers k, and $u_k(x)$ is a convex polyhedron.

We can choose the set functions $\mu_i(e)$, where

$$\mu_k(e) = \sum_{A_i^{(k)} \in e} \mu_i^{(k)}, \tag{11.34}$$

so that they weakly converge to the given set function $\mu(e)$, i.e.

$$\lim_{k \to \infty} \int_G f \mu_k(de) = \int_G f \mu(de) \tag{11.35}$$

for any continuous function f vanishing outside of a compact set $M_f \subset G$.

Let $u_k(x)$ be convex polyhedral solutions of the Dirichlet problem (11.30–31) in the convex polyhedral domains $P_\varepsilon^{(k)}$. We denote by L_k the graphs of their borders $h_{u_k}(x)$. Finally let L be the graph of the prescribed continuous function h: $\partial G \to R$ (see condition (11.27)).

Clearly we can choose the numbers $b_j^{(k)}$ so that:

1) L_k uniformly converge to L in E^{n+1};
2) the inequalities

$$-\infty < m = \inf_{\partial G} h(x) \leq b_j^{(k)} \leq \sup_{\partial G} h(x) = M < +\infty$$

hold.

Conditions (11.31) and (11.32) permit us to apply Theorem 10.4 to the sequence of convex functions $u_k(x)$. Thus we obtain the uniform estimates

$$m - T_R(\mu(G_\varepsilon))\text{diam } G \leq u_k(x) \leq M \tag{11.36}$$

for $k = 1, 2, 3, \ldots$. Note that $x \in \overline{P}_\varepsilon^{(k)}$ if a positive k is fixed.

Since all $A_i^{(k)}$, $i = 1, 2, \ldots, m_k$; $k = 1, 2, 3, \ldots$, are interior points of G, all functions $u_k(x)$ have the same Lipschitz constant of degree one in the closed polyhedral domains $\overline{P}_\varepsilon^{(k)}$, $k = 1, 2, 3, \ldots$. According to Theorems 10.5 and 10.6 we can take a subsequence $u_{k_p}(x)$ from the sequence $u_k(x)$ converging uniformly to some convex function $\tilde{u}(x)$. Clearly,

$$h_{\tilde{u}}(x) = h(x),$$

where $h_{\tilde{u}}(x)$ is the border of the convex function $\tilde{u}(x)$.

Since

$$\mu_k(G \backslash G_\varepsilon) = \mu(G \backslash G_\varepsilon) = 0, \qquad k = 1, 2, 3, \ldots$$

the set functions $\omega(R, u_{k_p}, e)$ converge weakly to the set function $\omega(R, \tilde{u}, e)$. Since

$$\omega(R, u_{k_p}, e) = \mu_{k_p}(e); \quad p = 1, 2, 3, \ldots$$

and the set functions $\omega(R, u_{k_p}, e)$ and $\mu_{k_p}(e)$ weakly converge to $\omega(R, \tilde{u}, e)$ and $\mu(e)$ respectively, we have

$$\omega(R, \tilde{u}, e) = \mu(e) \tag{11.37}$$

for all Borel subsets e of G. Thus the function $\tilde{u}(x)$ is a weak solution of the Dirichlet problem (11.26–27) satisfying the boundary condition in the classical sense, if we impose the additional condition (11.28).

The Dirichlet problem (11.26–27) cannot have more than one solution in $W^+(G)$. This follows directly from Theorem 10.3. Thus $\tilde{u}(x)$ is the unique solution of the Dirichlet problem (11.26–27), if conditions (11.28) and (11.29) are fulfilled. □

Remarks. 1) The inequality

$$\mu(G) \leq \int_{R^n} R(p)dp \tag{11.38}$$

is necessary for the solvability of the Dirichlet problem (11.26–27); see Subsection 11.1. But this condition is insufficient for the solvability of the same Dirichlet problem. Consider the following simple example: Let G be the unit n-ball $|x| < 1$ in E^n, $R(p) = (1 + |p|^2)^{-(n+2)/2}$ and

$$\mu(e) = \begin{cases} A \text{ if the subset } e \subset G \text{ is the point } x = 0; \\ A \text{ if the subset } e \subset G \text{ contains the point } x = 0; \\ 0 \text{ if the subset } e \subset G \text{ does not contain the point } x = 0, \end{cases}$$

where A is any non-negative number. Now we consider the Dirichlet problem

$$\omega(R, u, e) = \mu(e), \tag{11.39}$$

$$u|_{\partial G} = 0. \tag{11.40}$$

The solutions of this problem can only be convex cones of revolution whose vertices are projected on the same point $x = 0$ and bases coincide with the unit $(n-1)$-sphere $|x| = 1$. Clearly the Dirichlet problem (11.39–40) has solutions if and only if

$$0 \leq A < B(R) \equiv \int_{R^n} R(p)dp = \mu_n,$$

where μ_n is the volume of the unit n-ball.

Thus the sufficient condition (11.29) in Theorem 11.2 interlocks with the necessary one. This condition is sharp.

2) The condition (11.28) used in Theorem 11.2 is a restriction which we would like to remove. But if we remove it and use only the inequality

$$\mu(G) < B(R), \tag{11.41}$$

then the boundary condition (11.27) for the Dirichlet problem (11.26–27) may not be satisfied in the classical sense (see Subsection 11.1).

Therefore the problem of the generalized satisfaction of the boundary condition (11.27) arises, if we want to impose only one restriction

$$\mu(G) < B(R)$$

for the set function $\mu(e)$.

This problem will be considered below (see Theorem 11.3). *Theorem 11.3 is the main existence and uniqueness theorem* for solutions of the Dirichlet problem (11.26–27) in the class of weak and generalized solutions.

The second main theorem of existence and uniqueness for the same Dirichlet problem will be considered in Subsection 11.5. In this theorem we will give necessary and sufficient conditions that the boundary data be satisfied in the classical sense.

We now return to the Dirichlet problem

$$\omega(R, u, e) = \mu(e) \tag{11.42}$$

$$u|_{\partial G} = h(x). \tag{11.43}$$

We are concerned with the most general conditions of an open bounded convex domain G and functions $R(p)$, $\mu(e)$ and $h(x)$. These conditions can be stated in the following way.

Assumption 11.1. G is an open bounded convex domain in E^n and ∂G has a parabolic support of order not greater than τ, where $\tau = \text{const} \geq 0$ is any real fixed number.

Assumption 11.2. $R(p)$ is a positive and locally summable function in the gradient space R^n.

Assumption 11.3. $\mu(e)$ is a non-negative completely additive set function of Borel subsets of G such that

$$\mu(G) < B(R), \tag{11.44}$$

where

$$B(R) = \int_{R^n} R(p)dp. \tag{11.45}$$

Assumption 11.4. $h(x)$ is continuous on ∂G.

We denote by $V(R, \mu, h)$ the set of all weak solutions $z(x) \in W^+(G)$ of equation (11.42), whose borders $h_z(x)$ satisfy the inequality

$$h_z(x) \leq h(x) \tag{11.46}$$

for all $x \in \partial G$.

Lemma 11.1. *The set $V(R, \mu, h)$ is not empty if all of Assumptions 11.1–4 hold for the Dirichlet problem (11.42–43).*

Proof. Let

$$\mu_\varepsilon(e) = \mu(e \cap (G \backslash G_\varepsilon)) \tag{11.47}$$

be a new non-negative completely additive set function defined for all sufficiently small numbers $\varepsilon > 0$. This set function $\mu_\varepsilon(e)$ is defined for all Borel subsets of G; the set G_ε was introduced in Theorem 11.2.

Clearly the set functions $\mu_\varepsilon(e)$ converge weakly to $\mu(e)$ as $\varepsilon \to 0$. From Theorem 11.2 it follows that there exists only one weak solution $u_\varepsilon(x) \in W^+(G) \cap C(\overline{G})$ of the Dirichlet problem

$$\omega(R, u_\varepsilon, e) = \mu_\varepsilon(e); \tag{11.48}$$

$$u_\varepsilon(x)|_{\partial G} = h(x), \tag{11.49}$$

where (11.49) holds in the classical sense. Since all functions $u_\varepsilon(x)$ coincide for all $x \in \partial G$ and

$$\mu_{\varepsilon'}(e) \geq \mu_{\varepsilon''}(e)$$

for $\varepsilon' < \varepsilon''$, from Theorem 10.1 it follows that

$$u_{\varepsilon''}(x) \geq u_{\varepsilon'}(x)$$

everywhere in G.

Now the inequalities

$$\omega(R, u_\varepsilon, G) = \mu_\varepsilon(G) \leq \mu(G) = \text{const} < B(R)$$

lead to the uniform estimates

$$\sup_{\partial G} h(x) \geq u_\varepsilon(x) \geq \inf_{\partial G} h(x) - T_R(\mu(G))\text{diam } G \tag{11.50}$$

for all $x \in \overline{G}$. From (11.48) and (11.50) it follows that the convex functions $u_\varepsilon(x)$ converge to some convex function $u_0(x)$ in the open domain G.

Since $\omega(R, u, e)$ converge weakly to $\omega(R, u_0, e)$ inside G (see Theorem 9.1), $u_0(x)$ is a solution of the equation

$$\omega(R, u_0, e) = \mu(e)$$

and

$$h_{u_0}(x) \leq h(x)$$

for all $x \in \partial G$, where $h_{u_0}(x)$ is the border of the function $u_0(x)$.

At some points $x \in \partial G$ the values of $h_{u_0}(x)$ can be strictly less than the corresponding values of $h(x)$ (see Subsection 11.1). □

The convex function $u(x) \in V(R, \mu, h)$ is called a *weak solution of the Dirichlet problem (11.42–43)* if

$$h_u(x) \geq h_v(x), \quad x \in \partial G \tag{11.51}$$

for all weak convex solutions $v(x) \in V(R, \mu, h)$ of equation (11.42), where $h_u(x)$ and $h_v(x)$ are the borders of $u(x)$ and $v(x)$.

Theorem 11.3. *If Assumptions 11.1–4 hold, then the Dirichlet problem (11.42–43) has only one weak solution, i.e. there exists only one weak solution $u(x) \in V(R, \mu, h)$ of equation (11.42), for which inequality (11.51) holds. Moreover such a weak solution of the Dirichlet problem (11.42–43) is the convex function $u_0(x)$ constructed in Lemma 11.1.*

Proof. Let $v(x) \in V(R, \mu, h)$ be any weak solution of equation (11.42). Then

$$\omega(R, u_\varepsilon, e) = \mu_\varepsilon(e) \leq \mu(e) = \omega(R, v, e) \tag{11.52}$$

for any Borel subset e of G and for any sufficiently small number $\varepsilon > 0$.

From (11.52) it follows directly that

$$u_0(x) \geq v(x)$$

for all $x \in G$. This yields the inequality

$$h_{u_0}(x) \geq h_v(x), \quad x \in \partial G$$

for the borders of $u_0(x)$ and $v(x)$. If $h_{u_0}(x) = h_v(x)$ for all $x \in \partial G$, then from Theorem 10.2 it follows that

$$u_0(x) = v(x)$$

for all $x \in \overline{G}$. □

11.5 The Inverse Operator for the Dirichlet Problem

In this subsection we are concerned with the inverse operator $M_R(\mu)$ for the Dirichlet problem (11.42–43). The domain of $M_R(\mu)$ will be some convex set in the space of completely additive set functions $\mu(e)$, whose variations are bounded. The range of $M_R(\mu)$ consists of weak solutions of the Dirichlet problem (11.42–43), whose borders coincide with a prescribed continuous function $h(x)$ on ∂G.

Now we state the following assumptions with respect to the Dirichlet problem (11.42–43):

Assumption 11.4. $h(x)$ is continuous on ∂G.
Assumption 11.5. G is an open bounded convex domain in E^n and ∂G has a parabolic support of order not greater than $\tau = \text{const} \geq 0$.
Assumption 11.6. $R(p)$ is a positive locally summable function in the gradient space R^n and the inequality

$$R(p) \geq C_0 |p|^{-2k}, \quad p \neq (0, 0, \dots, 0) \tag{11.53}$$

holds for all $p \in P^n$, where $k \geq 0$ and $C_0 > 0$ are some constants.
Assumption 11.7. The set function $\mu(e)$ satisfies the following condition:

If x_0 is an arbitrary point of ∂G, then there exists some n-ball $U_\rho(x_0) \subset E^n$ such that the inequality

$$\mu(e \cap G) \leq a[\sup_e [\text{dist}(x, \partial G)]^\lambda] \text{mes } e \tag{11.54}$$

holds, where e is a Borel subset of $U_\rho(x_0)$, $a = \text{const} > 0$, and $\lambda = \text{const} \geq 0$.

Theorem 11.4. *Let the domain G and functions $h(x)$, $R(p)$ and $\mu(e)$ satisfy Assumptions 11.4–7 and let the inequalities*

$$\mu(G) < B(R) \tag{11.55}$$

and

$$\begin{cases} k \leq K & \text{if } \; k < 1 \;\text{ and }\; k \geq \frac{n}{2}, \\ k < K & \text{if } \; 1 \leq k < \frac{n}{2} \end{cases} \tag{11.56}$$

hold, where

$$K = \frac{n + \tau + 1}{\tau + 2} + \frac{\lambda}{2}.$$

Then the Dirichlet problem (11.42–43) has one and only one weak solution $u(x) \in W^+(G)$, and $h_u(x) = h(x)$, where $h_u(x)$ is the border of the function $u(x)$, i.e. $u(x)$ satisfies the boundary condition (11.43) in the classical sense.

The proof of this theorem follows directly from Theorems 11.2, 10.6 and 10.7.

We now turn to the analogy of Theorem 11.4 for generalized solutions of the Dirichlet problem

$$\det(u_{ij}) = \frac{\varphi(x)}{R(Du)}, \tag{11.57}$$

$$u|_{\partial G} = h(x). \tag{11.58}$$

We recall that generalized solutions $u(x)$ of equation (11.57) satisfy this equation almost everywhere, and their R-curvature $\omega(R, u, e)$ is an absolutely continuous set function.

Obviously Assumption 11.7 can be replaced with the following simpler one.

Assumption 11.7'. There exists some sufficiently small neighborhood U of ∂G such that

$$0 \le \varphi(x) \le a[\operatorname{dist}(x, \partial G)]^{\lambda} \tag{11.59}$$

for all $x \in G \cap U$, where $a = \text{const} > 0$ and $\lambda = \text{const} \ge 0$.

Theorem 11.5. *Let Assumptions 11.4–6 and 11.7' hold for the functions $h(x)$, $R(p)$, $\varphi(x)$ and domain G. Let the inequalities*

$$\begin{aligned} &k \le K \quad \text{if} \quad k < 1 \quad \text{and} \quad k \ge \tfrac{n}{2}, \\ &k < K \quad \text{if} \quad 1 \le k < \tfrac{n}{2} \end{aligned} \tag{11.60}$$

and

$$\mu(G) = \int_G \varphi(x)dx < B(R) = \int_{R^n} R(p)dp \tag{11.61}$$

hold, where

$$K = \frac{n+\tau+1}{\tau+2} + \frac{\lambda}{2}. \tag{11.62}$$

Then the Dirichlet problem (11.57–58) has one and only one generalized solution $u(x) \in W^{+}(G)$, and $h_u(x) = h(x)$ for all $x \in \partial G$, where $h_u(x)$ is the border of the convex function $u(x)$, i.e. $u(x)$ satisfies the boundary data (11.58) in the classical sense.

This theorem is a particular case of Theorem 11.4 and it does not need special proof.

We now fix a convex domain G and functions $R(p)$ and $h(x)$ such that all Assumptions 11.4–6 are fulfilled. Then according to Theorems 11.4 and 11.5 we obtain the inverse operator for the Dirichlet problem (11.42–43) or (11.57–58) respectively.

Let M_R be the inverse operator for the Dirichlet problem (11.42–43). Then M_R is defined on the set $D(M_R)$ consisting of all non-negative completely additive set functions $\mu(e)$ such that conditions (11.54), (11.55) and (11.56) hold. Clearly $D(M_R)$ is a convex subset of the cone M of all non-negative completely additive set functions $\mu(e)$, where e is any Borel subset of G.

Let $Q(M_R)$ be the range of M_R. Clearly $Q(M_R)$ consists of all weak solutions $u(x) \in W^+(G)$ of equation (11.42) whose borders coincide with prescribed continuous data on ∂G and for which

$$\omega(R, u, G) < B(R).$$

Let N_R be the inverse operator for the Dirichlet problem (11.57–58). Then N_R is defined on the set $D(N_R)$ consisting of all non-negative functions $\varphi(x)$ such that conditions (11.59), (11.60) and (11.61) hold. Clearly $D(N_R)$ is a convex subset of the cone of all non-negative summable functions in G. Let $Q(N_R)$ be the range of the operator N_R. Clearly $Q(N_R)$ consists of all generalized solutions $u(x) \in W^+(G)$ of equation (11.57), whose borders coincide with prescribed continuous data on ∂G, and for which $\omega(R, u, e)$ is absolutely continuous and

$$\int_G \varphi(x)dx < B(R).$$

11.6 Hypersurfaces with Prescribed Gaussian Curvature

Let S be a convex hypersurface in E^{n+1}, which is given by the equation $z = u(x)$ in the bounded convex domain G in E^n. According to the notation introduced in § 5 the integral Gaussian curvature σ_F of the convex hypersurface S can be expressed in terms of R-curvature in the following way:

$$\sigma_F(e^*) = \omega(R, u, e) = \int_{\chi_u(e)} (1 + |p|^2)^{-\frac{n+1}{2}} dp, \tag{11.63}$$

where $R(p) = (1 + |p|^2)^{-\frac{n+1}{2}}$, e^* is a Borel subset of the hypersurface S and e is the projection of e^* on E^n.

Thus for the case of the integral Gaussian curvature the Dirichlet problem (11.42–43) becomes

$$\omega(R, u, e) \equiv \int_{\chi_u(e)} (1 + |p|^2)^{-\frac{n+1}{2}} dp = \mu(e), \tag{11.64}$$

$$u|_{\partial G} = h(x). \tag{11.65}$$

Let all of the assumptions of Theorem 11.4 be fulfilled. Then the Dirichlet problem (11.64–65) has one and only one weak convex solution $u(x)$ for which (11.65) is satisfied in the classical sense.

We now consider the specific properties of the non-negative constants k, τ, λ and $B(R)$, which are contained in inequalities (11.55–65), providing the solvability of the Dirichlet problem (11.64–65), (see the statement of Theorem 11.4).

Clearly $k = \frac{n+1}{2}$ and

$$B(R) = \int_{R^n} (1 + |p|^2)^{-\frac{n+1}{2}} dp = \frac{1}{2}\sigma_n, \tag{11.66}$$

where σ_n is the area of the unit n-sphere in E^{n+1}. Since

$$k = \frac{n+1}{2} > \frac{n}{2},$$

inequality (11.56) becomes

$$(n-1)\frac{\tau}{\tau+2} \leq \lambda \tag{11.67}$$

after simplifications, where $\tau = \text{const} \geq 0$ and $\lambda = \text{const} \geq 0$. Clearly we can use the minimal admissible value of λ in Theorem 11.4 for the Dirichlet problem (11.64–65). Thus we can set $\lambda = (n-1)\frac{\tau}{\tau+2}$ in (11.54), see Assumption 11.7, Subsection 11.5.

Thus the statement of existence and uniqueness theorem for the Dirichlet problem (11.64–65) is as follows.

Theroem 11.6. *Let the following conditions be fulfilled for the Dirichlet problem (11.64–65):*

1. *G is an open bounded convex domain and ∂G has a parabolic support of order not greater than $\tau = \text{const} \geq 0$;*
2. *$h(x)$ is a continuous function on ∂G;*
3. *the set function $\mu(e)$ satisfies Assumption 11.7 with $\lambda = (n-1)\frac{\tau}{\tau+2}$ and*

$$\mu(G) < \frac{1}{2}\sigma_2.$$

Then the Dirichlet problem (11.64–65) has one and only one weak solution $u(x) \in W^+(G)$ for which (11.65) is fulfilled in the classical sense.

We do not need to give the proof of this theorem because it is a modification of Theorem 11.4 for the Dirichlet problem (11.64–65).

Remarks. 1. If $\tau = 0$, then ∂G has a parabolic support by the paraboloid $x_n = b_0\left(\sum_{i=1}^{n-1} x_i^2\right)$ at all points of ∂G, where we consider this paraboloid in the special Cartesian coordinates, depending on the point $x_0 \in \partial G$ (see § 10). In this case λ is also equal to zero. Thus Assumption 11.7 imposes a significantly less stringent restriction on the set function $\mu(e)$ for the case $\tau = 0$ as in Remark 2.

2. If $\tau > 0$, then λ is also positive and strictly increases to 1 as $\tau \to +\infty$.

The integral Gaussian curvature $\omega(R, u, e)$ is absolutely continuous for generalized solutions of equation (11.64). Thus for generalized solutions $u(x)$ the Dirichlet problem (11.64–65) becomes

$$\det(u_{ij}) = \varphi(x)(1 + |Du|^2)^{\frac{n+1}{2}} \tag{11.68}$$

$$u|_{\partial G} = h(x). \tag{11.69}$$

For equation (11.68) the function $\mu(e)$ is absolutely continuous and can be represented by the formula

$$\mu(e) = \int_e \varphi(x)dx. \tag{11.70}$$

If we apply Theorem 11.5 to the Dirichlet problem (11.68–69), then the following theorem will be obtained.

Theorem 11.7. *Let the following conditions be fulfilled for the Dirichlet problem (11.68–69):*

1. *G is an open bounded convex domain and ∂G has a parabolic support of order not greater than $\tau = const \geq 0$;*
2. *$h(x)$ is a continuous function on ∂G;*
3. *there exists some sufficiently small neighborhood U of ∂G such that*

$$0 \leq \varphi(x) \leq a[dist(x, \partial G)]^{(n-1)\frac{\tau}{\tau+2}} \tag{11.71}$$

for all $x \in G \cap H$, where $a = const > 0$.

4. $$\int_G \varphi(x)dx < \frac{1}{2}\sigma_n.$$

Then the Dirichlet problem (11.68–69) has one and only one generalized convex solution $u(x)$, for which the boundary condition (11.69) is fulfilled in the classical sense.

We now turn to the existence and uniqueness of generalized solutions of the Dirichlet problem

$$\det(u_{ij}) = \varphi(x)(1 + |Du|^2)^{\frac{n+2}{2}}, \tag{11.72}$$

$$u|_{\partial G} = h(x), \tag{11.73}$$

where $\varphi(x) \geq 0$ is a summable function in G.

First we consider the geometric aspect of the Dirichlet problem (11.72–73). Let $u(x)$ be a function defined in G. If $u(x)$ has first and second differentials at the point $x_0 \in G$, then the point x_0 is called *C^2-regular* for $u(x)$. It is well known that almost all points of G are C^2-regular for any convex (concave) function $u(x) \in W^+(G)$ ($u(x) \in W^-(G)$). It is also known that all definitions and properties of pointwise geometric quantities used in classical differential geometry are valid at all C^2-regular points of any convex (concave) function $u(x)$. Such geometric quantities are the first and second fundamental form of the hypersurface S_u: $z = u(x)$, the principal normal curvatures of S_u, the mean and Gaussian curvatures of this surface and others.

Thus we can use the formula

$$K = \frac{\det(u_{ij})}{(1 + |Du|^2)^{\frac{n+2}{2}}} \tag{11.74}$$

for the Gaussian curvature at any C^2-regular point of the hypersurface S_u.

Now let $u(x)$ be a convex generalized solution of equation (11.72). Let

$$R(p) = (1 + |p|^2)^{-\frac{n+2}{2}}.$$

Then the set function

$$\omega(R, u, e) = \int_{\chi_u(e)} (1 + |p|^2)^{-\frac{n+2}{2}} dp$$

is absolutely continuous. Hence

$$\omega(R, u, e) = \int_{\chi_u(e)} (1 + |p|^2)^{-\frac{n+2}{2}} dp = \int_e \varphi(x) dx. \tag{11.75}$$

From (11.74) and (11.75) it follows that

$$K = \varphi(x) \tag{11.76}$$

at any C^2-regular point of the generalized solution $u(x)$ of equation (11.72) and

$$\omega(R, u, e) = \int_e K \, dx \tag{11.77}$$

for any Borel subset e of the convex domain G.

Thus the R-curvature of the convex generalized solutions $u(x) \in W^+(G)$ of equation (11.72) can be restored by the integration of prescribed Gaussian curvature $K = \varphi(x)$ with respect to the Lebesgue measure in E^n.

Theorem 11.8. *Let the following conditions be fulfilled for the Dirichlet problem (11.72–73):*

1. *G is an open bounded convex domain in E^n and ∂G has a parabolic support of order not greater than $\tau = const \geq 0$;*
2. *$h(x)$ is a continuous function on ∂G;*
3. *there exists some sufficiently small neighborhood U of ∂G such that*

$$0 \leq \varphi(x) \leq a[dist(x, \partial G)]^{1+(n-1)\frac{\tau}{\tau+2}} \tag{11.78}$$

for all $x \in G \cap H$, where $a = const > 0$;

4. $$\int_G \varphi(x) dx < \mu_n,$$

where μ_n is the volume of the unit n-ball in E^{n+1}.

Then the Dirichlet problem (11.72–73) has one and only one generalized convex solution $u(x)$, for which the boundary data (11.73) is fulfilled in the classical sense.

The proof of this theorem follows from Theorem 11.5 if we consider the calculations used in the proof of Theorem 11.6.

Remark. The number λ in (11.78) is chosen as $1+(n-1)\frac{\tau}{\tau+2}$. Thus if $\tau=0$, then $\lambda=1$. This is the smallest possible value of λ which can be chosen in Theorem 11.8.

It is very useful to consider again the second example in Subsection 11.1 and compare it with Theorem 11.8.

Finally we consider the boundary value problem associated with the Minkowski problem (see Subsections 8.1 and 8.2). We call it briefly the Minkowski boundary value problem.

Let G^* be a spherical convex domain on the unit hypersphere S^n in E^{n+1}. We assume that ∂G^* is not empty and $\overline{G}^* = G^* \cup \partial G^*$ lies inside some open hemisphere S^n_+ of S^n. Clearly

$$\delta = \operatorname{dist}\{\overline{G}^*, \partial S^n_+\} > 0. \tag{11.79}$$

Let F be a convex hypersurface whose image under the Gaussian (spherical) mapping is the set G^*. If F is not smooth, then we consider the Gaussian mapping by means of supporting hyperplanes of F. The set of such convex hypersurfaces we denote by $W(G^*)$.

Now let $\varphi^*(m) > 0$ be a given function in G^*, where m is a point of the unit sphere S^n in E^{n+1}, and let $h^*(m)$ be a prescribed continuous function in ∂G^*.

We seek the convex hypersurfaces $F \in W(G^*)$ satisfying two conditions:

a) the Gaussian curvature K at the point of F with a unit outward normal m is equal to $\varphi^*(m)$;
b) the supporting function $p^*(m)$ of F (see Section 6) satisfies the boundary data

$$p^*(m) = h^*(m)$$

for all $m \in \partial G^*$.

This geometric problem is called the *Minkowski boundary value problem*. The statement of this problem is related to C^2-convex hypersurfaces with one-to-one Gaussian mapping. In Subsection 8.1 the statement of this problem was extended to all convex hypersurfaces by means of the spherical mapping and the surface function.

Let m_0 be some interior point of G^*. We consider the central projection g of S^n_+ on the tangential hyperplane α of S^n at the point m_0. Let $m_1, m_2, m_3, \ldots, m_n$, m_{n+1} be components of a unit vector m with respect to the Cartesian system with the origin 0 and the vector $0m_0$ as the basis vector of the $(n+1)$-st axis. From Subsection 6.2 it follows that Cartesian coordinates in α can be introduced by the formulas

$$x_1 = \frac{m_1}{m_{n+1}}, \ldots, x_n = \frac{m_1}{m_{n+1}}.$$

Now the function

$$p(x) = \frac{1}{m_{n+1}} p^*(m)$$

satisfies the differential equation

$$\det(p_{ij}) = \frac{1}{\varphi^*(m)[1 + x_1^2 + x_2^2 + \cdots + s_n^2]^{n+2/2}} \tag{11.80}$$

in the convex domain $G = g(G^*)$. Clearly

$$p|_{\partial G} = \frac{1}{m_{n+1}} h^*(m)|_{\partial G^*}. \tag{11.81}$$

Now from Theorem 11.5 follows the existence and uniqueness theorem for the Dirichlet problem (11.80–81).

Theorem 11.9. *Let the domain $G = g(G^*)$ and the function*

$$\varphi(x) = \frac{1}{[q(x)]^{n+2} \varphi^*(m(x))} \tag{11.82}$$

where $q(x) = [1 + |x|^2]^{\frac{1}{2}}$ and $m(x) = \left\{\frac{x_1}{q(x)}, \frac{x_2}{q(x)}, \dots, \frac{x_n}{q(x)}, -\frac{1}{q(x)}\right\}$, satisfy the following conditions

1. *G is an open bounded convex domain in E^n and ∂G has a parabolic support of order not greater than $\tau = const \geq 0$;*
2. *$p(x)$ on ∂G must coincide with any prescribed continuous function;*
3. *$\varphi(x)$ is any non-negative summable function in G.*

Then the Dirichlet problem (11.80–81) has one and only one generalized convex solution.

§12. The Dirichlet Problem for Elliptic Solutions of Monge–Ampere Equations $\mathrm{Det}(u_{ij}) = f(x, u, Du)$

In this section we are concerned with existence of one or several different generalized elliptic solutions of the Dirichlet problem

$$\det(u_{ij}) = f(x, u, Du), \tag{12.1}$$

$$u|_{\partial G} = h(x). \tag{12.2}$$

Generalized elliptic solutions of problem (12.1–2) are convex or concave functions. Without loss of generality, we consider only convex generalized solutions, which are defined in a prescribed bounded convex domain G, and which satisfy prescribed continuous data (12.2) in the classical sense.

12.1 The First Main Existence Theorem for the Dirichlet Problem (12.1–2)

We suppose that G is an open bounded convex domain which satisfies Assumption 11.5 (see § 11), i.e. ∂G has a parabolic support of order not greater than $\tau = \text{const} \geq 0$. In this subsection we also suppose that the following assumptions are fulfilled:

Assumption 12.1. The function $f(x,u,p)$ is continuous in $\overline{G} \times R \times R^n$ and the inequalities

$$0 \leq f(x,u,p) \leq \frac{\varphi(x)}{R(p)} \tag{12.3}$$

hold in the same domain $\overline{G} \times R \times R^n$, where $\varphi(x)$ is non-negative and summable in G, and $R(p)$ is positive and locally summable in the gradient space R^n. (Remember that $\overline{G} = G \cup \partial G$ is the closure of G.)

The function $R(p)$ satisfies Assumption 11.6, i.e. the inequality

$$R(p) \geq C_0 |p|^{-2k} \tag{12.4}$$

holds for all $p \in R^n$ except the point $p = (0,0,\ldots,0)$, where $C_0 = \text{const} > 0$, $k = \text{const} \geq 0$. (Clearly, the case $k < 0$ is reduced to the case $k = 0$.)

The function $\varphi(x)$ satisfies Assumption 11.7', i.e. there exists some sufficiently small neighborhood U of ∂G such that

$$0 \leq \varphi(x) \leq a[\text{dist}(x, \partial G)]^{\lambda} \tag{12.5}$$

for all $x \in G \cap U$, where $a = \text{const} > 0$, $\lambda = \text{const} \geq 0$.

Finally the function $h(x)$ satisfies Assumption 11.4, i.e. $h(x)$ is continuous on ∂G.

We will consider generalized solutions of the Dirichlet problem (12.1–2). We call the convex function $u(x) \in W^+(G)$ a *generalized solution of the Dirichlet problem (12.1–2)*, if the border of $u(x)$ coincides with $h(x)$ on ∂G, $u(x)$ satisfies equation (12.1) almost everywhere in G, and $\omega(R,u,e)$ is an absolutely continuous set function.

The following main theorem holds.

Theorem 12.1. *The Dirichlet problem*

$$\det(u_{ij}) = f(x,u,Du), \tag{12.6}$$

$$u|_{\partial G} = h(x) \tag{12.7}$$

has at least one generalized solution $u(x) \in W^+(G)$, if the following assumptions are fulfilled:

A) The convex domain G, functions $f(x,u,p)$, $\varphi(x)$, $R(p)$ and $h(x)$ satisfy Assumptions 11.4–6, 11.7' and 12.1;

B)

$$\int_G \varphi(x)dx < B(R) = \int_{R^n} R(p)dp. \tag{12.8}$$

C) If

$$K = \frac{n+\tau+1}{\tau+2} + \frac{\lambda}{2},$$

then

$$k \le K \quad \text{if} \quad k < 1 \quad \text{and} \quad k \ge \frac{n}{2}$$

and

$$k < K \quad \text{if} \quad 1 \le k < \frac{n}{2}.$$

Proof. We denote by $W_h^+(G)$ the set of all convex functions $u(x)$ satisfying the condition

$$h_u(x) = h(x),$$

where $h_u(x)$ is the border of $u(x)$. The set $W_h^+(G)$ is not empty, because the Dirichlet problem

$$\det(z_{ij}) = \frac{\varphi(x)}{R(Dz)}$$

$$z|_{\partial G} = h(x)$$

has the generalized solution $z(x) \in W_h^+(G)$. This follows directly from Theorem 11.5.

Clearly, $W_h^+(G)$ is a convex set in the space $C(\overline{G})$. Let

$$F_u(x) = f(x,u,Du)R(Du), \tag{12.9}$$

where $u(x) \in W^+(G)$. The function $F_u(x)$ is non-negative in G and

$$F_u(x) \le \varphi(x) \tag{12.10}$$

almost everywhere in G. If $u(x) \in W^+(G) \cap C^1(G)$, then $F_u(x) \in C(G)$ and $F_u(x) \in L(G)$, because

$$\int_G F_u(x)dx \le \int_G \varphi(x)dx.$$

If $u(x) \in W^+(G)$, then the same integral inequality can be obtained by approximating with functions $u_m(x) \in W^+(G) \cap C^1(G)$. Hence the Dirichlet problem

$$\det(z_{ij}) = \frac{F_u(x)}{R(Dz)} \tag{12.11}$$

$$z|_{\partial G} = h(x) \tag{12.12}$$

has only one generalized solution $z(x) \in W_h^+(G)$, where $u(x) \in W_h^+(G)$. This statement follows directly from Theorem 11.5, because all conditions of this theorem are fulfilled.

Thus we have constructed the operator

$$\widetilde{B}\colon W_h^+(G) \longrightarrow W_h^+(G)$$

and $z(x) = \widetilde{B}(u(x))$. Clearly, the fixed points of operator $\widetilde{B}$ are generalized solutions of our initial Dirichlet problem (12.1–2). The inequality

$$\int_G \varphi(x)dx < B(R)$$

yields the estimates

$$\inf_{\partial G} h(x) - T_R\left(\int_G \varphi(x)dx\right) \operatorname{diam} G \leq z(x) = \\ = \widetilde{B}(u(x)) \leq \sup_{\partial G} h(x)$$

for every function $u(x) \in W_h^+(G)$ (see Section 10). Thus the set $\widetilde{B}(W_h^+(G))$ is bounded in $C(\overline{G})$. Therefore we can take a subsequence

$$z_{i_m}(x) = \widetilde{B}(u_{i_m}(x))$$

for each sequence $u_1(x), u_2(x), \ldots, u_m(x), \ldots \in W_h^+(G)$ converging for all $x \in G$ (see the Blaschke Theorem, Section 3.5). But it is possible that it converges non-uniformly. From the conditions of Theorem 11.4 it follows that the sequence $z_{i_m}(x)$ converges uniformly to some functions $z_0(x) \in W_h^+(G)$. Thus the operator B is compact.

Let the functions $u_m(x) \in W_h^+(G)$ and let them converge to some function $u_0(x) \in W_h^+(G)$. Then the set of functions $v_m(x) = \widetilde{B}(u_m(x))$ is compact in $W_h^+(G)$, and considered a subspace of $C(\overline{G})$. This statement is proved in the same way as the compactness of the set $\widetilde{B}(W_h^+(G))$. Let $v_0(x) = \widetilde{B}(u_0(x))$. We take some uniformly convergent subsequence $v_{m_k}(x)$ in $C(\overline{G})$. Let $\bar{v}(x)$ be the limit of this subsequence. From the conditions of Theorem 12.1 and Section 10 it follows that $\bar{v}(x) \in W_h^+(G)$.

Using the properties of the convergent sequence of convex functions and their R-curvature, we obtain that $\bar{v}(x)$ and $v_0(x)$ are generalized solutions of the same equation

$$\det(v_{ij}) = \frac{F_{u_0}(x)}{R(Dv)},$$

and the borders of $\bar{v}(x)$ and $v_0(x)$ coincide. Therefore these functions coincide in G, and operator $\widetilde{B}$ is continuous.

Now all the conditions of Schauder's principal are fulfilled and the initial Dirichlet problem (12.1–2) has at least one generalized solution. □

12.2 Existence of at Least One Generalized Solution of the Dirichlet Problem for Equations $\det(u_{ij}) = f(x, u, Du)$

In Subsections 12.2 and 12.3 we are concerned with the development of Theorem 12.1 in two directions. The first one is the extension of Theorem 12.1 to the wider classes of non-negative functions $f(x, u, p)$ which can infinitely increase together with $|u| \to +\infty$. The second one is the problem of existence of several different generalized solutions for the Dirichlet problem (12.1–2).

First we present the statement of lemmas related to fixed points of positive operators acting in convex cones of Banach spaces.

Let B be a Banach space and K be a convex cone in B. We shall consider operators F: $K \to K$. Let S_ρ be the intersection of K with the sphere $\|x\|_B = \rho$, where ρ is a positive number.

Lemma 12.1. *Let F: $K \to K$ be a compact and continuous operator in the convex cone K of a Banach space B. If there exists some positive number ρ such that*

$$\|F(x)\|_B \leq \|x\|_B \tag{12.13}$$

for all $x \in S_\rho$, then operator F has at least one fixed point $x_0 \in K$ and $\|x_0\|_B \leq \rho$.

This lemma follows from the well known principle of fixed points in Banach spaces.

Lemma 12.2. *Let F: $K \to K$ be a compact and continuous operator in the convex cone K of a Banach space B and let ρ_1 and ρ_2 be two different positive numbers. If the inequalities*

$$\|F(x)\|_B \leq \|x\|_B \tag{12.14}$$

for all $x \in S\rho_1$ and

$$\|F(x)\|_B \geq \|x\|_B \tag{12.15}$$

for all $x \in S\rho_2$ hold, then the operator F has at least one fixed point $x_0 \in K$ such that either the inequalities

$$\rho_1 \leq \|x_0\|_B \leq \rho_2 \quad \textit{hold}$$

if $\rho_1 < \rho_2$, or the inequalities

$$\rho_1 \geq \|x_0\|_B \geq \rho_2 \quad \textit{hold}$$

if $\rho_1 > \rho_2$.

Now we consider the Dirichlet problem (12.1–2). We shall suppose that the following assumptions hold.

A) G is a bounded convex domain and ∂G has parabolic support of order not greater than $\tau = \text{const} \geq 0$.

B) We denote by q_G the infimum of distances between pairs of parallel supporting hyperplanes of G with opposite outward normals. Let $\varepsilon > 0$ be any number less than $\frac{q_G}{10}$. We introduce the functions

$$\varphi(x, \lambda, \varepsilon) = \begin{cases} 1 & \text{if } \operatorname{dist}(x, \partial G) \geq \varepsilon; \\ [\operatorname{dist}(x, \partial G)]^{\lambda}, & \text{if } \operatorname{dist}(x, \partial G) < \varepsilon \end{cases} \tag{12.16}$$

and

$$R_k(p) = \begin{cases} 1, & \text{if } |p| \leq 1; \\ \frac{1}{|p|^{2k}}, & \text{if } |p| > 1, \end{cases} \tag{12.17}$$

where $\lambda = \text{const} \geq 0$ and $0 \leq k = \text{const} < \frac{n}{2}$.

The statement of Assumption B is as follows: the function $f(x, u, p)$ is continuous in $\overline{G} \times R \times P^n$ and the inequalities

$$0 \leq f(x, u, p) \leq \frac{(a|u| + b)^{\alpha}}{R_k(p)} \varphi(x, \lambda, \varepsilon) \tag{12.18}$$

holds for all $x \in \overline{G}$, $u \leq 0$, $p \in P^n$, where $a = \text{const} \geq 0$, $b = \text{const} \geq 0$, $a^2 + b^2 > 0$ and $\alpha = \text{const} \geq 0$.

Theorem 12.2. *Let Assumptions A and B be fulfilled. Then the Dirichlet problem (12.1–2) has at least one convex generalized solution if the inequalities*

$$k < \frac{n + \tau + 1}{\tau + 2} + \frac{\lambda}{2} \tag{12.19}$$

and

$$0 \leq \alpha < n - 2k \tag{12.20}$$

hold.

Proof. Let K be the set of all convex functions satisfying the boundary condition (12.2) in the classical sense. Clearly, the set K is not empty and forms a convex cone in $C(\overline{G})$. All convex functions $u(x) \in K$ are non-positive in $\overline{G}$. Let

$$F_u(x) = f(x, u(x), Du(x)) \cdot R_k(Du(x)), \tag{12.21}$$

where $u(x) \in K$. From Assumption B it follows that $F_u(x)$ is non-negative and summable in G and

$$\int_G F_u(x) dx \leq \int_G (a|u(x)| + b)^{\alpha} dx \leq (a\|u\| + b)^{\alpha} \operatorname{mes} G, \tag{12.22}$$

where $\|u\| = \|u\|_{C(\overline{G})}$.

Now we consider the Dirichlet problem

$$\det(z_{ij}) = \frac{F_u(x)}{R_k(Dz)} \tag{12.23}$$

$$z|_{\partial G} = 0. \tag{12.24}$$

Since $k < \frac{n}{2}$, then

$$A(R_k) = \int_{|p|\le 1} dp + \int_{|p|>1} \frac{dp}{|p|^{2k}} = +\infty. \tag{12.25}$$

Therefore

$$\int_G F_u(x)dx \le (a\|u(x)\| + b)^\alpha \cdot \text{ mes } G < A(R_k). \tag{12.26}$$

Inequalities (12.26), (12.19), (12.20) together with Assumptions A and B permit us to apply Theorem 11.5 to the Dirichlet problem (12.23–24). Thus we obtain that this Dirichlet problem has only one generalized solution $z(x) \in K$. Hence the operator A: $K \to K$ arises such that

$$z(x) = A(u(x)), \tag{12.27}$$

where $z(x)$ is a generalized solution of the Dirichlet problem (12.23–24).

From Theorem 10.4 (see Subsection 10.2) it follows that

$$-T_{R_K}(\omega_u)\text{diam } G \le z(x) = A(u(x)) \le 0 \tag{12.28}$$

for all $x \in \overline{G}$, where

$$\omega_u = \int_G F_u(x)dx$$

and $T_R(\tau)$ is the inverse of the function

$$g_{R_K}(\rho) = \int_{|p|\le\rho} R_k(p)dp = \mu_n\left(1 + \frac{n}{n-2k}(p^{n-2k} - 1)\right) \tag{12.29}$$

(μ_n is the volume of the unit ball in R^n). Since $0 \le \frac{2k}{n} < 1$,

$$0 < \frac{n-2k}{n} \le 1 \quad \text{and} \quad g_{R_k}(+\infty) = A(R_k) = +\infty,$$

we obtain

$$T_{R_k}(\tau) = \left[1 + \frac{n-2k}{n}\left(\frac{\tau}{\mu_n} - 1\right)\right]^{\frac{1}{n-2k}} \le \left[\frac{\tau}{\mu_n} + 1\right]^{\frac{1}{n-2k}} \tag{12.30}$$

for all $\tau \in [0, +\infty)$. Thus it follows that

$$\|A(u(x))\|_{C(\overline{G})} \leq \left[\frac{\omega_u + \mu_n}{\mu_n}\right]^{\frac{1}{n-2k}} \cdot \text{ diam } G \tag{12.31}$$

$$\leq \left(\frac{1}{\mu_n}(a\|u(x)\| + b)^\alpha \text{ mes } G + \mu_n\right)^{\frac{1}{n-2k}}.$$

Therefore the operator $A\colon K \to K$ maps every bounded subset $Q \subset K$ in a bounded subset $A(Q)$ of K. Using the same considerations as in the proof of Theorem 12.1 we obtain that the operator $A\colon K \to K$ is compact and continuous.

Since $\alpha < n - 2k$, $a \geq 0, b \geq 0$ and $a^2 + b^2 > 0$, one can find a positive sufficiently large number r_0 such that the inequality

$$\left[\frac{1}{\mu_n}(a\|u(x)\| + b)^\alpha \text{ mes } G + \mu_n\right]^{\frac{1}{n-2k}} \text{ diam } G < r_0 \tag{12.32}$$

holds. Then from (12.31) we obtain

$$\|A(u(x))\| < \|u(x)\|$$

for all $u(x) \in S_{r_0}$, where S_{r_0} is the intersection of the cone K with the sphere $\|x\| = r_0$ in $C(\overline{G})$.

Now from Lemma 12.1 it follows that the Dirichlet problem (12.1–2) has at least one generalized solution $z(x) \in K$ and $\|z(x)\| \leq r_0$. □

Remarks. 1. Let $\alpha \geq 0$ and $2k \geq 0$ be the orders of the growth of the functions $(a\|u\| + b)^\alpha$ and

$$\frac{1}{R_k(p)} = \begin{cases} 1, & \text{if } \ |p| \leq 1; \\ |p|^{2k}, & \text{if } \ |p| > 1. \end{cases}$$

Then the inequalities

$$0 \leq \alpha + 2k < n$$

give the largest values of α and $2k$ in Assumption B of Theorem 12.2.

2. Theorem 12.2 is also true if the inequalities (12.31) hold for all $x \in G$, $|u| < r_0$, $p \in R^n$, where the positive number r_0 satisfies the inequality (12.32) and $a = \text{const} \geq 0$, $b = \text{const} \geq 0$, $a^2 + b^2 > 0$ and $0 \leq \alpha < n - 2k$ are given numbers.

12.3 Existence of Several Different Generalized Solutions for the Dirichlet Problem (12.23–24)

Let G be a bounded convex open domain in R^n, and let Q be a closed n-dimensional ball inside G.

Lemma 12.3. *If $\delta = dist(Q, \partial G) > 0$, then the inequality*

$$\inf_Q |u(x)| \geq \frac{\delta}{\text{diam } G} \|u(x)\|_{C(\overline{G})} \tag{12.33}$$

holds for all convex functions $u(x) \in K$, where K is the set of all convex functions, satisfying the boundary condition $u|_{\partial G} = 0$.

Proof. Lemma is obviously true for convex functions $v(x) \in K$, whose graphs are convex cones with the base ∂G. Now let $u(x)$ be any convex function belonging to K. Since the lemma is trivially true for the function $u(x) = 0$, we are concerned only with the case $\|u(x)\| > 0$, where $\|u(x)\|$ is the norm of $u(x)$ in the space $C(\overline{G})$. Let $x_0 \in G$ be the point where

$$|u(x_0)| = \|u(x)\|$$

and let $v(x) \in K$ be the function, whose graph is the convex cone with base ∂G and vertex $(x_0, u(x_0))$. Clearly,

$$\|v(x)\| = |u(x_0)| = \|u(x)\|$$

and

$$0 \geq v(x) \geq u(x) \tag{12.34}$$

for all $x \in \overline{G}$. Now we complete the proof of Lemma C by the following chain of equalities and inequalities

$$\begin{aligned} \inf_Q |u(x)| &= |u(x^*)| \geq |v(x^*) \geq \inf_Q |v(x)| \\ &\geq \frac{\delta}{\text{diam } G} \|v(x)\| = \frac{\delta}{\text{diam } G} \|u(x)\|, \end{aligned}$$

where x^* is the point of Q such that

$$\inf_Q |u(x)| = |u(x^*)|.$$

□

Here we present the assumption that we shall use below. They are as follows:

A) The boundary of a bounded convex open domain G has parabolic support of order not more than $\tau = \text{const} \geq 0$, i.e. we keep the same Assumption A which we used in Sections 12.1, 12.2, 12.3.

B') The function $f(x, u, p)$ is continuous in $\overline{G} \times R \times R^n$ and the inequalities

$$0 \leq f(x, u, p) \leq (a|u| + b)^\alpha \tag{12.35}$$

hold, where $a = \text{const} \geq 0$, $b = \text{const} \geq 0$, $a^2 + b^2 > 0$, $u \leq 0$, $0 \leq \alpha = \text{const} < n$.

Assumption B' can be considered as a particular case of Assumption B (see Subsection 12.2) when $k = 0$. Therefore we can use $\lambda = 0$ in Assumption B' because inequality (12.19) is reduced to inequality

$$0 < \frac{n+\tau+1}{\tau+2}$$

which holds for all $\tau \in [0, +\infty)$.

We shall suppose that the Assumptions A and B' hold in all our further considerations together with the new Assumption C.

C) The inequalities

$$\gamma_0 |u|^\beta \le f(x,u,p) \le (\gamma_1 |u| + \gamma_2)^\beta \tag{12.36}$$

hold for all $x \in \overline{G}, -u_0 \le u \le -u_1$, $p \in R^n$, where $\gamma_0 = \text{const} > 0$, $\gamma_1 = \text{const} > 0$, $\gamma_2 = \text{const} > 0$, $\beta = \text{const} > n$ and $u_0 = \text{const} > u_1 = \text{const} > 0$.

We pick a convex function $u(x) \in K$ such that

$$u_1 < \frac{\delta}{\text{diam } G} \|u(x)\| < \|u(x)\| \le u_0 \tag{12.37}$$

where $\|u(x)\| = \|u(x)\|_{C(\overline{G})}$, and consider two Dirichlet problems:

$$\det(v_{ij}) = \gamma_0 \left(\frac{\delta}{\text{diam } G}\right)^\beta \|u(x)\|^\beta, \tag{12.38}$$

$$v|_{\partial Q} = -\frac{\delta}{\text{diam } G} \|u(x)\| \tag{12.39}$$

in the ball Q and

$$\det(z_{ij}) = F_u(x), \tag{12.40}$$

$$z|_{\partial G} = 0 \tag{12.41}$$

in G, where $F_u(x) = f(x, u, (x), Du(x))$. From Theorem 11.5 it follows that there exists only one generalized solution $z(x)$ of the problem (12.40–41). We can also see that the function

$$v(x) = \frac{1}{2} B \left(\sum_{i=1}^n x_i^2 - r^2\right) - \frac{\delta}{\text{diam } G} \|u(x)\| \tag{12.42}$$

is the unique generalized solution of the problem (12.38–39), where r is the radius of the ball Q and

$$B = \gamma_0^{\frac{1}{n}} \left(\frac{\delta}{\text{diam } G}\right)^{\beta/n} \|u(x)\|^{\beta/n}.$$

From (12.37–39) it follows that

$$0 \geq -\frac{\delta}{\operatorname{diam} G}\|u(x)\| \geq v(x) \geq z(x)$$

for all $x \in Q$. Therefore

$$\|v(x)\|_{C(Q)} \leq \|z(x)\|_{C(\overline{G})}.$$

Thus we obtain

$$\|z(x)\|_{C(\overline{G})} \geq \|v(x)\|_{C(Q)} = \frac{r^2}{2}\gamma_0^{\frac{1}{n}}\left(\frac{\delta}{\operatorname{diam} G}\right)^{\beta/n} \|u(x)\|^{\beta/n}_{C(\overline{G})}$$
$$+\frac{\delta}{\operatorname{diam} G}\|u(x)\|_{C(\overline{G})}. \tag{12.43}$$

The inequality (12.43) is a non-trivial estimate $\|z(x)\|_{C(\overline{G})}$ from below, where $z(x)$ is the solution of the Dirichlet problem (12.40–41).

Now we prove the existence of at least two different generalized solutions of the Dirichlet problem

$$\det(u_{ij}) = f(x, u, Du) \tag{12.44}$$
$$u|_{\partial G} = 0 \tag{12.45}$$

under suitable conditions on the function $f(x, u, p)$.

Theorem 12.3. *Let the following conditions be fulfilled:*

1) The bounded convex open domain G satisfies Assumption A, and G contains a closed ball Q of radius r such that

$$\delta = dist(Q, \partial G) > 0.$$

2) The function $f(x, u, p)$ is non-negative and continuous for all $x \in \overline{G}$, $u \leq 0$, and $p \in P^n$, and the inequality

$$f(x, u, p) \leq (a|u| + b)^\alpha \tag{12.46}$$

holds for all $x \in \overline{G}$, $-r_0 \leq u \leq 0$, $p \in P^n$, where $a = const \geq 0$, $b = const \geq 0$, $a^2 + b^2 \geq 1$, $0 \leq \alpha < n$ and the inequality

$$\left(\left[\frac{(ar_0 + b)^\alpha}{\mu_n}\right] \operatorname{mes} G + \mu_n\right)^{1/n} \operatorname{diam} G < r_0 \tag{12.47}$$

holds for the number $r_0 > 0$.

3) Let $r_1 > 0$ be a number such that the inequalities

$$\frac{\delta}{\text{diam } G} r_1 > r_0 \tag{12.48}$$

and

$$r_1 < \frac{r^2}{2} \gamma_0^{1/n} \left(\frac{\delta}{\text{diam } G} \right)^{\beta/n} \left(\frac{r_1 \delta}{\text{diam } G} \right)^{\beta/n}, \tag{12.49}$$

where $\beta = \text{const} > n$, — hold.

4) The inequality

$$f(x, u, p) \leq \varphi(u)$$

holds for all $x \in \overline{G}$, $u \in (-\infty, 0]$, $p \in P^n$, where $\varphi(u)$ is a strictly positive and continuous function of u.If $r_0 \geq 1$, then we can take

$$\varphi(u) = (a|u| + b)^{\beta}.$$

If the conditions 1–4 are fulfilled then the Dirichlet problem (12.44–45) has at least two different convex generalized solutions.

Proof. First of all we note that the inequalities

$$\gamma_0 |u|^{\beta} \leq f(x, u, p) \leq (a|u| + b)^{\beta}$$

hold for all $x \in \overline{G}, -r_1 \leq u \leq -\frac{\delta}{\text{diam } G} r_1$, $p \in P^n$, where $\beta = \text{const} > n$. This is a consequence of conditions 3 and 4.

Now consider the Dirichlet problem (12.23–24), where $u(x)$ is any functions belonging to the convex cone $K \subset C(\overline{G})$. From the conditions of the present theorem it follows that the problem (12.23–24) has only one generalized solution $z(x)$. Therefore we can consider the operator A: $K \to K$ constructed in the proof of Theorem 12.2. This operator is compact and continuous. We establish this fact by means of the same methods used in the proof of Theorem 12.2. Using condition 2 together with Theorem 12.2 we obtain that the operator A has a fixed point $u_1(x) \in K$ such that

$$\|u_1(x)\|_{C(\overline{G})} \leq r_0.$$

Now from conditions 3 and 4, Lemma 12.3 and the inequality (12.43) we establish the existence of another fixed point $u_2(x) \in K$ such that

$$r_0 < \frac{\delta}{\text{diam } G} r_1 \leq \|u_2(x)\|_{C(\overline{G})} \leq r_1. \qquad \square$$

The methods used in Subsection 12.3 permit us to establish the existence of infinite number of different generalized solutions if the function $f(x, u, p)$ satisfies suitable conditions.

Comments to Chapter 3.

In Chapter 3 we are concerned with the Dirichlet problem for Monge–Ampere equations in the classes of weak and generalized convex solutions.

§ 9. This section is devoted to the concepts of the normal mapping and R-curvature of convex functions depending on n independent variables. These key concepts in the theory of weak and generalized solutions for Monge–Ampere equations were introduced for the first time in the author's paper [1]. The presentation of the normal mapping and R-curvature of convex functions together with their properties is given in detail. A brief presentation is given in the author's papers [2], [7], [20]. The detailed proof of Theorem 9.1 is published here for the first time.

§ 10. In Section 10 we are concerned with n-dimensional comparison, compactness and convergence theorems for convex functions in the terms of their R-curvature. The fundamental techniques in the theory of weak and generalized solutions are based on these theorems. The scheme of presentation corresponds to the author's papers [1], [7]. The proofs of all theorems are given in detail and all of them are proved in the author's papers [1], [7], [11], [12]. The detailed proof of the basic Theorems 10.6 and 10.7 related to the uniform convergence of convex functions in a closed bounded convex domain is published here for the first time. The scheme of the proof of Theorem 10.6 is presented in my paper [15].

§ 11. In this section we are concerned with existence, uniqueness and non-uniqueness theorems for weak and generalized solutions of the Dirichlet problem for n-dimensional Monge–Ampere equations

$$\det(u_{ij}) = \frac{\varphi(x)}{R(Du)}. \tag{$*$}$$

These equations arise from the well-known geometric problems of the reconstruction of convex hypersurfaces with prescribed Gaussian curvature as a function of the unit exterior normal (Minkowski problem) or as a function of the projection of the variable point of the desired hypersurfaces. The function $R(p) = 1$ corresponds to the Minkowski problem, and the function

$$R(p) = (1 + |p|^2)^{-\frac{n+2}{2}}$$

corresponds to the second problem. We shall significantly apply existence and uniqueness theorems for weak and generalized convex solutions of the Dirichlet problem for equation $(*)$ to the study of the Dirichlet problem for the general Monge–Ampere equations

$$\det(u_{ij}) = f(x, u, Du) \tag{$**$}$$

in the class of generalized convex solutions (see Section 12). Two obstructions showing that the Dirichlet problem for the n-dimensional equation $(*)$ can not be solved for arbitrary functions $\varphi(x)$ and $R(p)$ even if they are strictly positive and continuous were published by the author in [2] for the first time. These obstructions are described in Subsection 11.1.

In Subsections 11.2–11.6 the geometric theory of Monge–Ampere equations (*), relating to weak and generalized convex solutions of the Dirichlet problem, is presented. This theory was created by Bakelman [1], [2], [4], [7], [12], [13], [20]. The main definitions of weak and generalized solution of equations (*), presented in Subsection 11.2, were published by Bakelman [1] for the first time. In Subsections 11.3–11.5 we are concerned with existence and uniqueness theorems of the Dirichlet problem for equations (*) in the class of weak and generalized elliptic solutions. These solutions are convex functions.

In Subsection 11.3 we present the solution of the Dirichlet problem in the class of convex polyhedra. For convex polyhedra this Dirichlet problem can be solved by means of some extremal problem in the similar way as the Minkowski problem in Chapter 2. In Theorem 11.1 we obtain sharp necessary and sufficient conditions of existence of such a polyhedron. This polyhedron is also unique. The study of the Dirichlet problem (11.26–27) can be obtained by the suitable approximation by convex polyhedra. As we mentioned above this method was introduced by Minkowski [1] and later developed by Alexandrov [1], [5], Bakelman [1], [4], Pogorelov [5], [7] and others in different directions.

For weak solutions of the Dirichlet problem (11.26–27) there are two main existence theorems. One of them is Theorem 11.3, which is presented in Subsection 11.4. In this theorem it was established that the Dirichlet problem (11.26–27) has an unique weak elliptic (convex) solution in a bounded convex domain G, satisfying the prescribed continuous boundary data in a generalized meaning (see inequality (11.51) and the corresponding definition on page 193), if

$$\mu(G) < \int_{R^n} R(p)dp. \qquad (***)$$

This sufficient condition is sharp. The necessary condition is as follows

$$\mu(G) \leq \int_{R^n} R(p)dp. \qquad (****)$$

This theorem was first published by Bakelman [1]. Different detailed proofs of Theorem 11.3 were given by Bakelman [4], [7].

Sharp necessary and sufficient conditions, providing the classical satisfaction of the boundary data by the unique weak (convex) solution of the Dirichlet problem (11.43–44) are proved in Theorem 11.4. These conditions are stated in the terms of ∂G, the set function $\mu(e)$ and the positive locally summable function $R(p)$.

This theorem for the general n-dimensional case was published for the first time by Bakelman [7]. The essentially simpler two-dimensional case of Theorem 11.4 was studied by Bakelman [1], [2].

From Theorems 11.3 and 11.4 it follows that analogous theorems are valid for generalized elliptic (convex) solutions of the Dirichlet problem (11.57–58).

These analogies can be used for obtaining the inverse operator for the Dirichlet problem (11.57–58). The most convenient inverse operator for the Dirichlet problem (11.57–58) is constructed in Subsection 11.5. In Subsection 11.6 we present applications of Theorems 11.4 and 11.5 to the Dirichlet problem for convex hypersurfaces with prescribed Gaussian curvature. These results contain necessary and sufficient conditions for the existence of general convex hypersurfaces with positive constant Gaussian curvature (the Bernstein problem) and variable positive prescribed curvature (the Minkowski and other problems) with prescribed continuous Dirichlet boundary data. The smoothness of these solutions will be studied in Chapter 6. These results are published here for the first time.

§12. Bakelman's investigations related to weak and generalized solutions of the Dirichlet problem for Monge–Ampere equations $(*)$ (see § 9, 10, 11) and especially Theorem 11.3, proved in 1956, have stimulated interest in the Dirichlet problem for the equations

$$\det(u_{ij}) = f(x, u, Du). \qquad (**)$$

In 1958 Alexandrov [5] introduced the concept of weak solutions for the equations $(**)$ by means of the extension of the Monge–Ampere operator

$$\frac{1}{f(x, u, Du)} \det(u_{ij})$$

to the set of all convex functions, where the function $f(x, u, p)$ can be estimated from above and below by products $\varphi_1(x)R_1(p)$ and $\varphi_2(x)R_2(p)$. Alexandrov has proved that equation $(**)$ has at least one weak solution $u(x) \in W^+(G)$ assuming convexity and boundedness of the domain G and an inequality between the functions f, φ_i, $R_i(p)$, $i = 1, 2$. This inequality can be considered as an analog of the inequality $(***)$ for equations $(*)$. Note that the sufficient condition $(***)$ of Theorem 11.3 of the solvability of the Dirichlet problem for equations $(*)$ interlocks with the necessary condition $(****)$ of the solvability of the same problem. Alexandrov's assumption for the more general equation $(**)$ is a sufficient condition for existence of at least one weak solution of this equation.

First Alexandrov proved the existence of a solution of the Dirichlet problem in the class of convex polyhedra and then he used the polyhedral approximation for the general case. A detailed survey of Alexandrov's paper [5] has been presented by Bakelman [15], pages 89–92. The unsolved problems about the satisfaction of Dirichlet boundary data for two-dimensional Monge–Ampere equations $(**)$ were investigated by Bakelman [1], [2] and Pogorelov [2], [6], and for n-dimensional Monge–Ampere equations by Bakelman [4], [7] and Cheng and Yau [3], [4].

Cheng and Yau's papers are related to the Jörgens–Calabi–Pogorelov problem and will be considered in Chapters 5 and 10.

Bakelman introduced other methods to study the Dirichlet problem for equation (**) from 1959 up to present time. His main considerations are based on the reduction of the Dirichlet problem for equation (**) to some operator equation in Banach spaces and then the application of global fixed points theorems. The reduction of equation (**) to an operator equation is given by means of the inverse operator to the Dirichlet problem

$$\det(z_{ij}) = \frac{\varphi(x)}{R(Dz)}, \quad z|_{\partial G} = h(x) \in C(\partial G),$$

constructed in Subsection 11.5. In Subsections 12.1, 12.2, 12.3 different existence theorems for generalized convex solutions of the Dirichlet problem for n-dimensional Monge–Ampere equations (**) were presented. These theorems were proved by Bakelman in 1983–1984 and published in his paper [15] for the first time. Theorem 12.1 provides the existence of at least one generalized solution of the Dirichlet problem

$$\det(u_{ij}) = f(x, u, Du), \quad u|_{\partial G} = h(x) \in C(\partial G) \qquad (*****)$$

satisfying Dirichlet boundary data in the classical meaning, if the function $f(x, u, p)$ is continuous and satisfies the inequalities

$$0 \leq f(x, u, p) \leq \frac{\varphi(x)}{R(p)}$$

for all $(x, u, p) \in \overline{G} \times R \times R^n$, where the functions $\varphi(x)$ and $R(p)$ satisfy all the conditions of Theorem 11.5. The detailed statement and proof of Theorem 12.1 is contained in Subsection 12.1.

In Subsections 12.2 and 12.3, Theorem 12.1, is developed in two directions. The first one is an extension of Theorem 12.1 to a wider class of non-negative functions $f(x, u, p)$, which can infinitely increase as $|u| \to +\infty$. The second one is the problem of existence of several different generalized solutions of the Dirichlet problem (*****). The first theorems of this kind were obtained by Bakelman and Krasnoselskii [1] in 1960 (see also Bakelman [7] and Krasnoselskii [1]). A significant improvement of these theorems, to the general n-dimensional case was obtained by Bakelman [15] in 1984. Our presentation corresponds to Bakelman [15].

Chapter 4. Variational Problems and Generalized Elliptic Solutions of Monge–Ampere Equations

There is a profound connection between n-dimensional variational problems and the Dirichlet problem for n-dimensional Monge–Ampere equations. The absolute minimum of these variational problems turns out to be a generalized solution of the corresponding Monge–Ampere equations. In this chapter we study explicitly the main variational problem connected with the Monge–Ampere equation

$$\det(u_{ij}) = f(x_1, x_2, \ldots, x_n) \tag{13.1}$$

and also consider generalizations of it.

§13. Introduction. The Main Functional

13.1 Statement of Problems

The functional

$$E(u) = -\iint\limits_G [u_x^2 u_{yy} - 2u_x u_y u_{xy} + u_y^2 u_{xx} + 6fu]dxdy \tag{13.2}$$

is given by Courant and Hilbert [1] for the case of functions of two variables where G is an bounded open domain in the x, y-plane. Without losing generality we can assume that G is a convex domain. The Euler equation for functional (13.2) is

$$u_{xx}u_{yy} - u_{xy}^2 = f(x, y). \tag{13.3}$$

Unfortunately the functional (13.2) does not give any clue concerning the functional spaces in which this variational problem belongs, neither does it suggest the generalization for functions of n-variables.

Another functional, whose Euler equation is (13.3), was found by Bakelman [21], [7]. This functional is

$$I(u) = -\iint\limits_G [u(u_{xx}u_{yy} - u_{xy}^2) - 3fu]dxdy. \tag{13.4}$$

If ∂G is a C^2-curve and $u(x, y) \in C^2(\overline{G})$ and $u|_{\partial G} = 0$, then $E(u) = -2I(u)$. The functional (13.4) is closely connected with the Monge–Ampere operator

$u_{xx}u_{yy} - u_{xy}^2$ and admits a simple geometric interpretation by means of the tangential mapping constructed by the function $u(x, y)$. This functional also admits a simple natural generalization to functions of n variables:

$$I_n(u) = -\int_G [u(x)\det(u_{ij}(x)) - (n+1)f(x)u(x)]dx, \tag{13.5}$$

where G is an open convex bounded domain in Euclidean space E^n. If $x_1, x_2, \dots, x_n$ are Cartesian coordinates in E^n and $dx = dx_1 dx_2 \dots dx_n$, then the formal Euler equation for $I_n(u)$ is the Monge–Ampere equation

$$\det(u_{ij}) = f(x_1, x_2, \dots, x_n). \tag{13.6}$$

Bakelman [21], [7] investigated the variational problem for the functional (13.5) and proved that the absolute minimum for this problem is a generalized elliptic solution of equation (13.6).

Since the absolute minimum of the functional $I_n(u)$ was considered to be in the suitable class of functions satisfying the boundary condition $u|_{\partial G} = 0$, the solution of (13.6) considered above satisfies the boundary condition $u|_{\partial G} = 0$.

The results in Sections 13, 14, and 15 appeared in Bakelman [13].

13.2 Preliminary Considerations

Since the Euler equation (13.6) for the functional $I_n(u)$ is of second order instead of the fourth order, the variational problems for this functional are degenerate. Therefore there is only one boundary condition in variational problems for $I_n(u)$. (Note that there are two boundary conditions in nondegenerate variational problems of the second order.) This fact influences the semiboundedness, continuity and asymptotic behavior of $I_n(u)$ as $\|u\| \to \infty$. Here $\|u\|$ is the norm of the function u in the corresponding function set. We can see for example that $I_n(u)$ is unbounded from above and below if we consider it for all functions $u \in C^2(G) \cap C(\overline{G})$ and if $u|_{\partial G} = 0$. Therefore it is natural to construct the domain of definition for $I_n(u)$ by taking into account the properties of elliptic solutions of the corresponding boundary value problem for the Monge–Ampere equation (13.6). In this paper we consider elliptic solutions of the Dirichlet problem for the equation (13.6) with the zero boundary condition.

Evidently all C^2 elliptic solutions $u(x_1, \dots, x_n)$ have a fixed sign second differential everywhere in G. Hence they are either nonpositive convex or nonnegative concave functions in G. Therefore the variational problem corresponding to the absolute minimum (maximum) of $I_n(u)$ should be considered in the class of nonpositive convex (nonnegative concave) functions. It is clear that it is sufficient to consider only the problem of finding the absolute minimum of $I_n(u)$. The functional $I_n(u)$ can be extended to the class of all nonnegative convex functions, vanishing on ∂G, by means of the technique of generalized elliptic solutions for Monge–Ampere equations (see Chapter 3). Thus the main variational problem for the functional $I_n(u)$ is reduced to the establishment of the absolute minimum of the extension of $I_n(u)$ to the class of all convex functions vanishing on ∂G.

But the extension of $I_n(u)$ turns out to be discontinuous. This fact appears because the second boundary condition for the comparison functions is excluded. Moreover the problem of finding the first variation of $I_n(u)$ is connected with the extension of $I_n(u)$ to the class of all nonpositive continuous functions vanishing on ∂G. Therefore deeper and more refined techniques are required to extend the functional $I_n(u)$ and to proof the existence of its absolute minimum and to express the first variation of the extended functional $I_n(u)$ than we had briefly considered above. We present all these problems in Sections 13, 14, 15.

13.3 The Functional $I_H(u)$ and its Properties

In this subsection we construct the extension $I_H(u)$ of the functional $I_n(u)$ to the set of all nonpositive continuous functions vanishing on ∂G and establish the continuity of $I_H(u)$. Here H is any convex subdomain of G whose distance from ∂G is some positive number and G is a given convex bounded domain in E^n.

The Operator F_H and its Properties. Let G be an open convex bounded domain in E^n and H be a convex subdomain of G whose distance from ∂G is the positive number h_H. Let $\overline{H}$ and $\overline{G}$ be the closures of H and G. Denote by $C_0^-(\overline{G})$ the closed subset of the space $C(\overline{G})$ consisting of all nonpositive continuous functions vanishing on ∂G. The operator F_H considered in this subsection maps the set $C_0^-(\overline{G})$ in the special class of convex functions which will be introduced below. This operator will be used for the extension of the functional $I_n(u)$ to the set $C_0^-(\overline{G})$.

Now consider the construction of F_H and its properties. Let

$$v(x) = \begin{cases} u(x) & \text{if } \ x \in \overline{H}, \\ 0 & \text{if } \ x \in \overline{G} \backslash \overline{H} \end{cases} \tag{13.7}$$

for any function $u(x) \in C_0^-(\overline{G})$, i.e. $v(x) = u(x)\varphi_{\overline{H}}(x)$, where $\varphi_{\overline{H}}(x)$ is the characteristic function of the set $\overline{H}$. If S_v is the graph of $v(x)$, then the boundary of $\overline{Co}\{S_v\}$ consists of $\overline{G}$ and the graph S_w of some convex function $w(x)$. Clearly

$$w(x) \in W^+(\overline{G}) \cap C_0^-(\overline{G}).$$

We say in this case the function $w(x)$ spans $u(x)$ on the set $\overline{H}$ from below, and denote by F_H the operator mapping any function $u(x) \in C_0(\overline{G})$ in the corresponding convex function $w(x)$.

The Properties of F_H.

(1) Every supporting hyperplane β to S_w passing through the point $(x_0, w(x_0))$, where $x_0 \in G \backslash \overline{H}$, contains at least some line segment AB such that $AB \subset \beta \cap S_w$, $A \in \partial G$ and $B' \in \overline{H}$ is the projection of the point $B \in S_w \cap \beta$ onto the hyperplane E^n.

The proof follows directly from the definition of S_w and from the well-known properties of a convex hull.

(2) The equality

$$\text{mes}\ \chi_w(w, G\backslash\overline{H}) = 0 \tag{13.8}$$

holds for every function $w(x) = F_H(u(x))$.

Let β be any supporting hyperplane of S_w at the point $(x_0, w(x_0))$, where $x_0 \in G\backslash\overline{H}$. Then from property (1) it follows that β is a singular supporting hyperplane of S_w. Since the spherical image of all supporting hyperplanes of every convex hypersurface has measure zero,

$$\text{mes}\ \chi_w(w, G\backslash\overline{H}) = 0. \tag{13.9}$$

Now we denote by $W_H^+(\overline{H})$ the set $F_H(C_0^-(\overline{G}))$. It is evident that

$$W_H^+(\overline{G}) \subset W^+(\overline{G}) \cap C_0^-(\overline{G})$$

and the set $W^+(\overline{G}) \subset C_0^-(\overline{G}) \cap W_H^+(\overline{G})$ is not empty.

(3) The equality

$$\chi_w(G) = \chi_w(\overline{H}) \tag{13.10}$$

holds for all functions $w(x) \in W_H^+(\overline{G})$.

The proof follows directly from property(1).

(4) The equality

$$w(x) = F_H(w(x)) \tag{13.11}$$

holds if and only if $w(x) \in W_H^+(\overline{G})$.

Proof. If the function $w(x) \in C_0^-(\overline{G})$ satisfies equation (13.11), then from the definition of the operator F_H it follows that $w(x) \in W_H^+(\overline{G})$. The converse assertion follows from the properties of the convex hull.

(5) The set $W_H^+(\overline{G})$ is a closed subset of the space $C(\overline{G})$.

Proof. Let $w(x)$ be the limit of functions $w_1(x), w_2(x), \ldots, w_m(x), \ldots$ belonging to $W_H^+(\overline{G})$ in the space $C(\overline{G})$. Clearly $w(x)$ is a convex function belonging to $C_0^-(\overline{G})$. The considered property will be proved if we establish the equality

$$w(x) = F_H(w(x)) \tag{13.12}$$

(see property (4)). Let

$$v(x) = w(x)\varphi_{\overline{H}}(x), \qquad v_m(x) = w_m(x)\varphi_{\overline{H}}(x). \tag{13.13}$$

The restrictions $v_m(x)$ and $v(x)$ to the convex compact set $\overline{H}$ are convex functions and $\lim_{m\to\infty} v_m(x) = v(x)$ for all $x \in \overline{H}$ and $v_m(x) = v(x) = 0$ for all $x \in \overline{G}\backslash\overline{H}$. Therefore

$$\overline{Co}\{S_w\} = \lim_{m\to\infty} \overline{Co}\{S_{w_m}\} = \lim_{m\to\infty} \overline{Co}\{S_{v_m}\} = \overline{Co}\{S_v\}$$

because the equality
$$\overline{Co}\{S_{w_m}\} = \overline{Co}\{S_{v_m}\} \tag{13.14}$$
follows from the condition that $w_m(x) \in W_H^+(\overline{G})$ for all positive integers m. Property (5) is proved since (13.14) is equivalent to equality (13.12).

(6) The set $\chi_w(G)$ is contained in the n-dimensional ball $|p| \le \|w(x)\|/h_H$ for all functions $w(x) \in W_H^+(\overline{G})$.

Proof. Let α be the supporting hyperplane of the graph S_w of any function $w(x) \in W_H^+(\overline{G})$. Then there exists a point $x_0 \in \overline{H}$ such that the point $(x_0, w(x_0))$ belongs to α. Note that $\operatorname{dist}(x_0, \partial G)$ is not less than $h_H = \operatorname{dist}(\overline{H}, \partial G) > 0$.

Let K_{x_0} be the convex cone with the vertex $(x_0, w(x))$ and the base $U(x_0, h_H)$, where $U(x_0, h_H)$ is the closed n-ball with the center x_0 and the radius h_H. Let k_{x_0} be the convex function defining K_{x_0}. Then

$$\chi_w(\alpha) \subset \chi_{k_{x_0}}(U(x_0, h_H)).$$

The set $\chi_{k_{x_0}}(U(x_0, h_H))$ is the n-dimensional ball with the center $0(0, 0, \ldots, 0)$ and the radius $\rho = \|w(x)\|/h_H$. Therefore $\chi_w(G)$ is contained in the n-dimensional ball $|p| \le \|w(x)\|/h_H$ in R^n for all functions $w(x) \in W_H^+(\overline{G})$.

(7) From (6) it follows directly that any function $w(x)$ belonging to $W_H^+(\overline{G})$ satisfies the Lipschitz condition with the constant $\|w(x)\|/h_H$, i.e.

$$|w(x+q) - w(x)| \le \frac{\|w(x)\|}{h_H}|q|,$$

where x and $x + q$ are any points of G.

(8) The operator F_H: $C_0^-(\overline{G}) \to W_H^+(\overline{G})$ is continuous.

Proof. Let the functions $u_n(x) \in C_0^-(\overline{G})$ converge uniformly to the function $u(x) \in C_0^-(\overline{G})$. Take any number $\varepsilon > 0$ and consider two functions

$$v_\varepsilon^{(1)}(x) = \begin{cases} 0 & \text{if } \; x \in \overline{G} \backslash \overline{H} \text{ or if } u(x) \ge -\varepsilon, \\ u(x) + \varepsilon & \text{if } \; u(x) < -\varepsilon \end{cases}$$

and

$$v_\varepsilon^{(2)}(x) = \begin{cases} 0 & \text{if } \; x \in \overline{G} \backslash \overline{H} \\ u(x) - \varepsilon & \text{if } \; x \in \overline{H}. \end{cases}$$

Let $v(x) = u(x)\varphi_{\overline{H}}(x)$ and $v_n(x) = u_n(x)\varphi_{\overline{H}}(x)$ be the functions considered by definition of the operator F_H (note that $\varphi_{\overline{H}}(x)$ is the characteristic function of the set $\overline{H}$). Then $v^{(2)}(x) = v(x) - \varepsilon$ for all $x \in \overline{H}$ and

$$v_\varepsilon^{(1)}(x) = \begin{cases} v(x) + \varepsilon & \text{if } \; u(x) < -\varepsilon, \\ 0 & \text{if } \; x \in \overline{H} \text{ and } u(x) \ge -\varepsilon \end{cases}$$

also for all $x \in \overline{H}$.

Since $v_n(x)$ uniformly converge to $v(x)$ in G, there exists a natural number N such that

$$v_\varepsilon^{(2)}(x) \le v_n(x) \le v_\varepsilon^{(1)}(x)$$

for all $n \ge N$ and $x \in \overline{G}$. From the definition of the operator F_H it follows that

$$F_H(v^{(2)}(x)) \le w_n(x) = F_H(u_n(x)) \le F_H(v^{(1)}(x))$$

for all $n \ge N$ and $x \in \overline{G}$. Since

$$\lim_{\varepsilon \to 0} F_h(v_\varepsilon^{(2)}(x)) = \lim_{\varepsilon \to 0} F_H(v_\varepsilon^{(1)}(x)) = w(x),$$

we have

$$F_H(u(x)) = w(x) = \lim_{n \to \infty} w_n(x) = \lim_{n \to \infty} F(u_n(x)).$$

Property 8 is proved.

The Functional $I_H(u)$. Let H be a convex subdomain of a given convex bounded domain G in E^n such that $\operatorname{dist}(\overline{H}, \partial G) = h_H > 0$. Let $u(x) \in C_0^-(\overline{G})$ and $w(x)$ be the convex function constructed above by means of $u(x)$ (see § 13.3).

Now we define the functional

$$\Phi_H(u) = -\int_G u\omega(w, de) \tag{13.15}$$

on the set $C_0(\overline{G})$, where $\omega(w, e)$ is the measure of the normal mapping of the convex function $w(x) = F_H(u(x))$. From property (2) (see § 13.3) it follows that

$$\omega(w, G \backslash \overline{H}) = 0 \tag{13.16}$$

for all functions $w(x) = F_H(u(x))$ (see § 13.3, equality (13.8)).

Letting $\psi(e)$ be a nonnegative completely additive set function on the subsets of G and $\psi(G) < +\infty$, we define the new set function

$$\psi_H(e) = \psi(e \cap H). \tag{13.17}$$

Clearly, $\psi_H(e)$ is a nonnegative completely additive set function on Borel subsets of G and

$$\psi_H(G \backslash H) = 0. \tag{13.18}$$

Now we introduce the functionals

$$\tau_H(u) = \int_G u\psi_H(de) \tag{13.19}$$

and

$$I_H(u) = \Phi_H(u) + (n+1)\tau_H(u) \tag{13.20}$$

on the set $C_0^-(\overline{G})$.

The properties of the functionals Φ_H, τ_H and I_H.

Theorem 13.1. *The inequalities and equalities*

$$\Phi_H(w) = \Phi_H(u), \tag{13.21}$$
$$\tau_H(w) \leq \tau_H(u), \tag{13.22}$$
$$I_H(w) \leq I_H(u) \tag{13.23}$$

hold for all functions $u(x) \in C_0^-(\overline{G})$ *and convex functions* $w(x) = F_H(u(x))$.

Remark. Let $\psi(e) \geq C_0 \operatorname{mes}(e)$ for every Borel subset $e \subset G$, where $C_0 = \text{const} > 0$. Then the equality can hold in (13.22) and (13.23) if and only if the restriction of $u(x)$ on the n-dimensional convex body $\overline{H}$ is some convex function, i.e.

$$u(x)|_{\overline{H}} = w(x)|_{\overline{H}}, \tag{13.24}$$

where $w(x) = F_H(u(x))$.

Proof. From the equalities (13.16) and (13.18) it follows that

$$\Phi_H(u) = -\int_{\overline{H}} u(x)\omega(w(x), de), \tag{13.25}$$

$$\tau_H(u) = \int_{\overline{H}} u(x)\psi_H(de). \tag{13.26}$$

Since $u(x) \geq w(x)$ for all $x \in \overline{H}$, then from (13.26) we obtain

$$\tau_H(u) \geq \int_H w(x)\psi_H(de) = \int_G w(x)\psi_H(de) = \tau_H(w).$$

It is clear that

$$\tau_H(u(x)) = \tau_H(w(x)), \tag{13.27}$$

if $u(x)$ and $w(x)$ coincide on the set $\overline{H}$. The condition $\psi(e) \geq C_0 \operatorname{mes}(e)$ (see Remark) and equality (13.27) yield conversely $u(x) = w(x)$ for all $x \in \overline{H}$.

We denote by H_u the set of points $x \in \overline{H}$, where $u(x) = w(x)$ and by S_{H_u} the part of the graph of $u(x)$ for $x \in H_u$. Every supporting hyperplane a of the graph of the function $w(x)$ has at least one common point with the set S_{H_u}. Therefore $\chi_w(w, G\backslash H_u)$ consists only of the images of singular supporting hyperplanes to the graph of $w(x)$. Thus from property (2) (see § 13.2) it follows that $\omega(w, G\backslash H_u) = 0$. Hence

$$\Phi_H(u) = \int_G u\omega(w, de) = \int_{H_u} u\omega(w, de)$$
$$= \int_{H_u} w\omega(w, de) = \int_G \omega(w, de) = \Phi_H(w).$$

The inequality (13.23) now follows directly from (13.21) and (13.22). Theorem 13.1 is proved. □

Theorem 13.2. *The functionals $\Phi_H(e), \tau_H(e)$ and $I_H(e)$ are continuous on the set $C_0^-(\overline{G})$.*

Let the functions $u_1(x), u_2(x), \dots, u_n(x), \dots$ belong to $C_0^-(\overline{G})$ and uniformly converge to the function $u(x) \in C_0^-(\overline{G})$. Then the set functions $\omega(w_n, e)$ converge weakly to the set function $\omega(w, e)$ (see § 13.2), where $w_n = F_H(u_n)$ and $w = F_H(u)$ are convex functions belonging to $W_H^+(\overline{G})$. Now using the facts mentioned above and property (2) (see § 13.2) we obtain the proof of Theorem 13.2 by the standard considerations. □

Thus we can seek functions realizing the absolute minimum of the continuous functional $I_H(u)$ only in the set of convex functions $W_H^+(\overline{G})$.

§14. Variational Problem for the Functional $I_H(u)$

From § 13 it follows that the absolute minimum of the functional $I_H(u)$ can be reached only for the convex functions $w(x) \in W_H^+(\overline{G})$. In this section we establish the existence of absolute minimum $w(x)$ for the functional $I_H(u)$, where $w(x) \in W_H^+(\overline{G})$ (see Theorem 14.3). This is the first main result of the present chapter. The proof is based on the nontrivial bilateral estimates for the values of I_H considered only on all convex functions $w(x) \in W_H^+(\overline{G})$ (see Theorems 14.1 and 14.2).

14.1 Bilateral Estimates for $I_H(u)$

Lemma 14.1. *The inequality*

$$|w(x)| \geq \frac{h_H}{\textit{diam } G} \|w(x)\| \tag{14.1}$$

holds for every convex function $w(x) \in W_H^+(\overline{G})$.

Proof. The inequality(14.1) holds trivially for the function $w(x) = 0$ in G. Therefore we assume that $\|w(x)\| > 0$. Let $w(x)$ be any function from $W_H^+(\overline{G})$. From property (1) (see § 13.2) it follows that there exists the point $x_0 \in \overline{H}$ such that

$$\|w(x)\| = |w(x_0)|. \tag{14.2}$$

Now we consider the convex cone K with the vertex $(x_0, w(x_0))$ and the base G. Let K be the graph of the convex function $k(x)$. Then

$$w(x) < k(x) < 0 \tag{14.3}$$

for all $x \in \overline{G}$ and

$$w(x)|_{\partial G} = k(x)|_{\partial G} = 0. \tag{14.4}$$

The equality

$$|k(x)| = \frac{|xx'|}{|x_0x'|} |k(x_0)| \tag{14.5}$$

holds for any point $x \in \overline{H}$, where x' is the point of intersection of the ray x_0x (with origin x_0) and ∂G, and $|xx'| = \mathrm{dist}(x, x')$, $|x_0x'| = \mathrm{dist}(x_0, x')$. Since

$$|xx'| \geq h_H \tag{14.6}$$

and

$$|x_0x'| \leq \mathrm{diam}\ G \tag{14.7}$$

then from (14.6), (14.7), (14.8) we obtain

$$|k(x)| \geq \frac{h_H}{\mathrm{diam}\ G}|k(x_0)|. \tag{14.8}$$

But

$$|k(x_0)| = |w(x_0)| = \|w\|, \tag{14.9}$$

since the cone K has the vertex at the point $(x_0, w(x_0))$ and

$$\|w(x)\| = |w(x_0)|. \tag{14.10}$$

Now from (14.3), (14.8) and (14.9) we obtain the inequality (14.1). Lemma 14.1 is proved. □

Lemma 14.2. *The inequality*

$$\|w(x)\| \leq \left[\frac{\omega(w, G)}{\mu_n}\right]^{1/n} \mathit{diam}\ G \tag{14.11}$$

holds for every convex function $w(x) \in W^+(\overline{G}) \cap C_0^-(\overline{G})$, where μ_n is the volume of the n-unit ball.

This lemma is a special case of Theorem 10.3 (see § 10.2).

Theorem 14.1. *The inequality*

$$I_H(w) \geq \frac{\mu_h h_N}{(\mathrm{diam}\ G)^{n+1}}\|w(x)\|^{n+1} - \psi(G)(n+1)\|w(x)\| \tag{14.12}$$

holds for any $w(x) \in W_H^+(\overline{G})$.

Proof. From Lemma 14.1 we obtain

$$\int_G [-w(x)]\omega(w, de) \geq \frac{h_H}{\mathrm{diam}\ G}\|w(x)\|\omega(w, \overline{H}). \tag{14.13}$$

But from property (3) (see § 13) it follows that $\omega(w, \overline{H}) = \omega(w, G)$. Now from Lemma 14.2 we obtain

$$\omega(w, G) \geq \frac{\mu_n}{(\mathrm{diam}\ G)^n}\|w(x)\|^n. \tag{14.14}$$

Thus the inequalities (14.13) and (14.14) lead to the inequality

$$\int_G [-w(x)]\omega(w, de) \geq \frac{\mu_n h_H}{(\text{diam } G)^{n+1}} \|w(x)\|^{n+1}. \tag{14.15}$$

From (14.15) we obtain finally the inequality (14.12) for

$$I_H(w) = -\int_G w(x)\omega(w, de) + (n+1)\int_G w(x)\psi_H(de).$$

Theorem 14.1 is proved. □

Theorem 14.2. *The inequality*

$$I_H(w) \leq \frac{\mu_n}{h_H^n}\|w\|^{n+1} - \frac{(n+1)h_H}{\text{diam } G}\psi_H(G)\|w\| \tag{14.16}$$

holds for every convex function $w(x) \in W_H^+(\overline{G})$.

Proof. First we estimate from above the integral $\int_G [-w(x)]\omega(w, de)$. We have

$$0 \leq \int_G [-w(x)]\omega(w, de) \leq \|w\|\omega(w, G)$$

and from property (6) (see § 13) we obtain

$$0 \leq \int_G [-w(x)]\omega(w, de) \leq \frac{\mu_n}{h_H^n}\|w\|^{n+1}. \tag{14.17}$$

Now we estimate from below $\int_G |w(x)|\psi_H(de)$. Since $\psi_H(G\backslash\overline{H}) = 0$ then

$$\int_G |w(x)|\psi_H(de) = \int_{\overline{H}} |w(x)|\psi_H(de).$$

Now from Lemma 14.1 it follows that

$$\int_G |w(x)|\psi_H(de) \geq \frac{h_H}{\text{diam } G}\|w(x)\|\psi_H(G). \tag{14.18}$$

Thus from (14.17) and (14.18) we finally obtain

$$\begin{aligned} I_H(w) &= -\int_G w\omega(w, de) + (n+1)\int_G w\psi_H(de) \\ &\leq \frac{\mu_n}{h_H^n}\|w\|^{n+1} - \frac{(n+1)h_H\psi_H(G)\|w\|}{\text{diam } G}, \end{aligned}$$

because $w(x) \leq 0$ in G. Theorem 14.2 is proved. □

14.2 Main Theorem about the Functional $I_H(u)$

Let $U(H, m, M)$ denote the subset of functions $w(x) \in W_H^+(\overline{G})$ satisfying the condition

$$m \le \|w(x)\| \le M, \tag{14.19}$$

where $0 \le m < M < +\infty$ are constants. If $m = 0$, then $U(H, 0, M)$ consists of functions $w(x) \in W_H^+(\overline{G})$ satisfying the inequality

$$\|w(x)\| \le M. \tag{14.20}$$

Lemma 14.3. *Every set $U(H, m, M)$ is compact in $C(\overline{G})$.*

Proof. The set $U(H, m, M)$ is bounded and closed in $C(\overline{G})$ and any function $w(x) \in U(H, m, M)$ satisfies the Lipschitz condition of degree one and constant $M\mu_n^{1/n}(h_H)^{-1}$ Thus $U(H, m, M)$ is compact in $C(\overline{G})$. Lemma 14.3 is proved.□

Theorem 14.3 (Main Theorem About the Absolute Minimum for $I_H(u)$). *The functional $I_H(u)$ has at least one absolute minimum and a function $w_0(x)$ belonging to $W_H^+(\overline{G})$ and realizing this minimum satisfies the inequalities, $m_0 \le \|w_0(x)\| \le M_0$, where*

$$m_0 = \frac{1}{2}\left[\frac{h_H^{n+1}\psi_H(G)}{\mu_n(\text{diam } G)}\right]^{1/n},$$

$$M_0 = \max\left\{I, \frac{(n+1)\psi_H(G) + I}{\mu_n h_H}(\text{diam } G)^{n+1}\right\}.$$

Proof. From Theorem 4.1 it follows that $\lim_{k\to\infty} I_H(w_k) = +\infty$ if $w_k(x) \in W_H^+(\overline{G})$ and $\|w_k(x)\| \to +\infty$.

Therefore we can find a positive number M_0 such that $I_H(w) > 1$ if $\|w(x)\| > M_0$. For example we can take M_0 to be the number M_0 mentioned in Theorem 14.3. Now from the expression of $I_H(u)$ and Theorem 14.2 we see that $I_H(0) = 0$ and $I_H(w) < 0$ if $w \in W_H^+(\overline{G})$, $\|w\| > 0$ and $\|w\|$ is sufficiently small.

Therefore the functional $I_H(u)$ is bounded from below and $I_H(u)$ takes negative values.

Now we consider the function

$$\varphi(t) = \frac{\mu_n}{h_H^n}t^{n+1} - \frac{(n+1)_{h_H}}{\text{diam } G}\psi_H(G)t$$

for $t \in [0, +\infty)$. This function has only two roots 0 and some positive number t_0 and takes negative values only inside the interval $(0, t_0)$. Let t^* be such a point that

$$\varphi(t^*) = \inf \varphi(t), \qquad 0 \le t \le t_0.$$

Then $\varphi'(t^*) = 0$ and $t^* = [h_H^{n+1}\psi_H(G)/\mu_n(\text{diam } G)]^{1/n}$.

Now the function

$$\Phi(t) = \frac{\mu_n h_H}{(\text{diam } G)^{n+1}} t^{n+1} - (n+1)\psi_H(G)t$$

has only one negative minimum at the point

$$t^{**} = \left[\frac{\psi_H(G)[\text{diam } G]^{n+1}}{\mu_n h_H}\right]^{1/n}$$

(clearly t^{**} is the unique root of $\Phi'(t)$ and $\Phi(t^{**}) = \inf_{[0,\infty)} \Phi(t)$). Since

$$t^* = \left[\frac{h_H}{\text{diam } G}\right]^{\frac{n+2}{n}} t^{**} < t^{**}$$

and $\Phi'(t) < 0$ on $[0, t^{**})$, we can set

$$m_0 = \Phi^{-1}(\varphi(t^*)).$$

Recall that $\inf_{W_H^+(\overline{G})} I_H(w)$ is a finite negative number. It is clear that

$$\inf_{W_H^+(\overline{G})} I_H(w) = \inf_{U(H,m_0,M_0)} I_H(u)$$

where m_0 and M_0 were defined above.

From Lemma 14.3 it follows that there exists at least one function $w_0(x) \in U(H, m_0, M_0)$ such that

$$I_H(w_0(x)) = \inf_{U(H,m_0,M_0)} I_H(w).$$

Theorem 14.3 is proved. □

§15. Dual Convex Hypersurfaces and Euler's Equation

From Theorem 14.3 (see § 14) it follows that the absolute minimum of the functional $I_H(u)$ is achieved on some convex function $w_0(x) \in W_H^+(\overline{G})$. In the present section we establish that $w_0(x)$ is the general solution of the Dirichlet problem

$$\omega(w, e) = \psi_H(e), \qquad w|_{\partial G} = 0$$

(see Theorem 15.5). This is the second main result of this chapter and its proof is based on the special formula for the first variation of the functional $I_H(u)$. The fundamental technique used to establish this formula is the dual convex hypersurface for all non-positive continuous functions $u(x)$ defined in $\overline{G}$ and vanishing on ∂G. Here H is an open subdomain of G such that $\text{dist}(\overline{H}, \partial G) =$

$h_H > 0$, where $\overline{H}$ is the closure of H. If $\varphi_{\overline{H}}(x)$ is the characteristic function of $\overline{H}$, then the dual hypersurfaces $P_H(u)$ are constructed by means of the functions $v(x) = u(x)\varphi_{\overline{H}}(x)$. The dual hypersurfaces $P_H(u)$ generate some solid cones and the values of the functional $\Phi_H(u) = -\int_G u\omega(w, de)$ are exactly the volumes of these bodies. The mentioned relationship permits us to use methods and results from the theory of mixed volumes in the investigation of the first variation for the functional $I_H(u)$.

15.1 Special Map on the Hemisphere

Let $\overline{G}$ be an open convex domain in E^n. Let $\widetilde{R}^{n+1} = (p_1, p_2, \ldots, p_{n+1})$ be an $(n+1)$-dimensional Euclidean space and S^n_- be the unit n-hemisphere:

$$p_{n+1} < 0, \quad p_1^2 + p_2^2 + \cdots + p_{n+1}^2 = 1 \tag{15.1}$$

in $\widetilde{R}^{n+1}$. We consider the map γ: $S^n_- \to E^n$ defined by

$$x_1 = \frac{p_1}{|p_{n+1}|}, x_2 = \frac{p_2}{|p_{n+1}|}, \ldots, x_n = \frac{p_n}{|p_{n+1}|}, \tag{15.2}$$

where $x = (x_1, \ldots, x_n) \in E^n, p = (p_1, \ldots, p_{n+1}) \in S^n_-$, $x = \gamma(p)$. We can also consider γ as a diffeomorphism between the smooth manifolds S^n_- and E^n with natural differential structures. Then the diffeomorphism γ^{-1}: $E^n \to S^n_-$ maps any point $x = (x_1, x_2, \ldots, x_n) \in E^n$ to the point

$$p = \gamma^{-1}(x) = \left(\frac{x_1}{q}, \frac{x_2}{q}, \ldots, \frac{x_n}{q}, -\frac{1}{q}\right), \tag{15.3}$$

where $q = (I + x_1^2 + x_2^2 + \cdots + x^2 n)^{1/2}$. We denote γ^{-1} by γ_1. The set $\overline{G}^* = \gamma_1(\overline{G})$ is a closed convex domain in S^n_-, where $\overline{G}$ is the closure of G and

$$\operatorname{dist}(\partial S^n_-, \overline{G}^*) = \delta_0 > 0 \tag{15.4}$$

in the intrinsic spherical meaning.

15.2 Dual Convex Hypersurfaces

Let $u(x)$ be any continuous nonpositive function in $\overline{G}$ satisfying condition $u|_{\partial G} = 0$.

The function $u(x)$ defines the new function $u^*(p)$ in $\overline{G}^*$ by the formula

$$u^*(p) = (I - p_1^2 - \cdots - p_n^2)^{1/2} u(\gamma(p)) \tag{15.5}$$

for $p = (p_1, \ldots, p_n, p_{n+1}) \in \overline{G}^* \subset S^n_-$, where $x = \gamma(p)$ (see (16.2)). Conversely if we define

$$\bar{u}^*(x) = u^*(\gamma_1(x)), \tag{15.6}$$

where $p = \gamma_1(x)$, then

$$u^*(x) = \frac{I}{(I + x_1^2 + \cdots + x_n^2)^{1/2}} u(x). \tag{15.7}$$

We denote by H and $\overline{H}$ an open convex subdomain of G and its closure and assume that $\operatorname{dist}(\overline{H}, \partial G) = h_H > 0$. Then $H^* = \gamma_1(H)$ and its closure $\overline{H}^* = \gamma_1(\overline{H})$ are respectively open and closed spherical convex domains and the intrinsic distance h_{H^*} between $\overline{H}^*$ and ∂G^* is positive. Clearly h_{H^*} depends only on h_H.

The inequality

$$(p, z) \leq u^*(p) \tag{15.8}$$

defines the closed half-space $U_p \subset \widetilde{R}^{n+1}$ for each fixed vector $p \in \overline{G}^*$ and any vector $z \in \widetilde{R}^{n+1}$, satisfying the inequality (15.8). The set

$$Q_H(u) = \bigcap_{p \in \overline{H}^*} U_p \tag{15.9}$$

is a closed infinite convex body in $\widetilde{R}^{n+1}$. The sets

$$K(\partial G^*) = \bigcap_{q \in \partial G^*} V_q \quad \text{and} \quad K(\overline{G}^*) = \bigcap_{q \in \overline{G}^*} V_q \tag{15.10}$$

are one and the same convex cone in $\widetilde{R}^{n+1}$ with vertex $\tilde{0}(0, 0, \ldots, 0)$, where V_q is the closed half-space

$$(q, z) \leq 0 \tag{15.11}$$

for any fixed $q \in \partial G^*$ (or $\overline{G}^*$) and any vector $z \in \widetilde{R}^{n+1}$.

Now the sets

$$P_H(u) = \partial Q_H(u) \tag{15.12}$$

and

$$L(\partial G^*) = \partial K(\partial G^*) \tag{15.13}$$

are complete infinite convex hypersurfaces in $\widetilde{R}^{n+1}$ and the latter is a convex n-dimensional cone with the vertex $\tilde{0}(0, 0, \ldots, 0)$.

Theorem 15.1. *Let $w(x)$ be the convex function spanned by $u(x) \in C_0^-(\overline{G})$ from below on the set $\overline{H}$. Then*

$$Q_H(u) = Q_H(w) \quad \text{and} \quad P_H(u) = P_H(w). \tag{15.14}$$

Moreover the convex body $Q_H(u)$ and the convex hypersurface $P_H(u)$ have one and the same supporting function $w^*(p)$ defined on $\overline{H}^*$.

Proof. From definition of the function $w(x)$ it follows that $w(x) \leq u(x) \leq 0$ for any $x \in \overline{H}$. Therefore $w^*(p) \leq u^*(p) \leq 0$ for any $p \in \overline{H}^*$. Thus $W_p \subset U_p$

for any $p \in \overline{H}^*$, where W_p and U_p correspondingly are the closed half-spaces $(p, z) \leq w^*(p)$ and $(p, z) \leq u^*(p)$ for every fixed vector $p \in \overline{H}^*$ and any vector $z \in \widetilde{R}^{n+1}$. Therefore

$$Q_H(w) = \bigcap_{p \in \overline{H}^*} W_p \subset \bigcap_{p \in \overline{H}^*} U_p = Q_H(u). \tag{15.15}$$

From the theory of convex bodies it is well known that if M is an infinite closed convex body and $v^*(p) < 0$, $p \in \overline{H}^* \subset S^n_-$, the supporting function of M, that the function

$$v(x) = \left(I + \sum_{i=1}^{n} x_i^2\right)^{1/2} v^*(\gamma(x))$$

is a negative convex function for $x \in \overline{H}$, where

$$\gamma(x) = \left(\frac{x_1}{q}, \frac{x_2}{q}, \ldots, \frac{x_n}{q}, -\frac{1}{q}\right) \in \overline{H}^*$$

and $q = \left(I + \sum_{i=1}^{n} x_i^2\right)^{1/2}$. Now we apply this fact to the case $M = Q_H(u)$. Let $v^*(p)$ be the supporting function of he convex body $Q_H(u)$. Then clearly $0 \geq u^*(p) \geq v^*(p)$ for any $p \in \overline{H}^*$. Therefore we obtain for negative convex function $v(x)$ the inequality $0 \geq u(x) \geq v(x)$ for any $x \in \overline{H}$. From the definition of the convex function $w(x)$ spanned on $u(x)$ from below on $\overline{H}$ it follows that $u(x) \geq w(x) \geq v(x)$ for any $x \in \overline{H}$. Repeating our reasoning for the functions $w(x)$ and $v(x)$ we obtain

$$Q_H(w) \supset Q_H(v) = Q_H(u). \tag{15.16}$$

From (15.16) it follows that $Q_H(u) = Q_H(w)$, and hence $P_H(u) = P_H(w)$. Theorem 15.1 is proved. □

Now we consider the new convex body

$$Q_G(w) = \bigcap_{p \in \overline{G}^*} W_p \tag{15.17}$$

for every function $w(x) \in W_H^+(\overline{G})$, where the closed half-space W_p was defined above in this section.

Theorem 15.2.

$$Q_G(w) = Q_H(w) \cap K(\partial G^*). \tag{15.18}$$

Proof. It follows from definitions of the sets $Q_G(w)$ and $Q_H(w)$ that

$$Q_G(w) = Q_H(w) \cap Q_{G \setminus H}(w), \tag{15.19}$$

where

$$Q_{G\backslash H}(w) = \bigcap_{p \in \overline{G}^* \backslash \overline{H}^*} W_p.$$

First of all we note that the asymptotic solid cone $K_H(w)$ to $Q_H(w)$ has the set $\overline{H}^*$ as a spherical image. If the vertex of $K_H(w)$ lies inside $Q_H(w)$, then the whole cone $K_H(w)$ lies inside $Q_H(w)$. Let $L_H(w)$ be the boundary of $K_H(w)$. We suppose that the vertex of $K_H(w)$ coincides with the nearest point of $P_H(w)$ to the origin $\tilde{0}$ of $\widetilde{R}^{n+1}$.

Then the set

$$\lambda(w) = L_H(w) \cap L(\partial G^*)$$

is the $(n-1)$-dimensional hypersurface homeomorphic to $(n-1)$-sphere. Recall that $L(\partial G^*) = \partial K(\partial G^*)$, where $K(\partial G^*)$ is the convex solid cone.

Clearly $\sup\limits_{Z \in \lambda(w)} \{\text{dist}\{\tilde{0}, z\}\}$ can be estimated above by means of $\|w(x)\|$, $h_H = \text{dist}\{\overline{H}, \partial G\}$ and $\delta_0 = \text{dist}(\partial S^n_-, \overline{G}^*)$.

If

$$\nu(w) = P_H(w) \cap L(\partial G^*) \tag{15.20}$$

then $\nu(w)$ is also homeomorphic to $(n-1)$-sphere and $\nu(w)$ lies between $\lambda(w)$ and the origin of $\widetilde{R}^{n+1}$. Thus

$$\sup_{Z \in \nu(w)} \{\text{dis}\{\tilde{0}, Z\}\} \tag{15.21}$$

can also be estimated by means of $\|w(x)\|$, h_H and δ_0.

Now all supporting hyperplanes to the graph S_W of the convex function $w(x)$ of the points $(x, w(x))$ will be singular if x belongs to $\overline{G} \backslash \overline{H}$ (see the proof of Theorem 13.1). Let a be such a supporting hyperplane: then $a \in S_w$ is some closed bounded convex k-dimensional body, where $1 \leq k \leq n-1$. We denote by $\pi_a \subset \overline{G}$ the closed k-dimensional convex body which is the projection of the set $a \cap S_w$. Then π_a determines the singular point Y on $P_H(w)$ with k-dimensional set of supporting hyperplanes to $P_H(w)$, because $\pi_a \cap \overline{H} \neq \emptyset$. The spherical image of this set of supporting hyperplanes coincides with $\gamma_1(\pi_a) \subset S^n_-$. (The definition of the mapping γ_1 is in § 15.) Since a passes through some point $(x_0, 0)$, where $x_0 \in G$, then Y_a belongs to the cone $L(\partial G^*)$ or more precisely $Y_a \in \nu(w)$ (see (15.20)).

Clearly $\nu(w) = U_a Y_a$, where a runs through the set of all supporting hyperplanes to S_w having contact points $(x, w(x))$ with S_w, where $x \in \overline{G} \backslash \overline{H}$.

Therefore from (15.17), (15.18), (15.19) and from the last considerations it follows that $Q_G(w) = Q_H(w) \cap K(\partial G^*)$. Theorem 15.2 is proved. □

Thus the convex hypersurface

$$P_G(w) = \partial Q_G(w) \tag{15.22}$$

consists of two parts: the first one $S_H(w)$ lies inside the solid cone $k(\partial G^*)$ and the second one $T_{\partial G}(w)$ lies on the boundary $L(\partial G^*)$ of the one $K(\partial G^*)$. Both hypersurfaces have one and the same boundary $\nu(w) \subset P_G(w)$. Let us agree to include $\nu(w)$ as a part of $S_H(w)$ and $T_{\partial G}(w)$ and consider both hypersurfaces as closed hypersurfaces with a boundary.

We will call $S_H(w)$ the dual convex hypersurface (with respect to $\overline{H} \subset G$) of the convex function $w(x) \in W_H^+(\overline{G})$. The function

$$w^*(p) = (I - p_1^2 - \cdots - p_n^2)^{1/2} w(\gamma(p)) \tag{15.23}$$

is the supporting function for $S_H(w)$ for any $p \in \overline{G}^*$.

15.3 Expression of the Functional $I_H(u)$ by Means of Dual Convex Hypersurfaces

Let $w(x)$ be any convex function belonging to $W_H^+(\overline{H})$ and $S_H(w)$ be its dual convex hypersurface. We denote by $\sigma(S_H(w), e')$ the surface function of $S_H(w)$ (see § 7.4). The surface function $\sigma(S_H(w), e')$ is defined as the completely additive nonnegative function on the ring of Borel's subsets e' of the domain $\overline{G}^* \subset S_-^n$ and the values of this function equal to the area of the sets $\tilde{e} \subset S_H(w)$ such that $\tilde{e}$ consists of all points of $S_H(w)$ having supporting hyperplanes with unit outside normals belonging to e'. From our considerations it follows that

$$\sigma(S_H(w), \overline{G}^* \backslash \overline{H}^*) = 0. \tag{15.24}$$

It is well known (see § 7.4) that $\sigma(S_H(w_k), e')$ converges weakly to $\sigma(S_H(w_0), e')$ if $\lim_{k\to\infty} \|w_k - w_0\| = 0$. Let $V_H(w)$ be the volume of the part of the convex cone $K(\partial G^*)$ situated under the dual convex hypersurface $S_H(w)$.

Theorem 15.3. *The equality*

$$V_H(w) = -\frac{1}{n+1} \int_{\overline{H}} w(x) \omega(w, de) \tag{15.25}$$

holds for every convex function $w(x) \in W_H^+(G)$.

Proof. If $w(x) \in W_H^+(\overline{G})$, then for the volume $V_H(w)$ there is the formula

$$V_H(w) = -\frac{1}{(n+1)} \int_{\overline{G}^*} w^*(p) \sigma(S_H(w), de') \tag{15.26}$$

(see § 7.4).

But the surface function $\sigma(S_H(w), e')$ has the representation

$$\sigma(S_H(w), e') = \int_e \left(1 + \sum_{i=1}^n x_i^2\right)^{1/2} \omega(w, de), \tag{15.27}$$

in the map γ (see § 15.1). Formula (15.27) can be proved first for convex polyhedrons and extend for all class $W_H^+(\overline{G})$ of convex functions by approximation of polyhedrons.*) From (15.26) and (15.27) it follows that

$$V_H(w) = -\frac{1}{(n+1)}\int_{\overline{G}^*} w^*(\gamma_1(x))\left(1+\sum_{i=1}^n x_i^2\right)^{1/2}\omega(w,de)$$
$$= -\frac{1}{(n+1)}\int_{\overline{G}} w(x)\omega(w,de) = -\frac{1}{(n+1)}\int_{\overline{H}} w(x)\omega(w,de),$$

because $\omega(w,\overline{G}\setminus\overline{H}) = 0$. Theorem 15.3 is proved. □

Remark. Since any convex polyhedron can be approximated by C^2 convex hypersurfaces (function) with everywhere strictly principal normal curvatures, it is sufficient to establish (15.27) only of such a class of hypersurfaces (functions).

From the Gauss theorem it follows that

$$\sigma(S_H(w),e') = \int_{e'} \frac{ds_p}{K(p)}, \tag{15.28}$$

where ds_p is the element of area on S_-^n and $K(p)$ is the Gauss curvature of $S_H(w)$ at the point of $S_H(w)$ with the outside unit normal p. We find

$$\int_{e'} \frac{ds_p}{K(p)} = \int_e \left(I+\sum_{i=1}^n x_i^2\right)^{1/2} \det(w_{x_i x_j})dx, \tag{15.29}$$

where $e' = \gamma^{-1}(e)$ (see § 15.1 and also §§ 5, 6, 7). From (15.28) and (15.29) we obtain (15.27) for the C^2 convex functions (hypersurfaces) with strictly positive principal normal curvatures.

Theorem 15.4. *The functional $I_H(u)$ on $\overline{C}_0(\overline{G})$ has the representation*

$$I_H(u) = (n+1)\left[V_H(F_H(u)) + \int_{H^*} u^*(p)\psi_H^*(de)\right] \tag{15.30}$$
$$= -\int_{H_u^*} u^*(p)\sigma(S_H(w),de') + (n+1)\int_{H^*} u^*(p)\psi_H^*(de'),$$

where $w(x) = F_H(u(x))$ is the convex function spanned by $u(x)$ from below on $\overline{H} \subset G$; $\psi_H(e')$ is the nonnegative completely additive function of Borel subsets e' of G^* determined by the formula

$$\psi_H^*(e') = \int_{\gamma(e)} \left(I+\sum_{i=1}^n x_i^2\right)^{1/2} \psi_H(de); \tag{15.31}$$

*) Of course we use the weak convergence of surface functions.

$H_u^* = \gamma_1(H_u)$ and H_u is the closed subset of $\overline{H}$ where $u(x) = w(x)$.

Proof. From the definition of $I_H(u)$ it follows that

$$I_H(u) = -\int_G u\omega(w, de) + (n+1)\int_G u\psi_H(de). \tag{15.32}$$

Now

$$\int_G u(x)\psi_H(de) = \int_H u^*(x)\left(\ell + \sum_{i=1}^n x_i^2\right)^{1/2} \psi_H(de)$$

since $\psi_H(G \backslash H) = 0$. Therefore

$$\int_G u(x)\psi_H(de) = \int_{H^*} u^*(p)\psi_H^*(de') \tag{15.33}$$

if we use (15.31).

From Theorem 15.4 and Theorem 13.1 we obtain

$$\begin{aligned}(n+1)V_H(F_H(u)) &= -\int_{\overline{H}} w(x)\omega(w, dx) \\ &= -\int_{H_u} w(x)\omega(w, de) = -\int_{H_u} u(x)\omega(w, de) \\ &= -\int_{H_u^*} u^*(p)\sigma(S_H(w), de').\end{aligned} \tag{15.34}$$

From Theorem 13.1 we obtain

$$-\int_G u(x)\omega(w, de) = -\int_{H_u} u(x)\omega(de). \tag{15.35}$$

Now it follows from (15.34) and (15.35) that

$$\begin{aligned}-\int_G u(x)\omega(w, de) &= (n+1)V_H(F_H(u)) \\ &= \int_{H_u^*} u^*(p)\sigma(S_H(w), de').\end{aligned} \tag{15.36}$$

Thus from (15.32), (15.33) and (15.36) we obtain (15.30).
Theorem 15.4 is proved. □

15.4 Expression of the Variation of $I_H(u)$

First of all we study the variation of the functional

$$\Phi_H(u) = -\int_G u(x)\omega(w, de) \tag{15.37}$$

where $u(x) \in C_0^-(\overline{G})$ and $w(x)$ is the convex function spanned by $u(x)$ from below on the convex closed domain $\overline{H} \subset G$. From Theorem 15.4 it follows that

$$\Phi_H(u) = (n+1)V_H(w) = -\int_{H_u^*} w^*(p)\sigma(S_H(w), de') \qquad (15.38)$$

$$= -\int_{H_u^*} u^*(p)\sigma(S_H(w), de')$$

where H_u^* is a closed subset of $\overline{H}^* = \gamma_1(\overline{H})$, where $w^*(p) = u^*(p)$ and $V_H(w)$ is the volume of the part of the convex cone $K(\partial G^*)$ situated under the dual convex hypersurface $S_H(w)$.

Now we want to complement $S_H(w)$ to the whole closed convex hypersurface. The boundary $\nu(w)$ of $S_H(w)$ lies on the conic convex hypersurface $L(\partial G^*) = \partial\{K(\partial G^*)\}$ and homeomorphic to $(n-1)$-sphere. We can evidently find two numbers m_1 and m_2 depending on $\|w\|$, $\operatorname{dist}\{\overline{H}, \partial G\} = h_H > 0$ and $\operatorname{dist}\{\overline{G}^*, \partial S_-^n\}$ such that

$$0 < m_1 \le \operatorname{dist}\{\tilde{0}, \nu(w)\} \le m_2 < +\infty. \qquad (15.39)$$

We denote by $S_+^n(r)$ the hemisphere

$$p_1^2 + p_2^2 + \cdots + p_{n+1}^2 = r^2, \qquad p_{n+1} \ge 0 \qquad (15.40)$$

and by $U_+^n(r)$ the set

$$p_1^2 + p_2^2 + \cdots + p_{n+1}^2 \le r^2, \qquad p_{n+1} \ge 0. \qquad (15.41)$$

We'll only consider the functions $u(x) \in C_0^-(\overline{G})$ such that for the convex functions $w(x) = F_H(u(x))$ spanned on $u(x)$ from below the following inequalities

$$m_0 \le \|w(x)\| \le M_0 \qquad (15.42)$$

hold on the set $\overline{H}$.

Then there exist the common numbers $0 < m_1 < m_2 < +\infty$ such that for all functions $u(x) \in C_0^-(\overline{G})$ the following inequalities

$$0 < m_1 \le \operatorname{dist}\{\tilde{0}, \nu(w)\} \le m_2 < +\infty \qquad (15.43)$$

hold, if (15.42) are fulfilled $w(x) = F_H(u(x))$.

Thus we will be able to construct all the bounded convex bodies $\Pi_H(w)$.

Now consider the supporting function of $\Pi_H(x)$. We denote this function by $h_H(p)$ where p runs the whole unit sphere S^n: $p_1^2 + p_2^2 + \cdots + p_n^2 + p_{n+1}^2 = 1$.

The closed convex hypersurfaces $\Lambda_H(w)$ has at least two ribs $\nu(w)$ and $\nu(m_1 + 2m_2)$ which are the boundaries for three domains $S_H(w)$, $Z(m_1 + 2m_2)$ and $T(w, U_+^n(m_1 + 2m_2)) \subset L(\partial G^*)$.

Therefore

$$h_H^*(p) = \begin{cases} w^*(p) < 0 & \text{if } p \in G^* \\ 0 & \text{if } p \in \partial G^* \\ \begin{array}{c}\text{takes}\\ \text{positive}\\ \text{values}\end{array} & \text{if } p \in S^n \setminus \left(\overline{G}^* \cup \left\{\frac{1}{m_1+2m_2} Z(m_1+2m_2)\right\}\right) \\ m_1+2m_2 & \text{if } p \in \frac{1}{m_1+2m_2} Z(m_1+2m_2)) \subset S_+^n. \end{cases}$$

Note that $\sigma\left(\Lambda_H(w), S^n \setminus (\overline{G}^* \cup \frac{1}{m_1+2m_2} Z(m_1+2m_2))\right) = 0$.

Therefore the volume of $\Pi_H(w)$ can be found by the formula

$$\begin{aligned} V(\Pi_H(w)) &= \frac{1}{n+1} \int_{S^n} h_H^*(p) \sigma(\Lambda_H(w), de') \\ &= \frac{\sigma(K(\partial G^*))}{n+1} (m_1+2m_2)^{n+1} \\ &\quad + \frac{1}{n+1} \int_{\overline{G}^*} w^*(p) \sigma(S_H(w), de) \\ &= \frac{\sigma(K(\partial G^*))}{n+1} - \frac{1}{n+1} \int_{H_u^*} u^*(p) \sigma(S_H(w), de), \end{aligned} \tag{15.44}$$

where $\sigma(K(\partial G^*))$ is the solid angle of the convex cone $K(\partial G^*)$.

Thus

$$V(\Pi_H(w)) = \frac{\sigma(K(\partial G^*))}{n+1} + \frac{1}{n=1} \phi(u). \tag{15.45}$$

If we change the point of the reference of distances with the sign to supporting hyperplanes to any convex body, then the supporting function of this body changes its values. If such a point p_0 coincides with the inner point of $\Pi_H(w)$, then the supporting function takes only positive values and is some strictly positive function on S^n.

Minkowski [2], Alexandrov [3], Fenchel and Jessen [1] investigated the variation of the volume in the class of bounded convex bodies and established the formulas for the weak differential (the first variation) under different conditions (see also Bonnesen and Fenchel [1], and Busemann [1]). The main methods and techniques of these investigations were the theory of Minkowski mixed volumes and the Brunn–Minkowski inequality.

Alexandrov proved that if $h_0(p)$ is any strictly positive continuous function on the unit sphere S^n: $|p| = 1$ and H_0 is the closed convex body defined by intersection of all the halfspaces.

$$(p, z) \leq h_0(p), \qquad p \in S^n, \tag{15.46}$$

then

$$\lim_{t \to \infty} \frac{V(H_t) - V(H)}{t} = \int_{S^n} \eta(p) \sigma(H_0, de'), \tag{15.47}$$

where $\eta(p)$ is any continuous function on S^n, t is a real parameter converging to zero, $\sigma(H_0, e')$ is the surface function of H_0 and H_t is the closed bounded convex body defined by intersection of all the halfspaces

$$(p, z) \leq h_0(p) + t\eta(p).$$

Remark. Since $h_0(p)$ and $\eta(p)$ are continuous on S^n and $h_0(p)$ is strictly positive, $h_0(t) + t\eta(p)$ is also positive for sufficiently small t and the bodies H_t will be constructable.

Since 1) all terms of (15.46) are independent on the point of reference to the supporting hyperplanes and 2) it is possible to take any function $\eta \neq 0$ only on any closed set $\overline{H}_1 \subset H$, from (15.46), (15.45) and Theorem 15.4 it follows that

$$\lim_{t \to 0} \frac{I_H(u + t\eta) - I_H(u)}{t} = (n+1)\left[\int_{H_u} \eta[-\omega(w, de) + \psi_H(de)]\right]. \tag{15.48}$$

where $w = F_H(u)$.

From (15.48) and Theorem 15.3 it follows that the function

$$w_0(x) \in U(H, m_0, M_0) \subset W_H^+(\overline{G}),$$

realizing the absolute minimum of $I_H(u)$ in $C_0^-(G)$, is a generalized solution of the Dirichlet problem

$$\omega(w, e) = \psi_H(e) \tag{15.49}$$

$$w|_{\partial G} = 0. \tag{15.50}$$

Since the Dirichlet problem (15.49–50) has only one generalized solution and this solution belongs to $W_H^+(\overline{G})$ (see Theorem 11.4), there exists only one function realizing the absolute minimum of the functional $I_H(u)$ in the set $C_0^-(\overline{G})$ and this function belongs to $W_H^+(\overline{G})$.

Thus the proof of the following theorem is complete.

Theorem 15.5 (The Main Theorem for the Functional $I_H(u)$). *There exists only one convex function realizing the absolute minimum of the functional $I_H(u)$ in the set $C_0^-(\overline{G})$. This function $w_H(w) \in W_H^+(\overline{G})$ and $w_H(x)$ is the generalized solution of the Dirichlet problem (15.49–50).*

Chapter 5. Non-Compact Problems for Elliptic Solutions of Monge–Ampere Equations.

In this chapter we are concerned with elliptic generalized solutions of Monge–Ampere equations

$$\det(u_{ij}) = f(x, u, Du) \qquad (*)$$

in the entire n-dimensional Euclidean space E^n. Generalized elliptic solutions of equation $(*)$ are convex or concave functions and their graphs are complete convex hypersurfaces in the space $E^{n+1} = E^n \times R$. These hypersurfaces project one-to-one on E^n. Clearly it is sufficient to investigate only convex generalized solutions of equation $(*)$ and confine oneself only to nonnegative functions $f(x, u, p)$ for all $x \in E^n$, $u \in R$, $p \in R^n$. As we know, any convex generalized solution $u(x)$ of equation $(*)$ satisfies this equation almost everywhere in any compact subset of E^n and the set function

$$\omega(\ell, u, e) = \text{meas } \chi_u(e),$$

generated by $u(x)$, is absolutely continuous on the family of Borel subsets of E^n.

In Chapter 5 the main attention will be given to the boundary value problem for equation $(*)$ whose desired solutions have a prescribed asymptotic cone.

§16. Introduction. The Statement of the Second Boundary Value Problem

16.1 Asymptotic Cone of Infinite Complete Convex Hypersurfaces

We recall a few fundamental concepts from the theory of convex bodies (see Chapter 1, § 1). Let H be an infinite convex body in E^{n+1} and $S = \partial H$ be an infinite complete convex hypersurface. We shall consider only the case when S projects one-to-one on the hyperplane E^n. Let 0 be any point of H. The set of all rays starting from the point 0 and lying inside H forms some solid convex cone K with the vertex 0. The cone K does not depend on the choice of the vertex 0 in the body H. This cone can have dimensions $1, 2, 3, \ldots, n+1$. If we consider the cone K to within of the choice of its vertex in H, then K is called the *asymptotic cone* of the hypersurface $S = \partial H$.

The asymptotic cone K is called ***non-degenerate*** if ∂K projects one-to-one on the entire hyperplane E^n. For non-degenerate asymptotic cones it is convenient to replace them by their boundary. In this case the notation K will be used for the boundaries of a solid asymptotic cone and we will say that K is the asymptotic cone of the function $u(x)$.

Now we consider a few simple properties of the normal mapping of convex functions.

Property 1. Let $u(x)$ be any convex function; then the function

$$v(x) = u(x) + \sum_{i=1}^{n} a_i x_i + b,$$

where $a_1, a_2, \dots, a_n, b$ are constants, is convex and the set $\chi_v(e)$ can be obtained from $\chi_u(e)$ by parallel translation of the gradient space R^n along the vector $a = (a_1, a_2, \dots, a_n)$.

Property 2. The normal image of any convex cone K is a closed convex set in R^n, whose dimension can take the values $0, 1, 2, \dots, n$.

Property 3. If K is a non-degenerate convex cone, then $\chi_K(E^n)$ is a bounded closed n-dimensional convex domain with interior points.

The following remarks follow from Properties 1,2,3.

It is sufficient to consider only convex functions $u\colon E^n \to R$, whose normal images contain the origin of P^n. If the asymptotic cone of such a function is non-degenerate, then we can additionally assume that the origin θ' of R^n is an interior point of $\chi_K(E^n)$. If we also assume that the vertex of this non-degenerate convex cone K is the origin θ of E^n, then the equation of K is as follows

$$z = k(x),$$

where $k(x)$ is a nonnegative convex function in E^n and $k(x) = 0$ only at the point θ. Since any point of E^n can be taken as the origin, it is sufficient to consider asymptotic non-degenerate convex cones, whose vertices are projected onto the point $\theta \in E^n$. The equations of such cones are

$$z = k(x) + b,$$

where b is any constant.

16.2 The Statement of the Second Boundary Value Problem

We consider only convex generalized solutions $u(x)$ of the Monge–Ampere equation

$$\det(u_{ij}) = f(x, u, Du) \tag{16.1}$$

in the entire space E^n. Such solutions satisfy equation (16.1) almost everywhere in E^n and the set functions $\omega(\ell, u, e)$ are absolutely continuous. Note that

$$\omega(\ell, u, e) = \text{meas}\ \chi_u(e) \tag{16.2}$$

for any Borel subset e of E^n. Below we assume that $f(x,u,p)$ is continuous and nonnegative in $E^n \times R \times R^n$.

The statement of the second boundary value problem is as follows. Let K be a non-degenerate convex cone. Find broad sufficient conditions for equation (16.1) to have at least one generalized solution for which K is the asymptotic cone. First of all we present the solution of this problem for the special class of Monge–Ampere equations

$$\det(u_{ij}) = \frac{g(x)}{R(Du)}, \tag{16.3}$$

where we establish the necessary and sufficient conditions of solvability for the second boundary value problem. This necessary and sufficient condition for the second boundary value problem for equation (16.3) to have a generalized solution is as follows:

$$\int_{E^n} g(x)dx = \int_{\chi_k(R^n)} R(p)dp, \tag{16.4}$$

where $g(x) \geq 0, g(x) \in L(E^n)$, $R(p) > 0$, $R(p) \in L_{loc}(R^n)$ and $\chi_k(R^n)$ is the normal image of a prescribed non-degenerate convex asymptotic cone K. Note that the desired solution is defined to within an additive constant. This uniqueness theorem corresponds to the differential structure of equation (16.3).

The solution of the second boundary value problem for the general Monge–Ampere equation

$$\det(u_{ij}) = f(x,u,Du)$$

is essentially more difficult because the simple necessary and sufficient condition (16.4) for solvability of the same problem for equation (16.3) must be replaced by the complicated implicit necessary condition

$$\text{meas } \chi_k(P^n) = \int_{E^n} f(x,u(x),Du(x))dx. \tag{16.5}$$

Moreover, the application of fixed-point theorems offers difficulties because the set E^n is not compact. The proof of the existence theorem for the second boundary value problem, presented in this chapter, is based on the construction of a new Monge–Ampere equation in some specially introduced Banach space. Significantly, this construction is based upon asymptotic properties of the function $f(x,u,p)$ and its first derivatives as $|x| \to +\infty$ and $|u| \to +\infty$.

The second boundary value problem was studied by Alexandrov [8]; it is related to the special geometric problem of reconstruction of a general complete convex hypersurface with the prescribed area of Gaussian mapping. The existence and uniqueness theorem relating to generalized solutions of the second boundary value problem for Monge–Ampere equation (16.3) was established by Bakelman [2]. The particular case $R(p) = [1+(Du)^2]^{-(n+1)/2}$ corresponds to Alexandrov's Theorem.

The existence theorem for generalized solutions of the second boundary value problem for general Monge–Ampere equation (16.1) was established by Bakelman [16]. The presentation in Chapter 5 follows this paper.

§17. The Second Boundary Value Problem for Monge–Ampere Equations $\det(u_{ij}) = \frac{g(x)}{R(Du)}$

In §17 we establish an existence and uniqueness theorem for the second boundary value problem for the Monge–Ampere equations

$$\det(u_{ij}) = \frac{g(x)}{R(Du)}, \tag{17.1}$$

in the class of convex generalized solutions. The conditions of this theorem are necessary and sufficient. These results will be significantly used for the investigations of solvability of the second boundary value problem for general Monge–Ampere equations

$$\det(u_{ij}) = f(x, u, Du)$$

(see Chapter 5, §18).

17.1 The Necessary and Sufficient Conditions of Solvability of the Second Boundary Value Problem

Let $g(x) \geq 0$ be a summable function in E^n and let $R(p) > 0$ be a locally summable function in R^n. We denote by $E^{n+1} = E^n \times R$ the Euclidean space with Cartesian coordinates $x_1, x_2, \ldots, x_n, z$. We recall that E^n is an n-dimensional Euclidean space and $x_1, x_2, \ldots, x_n$ are Cartesian coordinates in E^n. Thus E^n is the hyperplane $z = 0$ in E^{n+1}. We shall use our traditional notation: x is a point of E^n, (x, z) is a point of E^{n+1} and $p = (p_1, p_2, \ldots, p_n)$ is a point of the gradient space R^n, where $p_1, p_2, \ldots, p_n$ are Cartesian coordinates in R^n.

Theorem 17.1. *Let K be a non-degenerate convex cone in E^{n+1} and let $z = k(x)$, $x \in E^n$, be the equation of K. Let*

$$\int_{E^n} g(x)dx = \int_{\chi_k(E^n)} R(p)dp. \tag{17.2}$$

Then the second boundary value problem for equation (17.1) has a generalized solution $u(x)$ and this solution is unique to within an additive constant.

Proof. First we extend the equation (17.1) to the equation

$$\omega(R, u, e) = \mu(e) \tag{17.3}$$

which is considered in the set of all general convex functions defined in R^n. In (17.3), the set function $\mu(e)$ is nonnegative, completely additive and satisfies the condition

$$\mu(E^n) = \int_{\chi_k(E^n)} R(p)dp. \tag{17.4}$$

According to § 11, solutions of equation (17.3) are called weak solutions of equation (17.1). If the set function $\mu(e)$ is absolutely continuous, i.e.

$$\mu(e) = \int_e g(x)dx,$$

then the R-curvature of weak solutions is also absolutely continuous and weak solutions turn out to be generalized solutions of equation (17.1).

The proof of the statement of uniqueness to within an additive constant for generalized solutions made in Theorem 17.1 is a simple corollary from Lemmas 10.1–2 (see § 10).

The proof of the existence statement for generalized solutions made in Theorem 17.1 can be proved by the approximation of convex polyhedra. If it is possible to impose the additional assumption that all constructed polyhedra pass through one and the same point in E^{n+1}. The normal images of all these polyhedra are uniformly bounded in the gradient space R^n. Hence all convex functions, whose graphs are convex polyhedra considered above, form a family of convex functions satisfying the Lipschitz condition of the first order with the common Lipschitz constant. This family of functions is compact in the space of continuous functions. Moreover R-curvatures of convergent sequences of convex functions converge weakly. From all these facts it follows that the statement of existence made in Theorem 17.1 can be proved, if this statement is proved for convex polyhedra.

17.2 The Second Boundary Value Problem in the Class of Convex Polyhedra

Let K be a polyhedral non-degenerate convex angle in E^{n+1} and let $a_1, a_2, \ldots, a_m$ be a system of prescribed fixed points in E^{n+1}. We consider the convex polyhedra whose vertices are projected only in the points $a_1, a_2, \ldots, a_m$ and whose asymptotic cones coincide with the prescribed convex polyhedral angle K introduced above. According to the Properties 1, 2, 3 (see Subsection 16.1) we can assume without losing generality that the origin θ' of R^n is an inner point of the set $\chi_K(E^n) \subset R^n$. Of course, we assume that some vertices of these polyhedra can degenerate. Then the measure of the normal images of such vertices together with values of R-curvature vanish. We denote by $W^+(a_1, a_2, \ldots, a_m, K)$ the set of all these convex polyhedra. The set $W^+(a_1, a_2, \ldots, a_m, K)$ is nonempty, because the convex polyhedral angle K with the vertex a_1 is an element of $W^+(a_1, a_2, \ldots, a_m, K)$.

Now let the vertices of a convex polyhedron P, $W^+(a_1, a_2, \ldots, a_m, K)$ be the points (a_1, A_1), $(a_2, A_2), \ldots, (a_m, A_m)$. Then P can be obtained as the boundary of the convex hull spanned on the vertices of P and the polyhedral angle K with the vertex (a_1, A_1) (see Theorem 4.6; Subsection 4.2; Chapter 1).

Now associate the number $\mu_i \geq 0$ with every point a_i, $i = 1, 2, \ldots, m$ and consider the following set function

$$\mu(e) = \sum_{a_i \in e} \mu_i \tag{17.5}$$

for any Borel subset e of E^n. If $P \in W^+(a_1, a_2, \ldots, a_m, K)$ and $z = u(x)$ is the equation of P, then

$$\omega(R, u, e) = \sum_{a_i \in e} \omega(R, u, a_i),$$

where e is again any Borel subset of E^n. Thus, the right parts of equations (17.3) can only be the set functions $\mu(e)$ which are constructed by means of formula (17.5) if we consider equation (17.3) in the class of convex polyhedra $W^+(a_1, a_2, \ldots, a_m, K)$.

Thus the statement of the second boundary value problem for equation (17.3) can be reformulated in the following way: prove the existence of convex polyhedra $P \in W^+(a_1, a_2, \ldots, a_m, K)$ such that

$$\omega(R, P, a_i) = \mu_i \tag{17.6}$$

$i = 1, 2, \ldots, m$, where $\mu_1 \geq 0$, $\mu_2 \geq 0, \ldots, \mu_m \geq 0$ are prescribed numbers.

Theorem 17.2. *If the numbers $\mu_1, \mu_2, \ldots, \mu_m$ are nonnegative and the equality*

$$\sum_{i=1}^{n} \mu_i = \int_{\chi_k(E^n)} R(p)dp \tag{17.7}$$

holds, then the second boundary value problem has solutions in the class of convex polyhedra $W^+(a_1, a_2, \ldots, a_m, K)$. Moreover if $u(x)$ is one of these solutions, then all others can be written in the form

$$v(x) = u(x) + C, \tag{17.8}$$

where C is an arbitrary constant.

Proof. First we establish the existence of a solution $u(x)$ satisfying the additional condition

$$u(a_1) = A_1,$$

where A_1 is an arbitrary real number. Let T be the set of convex polyhedra P; $z = u(x)$*, satisfying the following conditions:

*We do not make the distinction between a convex piecewise linear function $u(x)$ and its graph P.

a) T is a subset of $W^+(a_1, a_2, \ldots, a_m; K)$;
b) if $u(x) \in T$, then the inequalities

$$0 \le \omega(R, u, a_i) \le \mu_i \tag{17.9}$$

hold for all $i = 2, 3, \ldots, m$ and

$$\omega(R, u, a_1) = \int_{\chi_k(E^n)} R(p)dp - \sum_{i=2}^{m} \omega(R, u, a_i); \tag{17.10}$$

c)

$$u(a_1) = A_1. \tag{17.11}$$

The set T is not empty, because the convex cone K with the vertex (a_1, A_1) belongs to T.

The system of real numbers

$$\xi_1 = u(a_1), \xi_2 = u(a_2), \ldots, \xi_m = u(a_m) \tag{17.12}$$

taken from every convex polyhedron $u(x) \in W^+(a_1, a_2, \ldots, a_m, K)$, defines this polyhedron one-to-one. The metric

$$d(u(x), v(x)) = \left\{ \sum_{i=1}^{m} (u(a_i) - v(a_i))^2 \right\}^{1/2} \tag{17.13}$$

introduced in $W^+(a_1, a_2, \ldots, a_m, K)$ shows that the mapping (17.12) isometrically maps the set $W^+(a_1, a_2, \ldots, a_m, K)$ on some closed convex subset of the Euclidean space $R^m = \{\xi = (\xi_1, \xi_2, \ldots, \xi_m)\}$. Clearly $T \subset W^+(a_1, a_2, \ldots, a_m, K)$. Since $W^+(a_1, a_2, \ldots, a_m, K)$ is identified with a corresponding closed subset of R^m, then T is also a closed set in R^m. From (17.9–11) it follows that T is a bounded set in R^m. Thus T is a compact set in R^m.

The function f: $T \to R$ acting by the formula

$$f(u) = \sum_{i=1}^{m} u(a_i) \tag{17.14}$$

is continuous at any convex polyhedron $u(x) \in T$. Since T is a compact subset of R^m, then

$$\inf_T f(u) = f_0 > -\infty$$

and there exists the polyhedron $u_0(x) \in T$ such that

$$f(u_0(x)) = f_0. \tag{17.15}$$

Now we prove that $u_0(x)$ is the desired solution of the second boundary value problem, satisfying the condition (17.11):

$$u_0(a_1) = A_1.$$

If our assertion is incorrect, then at least at one point a_k, $(k = 2, 3, \dots, m)$

$$\omega(R, u_0, a_k) < \mu_k \tag{17.16}$$

since

$$\omega(R, u_0, a_k) \leq \mu_k$$

from the definition of the set T. Now consider the convex polyhedron

$$\widetilde{P}\colon\ z = \tilde{u}(x) \in W^+(a_1, a_2, \dots, a_k, K)$$

such that

$$\begin{aligned}\tilde{u}(a_1) = A_1, \tilde{u}(a_2) = u_0(a_2), \dots, \tilde{u}(a_{k-1})\\ = u_0(a_{k-1}), \tilde{u}(a_k) = u_0(a_k) - \varepsilon,\\ \tilde{u}(a_{k+1}) = u_0(a_{k+1}), \dots, \tilde{u}(a_m) = u_0(a_m),\end{aligned}$$

where ε is a sufficiently small number such that

$$\omega(R, \tilde{u}, a_k) < \mu_k. \tag{17.17}$$

Since

$$\sum_{i=1}^{m} \omega(R, \tilde{u}, a_i) = \int_{\chi_k(R^n)} R(p)dp$$

and

$$\omega(R, \tilde{u}, a_s) \leq \omega(R, u_0, a_s)$$

for $s = 1, 2, \dots, k-1, k+1, \dots, m$, then from (17.17–18) it follows that $\tilde{u}(x) \in T$. But

$$f(\tilde{u}(x)) = \sum_{i=1}^{m} \tilde{u}(a_i) = \left[\sum_{i=1}^{m} u_0(a_i)\right] - \delta = f_0 - \varepsilon < f_0.$$

The last inequality is incompatible with equality (17.15). Hence $u_0(x)$ is the desired solution of the second boundary value problem for the equation (17.6).

Now let convex polyhedra P_1 and P_2 be the graphs of convex solutions $u_1(x)$, $u_2(x) \in W^+(a_1, a_2, \dots, a_m; K)$ of the equation (17.6), satisfying the additional condition $u_1(a_1) = u_2(a_1) = A_1$. If $u_1(x)$ and $u_2(x)$ are different functions, then there exist at least one point a_j, where j is one of integers $1, 2, \dots, m$, such that the polyhedral angles $V_1 \subset P_1$ and $V_2 \subset P_2$ projected in the point a_j and $V_1 \supset V_2$ and $V_1 \backslash V_2 \neq \emptyset$. Hence

$$\text{mes}\ \chi_{u_1}(aj) > \text{mes}\ \chi_{u_2}(a_j).$$

Thus

$$\mu_j = \int_{\chi_{u_1}(a_j)} R(p)dp > \int_{\chi_{u_2}(a_j)} R(p)dp = \mu_j$$

and our assumption is incorrect. The proof of Theorem 17.2 is complete.

§18. The Second Boundary Value Problem for General Monge–Ampere Equations

In this section we investigate the second boundary value problem for the general Monge–Ampere equation

$$\det(u_{ij}) = f(x, u, Du) \tag{18.1}$$

in the class of convex generalized solutions. We divide the presentation into a few subsections.

18.1 The Main Assumptions

Assumption 18.1. Admissible Convex Cones. *) A non-degenerate convex cone in E^{n+1} is called admissible if the equation of K is

$$z = k(x), \tag{18.2}$$

where $k(x)$ is a continuous, convex function in E^n, homogeneous of order 1, satisfying the conditions

a) $k(\theta) = 0$,
b) $k(\theta) > 0$ for all $x \neq \theta$ in E^n.

It is sufficient to consider only admissible convex cones (see the final part of Subsection 16.1). We denote by K^* the normal image of a convex cone K, i.e.

$$K^* = \chi_k(E^n). \tag{18.3}$$

For every admissible convex cone K the set K^* is closed, bounded and convex in the gradient space R^n and the point θ' is an inner point of K^*, where θ' is the origin of R^n.

Assumption 18.2. The Properties of the Function $f(x,u,p)$ and its Derivatives. The function $f(x,u,p)$ is continuous in $E^n \times R \times R^n$ together with its derivatives $\frac{\partial f}{\partial u}$ and $\frac{\partial f}{\partial p_i}$, $(i = 1, 2, \ldots, n)$ and

$$f(x, u, p) > 0, \tag{18.4}$$

$$\frac{\partial f(x,u,p)}{\partial u} > 0 \tag{18.5}$$

---

*) Below we use notations A.1, A.2 and A.3 instead of Assumptions 18.1, 18.2 and 18.3.

if $E^n \times R \times R^n$.

We also assume that the inequalities

$$\left|\frac{\partial f}{\partial u}\right| \leq \frac{C_0}{|x|^{n+2+\alpha}}, \tag{18.6}$$

$$\left|\frac{\partial f}{\partial u}\right| \leq \frac{C_1}{|x|^{n+\alpha}} \tag{18.7}$$

hold for all $(x, u, p) \in E^n \times R \times K^*$ with $|x| \geq m_0$, where $\alpha = \text{const} > 0$, $C_0 = \text{const} \geq 0$, $m_0 = \text{const} \geq 1$.

Assumption 18.3. Estimators and Their Properties. For every admissible convex cone K there are two functions $\lambda_K(x, u)$ and $\phi_K(x, u)$, depending only on the prescribed cone K, such that:

a) $\lambda_K(x, u)$ and $\phi_K(x, u)$ are positive and continuous in $E^n \times R$ and increase with respect to u for every fixed $x \in E^n$;

b) the inequalities

$$\lambda_K(x, u) \leq f(x, u, p) \leq \phi_K(x, u) \tag{18.8}$$

hold for all $x \in E_n, u \in R, p \in K^*$.

18.2 The Statement of the Main Theorem and the Scheme of its Proof

Theorem 18.1. *Let K be an admissible convex cone described in Assumption 1 and let $\lambda_K(x, u)$, $\phi_K(x, u)$ be estimators satisfying Assumption 2. If there are two numbers*

$$-\infty < a_k < b_k < +\infty$$

such that

$$\text{a)} \qquad \int_{E^n} \phi_K(x, k(x) + q)dx < +\infty \tag{18.9}$$

for all $q \in [a_k, b_k]$;

$$\text{b)} \qquad \int_{E^n} \phi_K(x, k(x) + a_k)dx < \text{mes } K^*; \tag{18.10}$$

$$\text{c)} \qquad \inf_{\gamma \in K^*} \int_{E^n} \lambda_K(x, (x, \gamma) + b_k)dx > \text{mes } K^*, \tag{18.11}$$

where $z = k(x)$ is the equation of the cone K and $(x, \gamma) = \sum_{i=1}^{n} \gamma_i x_i$, $x = (x_1, x_2, \ldots, x_n) \in E^n$, $\gamma = (\gamma_1, \gamma_2, \ldots, \gamma_n) \in K^*$, then equation (18.1) has at least one generalized convex solution $u(x)$ with asymptotic cone K, and

$$a_k < u(\theta) < b_k.$$

Scheme of the Proof. Since we consider unbounded convex functions on the entire space E^n, application of fixed points principles requires us to construct a special function space in which the second boundary value problem can be investigated. In this special function space we study some modification of the Monge–Ampere operator $\det(u_{ij})$ and its inverse, which is induced by the original equation (18.1) and by prescribed admissible convex cone K. The final part of the proof for Theorem 18.1 is based on an application of the Schauder principle to the inverse of the modified Monge–Ampere operator.

18.3 The Function Space of the Second Boundary Value Problem

We denote by $C^0(E^n)$ the set of all continuous functions u: $E^n \to R$ and by $C^{0,1}(E^n)$ the subset of $C^0(E^n)$ consisting of all Lipschitz functions u: $E^n \to R$, i.e. $u(x) \in C^{0,1}(E^n)$ if and only if

$$L(u) = \sup_{x,y \in E^n} \frac{|u(y) - u(x)|}{|y - x|} < +\infty, \tag{18.12}$$

where $|y - x| = \operatorname{dist}(x, y)$ in E^n.

Let $U_1 \subset U_2 \subset \cdots \subset U_m \subset \cdots$ be the sequence of n-balls;

$$U_m\colon\ |x| \leq m, \tag{18.13}$$

$m = 1, 2, 3, \ldots$. We denote by $\|u\|_m$ the number $\sup\limits_{U_m} |u(x)|$ and by $\|u\|_A$ the number

$$\|u\|_A = |u(\theta)| + \sum_{m=1}^{\infty} \frac{\|u\|_m}{m^{2+\alpha}}, \tag{18.14}$$

where $\alpha = \text{const} > 0$ (see Assumption 2).

Let A be the subset of $C^0(E^n)$ consisting in all functions $u(x)$ such that

$$\|u\|_A < +\infty.$$

Clearly

$$\|u\|_A \leq |u(0)| + L(u) \sum_{m=1}^{\infty} \frac{1}{m^{1+\alpha}} < +\infty$$

for every $u(x) \in C^{0,1}(E^n)$. On the other hand $\|e^{x_1 + \cdots + x_n}\|_A = +\infty$. Thus A is a proper non-empty subset of $C^0(E^n)$.

The functional $\|u\|_A$ is a form of the set A. Clearly A is a Banach space with respect to this norm. We denote by A_1 the subspace of A consisting in all functions $u(x) \subset nC^{0,1}(E^n)$.

Now we introduce the equivalence relation Γ in A setting $u(X)\Gamma v(x)$ if and only if $u(x) - v(x) = \text{const}$ in E^n. Clearly the factor space $B = A/\Gamma$ is again a Banach space with respect to the induced form

$$\|\xi\|_B = \|\tilde{u}\| - A$$

where ξ is any element of B generated by the class of functions $\{u(x)+q\}$, where $u(x) \in A$ and q is any real number. We use the notation $\tilde{u}(x) = u(x) - u(\theta)$. We call the function $u(x) \in A$ a *basic representative* of the element $\xi \in B$, if $u(\theta) = 0$. We shall use the notation ξ_u for elements of the space B generated by basic representatives $u(x) \in A$. Clearly $\tilde{u}(x)$ is a basic representative for every $u(x) \in a$.

Every convex function $u(x)$ defined in E^n whose graph has an admissible asymptotic cone is an element of A_1. Therefore $u(x) \in A$. This statement follows directly from the compactness of $\chi_u(E^n)$ in the space P^n.

Let T_k be a subset of B consisting in all elements $\xi \in B$ such that $\xi = \xi_u$, where $u(x)$ is any convex function defined in E^n and satisfying two conditions:

a) $u(\theta) = 0$;
b) the asymptotic cone of $u(x)$ is a fixed admissible cone K, considered in the second boundary value problem for equation (18.1).

Lemma 18.1. *T_k is a convex subset of B.*

Lemma 18.2. *T_k is a closed subset of B.*

Lemma 18.3. *T_k is a compact subset of B.*

Proof of Lemma 18.1. Let λ and μ be any positive numbers such that $\lambda+\mu = 1$. If ξ_f and ξ_g are any elements of T_k generated by the basic representatives $f(x)$ and $g(x)$, then the functions $f(x)$ and $g(x)$ re convex in E^n, have one and the same asymptotic cone K, which is admissible, and $f(\theta) = g(\theta) = 0$.

Clearly the element $\lambda\xi_f + \mu\xi_g$ is generated by the function $\lambda f(x) + \mu g(x)$. Since $\lambda f(x) + \mu g(x)$ is convex and

$$\lambda f(\theta) + \mu g(\theta) = 0,$$

then

$$\lambda\xi_f + \mu\xi_g = \xi_{\lambda f+\mu g}. \tag{18.15}$$

If we prove that K is the asymptotic cone of $\lambda f(x) + \mu g(x)$, then from this fact and from equality (18.15) it follows that T_k is a convex set in B. Clearly the statement concerning the asymptotic cone of $\lambda f(x) + \mu g(x)$ needs to be established only for C^2 convex functions $f(x)$ and $g(x)$, because the case of general convex functions can be obtained by the simple approximation of corresponding C^2 convex functions.

Thus our convex functions $f(x)$ and $g(x)$, introduced in the beginning of the proof of Lemma 1, are twice differentiable in E^n.

Clearly

$$\operatorname{grad}(\lambda f + \mu g) = \lambda \operatorname{grad} f + \mu \operatorname{grad} g \tag{18.16}$$

at any point $x_0 \in E^n$. Thus

$$x(\gamma) = \lambda\chi(\gamma_1) + \mu\chi(\gamma_2),$$

where $\gamma, \gamma_1, \gamma_2$ are correspondingly the normal images of tangent hyperplanes of the graphs of the functions $\lambda f(x) + \mu g(x)$, $f(x)$ and $g(x)$ at the points

$$(x, \lambda f(x) + \mu g(x)), (x, f(x)), (x, g(x)).$$

Since $\chi(\gamma_1), \chi(\gamma_2)$ are points of the convex set $\chi_k(E^n)$, we have

$$\chi(\gamma) \in \chi_k(E^n). \tag{18.17}$$

If K' is the asymptotic cone of the convex function $\lambda f(x) + \mu g(x)$, then from (18.17) it follows that

$$\chi_{k'}(E^n) \subset \chi_k(E^n), \tag{18.18}$$

where $z = k'(x)$ is the equation of the asymptotic cone K_1, whose vertex is at the point $\theta \in E^n$. From (18.18) we obtain

$$0 \leqq k'(x) \leqq k(x) \tag{18.19}$$

for all $x \in E^n$.

Now we prove that

$$k'(x) = k(x)$$

for all $x \in E^n$. Let ℓ be any axis in E^n, passing through the point θ, and s be the Cartesian coordinate in ℓ such that

$$|s| = [x_1^2 + x_2^2 + \cdots + x_n^2]^{\frac{1}{2}}$$

for any point $x = (x_1, x_2, \ldots, x_n) \in \ell$.

Let $\tilde{k}'(s), \tilde{k}(s), \tilde{f}(s), \tilde{g}(s)$ be functions which are generated by $k'(x)$, $k(x)$, $f(x)$, $g(x)$ on the axis ℓ.

Then

$$k'(s) = \begin{cases} -k_1' s & \text{if} \quad s \leqq 0; \\ k_2' s & \text{if} \quad s > 0 \end{cases}$$

and

$$k(s) = \begin{cases} -k_1 s & \text{if} \quad s \leqq 0; \\ k_2 s & \text{if} \quad s > 0, \end{cases}$$

where $0 \leqq k_1' \leqq k_1$ and $0 \leqq k_2' \leqq k_2$. Thus we obtain the following chain of equalities

$$\begin{aligned} k_2' + k_1' &= \int_{-\infty}^{+\infty} \left(\lambda \frac{d^2 f}{ds^2} + \mu \frac{d^2 g}{ds^2} \right) ds = \lambda \int_{-\infty}^{+\infty} \frac{d^2 f}{ds^2} ds \\ &+ \mu \int_{-\infty}^{+\infty} \frac{d^2 g}{ds} ds \\ &= \lambda(k_2 + k_1) + \mu(k_2 + k_1) = k_2 + k_1, \end{aligned}$$

because $\lambda \geqq 0; \mu \geqq 0$ and $\lambda+\mu = 1$. Thus $k_2' + k_1' = k_2 + k_1$. Since $0 \leqq k_2' \leqq k_2$ and $0 \leqq k_1' \leqq k_1$, then

$$k_2' = k_2 \quad \text{and} \quad k_1' = k_1. \tag{18.20}$$

Hence

$$k'(x) = k(x) \tag{18.21}$$

for all $x \in E^n$, because ℓ is an arbitrary axis in E^n passing through θ. From (18.20) it follows that

$$K' = K$$

and the proof of Lemma 18.1 is complete. □

Proof of Lemma 18.2. Let T_k be the closure of T_k in the space B. If $\xi \in T_k$, then there exist the elements $\xi_1, \xi_2, \ldots, \xi_q, \ldots$ of the set T_k such that

$$\lim_{q\to\infty} \|\xi_q - \xi\|_B = 0.$$

The basic representatives $u_q(x)$ of ξ_q are convex functions on E^n with one and the same asymptotic cone K, which is admissible, and $u_q(\theta) = 0$, $q = 1, 2, 3, \ldots$. Clearly

$$\lim_{q\to\infty} \|u_q(x) - u(x)\|_A = 0,$$

where $u(x)$ is the basic representative of $\xi \in T_k$. Hence $u_q(x)$ converges uniformly to $u(x)$ in every closed ball U_m: $|x| \leqq M$. Thus $u(x)$ is a convex function with asymptotic cone K and $\xi \in T_k$. The proof of Lemma 18.2 is complete.

Proof of Lemma 18.3. Let ξ be any element of T_k and $u(x)$ be the basic representative of ξ. Then $u(\theta) = 0$ and the convex function $u(x)$ satisfies a Lipschitz condition in E^n with the constant $d_0 = \operatorname{diam} K^*$. Thus

$$\|\xi\|_B = \|u(x)\|_A = \sum_{m=1}^{\infty} \frac{\|u(x)\|_m}{m^{2+\alpha}}$$
$$\leqq d_0 \sum_{m=1}^{\infty} \frac{1}{m^{1+\alpha}} = \text{const} < +\infty.$$

Let $\{u_\gamma(x)\}$ be the collection of basic representatives of $\{u_\gamma(x)\}$. From the last estimate follows the existence of a sequence of convex functions $u_{\gamma_q}(x)$ which is a subsequence of $\{u_\gamma(x)\}$ and which is uniformly convergent to some convex function $u_0(x)$ in every ball U_m: $|x| \leqq m$. Clearly $u_0(\theta) = 0$ and K is the asymptotic cone of $u_0(x)$. Hence

$$\xi_{u_0} \in T_k.$$

If we establish that

$$\lim_{q\to\infty} \|u_{\gamma_q}(x) - u_(x)\|_A = 0,$$

then the proof of Lemma 18.3 will be complete.

Let $\varepsilon > 0$ be any number. We fix a positive integer m_0 such that

$$\sum_{m=m_0}^{\infty} \frac{1}{m^{1+\alpha}} < \frac{\varepsilon}{4 \text{ diam } K^*}.$$

Then

$$\begin{aligned} \|u_{\gamma_q} - u_0\|_A &= \sum_{m=1}^{m_0-1} \frac{\|u_{\gamma_q} - u_0\|_m}{m^{2+\alpha}} + \sum_{m=m_0}^{\infty} \frac{\|u_{\gamma_q} - u_0\|_m}{m^{2+\alpha}} \\ &\leqq (m_1 - 1)\|u_{\gamma_q} - u_0\|_m + 2 \text{ diam } K^* \sum_{m=m_0}^{m} \frac{1}{m^{1+\alpha}} \\ &< (m_0 - 1)\|u_{\gamma_q} - u_0\|_m + \frac{\varepsilon}{2}. \end{aligned}$$

Since $\lim \|u_{\gamma_q} - u_0\|_m = 0$, there is a positive integer Q_0 such that

$$(m_0 - 1)\|u_{\gamma_q} - u_0\|_m < \frac{\varepsilon}{2}$$

for $q > q_0$. Thus $\|u_{\gamma_q} - u_0\|_m < \varepsilon$ for $q > q_0$, where $\varepsilon > 0$ is any given number. Lemma 18.3 is proved. □

18.4 The Proof of Theorem 18.1

Now we return to the proof of Theorem 18.1, which is the main existence theorem of § 18.

Let ξ be any element of the set T_k and let $u(x)$ be its basic representative. We consider the collection of convex functions

$$u_a(x) = u(x) + a, \tag{18.22}$$

where $a \in (-\infty, +\infty)$. The function

$$\begin{aligned} F_{u_a}(x) &= f(x, u_a(x), Du_a(x)) \\ &= f(x, u(x) + a, Du(x)) \end{aligned}$$

is nonnegative for all $x \in E^n$. Let $z = k(x)$ be the equation of the cone K. Then

$$u_a(x) = u(x) + a \leq k(x) + a \tag{18.23}$$

for all $x \in E^n$. From Assumption A.3 it follows that

$$F_{u_a}(x) \leq \phi_k(x, u(x) + a) \leq \phi_k(x, k(x) + a) \tag{18.24}$$

for all $x \in E^n$ and all real values of a. From the conditions of Theorem 18.1 it follows that

$$\int_{E^n} F_{u_a}(x)dx \leq \int_{E^n} \phi_k(x, k(x) + a)dx, +\infty$$

for all $a \in [a_k, b_k]$. From Assumption A.3 we obtain the inequality

$$F_{u_a}(x) \geq \inf_{\gamma \in K^*} \lambda_k(x, (x, \gamma) + a) \tag{18.25}$$

for all $x \in E^n$ and all $x \in R$.

Now we introduce the function

$$\psi(a) = \int_{E^n} F_{u_a}(x)dx \tag{18.26}$$

for $a \in [a_k, b_k]$. Since

$$\psi(a) = \int_{E^n} f(x, u(x) + a, Du(x))dx \tag{18.27}$$

for $a \in [a_k, b_k]$, then from Assumptions A.2, A.3 and the conditions of Theorem 18.1 it follows that $\psi(a)$ is continuous, $\psi'(a)$ exists on $[a_k, b_k]$, and the inequalities

$$\psi(a_k) \leq \int_{E^n} \phi_k(x, k(x) + a_k)dx < \text{ meas } K^* \tag{18.28-a}$$

and

$$\psi(b_k) \geq \inf_{K^*} \int_{E^n} \lambda_k(x(x, \gamma) + b_k)dx > \text{ meas } K^* \tag{18.28-b}$$

hold. Since

$$\psi'(a) = \int_{E^n} \frac{\partial f(x, u(x) + a, Du(x))}{\partial u} dx > 0,$$

then from (18.28a–b) it follows that only one number $a^* \in [a_k, b_k]$ exists such that

$$\psi(a^*) = \int_{E^n} F_{u_{a^*}}(x)dx = \text{meas } K^*. \tag{18.29}$$

Now we consider the second boundary value problem for the equation

$$\det(z_{ij}) = F_{u_{a^*}}(x) \tag{18.30}$$

with prescribed asymptotic cone K.

Since all conditions of Theorem 18.1 are fulfilled, this boundary value problem has only one convex generalized solution $z(x)$ satisfying the condition $z(\theta) = a^*$ and having K as the asymptotic cone of its graph.

Let $\tilde{u}(x) = u_{a^*}(x) - a^*$ and $\tilde{z}(x) = z(x) - z(\theta)$ be the basic representatives of the elements ξ_u and η_z, where ξ_u and η_z are elements of the convex compact set T_k. Clearly the second boundary value problem for equation (18.30) and the prescribed admissible convex cone K generate some operator

$$G\colon T_k \to T_k$$

such that $\eta_z = G(\xi_u)$.

In the final part of this subsection we shall establish the continuity and compactness of the operator G on the convex set T_k. These facts will permit us to apply the Schauder fixed point theorem to the operator $G\colon T_k \to T_k$. The existence Theorem 18.1 is the final result of these investigations.

Theorem 18.4. *The functional $a^*\colon T_k \to R$ is continuous.*

Proof. Let ξ be any element of T_k and $u(x) \in A_1 \subset A$ be a basic representative of ξ. Then $u(x)$ is a convex function whose graph has K as its asymptotic cone, and $u(\theta) = 0$.

The real number $a^* = a^*(u)$ is the root of the equation

$$\int_{E^n} f(x, u(x) + a, Du(x))dx = \text{meas } K^*. \tag{18.31}$$

We proved above that this equation has only one root $a^* = a^*(u) \in [a_k, b_k]$. Now we have to prove that $a^*(u_q)$ converges to $a^*(u)$ if

$$\lim_{q\to\infty} \|u_q - u\|_A = 0$$

and $\xi_{u_q}, \xi_u \in T_k$. From (18.31) it follows that

$$\int_{E^n} f(x, u(x) + a^*(u), Du(x))dx =$$
$$\int_{E^n} f(x, u_q(x) + a^*(u_q), Du_q(x))dx = \text{meas } K^*.$$

Hence

$$\int_{E^n} \{[u(x) - u_q(x)] + [a^*(u) - a^*(u_q)]\} \left(\int_0^1 \frac{\partial f}{\partial u}\bigg|_{v_t} dt\right) dx \tag{18.32}$$
$$= -\sum_{i=1}^{n} \int_{E^n} \left[\frac{\partial u(x)}{\partial x_i} - \frac{\partial u_q(x)}{\partial x_i}\right] \cdot \left(\int_0^1 \frac{\partial f}{\partial u_i}\bigg|_{v_t} dt\right) dx,$$

where

$$v_t(x) = (1-t)u_q(x) + tu(x) + (1-t)a^*(u_q) + ta^*(u)$$

and

$$\frac{\partial v_t(x)}{\partial x_i} = (1-t)\frac{\partial u_q(x)}{\partial x_i} + t\frac{\partial u(x)}{\partial x_i},$$

$0 \leqq t \leqq 1, i = 1, 2, \ldots, n$.

Since $a^*(u)$ and $a^*(u_q)$ do not depend on x, from (18.32) we obtain

$$a^*(u) - a^*(u_q) = -\frac{\int_{E^n} \{u(x) - u_q(x)\} \left(\int_0^1 \frac{\partial f}{\partial u}\Big|_{v_t} dt\right) dx}{\int_{E^n} \left(\int_0^1 \frac{\partial f}{\partial u}\Big|_{v_t} dt\right) dx}$$
$$- \frac{\int_{E^n} \sum_{i=1}^{n} \left(\frac{\partial u(x)}{\partial x_1} - \frac{\partial u_q(x)}{\partial x_i}\right) \left[\int_0^1 \frac{\partial f}{\partial u}\Big|_{v_t} dt\right] dx}{\int_{E^n} \left(\int_0^1 \frac{\partial f}{\partial u}\Big|_{v_t} dt\right) dx}.$$

Since $\partial f/\partial u$ is positive and continuous in $E^n \times R \times R \times K^*$ (see Assumption A.2), for any compact set Q in $E^n \times R \times K^*$ there is a constant $h(Q) > 0$ such that

$$\frac{\partial f(x, u, p)}{\partial u} \geqq h(Q) > 0.$$

For our purpose it is sufficient to consider the compact set

$$Q_0 = U_1 \times [\delta_1, \delta_2] \times K^*,$$

where U_1 is the unit ball $|x| \leqq 1$ in E^n and the numbers δ_1, δ_2 are determined by conditions

$$\delta_1 \leqq v_1(x) \leqq \delta_2.$$

The finite values of δ_1, δ_2 depend only on $\|u(x)\|_A$, the numbers a_k, b_k and the integer $N_1 > 0$ such that

$$\|u_q(x) - u(x)\|_A < 1$$

if $q \geqq N_1$. If $x \in U_1$, then

$$(v_t(x), \nabla v_t(x)) \in [\delta_1, \delta_2] \times K^*$$

for all $t \in [0, 1]$, because* $\nabla v_t(x) = (1-t)\nabla u(x) + t\nabla u_q(x) \in K^*$. Here we take into account that $\nabla u(x)$ and $\nabla u_q(x)$ are points of the convex set K^*. Thus we obtain the inequality

$$\int_{E^n} \left(\int_0^1 \frac{\partial f}{\partial u}\Big|_{v_t} dt\right) dx \geqq h(Q_0) \text{ meas } Q_0 > 0 \tag{18.33}$$

for all $q \geqq N_1$.

---

* Since $u(x), u_q(x)$ and $v_t(x)$ are convex functions, the notation $\nabla u(x)$ etc. is used also for supporting hyperplanes if the graph of $u(x)$ has no tangent hyperplane at the point $(x, u(x))$.

According to Assumption A.2 the inequalities

$$0 < \frac{\partial f}{\partial u} \leqq \frac{C_0}{|x|^{n+2+\alpha}}$$

hold for all $x \in E^n$ with $|x| \geqq m_0, u \in R$ and $p \in K^*$, where $\alpha = \text{const} > 0$ and $m_0 = \text{const} \geqq 1$; without losing generality we can assume that m_0 is any positive integer greater than $(2^{\frac{1}{2+\alpha}} - 1)^{-1}$. Let

$$I_1 = \left| \int_{E^n} \left\{ [u(x) - u_q(x)] \int_0^1 \frac{\partial f}{\partial u}\bigg|_{v_t} dt \right\} dx. \right.$$

Then

$$I_1 \leqq \int_{|x| \leqq m_0} \left\{ |u(x) - u_q(x)| \int_0^1 \frac{\partial f}{\partial u}\bigg|_{v_t} dt \right\} dx \tag{2.34}$$
$$+ \int_{|x| > m_0} \left\{ |u(x) - u_q(x)| \cdot \int_0^1 \frac{\partial f}{\partial u}\bigg|_{v_t} dt \right\} dx.$$

Let

$$C_2 = \sup_{|x| \leqq m_0} \frac{\partial f}{\partial u}\bigg|_{v_t} < +\infty. \tag{18.35}$$

Clearly C_2 depends only on $\|u(x)\|_{m_0}$, K^* and the numbers a_k, b_k. Since $u(x)$ and $u_q(x)$ are basic representatives of elements of the set T_k,

$$u(\theta) = u_q(\theta) = 0 \tag{18.36}$$

and

$$|u(x)| \leqq |x| \cdot \text{ diam } K^*, \tag{18.37}$$
$$|u_q(x) \leqq |x| \cdot \text{ diam } K^*.$$

From (18.34–37) it follows that

$$I_1 \leqq C_2 \|u(x) - u_q(x)\|_{m_0}$$
$$+ \frac{C_0 \sigma_{n-1}}{2+\alpha} \sum_{m=m_0}^{\infty} \|u(x) - u_q(x)\|_{m+1} \left(\frac{1}{m^{2+\alpha}} - \frac{1}{(m+1)^{2+\alpha}} \right),$$

where σ_{n-1} is the area of the unit sphere S^{n-1}. Since

$$\|u(x) - u_q(x)\|_{m_0} \leqq m_0^{2+\alpha} \|u(x) - u_q(x)\|_A$$

and

$$\frac{1}{(m+1)^{2+\alpha}} > \frac{1}{m^{2+\alpha}} - \frac{1}{(m+1)^{2+\alpha}}$$

for $m \geqq m_0 > \left(\frac{1}{2^{2+\alpha}} - 1\right)^{-1}$,

$$I_1 \leqq \left[C_2 m_0^{2+\alpha} + \frac{C_0 \sigma_{n-1}}{2+\alpha}\right] \|u(x) - U_q(x)\|_A, \tag{18.38}$$

where the integer m_0 satisfies the inequality

$$m_0 > \left(\frac{1}{2^{2+\alpha}} - 1\right)^{-1}$$

and constants C_0 and C_2 are independent of q.

Let

$$I_2 = \left| \int_{E^n} \left\{ \sum_{i=1}^{n} \left[\frac{\partial u(x)}{\partial x_i} - \frac{\partial u_q(x)}{\partial x_i}\right] \left(\int_0^1 \left.\frac{\partial f}{\partial u_i}\right|_{v_t} dt \right) \right\} dx \right|.$$

Clearly

$$I_2 \leqq \sum_{i=1}^{u} \int_{R^n} \left\{ \left| \frac{\partial u(x)}{\partial x_i} - \frac{\partial u_q(x)}{\partial x_i} \right| \left(\int_0^1 \left.\frac{\partial f}{\partial u}\right|_{v_t} dt \right) \right\} dx. \tag{18.39}$$

According to Assumption A.2 the inequalities

$$\left|\frac{\partial f}{\partial u_i}\right| \leqq \frac{C_1}{|x|^{n+\alpha}}, \qquad i = 1, 2, \ldots, n \tag{18.40}$$

hold for all $x \in E^n$ with $|x| \geqq m_0$, $u \in R$, and $p \in K^*$. The functions $\frac{\partial u_q(x)}{\partial x_i}$ converge to $\frac{\partial u(x)}{\partial x_i}$ almost everywhere in E^n and

$$\nabla u_q(x) \in K^*, \quad \nabla u(x) \in K^* \tag{18.41}$$

for all $x \in K^*$.

From (18.39–41) it follows that

$$\begin{aligned} I_2 \leqq \sum_{i=1}^{n} \left(\sum_{|x| \leqq m} \int_0^1 \left| \left.\frac{\partial f}{\partial u_i}\right|_{v_t} \right| dt \right) \cdot \int_{|x| \leqq m} \left| \frac{\partial u(x)}{\partial x_i} - \frac{\partial u_q(x)}{\partial x_i} \right| dx \\ + 2C_1 n \operatorname{diam} K^* \int_{|x|>m} \frac{dx}{|x|^{n+\alpha}}, \end{aligned} \tag{18.42}$$

where $m \geqq m_0$ is an arbitrary positive integer. Since

$$\int_{|x|>m} \frac{dx}{|x|^{n+\alpha}} = \frac{\sigma_{n-1}}{\alpha m^\alpha},$$

we can find $m^* > 0$ such that for every integer $m \geqq \max\{m_0, m^*\}$ the inequality

$$2C_1 n \text{ diam } K^* \frac{\sigma_{n-1}}{\alpha m^\alpha} < \frac{\varepsilon}{2} \tag{18.43}$$

holds, where $\varepsilon > 0$ is a given positive arbitrary number. We fix some integer $m \geqq \max\{m_0, m^*\}$. Then

$$\sup_{|x| \leqq m} \int_0^1 \left| \frac{\partial f}{\partial u_i} \right|_{v_t} dt \leqq Cm < +\infty$$

($i = 1, 2, \ldots, n$), where the constant C_m depends only on m, a_k, b_k and diam K^*. Really $\frac{\partial f}{\partial u_i}$ are continuous functions according to Assumption A.2 and we consider the supremum of $\left|\frac{\partial f}{\partial u_i}\right|$ in the compact set:

$$\underset{p \in K^*}{|x|} \leqq m, a_k - |x| \cdot \text{ diam } K^* \leqq u \leqq b_k + |x| \text{ diam } K^*.$$

Since $u(x)$ and $u_q(x)$ are convex functions,

$$\lim_{q \to \infty} \int_{|x| \leqq m} \left| \frac{\partial u(x)}{\partial x_i} - \frac{\partial u_q(x)}{\partial x_i} \right| dx = 0.$$

Therefore we can find an N_2 such that

$$\sum_{i=1}^n \left(\sum_{|x| \leqq m} \int_0^1 \left| \frac{\partial f}{\partial u_i} \right|_{v_t} \left| dt \right) \int_{|x| \leqq m} \left| \frac{\partial u(x)}{\partial x_i} - \frac{\partial u_q(x)}{\partial x_i} \right| dx < \frac{\varepsilon}{2},$$

if $q \geqq N_2$. Thus

$$I_2 < \varepsilon, \tag{18.44}$$

if $q \geqq N_2$.

Now from (18.33), (18.38), (18.44) it follows that

$$\lim a^*(u_q) = a^*(u)$$

if $\|u_q - u\|_A \to 0$. Lemma 4 is proved.

Lemma 18.5. *The operator G: $T_k \to T_k$ is continuous.*

Proof. Let the sequence $\xi_q \in T_k$ converge to the element $\xi_0 \in T_k$ in the space B. We should prove that

$$\lim_{q \to \infty} \|\eta_q - \eta_0\|_B = 0, \tag{18.45}$$

where $\eta_2 = G(\xi_q)$ and $\xi_0 = G(\xi_0)$.

Since T_k is a compact subset of B (see Lemma 18.3), then there is a subsequence η_{q_i} convergent to some element $\tilde{\eta} \in B$, i.e.

$$\lim_{j\to\infty} \|\eta_{q_j} - \tilde{\eta}\|_B = 0.$$

Since T_k is closed in B, $\eta \in T_k$. It is well known that the set functions $\omega(\ell, \bar{v}_{q_j}, e)$ converge weakly* to the set function $\omega(\ell, v, e)$ in E^n, where $v_{q_j}(x)$ and $v(x)$ are representatives of η_{q_j} and η satisfies the conditions

$$v_{q_i}(\theta) = a^*(u_{q_i}), \quad v(\theta) = a^*(u_0).$$

On the other hand all the functions**

$$F_{u_{q_j}+a^*(x)} = f(x, u_{q_j} + a^*(u_{q_j}), \nabla u_{q_j})$$

are nonnegative and satisfy the inequality

$$F_{u_{q_j}+a^*(x)} \leqq \phi_k(x, k(x) + b_k)$$

for all $x \in E^n$, where $z = k(x)$ is the equation of admissible convex cone K (see Assumption A.1), prescribed for all functions $u_{q_j}(x)$.

Note that the following facts hold:

a) $\lim_{j\to\infty} \|u_{q_j} - u_0\|_A = 0$;
b) $\nabla u_{q_j}(x) \in K^*$ for all $x \in E^n$ and all integers q_j;
c) $\frac{\partial u_{q_j}}{\partial x_j}$ converges to $\frac{\partial u_0}{\partial x_i}$ almost everywhere in E^n, $(i = 1, 2, \dots, n)$;
d) $\phi_M(x, k(x) + b_k)$ is a nonnegative summable function in E^n;
e) The estimates (18.33) are correct for the functions u_q, u_0.

Now we use the Lebesgue Theorem and obtain

$$\begin{aligned}\omega(\ell, v, e) &+ \lim_{q_j\to\infty} \omega(\ell, u_{q_j}, e)\\ &= \lim_{q_j\to\infty} \int_e F_{u_{q_j}+a^*(u_{q_j})}(x)dx = \int_e F_{u_0+a^*(u_0)}(x)dx,\end{aligned}$$

where e is any Borel subset of E^n. Note that we used continuity of the functional $a^*\colon T_k \to R$ in deriving these equalities.

Thus $v(x)$ is a convex generalized solution of the equation

$$\det\left(\frac{\partial^2 \bar{v}}{\partial x_i \partial x_j}\right) = F_{u_0+a^*(u_0)}(x)$$

---

* $\omega(\ell, u, e)$ is the R-curvature of a convex function $u(x)$ with $R(p) \equiv 1$ for all $p \in P^n$.

** The functions $u_{q_j}(x)$ and $u_0(x)$ are basic representatives of the elements ξ_{q_j} and ζ_0.

and the function $v_0(x)$ is also a convex generalized solution of the equation

$$\det\left(\frac{\partial^2 v_0}{\partial x_i \partial x_j}\right) = F_{u_0+a^*(u_0)}(x).$$

Since

$$v(0) = v_0(\theta) = a^*(u_0)$$

and the admissible convex cone K is the asymptotic cone for both functions $\bar{v}(x)$ and $v_0(x)$, we have $\bar{v}(x) = v_0(x)$ for all $x \in E^n$. Thus (18.45) is correct.□

Now we can finish the proof of Theorem 18.1. Since T_k is a compact set in B, $G(T_k)$ is also compact in B. Moreover $G(T_k) = T_k$. Hence the operator G has at least one fixed point $\xi \in T_k$. But the function $u(x) + a^*(\tilde{u}(x))$ is the representative for both ξ and $G(\xi)$. Therefore $u(x)$ is the desired solution of the second boundary value problem for the equation

$$\det(u_{ij}) = f(x, u, Du). \qquad \square$$

Chapter 6. Smooth Elliptic Solutions of Monge–Ampere Equations

§19. The N-Dimensional Minkowski Problem

19.1 Introduction

In § 8 of Chapter 2 we presented in detail the classical Minkowski Theorem on the problem of existence and uniqueness of a closed convex hypersurface with prescribed Gaussian curvature $K(\eta)$ in $(n+1)$-dimensional Euclidean space E^{n+1}. Here $K(\eta)$ is a positive continuous function on the unit hypersphere $S^n \subset E^{n+1}$, which is centered at the origin of E^{n+1}. The Minkowski problem is the problem of existence and uniqueness of a closed convex hypersurface F with Gaussian curvature $K(\eta)$ at a point x with exterior unit normal η. Here we do not assume that F is a regular hypersurface. Therefore the Gaussian curvature of a hypersurface F at a point $x \in F$ is defined as the limit of the ratio $\frac{\omega(G)}{\sigma(G)}$ as domain G shrinks to the point x, where $\sigma(G)$ is the area of G and $\omega(G)$ is the area of the spherical image of G. Both set functions $\sigma(G)$ and $\omega(G)$ are defined in §§ 5, 8. This definition of Gaussian curvature does not assume the C^m-smoothness ($m \geq 2$) of a convex hypersurface.

If the Gaussian curvature of a hypersurface F is prescribed as a function of the unit exterior normal η, then the surface function of F (see § 8, Chapter 2) is defined by the formula

$$\mu(H) = \int_H \frac{d\omega}{K(\eta)},$$

where $d\omega$ is the element of area of S^n and H is any Borel subset of S^n. Conversely the assignment of the surface function as a set function on S^n is the same as the assignment of the Gaussian curvature as a function of the exterior unit normal.

This point of view permits us to consider the Minkowski problem in the class of all convex closed hypersurfaces. Clearly this class includes the class of all closed convex polyhedra as a particular case. First Minkowski solved the problem stated above for convex polyhedra. Then for any positive continuous function $K(\eta)$ on S^n he solved this problem by approximation with convex polyhedra. Both steps have been presented in detail in § 8.

Let us recall the classical Minkowski result, proved in 1903 (see Minkowski [2]).

Theorem 19.1 (Minkowski). *Let $K(\eta)$ be a prescribed positive continuous function on S^n, satisfying the condition*

$$\int_{S^n} \frac{\eta d\omega}{K(\eta)} = 0. \tag{19.1}$$

Then there exists a closed convex hypersurface F, which is unique up to translation and for which $K(\eta)$ is the Gaussian curvature at the point $x \in F$ with exterior unit normal η.

In 1937 A.D. Alexandrov [3] has developed the Minkowski Theorem by proving the existence of a closed convex hypersurface with arbitrary nonnegative completely additive surface function prescribed on S^n.

Theorem 19.2 (A.D. Alexandrov). *Let $\mu(H)$ be a completely additive set function on the unit sphere S^n in E^{n+1}, which satisfies the conditions*

$$\int_{S^n} \eta d\mu = 0, \quad \int_{S^n} |(\bar{e}, \eta)| d\mu \geq a = \text{ const } > 0$$

for any unit vector $\bar{e}$. Then there exists a closed convex hypersurface F, unique up to translation, for which $\mu(H)$ is its surface function.

Up to 1971 the following important conjecture had not been solved: If in Theorem 19.2 the strictly positive function $K(\eta)$ is sufficiently smooth on S^n, then a solution of the Minkowski problem must also be sufficiently smooth. This conjecture was positively solved by Pogorelov [3] in 1971 (see also Pogorelov [7]). In Subsections 19.2–4 we present Pogorelov's proof of the following theorem.

Theorem 19.3 (A.V. Pogorelov). *Let $K(\eta)$ be a prescribed function on the unit hypersphere S^n. Suppose that $K(\eta)$ is positive and smooth of class C^m, $m \geq 3$, and let it satisfy the condition*

$$\int_{S^n} \frac{\eta d\omega}{K(\eta)} = 0. \tag{19.1}$$

Then there exists a closed convex smooth hypersurface of class $C^{m+1,a}$, $0 < a < 1$ unique up to translation, whose Gaussian curvature is $K(\eta)$. If the function $K(\eta)$ is analytic, then this hypersurface is also analytic.

19.2 A Priori Estimates for the Radii of Normal Curvature of a Convex Hypersurface

Let F be a smooth closed convex hypersurface in E^{n+1}. Let $K(\eta)$ be the Gaussian curvature of F prescribed as a function of the unit exterior normal η. Let

$$\varphi(\eta) = \frac{1}{K(\eta)}.$$

The function $\varphi(\eta)$ can be considered on the spherical image of F.

Theorem 19.4 (Pogorelov). *The radii of normal curvature of the hypersurface F admit the a priori estimate*

$$R \leq \max_{\eta,\gamma} \varphi^{1/n} \left[1 + \frac{1}{n-1}\left(\frac{\varphi'^2}{\varphi^2} - \frac{\varphi''}{\varphi}\right)\right]^{1-\frac{1}{n}} \tag{19.2}$$

where φ is differentiated at point η with respect to length of arc of the great circle in the direction γ on the spherical image of F.

Proof. Let $R(\eta,\gamma)$ be the radius of normal curvature of F at the point with the unit exterior normal η in the direction γ. Let A be a point of F, where the function $R(\eta,\gamma)$ achieves its maximum with $\eta = \eta_0$ and $\gamma = \gamma_0$. We denote by $z, x_1, \ldots, x_n$ Cartesian coordinates in E^{n+1} which will be introduced in the following way. The origin 0 is a point on the interior normal of F at the point A, the z-axis is the straight line $0A$, and the axes $x_1, \ldots, x_n$ are parallel to the principal directions of F at the point A. Without loss of generality we can assume that the x_1-axis is chosen in the direction in which the radius of normal curvature has its maximum value. Thus the x_1-axis has the direction γ_0.

Let H be the support function of F and let

$$H_0 = R_1 \cdot (z^2 + x_1^2 + \cdots + x_n^2)^{1/2}$$

be the support function of a hypersphere of radius R_1, where $R_1 = R(\eta_0, \gamma_0)$. The radii of normal curvature of the hypersurface F are no greater than R_1. Therefore

$$d^2 H \leq d^2 H_0.$$

In particular

$$\frac{\partial^2 H}{\partial x_1^2} \leq \frac{\partial^2 H_0}{\partial x_1^2}.$$

Let

$$h(x_1, \dots, x_n) = H(1, x_1, \dots, x_n),$$
$$h_0(x_1, \dots, x_h) = H_0(1, x_1, \dots, x_n)$$

be the normalized support functions. Then

$$\frac{\partial^2 h}{\partial x_1^2} \leq \frac{\partial^2 h_0}{\partial x_1^2}$$

with equality sign at the point $A(x_1 = x_2 = \cdots = x_n = 0)$. Clearly the function

$$w_1 = (h_{x_1x_1} - h_{0,x_1x_1})\frac{(1 + x_1^2 + \cdots + x_n^2)^{3/2}}{(1 + x_2^2 + \cdots + x_n^2)},$$

achieves its maximum at the point A. This maximum is equal to zero. Since

$$R_1 = h_{0,x_1x_1}\frac{(1 + x_1^2 + \cdots + x_n^2)^{3/2}}{(1 + x_2^2 + \cdots + x_n^2)},$$

the function

$$w = h_{x_1x_1}\frac{(\ell + x_1^2 + \cdots + x_n^2)^{3/2}}{\ell + x_2^2 + \cdots + x_n^2}$$

achieves its maximum at the point A. The value of this maximum is R_1. Thus the estimate of the maximum of the radii of normal curvature of F is equivalent to the estimate of the maximum of the function w at the point A, where $x_1 = x_2 = \cdots = x_n = 0$.

The support function h, which defines the hypersurface F, satisfies the equation

$$(1 + x_1^2 + \cdots + x_n^2)^{\frac{n}{2}+1} \det(h_{ij}) = \varphi \tag{19.3}$$

(see Subsection 6.2). Clearly

$$w = h_{11}\frac{(1 + x_1^2 + \cdots + x_n^2)^{3/2}}{1 + x_2^2 + \cdots + x_n^2}. \tag{19.4}$$

At the point A we obtain

$$h_{ii} = R_i, \quad h_{ij} = 0 \quad \text{if} \quad i \neq j,$$

where R_i are the principal radii of the normal curvature of F at the point A. Differentiating relation (19.4) at the point A we obtain

$$w_i = h_{11i} = 0, \tag{19.5}$$

$$w_{11} = (h_{11})_{11} + 3R_1 \le 0, \quad w_{ii} = (h_{11})_{ii} + R_1, \quad i \ne 1, \tag{19.6}$$

because A is the maximum point of w. Differentiating equation (19.3) with respect to x_1 and using relations $h_{ij} = 0$ at the point A, we obtain the equation

$$\sum_i h_{11} \dots (h_{ii})_1 \dots h_{nn} = \varphi_1$$

at this point A. Now the last equation can be rewritten in the following form

$$\sum_i \frac{(h_{ii})_1}{h_{ii}} = \varphi_1 \tag{19.7}$$

at the point A, where $\varphi_1 = \frac{\partial \varphi}{\partial x_1}$. We now differentiate twice equation (19.3) with respect to x_1 to obtain the following relation at the point A:

$$(n+2)\varphi + \varphi \cdot \sum_i \frac{(h_{ii})_{11}}{h_{ii}} + \varphi \cdot \sum_{i \ne j} \frac{(h_{ii})_1}{h_{ii}} \cdot \frac{(h_{jj})_1}{h_{jj}} \tag{19.8}$$

$$- \varphi \cdot \sum_{i \ne j} \frac{(h_{ij})_1^2}{h_{ii} h_{jj}} = \varphi_{11},$$

where $\varphi_{11} = \frac{\partial^2 \varphi}{\partial x_1^2}$. Taking relation (19.7) into account we obtain

$$\sum_{i,j} \frac{(h_{ii})_1 (h_{jj})_1}{h_{ii} h_{ij}} \le \left(\frac{\varphi_1}{\varphi}\right)^2. \tag{19.9}$$

Now from inequalities (19.6) it follows that

$$\frac{(h_{11})_{11}}{h_{11}} \le -3, \quad \frac{(h_{ii})_{11}}{h_{ii}} \le -\frac{R_1}{R_i} \quad \text{for} \quad i \ne 1 \tag{19.10}$$

at the point A. Thus from equation (19.8) and inequalities (19.9) and (19.10) we obtain

$$(n-1)\varphi - \varphi R_1 \sum_{i>1} \frac{1}{R_i} + \frac{\varphi_1^2}{\varphi} \ge \varphi_{11} \tag{19.11}$$

at the point A. Since

$$\sum_{i>1} \frac{1}{R_i} \ge (n-1) \cdot \frac{1}{(R_2 \dots R_n)^{1/n-1}} = (n-1) \left(\frac{R_1}{\varphi}\right)^{\frac{1}{n-1}},$$

from inequality (19.11) we obtain the following inequality

$$(n-1)\varphi\left[1-R_1\left(\frac{R_1}{\varphi}\right)^{1/(n-1)}\right]+\frac{\varphi_1^2}{\varphi}\geq\varphi_{11} \tag{19.12}$$

at the point A. Thus we get the estimate

$$R_1\leq\varphi^{1/n}\cdot\left[1+\frac{1}{n-1}\left(\frac{\varphi_1^2}{\varphi^2}-\frac{\varphi_{11}}{\varphi}\right)\right]^{1-\frac{1}{n}} \tag{19.13}$$

at the same point A.

Let γ_0 be the great circle on the unit sphere

$$z^2+x_1^2+\cdots+x_n^2=1$$

in the x_1-direction. Then $x_1=\tan s$, where s is the arc of γ_0. Hence for $s=0$ (i.e. for $x_1=0$)

$$\varphi_1=\varphi'_s,\quad \varphi_{11}=\varphi''_{ss}.$$

Since R_1 is the maximum radius of normal curvature of the hypersurface F, we derive the estimate

$$R\leq\max_{\eta,\gamma}\varphi^{1/n}\left[1+\frac{1}{n-1}\left(\frac{\varphi'^2}{\varphi^2}-\frac{\varphi''}{\varphi}\right)\right]^{1-\frac{1}{n}}.$$

The proof of Theorem 19.4 is complete. □

19.3 Auxiliary Concepts and Formulas Obtained by E. Calabi [1] and A. Pogorelov [3]

Let $z=u(x_1,x_2,\ldots,x_n)$ be a convex hypersurface of positive Gaussian curvature. Then the quadratic form

$$dz^2=\sum_{i,j=1}^{n}u_{ij}dx_idx_j$$

is positive definite. According to Calabi [1] the following Riemannian metric

$$ds^2=\sum_{i,j=1}^{n}g_{ij}dx_idx_j \tag{19.14}$$

can be introduced, where

$$g_{ij}=\frac{\partial^2u}{\partial x_i\partial x_j}. \tag{19.15}$$

Let

$$\Phi = \det(g_{ij})$$

and

$$\varphi = \ln \sqrt{\Phi}$$

be invariants of metric (19.15). For metric (19.15) the Christoffel symbols of the first kind are as follows

$$\Gamma'_{ijk} = \frac{1}{2} \frac{\partial^2 u}{\partial x_i \partial x_j \partial x_k} = A_{ijk}.$$

The quantities A_{ijk} are symmetric with respect to all indices. Further the relation

$$g^{ij} A_{ijk} = \varphi_k \tag{19.16}$$

can be obtained by differentiating the equation

$$\det(g_{ij}) = \Phi. \tag{19.17}$$

Since

$$A_{ijk,\ell} = \frac{1}{2} \frac{\partial^4 u}{\partial x_i \partial x_j \partial x_k \partial x_\ell} + g^{hm}(A_{hjk}A_{mi\ell} + A_{ihk}A_{mj\ell} + A_{ijh}A_{mk\ell}),$$

the quantities $A_{ijk,\ell}$ are symmetric in each pair of indices, i.e.

$$A_{ijk,\ell} = A_{ij\ell,k}. \tag{19.18}$$

According to well-known formulas we obtain the following expression for the Riemann tensor

$$R_{ijk\ell} = g^{hm}(A_{hi\ell}A_{mjk} - A_{hik}A_{mj\ell}). \tag{19.19}$$

From (19.19) and (19.16) there follows the useful formulas for the Ricci tensor R_{ik} and the scalar curvature R:

$$\begin{aligned} R_{ik} &= g^{i\ell} R_{ijk\ell} g^{j\ell} g^{hm}(A_{hi\ell}A_{mjk} - A_{hik}A_{mj\ell}) \\ &= g^{j\ell} g^{hm} A_{hi\ell}A_{mjk} - g^{hm} A_{hik}\varphi_m. \end{aligned}$$

If we set

$$\overline{R}_{ik} = g^{j\ell} g^{hm} A_{hi\ell}A_{mjk},$$

then

$$R_{ik} = \overline{R}_{ik} - g^{hm} A_{hik}\varphi_m$$

and

$$R = g^{ik} R_{ik} = g^{ik}\overline{R}_{ik} - g^{hm}\varphi_h\varphi_m.$$

Setting

$$\overline{R} = g^{ik}\overline{R}_{ik} = A^{ijk}A_{ijk}$$

we obtain

$$R = \overline{R} - g^{hm}\varphi_h\varphi_m.$$

Lemma 19.1 (Calabi). *The following relation*

$$\frac{1}{2}\Delta\overline{R} = A^{ijk}\varphi_{i,jk} + \overline{R}_{ij}\overline{R}^{ij} + R_{ijk\ell}R^{ikj\ell} + A_{ijk,\ell}A^{ijk,\ell} \tag{19.20}$$

holds, where Δ is the Laplace–Beltrami operator of the metric (19.15).

Proof. First of all the following identities hold:

$$\begin{aligned}(\Delta A)_{ijk} &= g^{\ell m}A_{\ell jk,im} = g^{\ell m}A_{\ell jk,mi} + g^{\ell m}(A_{\ell jk,im} - A_{\ell jk,mi})\\ &= g^{\ell m}A_{\ell jk,mi} + g^{\ell m}(A_{jhk}R^h_{\ell im} + A_{\ell hk}R^h_{jim} + A_{\ell jh}R^h_{kim})\\ &= \varphi_{j,ki} + A^{h\ell m}(A_{h\ell i}A_{mjk} + A_{h\ell j}A_{mki} + A_{h\ell k}A_{mij})\\ &\quad - 2A^\alpha_{i\beta}A^\beta_{j\gamma}A^\gamma_{k\alpha}.\end{aligned}$$

The identical transformations made above are based on relations (19.16), the symmetric properties of the first covariant derivative $A_{ijk,\ell}$ of A_{ijk}, and the Ricci identity for the difference between the second covariant derivatives. Thus

$$\begin{aligned}\frac{1}{2}\Delta\overline{R} = \frac{1}{2}g^{\ell m}\overline{R}_{,\ell m} &= A^{ijk}\cdot(\Delta A)_{ijk} + A^{ijk,\ell}A_{ijk,\ell}\\ &= A^{ijk}\varphi_{i,jk} + 3A^{ijk}A^{h\ell m}A_{h\ell i}A_{mjk} - 2A^{ijk}A^r_{is}A^s_{jt}A^t_{kr}\\ &\quad + A^{ijk,\ell}A_{ijk,\ell}\\ &= A^{ij,k}\varphi_{i,jk} + \overline{R}_{ij}\overline{R}^{ij} + R_{ijk\ell}R^{ijk\ell} + A^{ijk,\ell}A_{ijk,\ell}.\end{aligned}$$

Thus the proof of Lemma 19.1 is complete. □

Let

$$B_{abcijk\ell} = \frac{1}{2}(A_{abc,i}A_{jk\ell} - A_{abc}A_{jk\ell,i}).$$

One can easily verify that

$$B_{abcijk\ell}B^{abcijk\ell} = \frac{1}{2}\overline{R}A_{abc,i}A^{abc,i} - \frac{1}{8}g^{ij}\overline{R}_{,i}\overline{R}_{,j}. \tag{19.21}$$

Now from Lemma 19.1 and equation (19.21) it follows that

$$\begin{aligned}\Delta\sqrt{\overline{R}} = \frac{\Delta\overline{R}}{2\sqrt{\overline{R}}} - \frac{g^{ij}\overline{R}_{,i}\overline{R}_{,j}}{4\overline{R}} &= \frac{1}{\overline{R}}(\overline{R}_{ij}\overline{R}^{ij} + R_{ijk\ell}R^{ijk\ell} + A^{ijk}\varphi_{i,jk})\\ &\quad + \frac{2}{\overline{R}^{3/2}}B_{abcijk\ell}B^{abcijk\ell}.\end{aligned}$$

Since the last term in the right-hand side is nonnegative, the following inequality

$$\Delta\sqrt{R} \geq \frac{1}{\sqrt{R}}(\overline{R}_{ij}\overline{R}^{ij} + R_{ijk\ell}R^{ijk\ell} + A^{ijk}\varphi_{i,jk}) \tag{19.22}$$

holds.

We now introduce the notation

$$\psi = \sqrt{R} = [A^{abc}A_{abc}]^{1/2}. \tag{19.23}$$

Lemma 19.2 (Pogorelov). *The following inequality*

$$\Delta\psi \geq \frac{n+1}{n(n-1)}\psi^3 + C_1\psi^2 + C_2\psi + C_3 + C_0|D\psi| \tag{19.24}$$

holds, where $|D\psi|$ is the maximum of the absolute values of the first derivatives of ψ with respect to the variables x_k; and C_0, C_1, C_2, C_3 can be estimated in terms of the second derivatives of the function $u(x)$ and the derivatives of order up to 3 of the function φ.

Proof. From (19.23) and expressions for R_{ik} and $\overline{R}_{ik}$ it follows that

$$R_{ij}R^{ij} = \overline{R}_{ij}\overline{R}^{ij} + a_1\psi^3 + a_2\psi^2, \tag{19.25}$$

where a_1, a_2 can be estimated in terms of the second derivatives of $u(x)$ and of the first derivatives of φ. We now prove that the expression

$$A^{ijk}\varphi_{i,jk} = b_0\psi \cdot |D\psi| + b_1\psi^3 + b_2\psi^2 + b_3\psi \tag{19.26}$$

can be estimated in terms of the second derivatives of $u(x)$ and the third derivatives of $\varphi(x)$. Actually

$$\varphi_{i,j} = \varphi_{ij} + \varphi_\alpha g^{\alpha s}A_{sij},$$
$$\varphi_{i,jk} = \varphi_\alpha g^{\alpha s}A_{sijk} + (*),$$

where $(*)$ denotes a quadratic expression with respect to quantities A. The coefficients of this expression admit a required estimate. Therefore

$$A^{ijk}\varphi_{i,jk} = \varphi_\alpha g^{\alpha s}A^{ijk}A_{ijks} + (**). \tag{19.27}$$

Clearly the first term in the right-hand side of equation (19.27) is estimated by $\psi|D\psi| + 0(\psi^3)$.

Calabi [1] established the following inequalities for a positive-definite Riemannian metric:

$$\frac{1}{n}R^2 \leq R_{ij}R^{ij}, \quad \frac{2}{n-1}R_{ij}R^{ij} \leq R_{ijk\ell}R^{ijk\ell}.$$

Hence

$$R_{ij}R^{ij} + R_{ijk\ell}R^{ijk\ell} \geq \frac{n+1}{n(n-1)}R^2.$$

Since $R = \overline{R} - g^{hm}\varphi_h\varphi_m$, from (19.25) it follows that

$$\overline{R}_{ij}\overline{R}^{ij} + R_{ijk\ell}R^{ijk\ell} \geq \frac{n+1}{n(n-1)}\psi^4 + d_1\psi^3 + d_2\psi^2.$$

Thus the desired inequality (19.24) follows from (19.22), (19.26) and (19.27).□

19.4 An A Priori Estimate for the Third Derivatives of a Support Function of a Convex Hypersurface

Let $z(x_1, \ldots, x_n)$ be a convex C^5-function, which is a solution of the Monge–Ampere equation

$$\det(z_{ij}) = \phi(x) > 0 \tag{19.28}$$

in the domain G. Here $\phi(x)$ is a positive C^3-function in G.

Theorem 19.5 (Pogorelov). *At any interior point of G the third derivatives of the solution $z(x_1, \ldots, x_n)$ of equation (19.28) admit an estimate which only depends on the second derivatives of $z(x_1, \ldots, x_n)$, the derivatives up to third order of the function ϕ, and the distance from this point to ∂G.*

Proof. Since the function $z(x_1, x_2, \ldots, x_n)$ is convex and $\det(z_{ij})$ is positive, the quadratic form d^2z is positive definite. Therefore we can introduce Calabi's metric

$$ds^2 = \sum_{i,j=1}^{n} g_{ij}dx_idx_j, \tag{19.29}$$

where $g_{ij} = \frac{\partial^2 z}{\partial x_i \partial x_j}$. Thus we define a Riemannian metric in G. We now set

$$\psi = \frac{1}{2}[g^{ia}g^{jb}g^{kc}z_{ijk}z_{abc}]^{1/2}.$$

According to Lemma 19.2 the following inequality holds:

$$\Delta\psi \geq \frac{n+1}{n(n-1)}\psi^3 + C_1\psi^2 + C_2\psi + C_3 + C_0 \cdot |D\psi|, \tag{19.30}$$

where $|D\psi|$ and the quantities C_0, C_1, C_2, C_3 have been explained in Subsection 19.3.

Let 0 be an arbitrary point of G and let $\rho > 0$ be the distance from 0 to ∂G. Without loss of generality we can assume that the point 0 is the origin of Cartesian coordinates $x_1, x_2, \ldots, x_n$ in E^n. In the ball ω: $\sum_{k=1}^{n}(x_k)^2 \leq \varepsilon^2\rho$ where $\frac{3}{4} \leq \varepsilon < 1$ and ε is a fixed number, we consider the function

$$w = \psi \cdot \lambda, \quad \lambda = \varepsilon^2\rho^2 - \sum_{k=1}^{n}(x_k)^2.$$

The function w is nonnegative and it vanishes on $\partial\omega$; hence w achieves its maximum w_0 at an interior point A of ω. Clearly

$$w_i = \frac{\partial w}{\partial x_i} = 0, \qquad i = 1, 2, \ldots, n$$

at the point A. Therefore

$$\psi_i = w\left(\frac{1}{\lambda}\right)_i, \quad \psi_{ij} = \frac{w_{ij}}{\lambda} + w\left(\frac{1}{\lambda}\right)_{ij}, \qquad i, j = 1, 2, \ldots, n$$

at the point A. We now substitute the values of ψ_i and ψ_{ij} in inequality (19.30). Then the left-hand side of inequality (19.30) becomes

$$\Delta\psi = g^{ij}\psi'_{ij} + g^{ij}\Gamma^k_{ij}\psi_j,$$

where the Γ^k_{ij} are the Christoffel symbols of the second kind for the Calabi metric (19.29) (see expressions for the Christoffel coefficients in Subsection 19.3). The important fact is that these expressions contain only the second derivatives of the function $z(x)$ and are linear in the third derivatives of ϕ. Since the function w achieves its maximum at the point A,

$$g^{ij}w_{ij} \leq 0$$

at this point. Thus

$$\Delta\psi \leq g^{ij}w\left(\frac{1}{\lambda}\right)_{ij} + C_i w\left(\frac{1}{\lambda}\right)_i \tag{19.31}$$

at the point A, where the coefficients C_i admit an estimate from above. From inequalities (19.30) and (19.31) it follows that there exists an inequality for w at the point A, i.e. inequality for w_0. This inequality is

$$\frac{n+1}{n(n-1)}w_0^2 + Q_2(w_0) \leq 0, \tag{19.32}$$

where Q_2 is a second-degree polynomial in w_0 with coefficients, which can be estimated in terms of the second derivatives of $z(x)$ and the derivatives of $\phi(x)$ up to third order.

From inequality (19.32) we obtain an estimate for w_0. Now it is possible to obtain an estimate for ψ at the point 0, which is the center of the ball ω, because

$$\psi \leq \frac{16w_0}{g\rho^2}$$

at the point 0. This estimate depends on the above-mentioned derivatives of the functions $z(x)$ and $\phi(x)$ and also on the distance ρ from 0 to ∂G. Since the form $d^2 z$ is positive definite and the second derivatives of $z(x)$ are bounded, the eigenvalues of this form and the eigenvalues of the form $g^{ij} dx_i dx_j$ admit an estimate in the terms mentioned in the statement of the present theorem. Thus an estimate for

$$\psi^2 = g^{ia} g^{jb} g^{kc} z_{ijk} z_{abc}$$

implies an estimate for the third derivatives z_{ijk}, which depends on the same quantities, i.e. on the second derivatives of the function $z(x)$, the derivatives of the function ϕ up to third order, $\inf_{S^n} \phi(x) = \phi_0 > 0$, and the distance from the point to ∂G.

The proof of Theorem 19.5 is complete. □

Theorem 19.6 (Pogorelov). *Let F be a closed convex hypersurface. Let F be of class C^5 and its Gaussian curvature $K(\eta)$ be strictly positive. Then for the support function H of this hypersurface and its derivatives up to third order one can obtain an estimate which only depends on the Gaussian curvature $K(\eta)$ and its derivatives up to third order.*

Proof. The eigenvalues of the quadratic form $d^2 H$ on S^n are the principal radii of curvature of F. From this fact and from Theorem 19.4 it follows that there is an estimate for the second derivatives H_{ij} of the support function H on S^n. The existence of an estimate for the function H and its first derivatives is a consequence of the facts that the diameter of F admits an estimate (see § 6 and 8) and that the origin of the coordinate system is inside the hypersurface F. All these estimates are not more than the diameter of the hypersurface F.

Let $x_0^2 + x_1^2 + \cdots + x_n^2 = 1$ be the equation of the unit hypersphere S^n. Let A be an arbitrary point of S^n. We now estimate the third derivatives of the support function H at the point A. Without loss of generality we can assume that the axis x_0 is identified with the straight line $0A$ and directed along the vector $\overrightarrow{0A}$. If

$$h(x_1, x_2, \ldots, x_n) = H(1, x_1, \ldots, x_n),$$

then the function h satisfies the differential equation

$$\det(h_{ij}) = \frac{1}{K}(x_1^2 + \cdots + x_n^2 + 1)^{-\frac{n}{2}-1}.$$

The estimate for the second derivatives of H guarantees an estimate for the second derivatives of h in the domain ω': $x_1^2 + \cdots + x_n^2 < 1$. The estimate for the second derivatives of h in ω' provides an estimate for the third derivatives of the function h at the center of ω', i.e. the point A. The third derivatives of the support function H at the point A can be estimated in the terms of the derivatives of h up to the third order. The proof of Theorem 19.6 is complete. □

19.5 The Proof of Theorem 19.3

Let $K(\eta)$ be a continuous strictly positive function class C^m, $m \geq 3$ on S^n, which satisfies the condition

$$\int_{S^n} \frac{\eta d\omega}{K(\eta)} = 0. \tag{19.33}$$

The main assertion of Theorem 19.1 is the existence of a closed convex hypersurface F of class $C^{m+1,a}$, $0 < a < 1$ for which $K(\eta)$ is the prescribed Gaussian curvature as function of the unit normal $\eta \in S^n$. If $\varphi(\eta) = \frac{1}{K(\eta)}$, then this existence problem can be reduced to the existence of a solution of differential equation

$$D(H, H, \ldots, H) = \varphi(\eta) \tag{19.34}$$

subject to the condition

$$\int_{S^n} \eta\varphi(\eta) d\omega = 0,$$

where $D(H, H, \ldots, H)$ is the sum of principal minors of the Hessian (H_{ij}) of a function H that is positively homogeneous of first degree on the unit hypersurface S^n. Set

$$\varphi(\eta, t) = 1 - t + t\varphi(\eta), \qquad 0 \leq t \leq 1$$

and introduce the equation with parameter t

$$D(H, H, \ldots, H) = \varphi(\eta, t) \tag{19.35}$$

subject to the condition

$$\int_{S^n} \eta\varphi(\eta, t) d\omega = 0.$$

To prove that equation (19.34) has a required solution, Pogorelov [7] applied the continuity method. This method provides the existence of a required solution if the two following assertions are proved:

1) If equation (19.35) has a solution for a certain t_0 then it has a solution for all t sufficiently close to t_0.

2) If equation (19.35) has a solution for each of a sequence of parameters $t_1, t_2, \ldots, t_n, \ldots$ converging to t_0, then it has a solution for $t = t_0$.

Indeed from these two assertions it follows that the set of all values of the parameter t for which equation (19.35) is solvable is both open and closed. This set is not empty, because there exists a trivial solution for $t = 0$. Hence the set of all t for which equation (19.35) is solvable is the segment [0,1]. Thus equation (19.35) is solvable for $t = 1$ and equation (19.34) has a required solution.

The Proof of Assertion 1). Suppose that H depends on the parameter t. The result of differentiation of $D(H, H, \ldots, H)$ with respect to t will be denoted by $D(H, H, \ldots, H, Z)$, where $Z = \frac{\partial H}{\partial t}$. Let

$$D(H, \ldots, H, Z) = \psi(\eta) \tag{19.36}$$

be the equation, where H is a support function of a convex hypersurface with positive Gaussian curvature and Z is an unknown function that is positively homogeneous of the first degree. It is well-known that equation (19.36) is a self-adjoint elliptic equation with respect to the function Z (see Hilbert [1]). It is also well known that this equation is linear and the homogeneous equation

$$D(H, \ldots, H, Z) = 0 \tag{19.37}$$

has exactly $n + 1$ linearly independent solutions

$$Z_0 = x_0, \quad Z_1 = x_1, \ldots, Z_n = x_n.$$

Therefore the solvability condition for the nonhomogeneous equation (19.36) is the system of equations

$$\int_{S^n} \eta_k \psi(\eta) d\omega = 0 \qquad (k = 0, 1, 2, \ldots, n), \tag{19.38}$$

where the numbers η_k are the components of the unit vector η. Equations (19.38) are equivalent to one vector equation

$$\int_{S^n} \eta \psi(\eta) d\omega = 0.$$

The solution itself can be represented in terms of the Green's function

$$H(\bar{q}) = \int_{S^n} G(\eta, \bar{q}) \psi(\eta) d\omega_\eta.$$

The solution of the nonlinear equation (19.35) can be obtained by the method of successive approximations. Below we present briefly the scheme of Pogorelov's considerations. Set

$$\begin{aligned} D(H, \ldots, H, Z) &= D(H + Z, \ldots, H + Z) - D(H, \ldots, H) \\ &\quad + R(H, Z) \\ &= D(H + Z) - D(H) + R(H, Z). \end{aligned}$$

Then for values of t close to t_0 we have to solve the equation

$$D(H, \ldots, H, Z) = \Delta t[\varphi(\eta) - 1] + R(H, Z).$$

Pogorelov solved this equation according to the special form of the method of successive approximations developed by Nirenberg [2]. The successive approximations can be found from the equation

$$D(H,\dots,H,Z_k) = \Delta t[\varphi(\eta) - 1] + R(H, Z_{k-1}). \tag{19.39}$$

The most important thing is to show that the solvability condition for equation (19.39) with respect to Z_k holds, i.e. that

$$\int_{S^n} (\eta)[\Delta t[\varphi(\eta) - 1] + R(H, Z_{k-1})]d\omega = 0. \tag{19.40}$$

The following identity

$$\int_{S^n} \eta D(H + \lambda Z_{k-1})d\omega = 0 \tag{19.41}$$

holds for the support function $H + \lambda Z_{k-1}$, where λ is a parameter. Since

$$D(H + \lambda Z_{k-1}) = \sum_s \lambda^s A_s(H, Z_{k-1}) = \lambda D(H, Z_{k-1}) + R_\lambda(H, Z_{k-1}),$$

it follows from (19.41), which is an identity with respect to λ, that

$$\int_{S^n} \eta A_s(H, Z_{k-1})d\omega = 0$$

for all A_s. (The definition of the mixed discriminants A_s is given in Chapter 1 of this book). Thus

$$\int_{S^n} \eta R(H, Z_{k-1})d\omega = 0.$$

Finally the equation

$$\int_{S^n} \eta\varphi(\eta)d\omega = 0$$

follows from the assumptions of the present theorem. Thus, equation (19.39) has a solution at each step of the successive approximations. The proof of assertion 1) is complete. □

The Proof of Assertion 2). We now assume the function $\varphi(\eta)$ is analytic. Then the $C^{2,a}$-solution for equation (19.35) obtained above is also analytic (see this well known fact in C. Miranda [1]). Let t_k be a sequence of values of the parameter t converging to t_0 and let, for any t_k, equation (19.35) be solvable. Let H_k denote the solution corresponding to the value t_k. In Subsection 19.3 it was shown that for the solution H of equation (19.35) and its derivatives up to the third order there exists an a priori estimate depending only on the function

φ and its derivatives of order up to three. Thus the sequence of solutions H_k contains a subsequence which converges uniformly along with its second derivatives on S^n. The limiting function H of this subsequence is of class $C^{2,a}$, $0 < a < 1$ and satisfies equation (19.35) when $t = t_0$. From the theorem on analyticity of solutions of elliptic equations it follows that H is analytic. Thus assertion 2) is proved for analytic functions $\varphi(\eta)$. Hence Theorem 19.3 is proved for the case when the function $\varphi(\eta)$ is analytic.

This result can be extended to the class of a three times differentiable functions $\varphi(\eta)$, which are strictly positive on S^n and satisfy the condition

$$\int_{S^n} \eta\varphi(\eta)d\omega = 0. \tag{19.42}$$

We approximate $\varphi(\eta)$ by strictly positive analytic functions $\varphi_k(x)$ in the C^3-norm on S^n, which satisfy the conditions:

a)
$$\inf_{S^n} \varphi_k(x) \geq \varphi_0 = \text{const} > 0$$
for all $k = 1, 2, \ldots, m, \ldots$;

b)
$$\int_{S^n} \eta\varphi_k(\eta)d\omega = 0.$$

For any analytic function $\varphi_k(\eta)$ the Minkowski problem can be solved and we shift to the limiting solution as $\varphi_k(\eta)$ converges to $\varphi(\eta)$. The a priori estimates obtained in Subsection 19.3 ensure the convergence of the sequence of analytic $H_k(\eta)$ to a function $H(\eta)$ of class $C^{2,a}$ on S^n. The functions $H_k(\eta)$ and $H(\eta)$ are solutions of the Minkowski problem for the analytic functions $\varphi_k(\eta)$ and $\varphi(\eta) \in C^3(S^n)$ respectively. From the theorem on smoothness of solutions of elliptic equations it follows that, if $\varphi(\eta) \in C^m(S^n)$, $m \geq 3$, then the solution is of class $C^{m+1,a}(S^n)$. This completes the Pogorelov proof of Theorem 19.3. □

Pogorelov's papers [3], [4], [5] devoted to the Minkowski multidimensional problem aroused a great interest. Since a few basic results in his papers were very briefly presented some of Pogorelov's proofs are difficult to follow. S.Y. Cheng and S.T. Yau [1], [3] then gave complete proofs for the Minkowski problem on S^n and topics related to the Dirichlet problem for multidimensional Monge–Ampere equations.

§20. The Dirichlet Problem for Smooth Elliptic Solutions of N-Dimensional Monge–Ampere Equations

In this section we study the Dirichlet problem for Monge–Ampere equations

$$\det(u_{ij}) = f(x, u, Du) \quad \text{in} \quad \overline{G}, \tag{20.1}$$
$$u = \psi(x) \quad \text{on} \quad \partial G, \tag{20.2}$$

where G is a convex domain in the Euclidean space $E^n = \{x = (x_1, \ldots, x_n)\}$. Here as usual $x_1, x_2, \ldots, x_n$ are Cartesian coordinates in E^n.

The Monge–Ampere operator $\det(u_{ij})$ is elliptic on a function $u(x) \in C^2(\overline{G})$ if and only if the function $u(x)$ is either strictly convex or strictly concave, i.e. the quadratic form d^2u is either positive-definite or negative-definite in $\overline{G}$ respectively. In this section we confine ourselves primarily to convex solution $u(x) \in C^2(\overline{G})$, $n \geq 2$, for equation (20.1) in a bounded convex domain G. We recall that $\overline{G} = G \cup \partial G$ is the closure of G. Thus we must require that $f(x, u, p)$ be a positive function for all $x \in \overline{G}$, $u \in R$, $p \in R^n$.

The theory of concave solutions of Monge–Ampere equations (20.1) is quite similar to the theory of convex solution for the same equation. Therefore we only study the Dirichlet problem for convex solutions of equation (20.1).

In Section 20 two main topics will be studied. The first one is a priori estimates for solutions of equation (20.1) and their derivatives up to third order in the spaces of continuous and Hölder functions. The second one is existence theorems for smooth convex solutions of the Dirichlet problem (20.1–2).

20.1 The Uniqueness and Comparison Theorems

A convex function $u(x) \in C^2(\overline{G})$ is called strictly convex if d^2u is a positive-definite form in $\overline{G}$.

Below we assume that the function $f(x, u, p)$ satisfies the following

Assumption M.A.-1. a) $f(x, u, p)$ is continuous together with its first derivatives $f_u(x, u, p)$ and $f_{p_i}(x, u, p)$, $i = 1, 2, \ldots, n$, in $\overline{G} \times R \times R^n$.

b) The inequalities

$$f(x, u, p) > 0, \tag{20.3}$$
$$f_u(x, u, p) \geq 0 \tag{20.4}$$

hold for all $x \in \overline{G}, u \in R, p = (p_1, \ldots, p_n) \in R^n$.

Theorem 20.1 (The Comparison Theorem). *Let $u(x)$ and $v(x)$ be two strictly convex solutions of class $C^2(\overline{G})$ of equation (20.1) and let the following conditions hold:*

a)

$$\det(u_{ij}) \geq \det(v_{ij}) \quad \text{in} \quad G; \tag{20.5}$$

b)

$$u(x) \leq v(x) \quad \text{on} \quad \partial G; \tag{20.6}$$

c) the assumption M.A.-1 holds.

Then

$$u(x) \leq v(x) \tag{20.7}$$

for all $x \in \overline{G}$.

The proof of this theorem is based on the following lemmas.

Lemma 20.1. *Let $u_t(x) = (1-t)u(x)+tv(x)$, $0 \le t \le 1$. We denote by $U_{ij}(t)$ the cofactor of the element $\frac{\partial^2 u_t(x)}{\partial x_i \partial x_j}$ of the det $\left(\frac{\partial^2 u_t(x)}{\partial x_i \partial x_j}\right)$. Then*

$$\det(v_{ij}) - \det(u_{ij}) = \sum_{i,j=1}^{n} A_{ij}(x) \cdot \delta_{ij}(x), \tag{20.8}$$

where

$$A_{ij}(x) = \int_0^1 U_{ij}(t)dt, \qquad i,j = 1,2,\ldots,n, \tag{20.9}$$

and

$$\delta(x) = v(x) - u(x). \tag{20.10}$$

Moreover the functions $A_{ij}(x)$ are continuous in $\overline{G}$ and the quadratic form $\sum_{i,j} A_{ij}(x)\xi_i\xi_j$ is positive-definite in $\overline{G}$ for all $x \in \overline{G}$, where $\xi = (\xi_1,\ldots,\xi_n)$ is any vector of R^n.

Proof. Since $u(x)$ and $v(x)$ are strictly convex in $\overline{G}$, all the functions $u_t(x) = (1-t)u(x) + tv(x)$, $0 \le t \le 1$ are also strictly convex in $\overline{G}$. Thus the matrix

$$U(t) = \begin{pmatrix} \frac{\partial^2 u_t(x)}{\partial x_1 \partial x_1} & \cdots & \frac{\partial^2 u_t(x)}{\partial x_1 \partial x_n} \\ \cdots & \cdots & \cdots \\ \cdots & \cdots & \cdots \\ \frac{\partial^2 u_t(x)}{\partial x_n \partial x_1} & & \frac{\partial^2 u_t(x)}{\partial x_n \partial x_n} \end{pmatrix}, \tag{20.11}$$

$0 \le t \le 1$, is positive-definite in $\overline{G}$. Clearly

$$\begin{aligned} \det(v_{ij}) - \det(u_{ij}) &= \sum_{i,j=1}^{n} U_{ij}(t) \cdot \frac{d\left(\frac{\partial^2 u_t(x)}{\partial x_i x_j}\right)}{dt} dt \\ &= \sum_{i,j=1}^{n} \left(\int_0^1 U_{ij}(t)dt\right) \cdot (v_{ij}(x) - u_{ij}(x)) \\ &= \sum_{i,j=1}^{n} A_{ij}(x) \cdot \delta_{ij}(x). \end{aligned} \tag{20.12}$$

From (20.9), (20.11) and (20.12) it follows that $A_{ij}(x)$ are continuous in $\overline{G}$ and the quadratic form

$$\sum_{i,j=1}^{n} A_{ij}(x)\xi_i\xi_j$$

is positive-definite in $\overline{G}$. □

Lemma 20.2. *The following identity holds:*

$$f(x, v(x), Dv(x)) - f(x, u(x), Du(x))$$
$$= \sum_{i=1}^{n} B_i(x) \cdot \delta_i(x) - C(x) \cdot \delta(x), \tag{20.13}$$

where

$$B_i(x) = \int_0^1 \frac{\partial f(x, u_t(x), Du_t(x))}{\partial p_i} dx, \; i = 1, 2, \ldots, n, \tag{20.14}$$
$$C(x) = \int_0^1 \frac{\partial f(x, u_t(x), Du_t(x))}{\partial u} dx \tag{20.15}$$

are continuous functions in $\overline{G}$ *and* $C(x) \geq 0$ *for all* $x \in \overline{G}$.

We omit the proof of this lemma, because it is similar to the proof of Lemma 20.1.

Now from Lemmas 20.1 and 20.2, and the conditions of Theorem 20.1 it follows that the function $\delta(x)$ satisfies the inequalities

$$\sum_{i,j=1}^{n} A_{ij}(x) \cdot \delta_{ij}(x) - \sum_{i=1}^{n} B_i(x) \cdot \delta_i(x) - C(x)\delta(x) \leq 0 \text{ on } \overline{G}$$

and

$$\delta(x) \geq 0 \quad \text{on} \quad \partial G.$$

From the Hopf maximum principle (see Subsection 8.9) it follows that $\delta(x) \geq 0$ for all $x \in \overline{G}$. Thus $u(x) \leq v(x)$ everywhere in $\overline{G}$.

The proof of Theorem 20.1 is complete.

Theorem 20.2 (The Uniqueness Theorem). *Let all conditions of Theorem 20.1 be fulfilled except (20.5) and (20.6), which are replaced by*

a)
$$\det(u_{ij}) = \det(v_{ij}) \quad \text{on} \quad G, \tag{20.16}$$

b)
$$u(x) = v(x) \quad \text{on} \quad \partial G. \tag{20.17}$$

Then $u(x) = v(x)$ *everywhere in* $\overline{G}$.

The proof of this theorem follows directly from Theorem 20.1.

20.2 C^0-Estimates for Solutions $u(x) \in C^2(\overline{G})$ of the Dirichlet Problem (20.2) by Subsolutions

In this subsection we consider the Dirichlet problem

$$\det(u_{ij}) = f(x, u, Du) \quad \text{on} \quad G, \tag{20.1}$$
$$u = \psi(x) \quad \text{on} \quad \partial G, \tag{20.2}$$

where G is a bounded convex domain in E^n and ∂G is a closed convex hypersurface in E^n of class C^2. We consider a strictly convex solution $u(x) \in C^2(\overline{G})$, which satisfies the boundary data (20.2) with $\psi(x) \in C^2(\partial G)$.*)

We recall a convex function $w(x) \in C^2(\overline{G})$ is a *weak subsolution* of the Dirichlet problem (20.1–2) if the inequality

$$\det(w_{ij}) \geq f(x, w(x), Dw(x)) \tag{20.18}$$

holds in G and the inequality

$$w(x) \leq \psi(x) \tag{20.19}$$

holds on ∂G.

If for the same function $w(x)$ the inequality (20.19) is replaced by the relation

$$w(x) = \psi(x)$$

for all $x \in \partial G$, then we call $w(x)$ a *strong subsolution* of the Dirichlet problem (20.1–2).

The following theorem, which we are going to prove, is useful for obtaining an a priori estimate of C^0-norms for convex solutions of the Dirichlet problem (20.1–2).

Theorem 20.3. *Let $u(x) \in C^2(\overline{G})$ be any solution of the Dirichlet problem (20.1–2) and let $w(x) \in C^2(\overline{G})$ be a weak convex subsolution of the same Dirichlet problem. If the function $f(x,u,p)$ satisfies Assumption M.A.-1 then the inequalities*

$$W_0 \leq u(x) \leq M \tag{20.20}$$

*hold for all $x \in \overline{G}$, where $M = \sup_{\partial G} \psi(x)$ and $W_0 = \inf_{\overline{G}} w(x)$.**)

Proof. The inequality

$$u(x) \leq M \tag{20.21}$$

is the direct consequence of convexity of the function $u(x)$. Since the strictly convex function $w(x)$ is a weak subsolution of the Dirichlet problem (20.1–2),

$$\delta(x) = w(x) - u(x) \tag{20.22}$$

---

*)Of course the requirements on ∂G, $\psi(x)$ and $u(x)$ can be significantly weakened: ∂G can be only a closed convex hypersurface and $u(x) \in C^2(G) \cap C(\overline{G})$. Since in § 20 we confine ourselves primarily to estimates for existence theorems, we prefer to use the requirements stated above.

*) According to Theorem 20.2 the Dirichlet problem (20.1–2) can not have more than one solution $u(x)$ in $C^2(\overline{G})$.

is nonpositive for all $x \in \partial G$. From Lemmas 20.1 and 20.2 it follows that inequality

$$\begin{aligned}\sum_{i,j=1}^{n} A_{ij}(x)\cdot\delta_{ij}(x) &= \det(w_{ij}) - \det(u_{ij}) \\ &\geq \psi(x, w(x), Dw(x) - \psi(x, u(x), Du(x)) \\ &= \sum_{i=1}^{n} B_i(x)\cdot\delta_i(x) + C(x)\cdot\delta(x) \end{aligned} \tag{20.23}$$

holds for all $x \in G$. Since the quadratic form $A_{ij}(x)\xi_i\xi_j$ is positive definite in $\overline{G}$, $C(x) \geq 0$ in G, and $\delta(x) \leq 0$ for all $x \in \partial G$, according to Hopf's maximum principle $\delta(x) \leq 0$ for all $x \in \overline{G}$. Hence $w(x) \leq u(x)$ for all $x \in \overline{G}$. Thus

$$W_0 = \inf_{\overline{G}} w(x) \leq u(x) \tag{20.24}$$

for all $x \in \overline{G}$. □

Examples. 1) We consider a priori estimates for solutions $u(x) \in C^2(\overline{G})$ of the Dirichlet problem:

$$\det(u_{ij}) = f(x) \quad \text{in} \quad G, \tag{20.25}$$

$$u = \psi(x) \quad \text{on} \quad \partial G. \tag{20.26}$$

According to our assumptions formulated at the beginning of Subsection 20.2 we have

$$0 < f(x) \leq \sup_{\overline{G}} f(x) = F_0 < +\infty, \tag{20.27}$$

$$-\infty < m = \inf_{\partial G} \psi(x) \leq \sup_{\partial G} \psi(x) = M < +\infty. \tag{20.28}$$

Let $x_0 = (x_1^0, \ldots, x_n^0)$ be any inner point of G and $d_0 = \operatorname{diam} G$. Then

$$\overline{G} \subset \overline{U},$$

where $\overline{U}$: $|x - x_0| \leq d_0$ is the closed n-ball with center x_0 and radius d_0. Let

$$\begin{aligned} w(x) &= m + \frac{1}{2}F_0^{1/n}[|x - x_0|^2 - d_0^2] \\ &= m + \frac{1}{2}F_0^{1/n}\left[\sum_{i=1}^{n}(x_i - x_i^0)^2 - d_0^2\right] \end{aligned} \tag{20.29}$$

be a function defined in $\overline{U}$. Clearly

$$\sum_{i=1}^{n}(x_i - x_i^0)^2 - d_0^2 < 0$$

in the open ball U and $\sum_{i=1}^{n}(x_i - x_i^0)^2 - d_0^2 = 0$ on ∂U. Since $\partial G \subset \overline{U}$, for any $x \in \partial G$

$$w(x) \leq m = \inf_{\partial G} \psi(x) \leq \psi(x).$$

Since

$$d^2 w = F_0^{1/n} \cdot \sum_{i=1}^{n} (dx_i)^2 \quad \text{in} \quad \overline{U},$$

$w(x)$ is a strictly convex function in $\overline{U}$. Finally

$$\det(w_{ij}) = \begin{vmatrix} F_0^{1/n} & 0 & 0 & \dots & 0 \\ 0 & F_0^{1/n} & 0 & \dots & 0 \\ \cdots & \cdots & \cdots & \cdots & \cdots \\ \cdots & \cdots & \cdots & \cdots & \cdots \\ 0 & 0 & 0 & \dots & F_0^{1/n} \end{vmatrix} = F_0.$$

Therefore the inequality

$$\det(w_{ij}) = F_0 = \sup_{\overline{G}} f(x) \geq f(x) > 0$$

holds for all $x \in \overline{G}$. Thus $w(x)$ is a weak subsolution for the Dirichlet problem (20.25–26). Now from Theorem 20.3 it follows that the inequalities

$$m - \frac{1}{2} F_0^{1/n} d_0^2 \leq u(x) \leq M \tag{20.30}$$

hold for all $x \in \overline{G}$, where $u(x) \in C^2(\overline{G})$ is any solution of the Dirichlet problem (20.25–26).

2) Bakelman [1], [18], [20] developed various construction for estimating solutions for the Dirichlet problem

$$\det(u_{ij}) = f(x, u, Du) \quad \text{in} \quad G, \tag{20.31}$$
$$u = \psi(x) \quad \text{on} \quad \partial G. \tag{20.32}$$

Below we present one of them connected with weak subsolutions. A few other constructions related to my recent investigations will be presented in Subsection 20.3. We assume that all requirements with respect to bounded convex domain G and its boundary ∂G, functions $f(x, u, p)$ and $\psi(x)$, and solutions $u(x)$ of the problem (20.31–32) are fulfilled. We formulated these requirements at the beginning of Subsection 20.2.

Let x_0 be any interior point of the convex domain G. Without loss of generality we can assume that x_0 is the origin of Cartesian coordinates in E^n. We denote by $\overline{U}$ the closed ball $|x - x_0| \leq d_0$, where $d_0 = \operatorname{diam} G$. Clearly $\overline{G} \subset \overline{U}$. Let $(\rho, \theta) = (\rho, \theta_1, \theta_2, \dots, \theta_{n-1})$ be the spherical coordinates in E^n with pole at the point x_0. We now extend the function $f(x,u,p)$ from $\overline{G} \times R \times R^n$ to $\overline{U} \times R \times R^n$ and keep the same notation $f(x,u,p)$ for the extended function.

Below we assume that $f(x,u,p)$ satisfies the following

Assumption M.A.-2. The function $f(x,u,p)$ is continuous and satisfies the inequalities

$$0 < f(x,u,p) \leq C \cdot (1+|p|^n)^{k/n} \tag{20.33}$$

for all $x \in \overline{U}$, $u \in R$, $p \in R^n$, where $C = \text{const} > 0$ and $k = \text{const} \geq 0$.

Assumption M.A.-2 differs from Assumption M.A.-1. Namely the condition $f_u(x,u,p) \geq 0$, contained in Assumption M.A.-1, provides the validity of comparison and uniqueness theorems for the Dirichlet problem (20.1–2). This inequality is replaced by inequality (20.33) in Assumption M.A.-2. Inequality (20.33) provides neither comparison nor uniqueness theorems. This inequality only describes the order of growth of a positive function $f(x,u,p)$ as $|p| \to +\infty$. Nevertheless inequality (20.33) permits us to construct weak and strong subsolutions for the problem (20.1–2).

First of all we consider the Dirichlet problem

$$\det(w_{ij}) = C \cdot (1+|p|^n)^{k/n} \quad \text{in} \quad U, \tag{20.34}$$

$$w = 0 \quad \text{on} \quad \partial U, \tag{20.35}$$

where U is the ball $|x| < d_0$ in E^n and $d_0 = \operatorname{diam} G$. We are looking for strictly convex solutions $w(x) \in C^2(\overline{U})$. According to Theorem 20.2 the Dirichlet problem (20.34–35) has not more than one solution of class $C^2(\overline{U})$.

We want to find the desired solution in the form $w = w(\rho)$ in $\overline{U}$, where

$$\rho = (1 + x_1^2 + \cdots + x_n^2)^{1/2}.$$

It turns out that for $w(\rho)$ an explicit formula can be found. The Dirichlet problem (20.1–2) for the function $w(\rho)$ becomes:

$$\frac{1}{\rho^{n-1}}(w')^{n-1}w'' = C \cdot (1+|w'|^n)^{k/n} \quad \text{in} \quad [0,d_0], \tag{20.36}$$

$$w'(0) = 0, \quad w(d_0) = 0, \tag{20.37}$$

where $w(\rho)$ is a strictly convex function and $w(\rho) \in C^2[0,d_0]$. Hence $w'(\rho)$ is a strictly increasing positive function in $(0,d_0]$ and $w'(d_0) < +\infty$. Thus $w(\rho)$ is strictly increasing convex function in $[0,d_0]$. Clearly $w(\rho) < 0$ for $\rho \in [0,d_0)$ and $w(d_0) = 0$. Moreover $w(\rho)$ achieves its least value at $\rho = 0$.

First of all we find conditions when problem (20.36–37) is solvable. From (20.36) we obtain

$$\int_{\overline{U}} \frac{1}{\rho^{n-1}} \frac{(w')^{n-1}w''dx}{[1+(w')^n]^{k/n}} \equiv \int_{\overline{U}} \frac{\det(w_{ij})dx}{(1+|\nabla w|^n)^{k/n}} = \int_{\overline{U}} C dx. \tag{20.38}$$

Therefore

$$\sigma_n \int_0^{d_0} \frac{(w')^{n-1} w'' d\rho}{(1+(w')^n)^{k/n}} = C\mu_n d_0^n, \tag{20.39}$$

where σ_n is the area of the unit hypersphere in E^n, and μ_n is the volume of the n-unit ball in E^n. It is well known that $\sigma_n = n\mu_n$. Thus (20.39) becomes

$$\int_0^{d_0} \frac{(w')^{n-1} w'' d\rho}{[1+(w')^n]^{k/n}} = \frac{C}{n} d_0^n. \tag{20.40}$$

Since $w'(\rho) > 0$ for all $\rho \in [0, d_0]$, $t = w'(\rho)$ is strictly increasing function of $\rho \in [0, d_0]$, for which $t(0) = w'(0) = 0$ and $t(d_0) = w'(d_0) < +\infty$. Thus equation (20.40) becomes

$$\frac{C}{n} d_0^n = \int_0^{w'(d_0)} \frac{t^{n-1} dt}{(1+t^n)^{k/n}} \tag{20.41}$$

$$= \begin{cases} \frac{1}{n}(w'(d_0))^n & \text{if } \ k = 0; \\ \frac{1}{n-k}([1+(w'(d_0)^n]^{\frac{n-k}{n}} - 1) & \text{if } \ 0 < k < n; \\ \frac{1}{n}\ln[1+(w'(d_0))^n] & \text{if } \ k = n; \\ \frac{1}{k-n}[1-(1+(w'(d_0))^n)^{-\frac{k-n}{n}}] & \text{if } \ k > n. \end{cases}$$

From (20.41) it follows that the boundary value problem (20.36–37) is solvable if and only if

$$\begin{aligned} &d_0 < +\infty && \text{for} \quad 0 \le k \le n; \\ &C\left(\tfrac{k-n}{n}\right) d_0^n < 1 && \text{for} \quad k > n. \end{aligned} \tag{20.42}$$

We now find an explicit formula for the function $w(\rho)$. First of all we integrate equation (20.36) in the ball $\overline{U}_\rho$: $|x| \le \rho$, where $\rho \le [0, d_0]$. After elementary calculations, similar to calculations leading to equations (20.40) and (20.41), we obtain the formula

$$\frac{1}{n} C\rho^n = \begin{cases} \frac{1}{n}[w'(\rho)]^n & \text{if } \ k = 0; \\ \frac{1}{n-k}([1+[w'(\rho)]^n]^{\frac{n-k}{n}} - 1) & \text{if } \ 0 < k < n; \\ \frac{1}{n}\ln(1+[w'(\rho)]^n) & \text{if } \ k = n; \\ \frac{1}{k-n}\left[1-\left(\frac{1}{1+(w'(\rho))^n}\right)^{\frac{k-n}{n}}\right] & \text{if } \ k > n. \end{cases} \tag{20.43}$$

Now four cases depending on relations between k and n should be considered.

1) $k = 0$. In this case

$$w'(\rho) = (C)^{1/n}\rho, \qquad 0 \le \rho \le d_0.$$

Thus

$$w(\rho) = w(0) + \frac{1}{2}C^{1/n}\rho^2, \qquad 0 \le \rho \le d_0.$$

Since $w(d_0) = 0$ (see condition (20.37)),

$$w(0) = -\frac{1}{2}C^{1/n}d_0^2.$$

Thus the final formula is

$$w(\rho) = \frac{1}{2}C^{1/n}[\rho^2 - d_0^2]. \tag{20.44}$$

(Compare formula (20.44) with the formula (20.29).)

2) $0 < k < n$. In this case

$$w'(\rho) = \left\{\left[1 + \frac{n-k}{n}C\rho^n\right]^{\frac{n}{n-k}} - 1\right\}^{\frac{1}{n}}, \qquad 0 \le \rho_0 \le d_0.$$

Thus

$$w(\rho) = w(0) + \int_0^\rho \left[\left(1 + \frac{n-k}{n}Cs^n\right)^{\frac{n}{n-k}} - 1\right]^{\frac{1}{n}} ds, \qquad 0 \le \rho \le d_0,$$

where

$$w(0) = -\int_0^{d_0} \left[\left(1 + \frac{n-k}{k}C \cdot s^n\right)^{\frac{n}{n-k}} - 1\right]^{\frac{1}{n}} ds.$$

3) $k = n$. In this case

$$w'(\rho) = [e^{C\rho^n} - 1]^{1/n}, \qquad 0 \le \rho \le d_0.$$

Thus

$$w(\rho) = w(0) + \int_0^\rho [e^{Cs^n} - 1]^{1/n} ds, \qquad 0 \le \rho \le d_0,$$

where

$$w(0) = -\int_0^{d_0} [e^{C \cdot s^n} - 1] ds.$$

4) $k > n$. In this case

$$w'(\rho) = \left(\frac{1}{[1 - \frac{k-n}{n}C\rho^n]^{n/k-n}} - 1\right)^{\frac{1}{n}}, 0 \le \rho \le d_0 < \frac{1}{C}\left(\frac{n}{k-n}\right)^{\frac{1}{n}}.$$

Thus

$$w(\rho) = w(0) + \int_0^\rho \left[\frac{1}{\left[1 - \frac{k-n}{n} C s^n\right]^{n/k-n}} - 1 \right]^{\frac{1}{n}} ds,$$

$$0 \le \rho \le d_0 < \frac{1}{C} \cdot \left(\frac{n}{k-n}\right)^{\frac{1}{n}},$$

where

$$w(0) = -\int_0^{d_0} \left[\frac{1}{\left[1 - \frac{k-n}{n} C s^n\right]^{n/k-n}} - 1 \right]^{\frac{1}{n}} ds.$$

Thus an explicit formula for $w(\rho)$ has been constructed. If $0 \le k \le n$ then the radius $d_0 = \operatorname{diam} G$ is not subjected to any limitations depending on the value of the constant $C > 0$. If $k > n$ then we have the restriction

$$d_0 < \left(\frac{n}{k-n}\right)^{\frac{1}{n}} \cdot \frac{1}{C}.$$

We now consider the function

$$v(x) = m + w(x),$$

where $m = \inf_{\partial G} \psi(x)$. Since $\overline{G} \subset \overline{U}$, $v(x) \le m \le \psi(x)$ for all $x \in \partial G$. On the other hand

$$\frac{\det(v_{ij}(x))}{(1 + |\nabla v|^n)^{k/n}} = \frac{\det(w_{ij}(x))}{(1 + |\nabla w|^n)^{k/n}} = C \ge \frac{\det(u_{ij})}{(1 + |\nabla u|^n)^{k/n}} > 0. \tag{20.45}$$

Let

$$\delta(x) = v(x) - u(x)$$

and $u_t(x) = (1-t)u(x) + tv(x)$, $0 \le t \le 1$. Clearly $\delta(x)$ and $u_t(x)$ are functions of class $C^2(\overline{G})$, $\delta(x) \le 0$ on ∂G, and $u_t(x)$ is strictly convex in $\overline{G}$ for all $t \in [0,1]$.

Since

$$0 \le \frac{\det(v_{ij})}{(1 + |\nabla v|^n)^{k/n}} - \frac{\det(u_{ij})}{(1 + |\nabla u|^n)^{k/n}} = \sum_{i,j=1}^{n} A_{ij}(x)\delta_{ij}(x)$$

$$+ \sum_{i=1}^{n} B_i(x)\delta_i(x)$$

where $A_{ij}(x)$ and $B_j(x)$ are continuous in $\overline{G}$ and the form $A_{ij}(x)\xi_i\xi_j$ is positive definite in $\overline{G}$*), from Hopf's maximum principle it follows that $\delta(x) \le 0$ in $\overline{G}$. Thus the function

$$v(x) = m + w([x_1^2 + \cdots + x_n^2]^{1/2})$$

*) This assertion can be proved in the same way as Lemma 20.1 and 20.2.

is a weak subsolution of the Dirichlet problem (20.34–35).

In all constructions and formulas used in example b) we operate with the number $d_0 = \operatorname{diam} G$. There will be no changes if we replace the ball $\overline{U}$: $|x| \le d_0$ by a smaller ball $\overline{V}$: $|x| \le r$ such that $\overline{G} \subset \overline{V} \subset \overline{U}$. It is possible to make our results even stronger if we choose the point $x_0 \in \operatorname{int} G$ and the radius r_0 of the admissible ball $\overline{V}_0$: $|x - x_0| \le r_0$ under the condition $r_0 = \inf r$, where the infimum is taken on the set of all admissible balls $\overline{V}$, i.e. for any $\overline{V}$ such that $\overline{G} \subset \overline{V}$. We call the number r_0 the *exterior radius* of $\overline{G}$. In all formulas derived above the number d_0 can be replaced by r_0.

All conclusions made above can be presented in the following theorem.

Theorem 20.4. *Let the function $f(x,u,p)$ be continuous and satisfy the inequalities*

$$0 < f(x,u,p) \le C \cdot (1 + |p|^n)^{k/n}$$

for all $x \in \overline{U}, u \in R, p \in R^n$, where $\overline{U}$ is the ball $|x| \le d_0$ in R^n, $C = \text{const} > 0$, $k = \text{const} \ge 0$. Let G be a convex domain in E^n such that $\overline{G} \subset \overline{U}$ and ∂G is of class C^2. Then for any solution $u(x) \in C^2(\overline{G})$ for the Dirichlet problem (39.1–2) the following inequalities hold:

$$m + w(0) \le m + w(x) \le u(x) \le M \tag{20.46}$$

for all $x \in \overline{G}$, where $m = \inf_{\partial G} \psi(x)$, $M = \sup_{\partial G} \psi(x)$ and $w(x)$ is a convex function defined by formulas presented above.

If $0 \le k \le n$, then d_0 can be any finite positive number. If $k > n$, then d_0 must satisfy the restriction

$$d_0 < \left(\frac{n}{k-n}\right)^{1/n} \cdot \frac{1}{C}. \tag{20.47}$$

20.3 Geometric Estimates of Convex Solutions for Monge–Ampere Equations

In this subsection we present geometric estimates for convex solutions of the Dirichlet problem

$$\det(u_{ij}) = f(x,u,Du) \quad \text{in} \quad G, \tag{20.48}$$

$$u = 0 \quad \text{on} \quad \partial G,^{*)} \tag{20.49}$$

where G is a bounded convex domain in E^n and ∂G is a closed convex hypersurface. These estimates are related to both generalized and smooth solutions

*) For brevity we consider the boundary condition $u = 0$ on ∂G.

of class C^m, $m \geq 2$, of the Dirichlet problem (20.48–49). They were recently obtained by Bakelman [18].

Assumption M.A.-3. The function $f(x,u,p)$ satisfies the following conditions:

1) $f(x,u,p)$ is nonnegative and continuous for all $x \in \overline{G}$, $u \leq 0$ and $p \in R^n = \{p = (p_1, \ldots, p_n)\}$;
2) the inequalities

$$0 \leq [f(x,u,p)]^{1/n} \leq \frac{\varphi(x) + q(x)|u| + d(x)|p|}{[R(|p|)]^{1/n}} \tag{20.50}$$

hold for all $x \in \overline{G}$, $u \leq 0$ and $p \in R^n$, where

a) $\varphi(x) \geq 0$ in G and $\varphi(x) \in L^n(G)$;
b) $q(x) \geq 0$ in G and $q(x) \in L^n(G)$;
c) $d(x) \geq 0$ in G and $d(x) \in L^n(G)$;
d) $R(|p|)$ is positive and continuous in R^n, and also $R(|p|)$ is decreasing as function of $|p| \in [0, +\infty)$;
e) $\lim\limits_{|p| \to +\infty} R(|p|) = R(+\infty) > 0$;

First of all we apply the Hölder inequality to the right-hand side of inequality (20.50). Then we obtain

$$\begin{aligned} 0 &\leq \frac{\varphi(x) + q(x)|u| + d(x)|p|}{[R(|p|)]^{1/n}} \\ &\leq \frac{[\alpha^n \cdot (\varphi(x) + q(x)|u|)^n + [d(x)]^n]^{1/n}}{[R(|p|)]^{1/n}} \\ &\quad \times [(\alpha)^{-\frac{n}{n-1}} + |p|^{\frac{n}{n-1}}]^{\frac{n-1}{n}}. \end{aligned} \tag{20.51}$$

Let $u(x) \in C(\overline{G})$ be a convex generalized solution of the Dirichlet problem (20.48). According to properties of such solutions for equations (20.48) (see Chapter 3, §§ 11, 12) we obtain the following formula

$$\omega(Q_\alpha, u, G) \equiv \int_{\chi_u(G)} Q_\alpha(|p|)dp = \int_G Q_\alpha(|\nabla u|) \det(u_{ij})dx, \tag{20.52}$$

where $\omega(Q_\alpha, u, G)$ is the Q_α-curvature of the convex function $u(x)$ on the set G generated by the function

$$Q_\alpha(|p|) = \frac{R(|p|)}{[\alpha^{-\frac{n}{n-1}} + |p|^{\frac{n}{-1}}]^{n-1}} \tag{20.53}$$

**)Here we omit the study of more general classes of admissible functions $R(|p|)$ because it is connected with additional complicated considerations. See my paper [20] and also my results related to geometric maximum principles for quasilinear equations, Chapter 8, § 27.

where $\alpha > 0$ is any real number. Since the convex function $u(x) \in C(\overline{G})$, $q(x)|u(x)| \in L^n(G)$. Now from inequalities (20.50) and (20.51) it follows that

$$\omega(Q_\alpha, u.G) \leq \alpha^n[\|\varphi(x)\|_{L^n(G)} + \|q(x)u(x)\|_{L^n(G)}]^n + \|d(x)\|^n_{L^n(G)}, \quad (20.54)$$

where $\alpha > 0$ is any real number.

Let $K(x_0)$ be the convex cone with vertex at the point $(x_0, u(x_0))$ and base ∂G, where x_0 is an arbitrary interior point of G. As we know (see Chapter 3, § 9) the inequality

$$\omega(Q_\alpha, k_{x_0}, G) \leq \omega(Q_\alpha, u, G) \quad (20.55)$$

holds. If $\widetilde{K}(x_0)$ is the normalized convex cone with vertex at the point $(x_0, -1)$ and base ∂G, then

$$k_{x_0}(x) = |u(x_0)|\tilde{k}_{x_0}(x) \quad (20.56)$$

for any $x \in \overline{G}$. In (20.55) and (20.56), $k_{x_0}(x)$ and $\tilde{k}_{x_0}(x)$ are convex functions in $\overline{G}$, whose graphs are $K(x_0)$ and $\widetilde{K}(x_0)$ respectively.

If $K^*(x_0)$ and $\widetilde{K}^*(x_0)$ are the normal (tangential) images of the cones $K^*(x_0)$ and $\widetilde{K}^*(x_0)$, then

$$K^*(x_0) = |u(x_0)| \cdot K^*(x_0). \quad (20.57)$$

Thus from the definition of the Q_α-curvature of convex functions and relations (20.55–57) it follows that

$$\begin{aligned}\omega(Q_\alpha, k_{x_0}, G) &= \int_{|u(x_0)|K^*(x_0)} Q_\alpha(|p|)dp \quad (20.58)\\ &= \int_{S^{n-1}} \left[\int_0^{|u(x_0)|\rho_{x_0}(\theta)} Q_\alpha(s)s^{n-1}ds\right] d\sigma_\theta,\end{aligned}$$

where S^{n-1} is the unit sphere $|p| = 1$ in the Euclidean space $R^n = \{p = (p_1, \ldots, p_n), \theta \in S^{n-1}$, $d\sigma_\theta$ is the element of the area of S^{n-1}, (ρ, θ) is a spherical system of coordinates in R^n with the beginning at the point $p = (0, 0, \ldots, 0)$, and finally $\rho = \rho_{x_0}(\theta)$ is the equation of $\partial\widetilde{K}^*(x_0)$ in this spherical system.

We now estimate from below the quantity $\omega(Q_\alpha, k_{x_0}, G)$. First of all we study the properties of the intrinsic integral in (20.58).

Below we use the inequality

$$\begin{aligned}&\int |u(x_0)|\rho_{x_0}(\theta)_0 \frac{R(s)s^{n-1}ds}{[\alpha^{-\frac{n}{n-1}} + s^{\frac{n}{n-1}}]^{n-1}}\\ &\geq (R + \infty)\int_0^{|u(x_0)|\rho_{x_0}(\theta)} \frac{s^{n-1}ds}{[\alpha^{-\frac{n}{n-1}} + s^{\frac{n-}{n-1}}]^{n-1}} \quad (20.59)\end{aligned}$$

and set $\mu = \ln \rho_{x_0}(\theta)$. Then

$$|u(x_0)|\rho_{x_0}(\theta) = e^{\ln \rho_{x_0}(\theta)} = e^\mu.$$

The function

$$F(\mu) = \int_0^{|u(x_0)|e^\mu} \frac{s^{n-1}ds}{[\alpha^{-\frac{n}{n-1}} + s^{\frac{n}{n-1}}]^{n-1}} \tag{20.60}$$

is strictly convex with respect to μ. This fact follows directly from the formula

$$\frac{dF}{d\mu} = [(\alpha|u(x_0)|e^\mu)^{-\frac{n}{n-1}} + 1]^{-(n-1)},$$

which shows that $\frac{dF}{d\mu}$ is a strictly increasing function. According to the concept of mapping's mean and Lemma 27.1 (see Chapter 8, § 27) we obtain the following inequality

$$\frac{1}{\sigma_n} \int_{S^{n-1}} F(\ln \rho_{x_0}(\theta))d\sigma_\theta \geq F\left(\frac{1}{\sigma_n} \int_{S^{n-1}} \ln \rho_{x_0}(\theta)d\sigma_\theta)\right). \tag{20.61}$$

Let

$$\phi(x_0) = e^{\frac{1}{\sigma_n} \int_{S^{n-1}} \ln \rho_{x_0}(\theta)d\sigma_\theta}. \tag{20.62}$$

Then from (20.58–62) *we obtain the first important inequality*

$$\omega(Q_\alpha, k_{x_0}, G) \geq R(+\infty)\sigma_n \int_0^{|u(x_0)|\phi(x_0)} \frac{s^{n-1}ds}{[\alpha^{-\frac{n}{n-1}} + s^{\frac{n}{n-1}}]^{n-1}}. \tag{20.63}$$

Our next step is devoted to calculations with the right-hand side of inequality (20.63). We introduce the new variable

$$t = 1 + (\alpha s)^{\frac{n}{n-1}}. \tag{20.64}$$

Then

$$dt = \frac{n}{n-1}(\alpha s)^{\frac{1}{n-1}}\alpha ds$$

and inequality (20.63) becomes

$$\rho_{x_0}(\theta) = \frac{1}{d_0} e^{e^{\ln \ln[d_0 \cdot \rho_{x_0}(\theta)]}}. \tag{20.65}$$

Let $\mu = \ln \ln(d_0 \cdot \rho x_0(\theta))$ and

$$F(\mu) = \int_0^{\frac{|u(x_0|}{d_0} e^{e^\mu}} \frac{R(s)s^{n-1}ds}{(\alpha^{-\frac{n}{n-1}} + s^{\frac{n-}{n-1}})^{n-1}}. \tag{20.66}$$

Then

$$\frac{dF}{d\mu} = R\left(\frac{|u(x_0)|}{d_0} e^{e^\mu}\right) \ln(e^{e^\mu} + e) \cdot \frac{\ln e^{e^\mu}}{\ln(e^{e^\mu} + e)}$$
$$\cdot \left(1 + \left(\frac{\alpha|u(x_0)|}{d_0} e^{e^\mu}\right)^{-\frac{n}{n-1}}\right)^{n-1}.$$

From condition f) it follows that $\frac{dF}{d\mu}$ is a nondecreasing function of μ as the product of three nondecreasing functions of μ. Hence $F(\mu)$ is a convex function with respect to μ. Now from convexity of the function $F(\mu)$ *we obtain the second important inequality*

$$\omega(Q_\alpha, k_{x_0}, G) \geq \sigma_n \int_0^{\frac{|u(x_0)|}{d_0}\psi(x_0)} \frac{R(s)s^{n-1}ds}{\left(\alpha^{-\frac{n}{n-1}} + s^{\frac{n}{n-1}}\right)^{n-1}}, \tag{20.67}$$

where

$$\psi(x_0) = e^{e^{\frac{1}{\sigma_n}} \int_{S^{n-1}} \ln\ln(d_0 \cdot \rho_{x_0}(\theta)) d\sigma_\theta}. \tag{20.68}$$

Our next step is devoted to computations with integrals at the right-hand sides of inequalities (20.63) and (20.67). For both integrals we introduce the same new variable

$$t = 1 + (\alpha s)^{\frac{n}{n-1}}. \tag{20.69}$$

Then

$$dt = \frac{n}{n-1}(\alpha s)^{\frac{1}{n-1}} ds \tag{20.70}$$

and inequality (20.63) becomes

$$\omega(Q_\alpha, u, G) \geq R(+\infty)\sigma_n \cdot \int_1^{1+L(x_0)} \frac{(t-1)^{n-2}dt}{t^{n-1}} \tag{20.71}$$

$$= \frac{n-1}{n} R(+\infty)\sigma_n \left[\int_1^{1+L(x_0)} \frac{dt}{t} - \int_0^{1+L(x_0)} \frac{t^{n-2} - (t-1)^{n-2}}{t^{n-1}} dt\right],$$

where $L(x_0) = [\alpha|u(x_0)|\phi(x_0)]^{\frac{n}{n-1}}$. Since $t^{n-2} - (t-1)^{n-2} > 0$ for all $t \in [1, +\infty)$, from (20.54) and (20.71) *we obtain the first main inequality.*

$$\begin{aligned} \frac{n-1}{n} R(+\infty)\sigma_n \ln[1 + (\alpha|u(x_0)|\phi(x_0))^{\frac{n}{n-1}}] &\leq \alpha^n [\|\varphi(x)\|_{L^n(G)} \\ &+ \|q(x)u(x)\|_{L^n(G)}]^n \\ &+ \|d(x)\|^n_{L^n(G)} + \frac{n-1}{n}\sigma_n \cdot \tau_n \cdot R(+\infty), \end{aligned} \tag{20.72}$$

where

$$\tau_n = \int_1^{+\infty} \frac{t^{n-2} - (t-1)^{n-2}}{t^{n-1}} dt$$

is a finite positive number depending only on n.

We now are ready to prove the main results of this subsection.

Theorem 20.6. *Let $u(x)$ be a convex nonzero solution of the Dirichlet problem*

$$\det(u_{ij}) = f(x, u, Du) \quad in \quad G, \tag{20.48}$$

$$u = 0 \quad on \quad \partial G, \tag{20.49}$$

where G is a bounded convex domain in E^n, and ∂G is a closed convex hypersurface. Let Assumption M.A.-3 hold in the following form: condition 1) and conditions 2a, 2b, 2c, 2d, 2e are fulfilled. We also assume that

$$\|\varphi(x)\|_{L^n(G)} + \|q(x)\|_{L^n(G)} > 0^{*)}. \tag{20.73}$$

Then the following estimate

$$|u(x_0)| \leq \frac{[\|\varphi(x)\|_{L^n(G)} + \|q(x)u(x)\|_{L^n(G)}]}{\phi(x_0)(\sigma_n)^{1/n}} \exp\left(\frac{n-1}{n}\tau_n + \frac{1+(\sigma_n)^{-1}\|d(x)\|^n_{L^n(G)}}{R(+\infty)}\right) \tag{20.74}$$

holds for any interior point $x_0 \in G$.

---

*) If $\|\varphi(x)\|_{L^n(G)} + \|q(x)\|_{L^n(G)} = 0$, then inequality (20.50) becomes

$$0 \leq [f(x,u,p)]^{1/n} \leq d(x)\frac{|p|}{R(|p|)^{1/n}} \leq \frac{d(x)\cdot(|p|^n+1)^{1/n}}{R(|p|)^{1/n}}.$$

Thus both generalized and smooth of class C^m, $m \geq 2$, convex solutions $u(x)$ of the Dirichlet (20.48–49) satisfy the inequality

$$0 \leq \frac{R(|Du|)}{1+|Du|^n}\det(u_{ij}) \leq [d(x)]^n$$

for all $x \in G$. Since $R(|p|) \geq (R+\infty) = \text{const} > 0$,

$$\int_{R^n}\frac{R(|p|)dp}{1+|p|^n} = +\infty.$$

According to condition 2c we have the inequality

$$\int_G [d(x)]^n dx < \int_{R^n}\frac{R(|p|)dp}{1+|p|^n} = +\infty,$$

which guarantees the estimate

$$0 \geq u(x) \geq -g(\|d(x)\|_{L^n(G)})\operatorname{diam} G$$

for any generalized (smooth) convex solution of the Dirichlet problem (20.48–49). Here $\rho = g(t)$, $t \in [0,+\infty)$, is the inverse for a continuous strictly increasing function

$$t = w(\rho) = \int_{|p|\leq\rho}\frac{R(|p|)dp}{1+|p|^n}, \qquad 0 \leq \rho < +\infty$$

(see details in Chapter 3, §§ 10, 11, 12). Therefore only the case (20.73) is interesting for considerations.

Proof. If $u(x)$ is any generalized convex nonzero solution of the Dirichlet problem (20.48–49), then $u(x) < 0$ at every interior point of G. Thus from this fact and assumption (20.73) it follows that

$$m = (\sigma_n)^{-1/n}[\|\varphi(x)\|_{L^n(G)} + \|q(x)u(x)\|_{L^n(G)}] > 0. \tag{20.75}$$

We can assume that

$$|u(x_0)| \cdot \phi(x_0) > m, \tag{20.76}$$

since in the opposite case inequality (20.74) is clearly valid.

Let

$$\alpha^{\frac{n}{n-1}} = m^{-\frac{n}{n-1}} - [|u(x_0)| \cdot \phi(x_0)]^{-\frac{n}{n-1}}. \tag{20.77}$$

This choice of α is admissible, since from (20.76) it follows that $\alpha^{\frac{n}{n-1}} > 0$. Clearly

$$1 + (\alpha|u(x_0)|\phi(x_0))^{n/n-1} = \left(\frac{|u(x_0)| \cdot \phi(x_0)}{m}\right)^{\frac{n}{n-1}}. \tag{20.78}$$

If we use relations (20.77) and (20.78), then the main inequality (20.72) becomes

$$R(+\infty)\ln\frac{|u(x_0)|\phi(x_0)}{m} \leq 1 + \frac{1}{\sigma_n}\|d(x)\|^n_{L^n(G)} + \frac{n-1}{n}R(+\infty)\tau_n.$$

Thus we obtain the desired inequality

$$|u(x_0)| \leq \frac{[\|\varphi(x)\|_{L^n(G)} + \|q(x)u(x)\|_{L^n(G)}]}{(\sigma_n)^{1/n}\phi(x_0)} \exp\left(\frac{n-1}{n}\tau_n + \frac{1 + (\sigma_n)^{-1}\|d(x)\|^n_{L^n(G)}}{R(+\infty)}\right).$$

Theorem 20.6 is proved. □

Remarks to Theorem 20.6.

1) For brevity we introduce the following notation:

$$\|g\| = \|g(x)\|_{L^n(G)} \tag{20.79}$$

for any function $g(x) \in L^n(G)$;

$$A(d, R, n) = \frac{1}{\sigma_n^{1/n}}\exp\left\{\frac{n-1}{n}\tau_n + \frac{1 + \sigma_n^{-1}\|d\|^n}{R(+\infty)}\right\}, \tag{20.80}$$

where

$$\tau_n = \int_1^{+\infty}\frac{t^{n-2} - (t-1)^{n-2}}{t^{n-1}}dt.$$

With these notation inequality (20.74) becomes

$$|u(x_0)| \le \frac{\|\varphi\| + \|qu\|}{\phi(x_0)} \cdot A(d,R,n). \tag{20.81}$$

From the definition of the function $\phi(x_0)$ it follows that

$$0 < \frac{1}{\phi(x_0)} \le \operatorname{diam} G. \tag{20.82}$$

Thus (20.81) and (20.82) lead to the uniform estimate

$$|u(x)| \ge [\|\varphi\| + \|qu\|] \cdot A(d,R,n) \cdot \operatorname{diam} G \tag{20.83}$$

for all $x \in \overline{G}$.

2) It is important to exclude $\|qu\|$ from the right-hand side of inequalities (20.81) and (20.83).

If $\|q\| = 0$, then we obtain directly from (20.81) and (20.83) estimates of such kind:

$$|u(x_0)| \le \frac{\|\varphi\|}{\phi(x_0)} \cdot A(d,R,n) \tag{20.84}$$

and

$$|u(x_0)| \le \|\varphi\| \cdot A(d,R,n) \cdot \operatorname{diam} G \tag{20.85}$$

for any generalized (smooth) convex solution, vanishing on ∂G.

We will study the case when $\|q\| \neq 0$ and find the condition, which allows us to exclude $\|qu\|$ from estimates (20.81) and (20.83).

Theorem 20.7. *If all conditions of Theorem 20.6 are fulfilled and the inequality*

$$\left\| \frac{q(x)}{\phi(x)} \right\|_{L^n(G)} < \frac{1}{A(d,R,n)} \tag{20.86}$$

holds, then

$$|u(x_0)| \le \frac{\|\varphi\| \cdot A(d,R,n)}{\left(1 - A(d,R,n)\left\|\frac{q}{\phi}\right\|\right) \cdot \phi(x_0)} \tag{20.87}$$

for all interior points x_0 of G.

Proof. From inequality (20.81) it follows that

$$\|qu\| \le [\|\varphi\| + \|qu\|] \cdot A(d,R,n) \left\| \frac{q}{\phi} \right\|.$$

According to Remark 1 all norms in this inequality are considered in $L^n(G)$. Hence

$$\|qu\| \cdot \left(1 - A(d,R,n)\left\|\frac{q}{\phi}\right\|\right) \le \|\varphi\| \cdot A(d,R,n) \left\|\frac{q}{\phi}\right\|. \tag{20.88}$$

From condition (20.86) we obtain the inequality

$$1 - A(d,R,n)\left\|\frac{q}{\phi}\right\| > 0. \tag{20.89}$$

Now from (20.88) and (20.89) we obtain the following important inequality

$$\|qu\| \le \frac{\|\varphi\| \cdot A(d,R,n)\left\|\frac{q}{\phi}\right\|}{1 - A(d,R,n)\left\|\frac{q}{\phi}\right\|}. \tag{20.90}$$

Finally we apply estimate (20.90) to inequality (20.81) and derive the desired estimate

$$|u(x_0)| \le \frac{\|\varphi\| \cdot A(d,R,n)}{\left(1 - A(d,R,n)\left\|\frac{q}{\phi}\right\|\right)\phi(x_0)}. \tag{20.91}$$

The proof of Theorem 20.7 is complete. □

3) From (20.91) and (20.82) we obtain the uniform estimate

$$|u(x)| \le \frac{\|\varphi\| \cdot A(d,R,n) \cdot \operatorname{diam} G}{\left(1 - A(d,R,n) \cdot \left\|\frac{q}{\phi}\right\|\right)}$$

for all $x \in \overline{G}$, where $u(x)$ is any generalized (smooth of class C^m, $m \ge 2$) convex solution of the Dirichlet problem (20.48–49).

20.4 Geometric Estimates of the Gradient of Convex Solutions for Monge–Ampere Equations

In this subsection we present geometric estimates of the gradients of convex solutions $u(x) \in C^2(\overline{G})$ of Monge–Ampere equations

$$\det(u_{ij}) = f(x,u,p) \tag{20.92}$$

for all $x \in \overline{G}$. Below we assume that $f(x,u,p) > 0$ in $\overline{G} \times R \times R^n$ and $f(x,u,p)$ is continuous in the same domain $\overline{G}$. These estimates were recently developed by the author on the base of his results [7], related to convex solutions of class C^2 for two-dimensional elliptic Monge–Ampere equations

$$u_{11}u_{22} - u_{12}^2 = f(x_1,x_2,u,u_1,u_2). \tag{20.93}$$

Below we assume that ∂G is a convex hypersurface of class $C^{2,a}$ and all principal curvatures $k_1(x),\ldots,k_{j-1}(x)$ of ∂G at any point $x \in \partial G$ are uniformly bounded from below by a number $\chi_0 = \text{const} > 0$, i.e.

$$k_i(x) \ge \chi_0, \tag{20.94}$$

for all $x \in \partial G$, $i = 1, 2, \ldots, n-1$. We consider convex solutions $u(x) \in C^m(\overline{G})$, $m \geq 2$, of the Dirichlet problem

$$\det(u_{ij}) = f(x, u, p) \quad \text{in} \quad G, \tag{20.95}$$

$$u = \psi(x) \quad \text{on} \quad \partial G, \tag{20.96}$$

where $\psi(x) \in C^{2,a}(\partial G), 0 < a < 1$.

Let S be the $(n-1)$-dimensional surface in E^{n+1} defined by the equation

$$S\colon\ z = \psi(x), \qquad x \in \partial G. \tag{20.97}$$

Then the *lower twisting* $M_1(S)$ of S (see Subsection 28.3) is a finite nonnegative number, which is estimated from above by $C^{2,a}$-norms of the function $\psi(x)$ and of functions defining ∂G. Therefore for any point $X_0(x_0, \psi(x_0)) \in S$, $x_0 \in \partial G$, there exists a hyperplane

$$\gamma_{X_0}\colon\ z = a_1 x_1 + a_2 x_2 + \cdots + a_n x_n + b \tag{20.98}$$

such that:

1) $\sum\limits_{i=1}^{n} a_i^2 \leq M_L(S)$;
2) γ_{X_0} passes through the tangent plane of S at the point $X_0 \in S$;
3) the surface S lies over the hyperplane γ_{X_0}.

From conditions imposed on ∂G it follows that there exists the least closed n-ball $\overline{U}(x_0)$ for any point $x_0 \in \partial G$ such that:

a) $\overline{G} \subset \overline{U}(x_0)$,
b) $\overline{U}(x_0)$ touches ∂G from outside at the point x_0;
c) if $r(x_0)$ is the radius of $\overline{U}(x_0)$, then

$$r(x_0) \leq \frac{1}{\chi_0}. \tag{20.99}$$

Let $r(\partial G) = \sup\limits_{x_0 \in \partial G} r(x_0)$, then

$$r(\partial G) \leq \frac{1}{\chi_0}. \tag{20.100}$$

Let $x_0 \in \partial G$ be any fixed point of ∂G. Without loss of generality we can assume that the origin of Cartesian coordinates is chosen at the center x_0 of the ball $\overline{U}(x_0)$. Then

$$\sum_{i=1}^{n} x_i^2 \leq [r(x_0)]^2.$$

Let $u(x) \in C^m(\overline{G})$, $m \geq 2$, be a convex solution of the Dirichlet problem (20.95–96) and let $v(x) \in C^2(\overline{U}(x_0))$ be a convex function satisfying the following conditions:

$$1) \qquad v = a_1x_1 + \cdots + a_nx_n + b \quad \text{on} \quad \partial\overline{U}(x_0);$$
$$2) \qquad v(x) \leq u(x) \quad \text{in} \quad U(x_0),$$

where $z = \sum_{i=1}^{n} a_ix_i + b$ is the equation of the hyperplane γ_{X_0}. We recall that $X_0 = (x_0, \psi(x_0))$. Then for the exterior normal derivatives $\frac{\partial u}{\partial n}$ and $\frac{\partial v}{\partial n}$ the following inequalities

$$0 \leq \frac{\partial u}{\partial n} \leq \frac{\partial v}{\partial n}$$

hold at the point $x_0 \in \partial G$. Therefore the estimate of $|\nabla u|$ at the point $x_0 \in \partial G$ can be reduced to reconstruction of an auxiliary convex function $v(x)$, for which $\frac{\partial v}{\partial n}$ at the point $x_0 \in \partial G \cap \partial\overline{U}(x_0)$ can be estimated from above by prescribed data of the Dirichlet problem (20.95–96).

Let the arbitrary point $x_0 \in \partial G$ be fixed. First we extend the function $f(x, u, p)$ into the set $\overline{U}(x_0) \times R \times R^n$. We will use the same notation $f(x, u, p)$ for this extension. Below we assume that $f(x, u, p)$ is continuous in $\overline{U}(x_0) \times R \times R$ and satisfies the following inequalities

$$0 < f(x, u, p) \leq \phi_0 \cdot Q(|p|) \tag{20.101}$$

on the same set $\overline{U}(x_0) \times R \times R^n$, where $0 < \phi_0 = \text{const} < +\infty$ and $Q(|p|)$ is a continuous function in R^n. According to the convex majorants method (see Subsection 28.3) we introduce the function

$$N\left(|p|, S, \frac{1}{Q}\right) = [\sup Q(|p|)]^{-1},$$

where "sup" is taken in the n-ball $|p - \bar{p}|^2 \leq M_L(S)$. We recall that the *lower twisting* $M_L(S)$ was introduced in Subsection 28.3. The function N is continuous in R^n and depends only on $|p|$. Below we denote this function by $N\left(\frac{1}{Q}\right)$.

We now prove that the inequality

$$\mu_n\phi_0 \cdot [r(\partial G)]^n < \int_{R^n} N\left(\frac{1}{Q}\right) dp \tag{20.102}$$

guarantees an upper finite estimate for $|\nabla u|$ on ∂G. If such an estimate is obtained, then we also obtain a finite estimate from above for $|\nabla u|$ in the closed convex domain $\overline{G}$. The last assertion follows from the fact that for any convex function $w(x) \in C^1(\overline{G})$ the function $|\nabla w|$ takes its maximum on ∂G.

The Dirichlet problem

$$\det(w_{ij}) = \frac{\phi_0}{N\left(\frac{1}{Q}\right)} \quad \text{in} \quad U(x_0); \tag{20.103}$$

$$w = 0 \quad \text{on} \quad \partial U(x_0), \tag{20.104}$$

has exactly one generalized convex solution if and only if inequality (20.102) holds (see § 11). In our case this solution w is a convex function of $\rho = |x| = (x_1^2 + \cdots + x_n^2)^{1/2}$, since $\frac{1}{N(\frac{1}{Q})}$ is a function of $|p|$, $\phi_0 = \text{const} \ > 0$ and $\overline{U}(x_0)$ is a ball in R^n. Thus $w(\rho)$ is a solution of the boundary value problem

$$\det(u_{ij}) \equiv \frac{1}{n-1}\left(\frac{dw}{d\rho}\right)^{n-1}\frac{d^2w}{d\rho^2} = \frac{\phi_0}{N\left(\frac{1}{Q}\right)}, \tag{20.105}$$

$$w(r(x_0)) = 0, \quad \frac{dw(0)}{d\rho} = 0. \tag{20.106}$$

This problem is exactly the previous Dirichlet problem (20.103–104), which is especially written for radial solutions $w(\rho)$ in spherical coordinates, whose pole is placed at the point $x_0 \in \partial G$. We can find the solution $w(\rho) \in C^2(\overline{G})$ of the problem (20.105–106) by quadratures, if condition (20.102) holds.

From (20.105) and (20.106) it follows that

$$\begin{aligned}\int_{|p|\le\frac{dw(r)}{d\rho}} N\left(\frac{1}{Q}\right)dp &= \int_{|x|\le r} N\left(\frac{1}{Q}\right)\det(w_{ij})dx \\ &= \sigma_n\int_0^r\left(\frac{dw}{d\rho}\right)^{n-1} N\left(\frac{dw}{d\rho}, S, \frac{1}{Q}\right)\frac{d^2w}{d\rho^2}d\rho \\ &= \sigma_n\int_0^r \phi_0\rho^{n-1}d\rho = \mu_n\phi_0 r^n\end{aligned}$$

for any $0 < r < r(x_0)$. According to inequality (20.102) we obtain the basic inequality

$$\begin{aligned}\int_{|p|\le\frac{dw(r(x_0))}{d\rho}} N\left(\frac{1}{Q}\right)dp &= \mu_{n_n}\phi_0[r(x_0)]^n \\ &\le \mu_n\phi_0[r(\partial G)]^n < \int_{R^n} N\left(\frac{1}{Q}\right)dp.\end{aligned} \tag{20.107}$$

From inequality (20.107) it follows that there exists an uniform estimate for the exterior normal derivative $\frac{\partial w}{\partial n}$ at all points $x_0 \in \partial G$.

We now set

$$v(x) = w(x) + \sum_{i=1}^n a_i x_i + b,$$

where $z = \sum_{i=1}^{n} a_i x_i + b$ is the equation of the hyperplane γ_{X_0}. Clearly $v(x) \in C^2(\overline{G})$. We now prove that $v(x)$ is the desired convex function. First of all

$$v(x) = w(x) + \sum_{i=1}^{n} a_i x_i + b = \sum_{i=1}^{n} a_i x_i + b$$

for all $x = (x_1, \ldots, x_n) \in \partial\overline{U}(x_0)$, where x_0 is an arbitrary point of ∂G. Secondly

$$\det(v_{ij}) = \det(w_{ij}) = \frac{\phi_0}{N\left(|\nabla w|^2, S, \frac{1}{Q}\right)} = \phi_0 \cdot \sup Q(|p|),$$

where "sup" is taken in the ball $|\bar{p} - \nabla u|^2 \leq M_L(S)$. Since $\nabla v = \nabla w + \nabla z$ and $|\nabla z|^2 = \sum_{i=1}^{n} a_i^2 \leq M_L(S)$, ∇v is a point of the ball $|\bar{p} - \nabla w|^2 \leq M_L(S)$. Therefore

$$\sup Q(|\bar{p}|) \geq Q(|\nabla v|).$$

Thus

$$\frac{\det(v_{ij})}{Q(|\nabla v|)} \geq \phi_0$$

and $v(x)$ is the desired function. We can use $v(x)$ for the estimate of the exterior normal derivative $\frac{\partial u}{\partial n}$ at an arbitrary fixed point $x_0 \in \partial G$. We proved above that by this construction we can establish a finite estimate for the C^1-norm of any solution of class $C^2(\overline{G})$ of the Dirichlet problem (20.95–96) in the closed convex domain $\overline{G}$, which satisfies the conditions imposed in the beginning of Subsection 20.4.

If we consider functions, that satisfy the condition

$$\int_{R^n} N\left(|p|S, \frac{1}{Q}\right) dp = +\infty,$$

then inequality (20.102) becomes

$$\sigma_n \phi_0 \cdot [r(\partial G)]^n < +\infty.$$

Thus $\phi_0 > 0$ can be an arbitrary positive number. Finally we note that in the case when $Q(|p|) = (1 + |p|^n)^{k/n}$, $k \geq 0$ is any number, the integrals

$$\int_{R^n} N\left(\frac{1}{Q}\right) dp \quad \text{and} \quad \int_{R^n} \frac{1}{Q(|p|)} dp \tag{20.108}$$

converge or diverge simultaneously. Thus if $0 \leq k \leq n$, then both integrals (20.108) diverge.

20.5 The Dirichlet Problem for the Monge–Ampere Equation $\det(u_{ij}) = \psi(x)$

In Subsections 20.5–20.10 we present a complete proof of the following theorem.

Theorem 20.8. *There exists a unique strictly convex solution $u \in C^\infty(\overline{\Omega})$ of the Dirichlet problem*

$$\det(u_{ij}) = \psi(x) \quad in \quad \Omega, \tag{20.109}$$
$$u = \phi(x) \quad on \quad \partial\Omega, \tag{20.110}$$

where $\psi \in C^\infty(\overline{\Omega}), \psi > 0$ in $\overline{\Omega}$, and $\phi \in C^\infty(\partial\overline{\Omega})$.

Theorem 20.8 is due to Caffarelli, Nirenberg, and Spruck [1] and also to Krylov [3] and Ivockina [1,2].

Below we mainly follow Caffarelli, Nirenberg, Spruck [1]. First of all we recall the continuity method adapted to the Dirichlet problem (20.109–110). Let $u^0 \in C^\infty(\overline{\Omega})$ be a strictly convex function which equals 0 on $\partial\overline{\Omega}$. It is easy to find such a function u^0 satisfying, in addition, the following inequality

$$\psi^0 = \det(u^0_{ij}) \geq \psi \quad \text{in} \quad \Omega.$$

Now for each $t \in [0,1]$ we want to find a strictly convex solution $u^t \in C^{2+\alpha}(\overline{\Omega})$ of the Dirichlet problem

$$\det(u^t_{ij}) = t\psi + (1-t)\psi^0 \quad \text{in} \quad \Omega, \tag{20.109^t}$$
$$u^t = t\phi \quad \text{on} \quad \partial\Omega. \tag{20.110^t}$$

For $t = 0$ this Dirichlet problem has a solution. This solution is the function u^0. Using the implicit function theorem and a classical Schauder theory for second order linear elliptic equations with coefficients in $C^\alpha(\overline{\Omega})$, one finds that the set of all t, for which the Dirichlet problem (20.109^t–110^t) has a solution, is open. Assume that the a priori estimate

$$\|u^t\|_{2+\alpha} \leq \overline{K} \quad \text{independent of} \quad t \tag{20.111}$$

can be obtained. Then the set of such t is also closed. Hence solutions $u^t \in C^\infty(\overline{\Omega})$ of problem (20.109^t–20.110^t) exist for all $t \in [0,1]$. The function $u^1(x)$ is then the desired solution of problem (20.109–110).

A priori estimates $\|u\|_{2+\alpha} \leq \overline{K}$, which are derived for solutions $u(x)$ of equations (20.109) by Caffarelli, Nirenberg, and Spruck [1], can be applied to the solutions of problem (20.109^t–110^t), since the constant $\overline{K}$ depends only on Ω, on the C^3-norm $\|\psi\|_3$ of $\psi(x)$, $\max\psi^{-1}$, and on $\|\phi\|_4$. The following a priori estimates are derived

$$\|u\|_2 \leq K, \tag{20.112}$$
$$\sum_{i,j} |u_{ij}(x) - u_{ij}(y)| \leq \frac{K}{1 + |\ln|x-y||}, \quad x, y \in \overline{\Omega}, \tag{20.113}$$

using, essentially, only the maximum principle.

Once one has an estimate of $\|u\|_2$ and of modulus continuity of the second derivatives of $u(x)$ it follows from standard elliptic theory (see Gilbarg and Trudinger [1]) that one can estimate $\|u\|_{2+\alpha}$ and, in fact, it follows that $u \in C^\infty(\overline{\Omega})$. Thus the desired estimate

$$\|u\|_{2+\alpha} \le \overline{K} \tag{20.114}$$

is obtained.

20.6 A Priori Estimates for Derivatives up to Second Order

A priori estimates for $|u|$ and $|\nabla u|$ in $\overline{\Omega}$ for solutions of the Dirichlet problem (20.109–110) can be obtained using results established in Subsections 28.2 and 28.4. We note that this problem is the simplest one, when results of these subsections can be applied.

We would like to emphasize that $\sup_{\overline{\Omega}} |u|$ can be estimated by $m = \inf_{\partial\Omega} \phi(x)$, $M = \sup_{\partial\Omega} \phi(x)$, diam Ω and $\sup_{\overline{\Omega}} \psi(x)$, while $\sup_{\overline{\Omega}} |\nabla u|$ can be estimated by $C^{2+\alpha}$ norms of $\phi(x)$ and of functions defining $\partial\Omega$, $\chi_0 = \inf_{\partial\Omega}\{\min k_i(x)\}$, $\sup \psi(x)$; where $k_i(x)$, $(i = 1, 2, \dots, n-1)$ are the principle normal curvatures of $\partial\Omega$ at the point $x \in \partial\Omega$.

Turning to the second derivatives of u we will first estimate them on the boundary with the aid of suitable barrier functions.

If we take the logarithms of both sides of the equation and differentiate with respect to x_k we find, using summation convention,

$$Lu_k = u^{ij} u_{kij} = (\log\psi)_k \tag{20.115}$$

where $\{u^{ij}\}$ is the inverse of the Hessian matrix $H = \{u_{ij}\}$. We will denote by $A = \{A^{ij}\}$ the cofactor matrix of the Hermitian matrix $H = \{u_{ij}\}$ i.e. $A = (\det H) \cdot H^{-1}$. L is the linearized operator

$$L = u^{ij} \partial_i \partial_j .$$

Note that

$$L(x_\ell u_k - x_k u_\ell) = (x_\ell \partial_k - x_\ell \partial_k) \log\psi . \tag{20.116}$$

This simply reflects the fact that the operator $x_\ell \partial_k - x_k \partial_\ell$ is an angular derivative (on $|x| =$ constant) and the expression $\det(u_{ij})$ is invariant under rotation of coordinates.

Consider any boundary point; without loss of generality we may take it to be the origin and the x_n axis to be interior normal. Then near the origin, $\partial\Omega$ is represented by, here $x' = (x_1, \dots, x_{n-1})$,

$$x_n = \rho(x') = \frac{1}{2} B_{\alpha\beta} x_\alpha x_\beta + 0(|x'|^3). \tag{20.117}$$

In the summation, Greek letters α, β etc. go from 1 to $n-1$. On $\partial\Omega$ we have $u - \phi = 0$ (recall ϕ is defined in $\overline{\Omega}$), so also

$$(\partial_\alpha + \rho_\alpha \partial_n)(u - \phi) = 0 \quad \text{for} \quad \alpha < n$$

i.e.

$$|(u - \phi)_\alpha + (u - \phi)_n B_{\alpha\beta} x_\beta| \leq C|x|^2.$$

Since $|\nabla u| \leq K_1$ on $\partial\Omega$ we have control of the constant C. Furthermore on $\partial\Omega$ we have $(\partial_\beta + \rho_\beta \partial_n)(\partial_\alpha + \rho_\alpha \partial_n)(u - \phi) = 0$. In particular, at the origin we have, of course,

$$|u_{\alpha\beta}| \leq C, \qquad \alpha, \beta < n. \tag{20.118}$$

We wish, in addition to establish the estimate

$$\sum_{\alpha,\beta<n} u_{\alpha\beta}(0)\xi_\alpha \xi_\beta \geq c_0 > 0 \tag{20.119}$$

for any unit vector $\xi = (\xi_1, \ldots, \xi_{n-1})$. There is no loss of generality in assuming $\xi_1 = 1$; we wish then to show

$$u_{11}(0) \geq c_0 > 0. \tag{20.119$'$}$$

We may suppose, furthermore, that

$$u(0) = u_\alpha(0) = 0, \qquad \alpha < n.$$

To prove (20.119′) we will make use of a more carefully constructed barrier function. On $\partial\Omega$ we have (20.117) and

$$u = \phi = \frac{1}{2}\gamma_{\alpha\beta} x_\alpha x_\beta + 0(|x|^3).$$

With $\lambda = \gamma_{11}/B_{11}$, the function

$$\tilde{u} = u - \lambda x_n \tag{20.120}$$

still satisfies (20.109). We claim

$$\tilde{u}|_{\partial\Omega} \leq \sum_{1<j\leq n} a_{1j} x_1 x_j + C\left(\sum_{1<\beta<n} x_\beta^2 + |x|^4\right). \tag{20.120$'$}$$

This is because in the error term, x_1^3 may be replaced by $a_{1n}x_1 x_n$ (with suitable a_{1n}) modulo an error controlled by the last term in (20.120).

Now choose as barrier function

$$h = -\varepsilon x_n + \delta |x|^2 + \frac{1}{2B} \sum_{1<j\leq n} (a_{1j} x_1 + B x_j)^2.$$

By taking $G \gg C$, and then fixing $\delta > 0$ small, and $\varepsilon \ll \delta$, we can ensure that

$$h \geq \tilde{u} \quad \text{on} \quad \partial\Omega.$$

Observe that 2δ is the smallest eigenvalue of $\{h_{ij}\}$ and that all other eigenvalues are uniformly bounded from above, independent of δ. Choosing δ small we have

$$\det(h_{ij}) < \psi \quad \text{in} \quad \Omega.$$

Thus h is an upper barrier for $\tilde{u}$, i.e. by the maximum principle,

$$\tilde{u} \leq h.$$

Consequently

$$\tilde{u}_n(0) \leq h_n(0) = -\varepsilon.$$

By construction we have

$$\frac{\partial^2}{\partial x_1^2}\tilde{u}(x', \rho(x')) = 0$$

at the origin, i.e.

$$\tilde{u}_{11} + \tilde{u}_n\rho_{11} = 0 \quad \text{at} \quad 0.$$

Thus

$$\begin{aligned} u_{11}(0) = \tilde{u}_{11}(0) &= -\tilde{u}_n(0)\rho_{11}(0) \\ &\geq \varepsilon\rho_{11}(0); \end{aligned}$$

(20.119′) and (20.119) are proved. The constant c_0 depends only on $\max\psi$, $\max\psi^{-1}$, Ω and $|\phi|_3$.

Next we will estimate the mixed derivative $u_{\alpha n}(0)$. Consider the vector field (directional derivative)

$$T = \partial_\alpha + \sum_{\beta<n} B_{\alpha\beta}(x_\beta\partial_n - x_n\partial_\beta), \qquad \alpha < n. \tag{20.121}$$

In view of (20.121) we have

$$L(Tu) = T(\log\psi). \tag{20.121′}$$

Since u^{ij} is positive definite,

$$|L(T(u-\phi))| \leq C\left(1 + \sum u^{ii}\right) \tag{20.122}$$

and

$$|T(u-\phi)| \leq C|x|^2 \quad \text{on} \quad \partial\Omega. \tag{20.123}$$

We will use as barrier function

$$w = -a|x|^2 + bx_n,$$

for suitable positive constants a, b. We have

$$Lw = -2a \sum u^{ii}$$

and hence for a large, $|L(T(u-\phi))| + Lw \leq -a \sum u^{ii} + C$. By the theorem of the arithmetic, geometric mean,

$$\frac{1}{n} \sum u^{ii} \geq (\det(u^{ij}))^{1/n} = \psi^{-1/n}.$$

Choosing a large, we may then assure that

$$|L(T(u-\phi))| + Lw \leq 0 \quad \text{in} \quad \Omega.$$

By (20.123) since Ω is strictly convex, we may then choose b so large that

$$|T(u-\phi)| \leq w \quad \text{on} \quad \partial\Omega.$$

Applying the maximum principle we infer that

$$|T(u\phi)| \leq w \quad \text{in} \quad \Omega$$

and hence

$$|\partial_n T(u-\phi)| \leq b \quad \text{at} \quad 0.$$

Thus we have proved the estimate

$$|u_{n\alpha}(0)| \leq K. \tag{20.124}$$

Finally we use the equation to estimate $u_{nn}(0)$. We have

$$\psi(0) = \det(u_{ij}) = \sum A^{ni} u_{in}.$$

By the estimates already established we see that the first $(n-1)$ terms in the sum are bounded so that $A^{nn}u_{nn} \leq C$ at the origin. From (20.119) we obtain for $A^{nn}(0)$ a bound from below. Hence we infer that

$$u_{nn}(0) \leq C. \tag{20.124'}$$

Having an upper bound for all the eigenvalues of $H = \{u_{ij}\}$, we obtain also a lower bound for each, since their product equals ψ. Thus we also have

$$u_{nn}(0) \geq c_0 > 0 \tag{20.124''}$$

for some c_0 under control.

The last step is to estimate the second derivatives in the interior. We do this by showing that, essentially, u_{rr} satisfies a maximum principle. Write the equation in the form

$$F(\partial^2 u) := \log\det(u_{ij}) = \log\psi. \tag{20.125}$$

As we remarked, the linearized operator is

$$L = F_{u_{ij}}\partial_i\partial_j = u^{ij}\partial_i\partial_j.$$

Claim. $F(H)$ a concave function of its arguments, the positive definite symmetric matrices $H = \{u_{ij}\}$.

In fact, a computation yields

$$\begin{aligned}\frac{\partial^2 F}{\partial u_{ij}\partial u_{k\ell}} &= \frac{\partial u^{ij}}{\partial u_{k\ell}} \\ &= -u^{ik}u^{\ell j}\end{aligned}$$

and the corresponding quadratic form on symmetric matrices $M = \{m_{ij}\}$

$$\frac{\partial^2 F}{\partial u_{ij}\partial u_{k\ell}} m_{ij}m_{k\ell},$$

is negative definite, as is easily seen by diagonalizing H.

Now differentiate (20.125) twice with respect to x_r. We find first

$$F_{u_{ij}}u_{rij} = \frac{\psi_r}{\psi},$$

then

$$F_{u_{ij}}u_{rrij} + F_{u_{ij}u_{k\ell}}u_{rij}u_{rk\ell} = \frac{\psi_{rr}}{\psi} - \frac{\psi_r^2}{\psi^2}.$$

By concavity established above we have

$$\begin{aligned}Lu_{rr} = F_{u_{ij}}u_{rrij} &\geq (\log\psi)_{rr} \\ &\geq -nC\end{aligned} \tag{20.125'}$$

for some constant C depending only on ψ. This same estimate holds for any constant directional derivative $T = \sum c_j\partial_j$, $\sum c_j^2 = 1$, i.e.

$$LT^2u \geq -C. \tag{20.126}$$

Since $Lu = n$ we see that

$$L(u_{rr} + Cu) \geq 0$$

and hence $u_{rr} + Cu$ achieves its maximum on the boundary. We conclude therefore that

$$u_{rr} \le K \quad \text{in} \quad \Omega.$$

Since the equation is invariant under orthogonal change of coordinates we see that for any directional derivative operator $T = \sum c_i \partial_i$, c_i constant, $\sum c_i^2 = 1$ we have

$$T^2 u \le K \quad \text{in} \quad \Omega. \tag{20.127}$$

Taking $T = \frac{1}{\sqrt{2}}(\partial_i \pm \partial_j)$ in recalling that $u_{ii} > 0$ we find also

$$|u_{ij}| \le K \quad \text{in} \quad \Omega \tag{20.127$'$}$$

and (20.112) is proved.

Since the eigenvalues of H are bounded from above by K and their product $= \psi$, we obtain positive lower bounds for each one.

Having established $|u|_2 \le K$, we know that the linearized operator L is uniformly elliptic. Furthermore, returning to the preceding calculation we see that for any T as above

$$\begin{aligned} LT^2 u &= u^{ik} u^{j\ell} T u_{ij} T u_{k\ell} + T^2(\log \psi) \\ &\ge c_0 \sum_{i,j} |T u_{ij}|^2 - C \end{aligned} \tag{20.128}$$

where c_0 and C are positive constants under control; in particular one has a positive lower bound for c_0.

20.7 Calabi's Interior Estimates for the Third Derivatives

We will first describe the Calabi computation, used by Pogorelov, to estimate the third derivatives in the interior. With $\{u^{ij}\} = H^{-1}$, $H = \{u_{ij}\}$, the following expression

$$\sigma = u^{k\ell} u^{pq} u^{rs} u_{kpr} u_{\ell qs} \tag{20.129}$$

measures the square of the third derivatives in terms of the Riemannian metric $ds^2 = u_{ij} dx_i dx_j$. (To be more consistent geometrically, we should use covariant derivatives.) We will show that for some positive constants c_1, c_2 (depending only on the estimates already established),

$$L\sigma = u^{ij} \sigma_{ij} \ge c_1 \sigma^2 - c_2. \tag{20.130}$$

Since L is uniformly elliptic it will then follow with the aid of the maximum principle that

$$\sigma(x) \le \frac{c_3}{d^2(x)} \tag{20.130$'$}$$

where $d(x) =$ distance of x to $\partial\Omega$. (For completeness we include the derivation of (20.130$'$) at the end of this section.)

Here is a concise presentation of the ingenious computation of Calabi [1]; the reader may wish to skip this.

$$
\begin{aligned}
u^{ij}u_{kij} &= (\log\psi)_k \\
u^{ij}u_{kpij} &= u^{ia}u_{abp}u^{bj}u_{kij} + (\log\psi)_{kp} \\
u^{ij}u_{kprij} &= u^{ia}u_{abr}u^{bj}u_{kpij} + u^{ia}u_{abpr}u^{bj}u_{kij} \\
&\quad + u^{ia}u_{abp}u^{bj}u_{krij} - 2u^{ic}u_{cdr}u^{da}u_{abp}u^{bj}u_{kij} \\
&\quad + (\log\psi)_{kpr}.
\end{aligned}
\tag{20.131}
$$

Setting $u^{k\ell}u^{pq}u^{rs}u_{\ell qs} = u^{kpr}$, this is symmetric in k, p, r, we find

$$
\begin{aligned}
u^{kpr}u^{ij}u_{kprij} = u^{kpr}[&(\log\psi)_{kpr} + 3u^{ia}u^{bj}u_{abr}u_{kpij} \\
&- 2u^{ic}u^{da}u^{bj}u_{cdr}u_{abp}u_{kij}].
\end{aligned}
$$

Next

$$
\begin{aligned}
u^{ij}\sigma_{ij} &= 2u^{k\ell}u^{pq}u^{rs}u_{\ell qs}u^{ij}u_{kprij} + 2u^{ij}u^{k\ell}u^{pq}u^{rs}u_{kpri}u_{\ell qsj} \\
&\quad - 12u^{ij}u^{ka}u_{abi}u^{b\ell}u^{pq}u^{rs}u_{kprj}u_{\ell qs} \\
&\quad + 6u^{ij}u_{kpr}u_{\ell qs}u^{ka}u_{abi}u^{b\ell}u^{pc}u_{cdj}u^{dq}u^{rs} \\
&\quad - 3u^{ij}u_{kpr}u_{\ell qs}u^{ka}u_{abij}u^{b\ell}u^{pq}u^{rs} \\
&\quad + 3u^{ij}u_{kpr}u_{\ell qs}u^{kg}u_{gmj}u^{ma}u_{abc}u^{b\ell}u^{pq}u^{rs} \\
&\quad + 3u^{ij}u_{kpr}u_{\ell qs}u^{ka}u_{abi}u^{bg}u_{gmj}u^{m\ell}u^{pq}u^{rs}.
\end{aligned}
$$

At any point x where we are evaluating this, we may assume after a suitable rotation, that u_{ij} is diagonal at the point. Thus we find, at this point, using summation convention,

$$
\begin{aligned}
u^{ij}\sigma_{ij} &= 6\frac{u_{kpr}u_{ijr}u_{kpij}}{u_{kk}u_{pp}u_{rr}u_{ii}u_{jj}} - 4\frac{u_{kpr}u_{iar}u_{ajp}u_{kij}}{u_{kk}u_{pp}u_{rr}u_{ii}u_{aa}u_{rr}} \\
&\quad + 2\frac{u_{kpr}(\log\psi)_{kpr}}{u_{kk}u_{pp}u_{rr}} + 2\sum\frac{u_{kpri}^2}{u_{kk}u_{pp}u_{rr}u_{ii}} \\
&\quad - 12\frac{u_{k\ell i}u_{p\ell r}u_{kpri}}{u_{kk}u_{\ell\ell}u_{pp}u_{rr}u_{ii}} + 6\frac{u_{kpr}u_{\ell qr}u_{k\ell i}u_{pqi}}{u_{kk}u_{pp}u_{rr}u_{\ell\ell}u_{qq}u_{ii}} \\
&\quad - 3\frac{u_{kpr}u_{\ell pr}u_{k\ell ii}}{u_{kk}u_{pp}u_{rr}u_{\ell\ell}u_{ii}} + 6\frac{u_{kpr}u_{\ell pr}u_{kai}u_{a\ell i}}{u_{kk}u_{pp}u_{rr}u_{\ell\ell}u_{ii}u_{aa}}.
\end{aligned}
$$

Also

$$
\frac{u_{kpr}u_{\ell pr}u_{k\ell ii}}{u_{kk}u_{pp}u_{rr}u_{\ell\ell}u_{ii}} = \frac{u_{kpr}u_{\ell pr}}{u_{kk}u_{pp}u_{rr}u_{\ell\ell}}\left(\frac{u_{kab}u_{\ell ab}}{u_{aa}u_{bb}} + (\log\psi)_{k\ell}\right).
$$

Insert this in the preceding and complete the square *cleverly*; we find, after some tedious calculation,

$$u^{ij}\sigma_{ij} = 2\sum \frac{1}{u_{ii}u_{kk}u_{pp}u_{rr}} \Big(u_{kpri} - \frac{1}{2u_{\ell\ell}}(u_{k\ell i}u_{p\ell r} + u_{p\ell i}u_{k\ell r}$$
$$+ u_{r\ell i}u_{kp\ell})\Big)^2 - \frac{1}{2}\frac{1}{u_{ii}u_{kk}u_{pp}u_{rr}}$$
$$\cdot \left| \sum_\ell \frac{1}{u_{\ell\ell}}(u_{k\ell i}u_{p\ell r} + u_{p\ell i}u_{k\ell r} + u_{r\ell i}u_{kp\ell}) \right|^2 \tag{20.132}$$
$$- 4A + 6A + \frac{2u_{kpr}(\log\psi)_{kpr}}{u_{kk}u_{pp}u_{rr}} - \frac{3u_{kpr}u_{\ell pr}}{u_{kk}u_{pp}u_{rr}u_{\ell\ell}}$$
$$\times \left(\frac{u_{kab}u_{\ell ab}}{u_{aa}u_{bb}} + (\log\psi)_{k\ell} \right) + 6B$$

where

$$A = \frac{u_{kpr}u_{\ell qr}u_{k\ell i}u_{pqi}}{u_{kk}u_{pp}u_{rr}u_{\ell\ell}u_{ii}u_{qq}}, \quad B = \frac{u_{kpr}u_{\ell pr}u_{kai}u_{\ell ai}}{u_{kk}u_{pp}u_{rr}u_{\ell\ell}u_{aa}u_{ii}}.$$

Thus

$$u^{ij}\sigma_{ij} \geq -C - C\sigma - \frac{1}{2}\frac{1}{u_{ii}u_{kk}u_{pp}u_{rr}} \left| \sum_\ell \frac{1}{u_{\ell\ell}}(u_{k\ell i}u_{p\ell r} \right.$$
$$\left. + u_{p\ell i}u_{k\ell r} + u_{r\ell i}u_{kp\ell}) \right|^2 + 3B + 2A$$
$$= -C - C\sigma - \frac{3}{2}B - 3A + 3B + 2A,$$

as direct calculation of the length square shows,

$$= \frac{1}{2}B - C - C\sigma + \frac{1}{2}(B - A). \tag{20.133}$$

Claim. $B \geq A$.

Proof. For convenience, set

$$u_{kpr}(u_{kk}u_{pp}u_{rr})^{-1/2} = v_{kpr}.$$

Then we have to show

$$B = v_{kpr}v_{\ell pr}v_{kai}v_{\ell ai} \geq v_{kpr}v_{\ell qr}v_{k\ell i}v_{pqi} = A$$

for v symmetric in all indices. Direct computation shows

$$\sum_{jik\ell} \left| \sum_r (v_{jri}v_{rk\ell} + v_{jrk}v_{ri\ell} - 2v_{jr\ell}v_{rik}) \right|^2 = 6B - 6A$$

and the claim follows.

Thus from (20.133) we find

$$u^{ij}\sigma_{ij} \geq \frac{1}{2}B - C\sigma - C, \tag{20.134}$$

for suitable (controlled) constant C. Finally we see that

$$\begin{aligned}
B &= \sum_{k,\ell}\left|\sum_{p,r} v_{kpr}v_{\ell pr}\right|^2 \\
&\geq \sum_{k}\left|\sum_{p,r} v_{kpr}^2\right|^2 \\
&\geq \frac{1}{n}\left(\sum_{k,p,r} v_{kpr}^2\right)^2 \qquad \text{by Schwarz inequality} \\
&= \frac{1}{n}\left(\sum \frac{1}{u_{kk}u_{pp}u_{rr}}u_{kpr}^2\right)^2 \\
&= \frac{1}{n}\sigma^2.
\end{aligned}$$

Inserting this in (20.134) we find, for suitable constant C,

$$u^{ij}\sigma_{ij} \geq \frac{1}{4n}\sigma^2 - C.$$

Inequality (20.130) is proved. We conclude this section with a proof of (20.130′) due to H. Brézis.

We formulate it in a more general form.

Lemma 20.3. *In a domain Ω let σ satisfy an elliptic inequality*

$$A\sigma = a^{ij}(x)\sigma_{ij} + a^i(x)\sigma_i \geq c_1\sigma^2 - c_2$$

where A is uniformly elliptic operator with bounded coefficients and $c_1 > 0$, c_2 are constants. Then for some constant c_3 depending only on c_1, c_2 the uniform ellipticity, and bounds on the coefficients of A, and Ω, we have

$$\sigma(x) \leq \frac{c_3}{d^2(x)} \tag{20.130$''$}$$

where $d(x)$ = distance of x to $\partial\Omega$.

Proof. To estimate σ at any point $y \in \Omega$ which we take to be the origin, whose distance to $\partial\Omega$ is $2R$, we make use of the function $\zeta = R^2 - |x|^2$ in $|x| < R$, $\zeta = 0$ for $|x| \geq R$. Set

$$\tau = \zeta^2\sigma.$$

Then for $|x| \le R$ we have

$$\begin{aligned} c_1|\zeta^{-2}\tau|^2 &\le c_2 + L\sigma \\ &= c_2 + \zeta^{-2}L\tau - 2\zeta^{-3}a^{ij}\zeta_i\tau_j + \tau L(\zeta^{-2}). \end{aligned}$$

At a point $\bar{x}$ in $|x| \le R$ where τ takes its maximum, $\tau_i = 0$ and $L\tau \le 0$ and so at $\bar{x}$:

$$\begin{aligned} c_1\tau^2 &\le c_2\zeta^4 + \tau\zeta^4 L(\zeta^{-2}) \\ &= c_2\zeta^4 - 2\tau\zeta a^{ij}\zeta_{ij} + 6\tau a^{ij}\zeta_i\zeta_j - 2\tau\zeta a^i\zeta_i \\ &\le C(R^8 + R^2\tau + R^3\tau) \end{aligned}$$

with C under control. Since $R \le \operatorname{diam} \Omega$ it follows that

$$\tau(\bar{x}) \le CR^2$$

with a different C. The same bound must hold for $\tau(0)$. Hence

$$R^4\sigma(0) \le CR^2$$

which yields (20.130″).

20.8 One-Sided Estimates at the Boundary for some Third Derivatives

In this section we argue locally near a point on $\partial\Omega$; we take the point to be the origin and suppose Ω is described near 0 as in Subsection 20.2, i.e.

$$\Omega\colon\ \{x_n > \rho(x')\} \quad \text{locally.}$$

Let T denote the differential operator

$$T = \partial_\alpha + \rho_\alpha\partial_n, \qquad \alpha < n,$$

which on $\partial\Omega$ acts tangentially to the boundary. In this section, we will compute LT^2u and construct a suitable barrier function in order to obtain an estimate, *from above*:

$$-\partial_\nu T^2 u \le K \quad \text{on} \quad \partial\Omega. \tag{20.35}$$

Here ν is the unit exterior normal. We will carry out the estimate at the origin. A similar estimate holds for any point nearby. Thus we are going to establish

$$(\partial_n T^2 u)(0) \le K. \tag{20.135'}$$

Applying T to the equation $\log det(u_{ij}) = \log\psi$, we find

$$u^{ij}Tu_{ij} = T\log\psi.$$

Now the commutator

$$[\partial_i\partial_j, T] = \rho_{\alpha i}\partial_n\partial_j + \rho_{\alpha j}\partial_n\partial_i + \rho_{\alpha ji}\partial_n \tag{20.136}$$

and so

$$u^{ij}\partial_i\partial_j(Tu) = T\log\psi + \rho_{\alpha i}u_{nj} + \rho_{\alpha j}u_{ni} + \rho_{\alpha ji}u_n.$$

Applying T once more we find

$$\begin{aligned} u^{ij}T\partial_i\partial_j Tu &= u^{ia}(Tu_{ab})u^{bj}(\partial_i\partial_j Tu) + T^2\log\psi \\ &\quad + \rho_{\alpha i}Tu_{nj} + \rho_{\alpha j}Tu_{ni} + f \\ &= u^{ia}(Tu)_{ab}u^{bj}(Tu)_{ij} + \rho_{\alpha i}(Tu)_{nj} + \rho_{\alpha j}(Tu)_{ni} + f \end{aligned}$$

where f represents bounded functions (under control) since $|u|_2 \le K$. Using (20.136) again we find

$$LT^2u \ge u^{ia}u^{bj}(tu)_{ab}(Tu)_{ij} - C\sum_{i,j}|(Tu)_{ij}| - C.$$

Thus

$$LT^2u \ge -C \quad \text{in} \quad \Omega \quad \text{near} \quad 0,$$

for a suitable constant C under control.

In a region

$$\Omega_\varepsilon = \Omega \cap \{x_n < \varepsilon\},$$

with fixed, small, $\varepsilon > 0$, we will take as barrier function

$$w = T^2\phi + ax_n - b(u - u(0) - \sum x_j u_j(0)).$$

Recall that $T^2u = T^2\phi$ on $\partial\Omega$. We have

$$Lw = LT^2\phi - nb \le -C \le LT^2u \quad \text{in} \quad \Omega_\varepsilon$$

if we fix b sufficiently large. Having thus fixed b, we may choose a so large that

$$w \ge T^2u \quad \text{on} \quad \partial\Omega_\varepsilon.$$

Thus w is an upper barrier and we have

$$T^2u \le w \quad \text{in} \quad \overline{\Omega}_\varepsilon$$

while equality holds at the origin. Hence

$$\begin{aligned} \partial_n T^2u(0) &\le \partial_n w(0) \\ &\le a + \partial_n T^2\phi(0) \\ &\le K, \end{aligned}$$

establishing (20.135′).

We have described all the estimates that were proved by Calabi and Nirenberg in 1974 (some proofs have been modified). In the next section we will see how to exploit (20.135). First we will reformulate it. From (20.135) it is clear that also

$$T^2 u_n \le K \quad \text{on} \quad \partial\Omega \quad \text{near} \quad 0,$$

and $w = u_n$ satisfies the linear equation

$$Lu_n = (\log \psi)_n.$$

It is convenient, for direct application of the crucial lemma of the next section, to straighten the boundary $\partial\Omega$ near Ω. Let us do so by introducing new variables y: $y_\alpha = x_\alpha$ for $\alpha < n$, $y_n = x_n - \rho(x')$. Then if we set

$$v = K \sum_1^{n-1} y_\alpha^2 - w, \qquad (20.137)$$
$$w = u_n,$$

for K a large constant, we see that in a half ball $|x| \le \delta$, $x_n \ge 0$,

$$|Lv| = \left|\sum a^{ij} v_{ij}\right| \le C \qquad (20.138)$$

where L is a uniformly elliptic operator; also

$$|v| + |\nabla v| \le K, \qquad (20.139)$$

and

$$v(y', 0) \text{ is } \textit{convex} \qquad (20.140)$$

provided K is chosen sufficiently large.

Formula (20.140) is the key for our next argument.

20.9 An Important Lemma

In this section we present a useful result. When applied to v given by (20.137) which satisfies (20.138–140) it will yield a logarithmic modulus of continuity of v_α on $x_n = 0$ for $\alpha < n$.

We formulate it for a general uniformly elliptic operator

$$L = a^{ij}(x)\partial_{ij} + a^i(x)\partial_i + a(x), \text{ with } a \le 0$$
$$M^{-1}|\xi|^2 \le a^{ij}\xi_i\xi_j \le M|\xi|^2, \quad |a^i|, |a| \le M, \qquad (20.141)$$

in a half ball

$$B_R^+ \text{ in } R^n = \{|x| < R, x_n > 0\}.$$

Lemma 20.4. *Let $v \in C^2(B_R^+) \cap C^1(\overline{B_R^+})$, satisfy*

(i) $Lv \le C$
(ii) $|v| \le C$, in B_R^+
(iii) $|\nabla_{x'} v(x',0)| \le C$
(iv) $v_n(x',0) \le C$
(v) $v(x',0)$ is convex.

Then

$$|\nabla_{x'} v(x',0) - \nabla_{\overline{x'}} v(\overline{x'},0)| \le \frac{\overline{C}}{1 + |\log|x' - \overline{x'}|\,|} \quad \text{for } |x'|, |\overline{x'}| \le \frac{R}{2}, \tag{20.142}$$

where $\overline{C}$ depends only on n, M, R and C.

We remark that, even for a bounded harmonic function v in B_R^+, in general, if (iii) holds and $\nabla_{x'} v(x',0)$ is continuous, then the function $v_n(x',0)$ need not be bounded. It is, provided $\nabla_x \cdot v(x',0)$ is Dini continuous. This lemma is a kind of converse of that fact. It says if (iv) *and* (v) hold, (20.142) necessarily holds — somewhat surprising.

Proof. We may suppose $R < 1$. It suffices to prove the lemma for $x' = 0$ and $|\overline{x'}|$ small, say $|\overline{x'}| \le \delta/2 < R/2$. We may suppose $v(0) = |\nabla_{x'} v(0)| = 0$ — after subtracting an affine function.

Furthermore, after rotating coordinates we may suppose

$$\nabla_{x'} v(\overline{x'},0) = (\alpha, 0, \ldots, 0), \quad \alpha > 0.$$

We have $\alpha \le C$. If $\alpha = 0$ there is nothing to prove; we wish to prove

$$\alpha \le \frac{\overline{C}}{|\log|\overline{x'}|\,|}. \tag{20.142'}$$

By convexity we have $v(x,0) \ge 0$ and

$$v(x',0) \ge v(\overline{x'},0) + \alpha(x_1 - \overline{x_1}) := \alpha(x_1 - \beta) \tag{20.143}$$

where β is here defined. Setting $x = 0$ and $x = x'$ we find

$$0 \le \beta \le \overline{x_1}. \tag{20.143'}$$

Consider the barrier

$$\begin{aligned} h(x', x_n) = {} & \frac{\alpha}{2}[(x_1 - \beta)^2 + x_n^2]^{1/2} + \frac{\alpha}{2}(x_1 - \beta) \\ & - \frac{\alpha\varepsilon}{2} x_n \log[(x_1 - \beta)^2 + x_n^2] - D(x_n + |x'|^2 - E x_n^2) \end{aligned} \tag{20.144}$$

where we will choose $0 < \varepsilon, \delta < 1 < D$, and E, depending only on n, M, R, C, in such a way as to guarantee

$$Lh \ge C \quad \text{(see (i))} \tag{20.145}$$

and

$$h \le v \quad \text{on} \quad \partial(B_\delta^+). \tag{20.146}$$

By the maximum principle we will have $h \le v$ in B_δ^+. Setting $x' = 0$, dividing by x_n and letting $x_n \to 0$ we conclude $\beta > 0$ and

$$h_n(0) \le v_n(0) \le C,$$

i.e.

$$-D - \alpha\varepsilon \log \beta \le C.$$

Thus

$$\alpha \le \frac{C+D}{\varepsilon|\log \beta|} \le \frac{C+D}{\varepsilon|\log |\overline{x'}|\,|} \quad \text{by (20.143')}$$

i.e. (20.142') holds.

Next we observe that (20.146) holds on $x_n = 0$, $|x'| \le \delta$. On $x_n = 0$ we have $h = \alpha(x_1 - \beta)_+ - D|x'|^2$. There are two cases.

Case (i) $x_1 - \beta > 0$. Then

$$\begin{aligned} h &\le \alpha(x_1 - \beta) \\ &\le v(x', 0) \quad \text{by (20.143)}. \end{aligned}$$

Case (ii) $x_1 - \beta \le 0$. Then

$$h(x', 0) \le 0 \le v(x', 0).$$

We have only to verify (20.146) on S, the curve part of ∂B_δ^+, with suitable choice of δ and D. First fix δ so small that

$$x_n + |x'|^2 - Ex_n^2 > \frac{\delta^2}{2} \quad \text{on} \quad S.$$

Then we may choose D so large that on S.

$$h \le -C \le v. \qquad \square$$

We remark that Lemma 20.4 is a local form of a result of the type of Liouville's theorem. We describe a simple case of this. Here R_+^n represents the upper half space $x_n > 0$ in R^n.

Lemma 20.5. *Let $w \in C^2(R_+^n)$ satisfy*

$$Lw = a^{ij}(x) w_{ij} \le 0$$

where L is uniformly elliptic, i.e.

$$\frac{1}{m}|\xi|^2 \le a^{ij}\xi_i\xi_j \le m|\xi|^2, \qquad m > 0.$$

Assume w is Lipschitz continuous in R^n_+ and

$$w(x',0) \geq v(x') \text{ where } v \text{ is convex.}$$

Then v is an affine function.

Proof. If v is not affine, since it is convex, we may add a suitable affine function to v and w and rotate coordinates so that

$$w(x',0) \geq v \geq \max\{\alpha x_1, 0\} \quad \text{for some} \quad \alpha > 0.$$

With $s = (x_1^2 + x_n^2)^{1/2}$, set

$$h_M(x) = \frac{\alpha}{2}s + \frac{\alpha}{2}x_1 - \alpha\varepsilon x_n \log|x| + Mx_n.$$

We may first choose $\varepsilon > 0$ sufficiently small (independent of M), as a computation shows, so that

$$Lh_M \geq 0.$$

On $x_n = 0$ we have

$$h_M = \max\{\alpha x_1, 0\} \leq w.$$

On $|x| = R, x_n > 0$ we have

$$\begin{aligned} h_M &= \frac{\alpha}{n}(s + x_1) + x_n(M - \alpha\varepsilon \log R) \\ &\leq \max\{\alpha x_1, 0\} + x_n(M - \alpha\varepsilon \log R + \alpha) \\ &\leq w + x_n(M - \alpha\varepsilon \log R + \alpha - \bar{L}) \end{aligned}$$

where $\bar{L}$ is the Lipschitz constant for w. Thus

$$h_M \leq w \text{ on } |x| = R, \quad x > 0 \text{ for } R \text{ sufficiently large.}$$

By the maximum principle

$$h_M \leq w \text{ in } R^n_+ \qquad \forall M.$$

Letting $M \to +\infty$ we obtain a contradiction. □

20.10 Completion of the Proof of Theorem 20.8

Return to Subsection 20.9 and consider the function v defined (20.137); in view of (20.138–140), Lemma 20.4 may be applied and we infer that for $|y'|, |\overline{y'}| \leq c$

$$\left| \frac{\partial y}{\partial y_\alpha}(y',0) - \frac{\partial v}{\partial y_\alpha}(\overline{y'},0) \right| \leq \frac{\overline{C}}{|\log|y' - \overline{y'}||}. \tag{20.147}$$

In the new coordinates we have

$$\begin{aligned} u_{x_\alpha x_\beta} &= u_{y_\alpha y_\beta} - u_{y_\alpha y_n}\rho_\beta - u_{y_\beta y_n}\rho_\alpha + u_{y_n y_n}\rho_\alpha\rho_\beta - u_{y_n}\rho_{\alpha\beta} \\ u_{x_\alpha x_n} &= u_{y_\alpha y_n} - u_{y_n y_n}\rho_\alpha \\ u_{x_n x_n} &= u_{y_n y_n}. \end{aligned}$$

Since $u = \phi$ on $\partial\Omega$, i.e. on $y_n = 0$ we have of course

$$u_{y_\alpha y_\beta} \text{ is Lipschitz on } y_n = 0, \quad \alpha, \beta < n$$

and, from (20.147)

$$u_{y_\alpha y_n} \text{ satisfies a logarithmic modulus of continuity}$$

which is under control. It is clear from the equation $\det(u_{x_i x_j}) = \psi$, expressed in terms of the y derivatives, near the origin, that we may solve for the missing second derivative $u_{y_n y_n}$ and conclude that it too satisfies a logarithmic modulus of continuity.

Returning to our original domain Ω we conclude that on $\partial\Omega$, the second derivatives have a fixed logarithmic modulus of continuity i.e. we have proved (20.113) for $x, y \in \partial\Omega$.

To go from here to the full inequality (20.113) is not difficult. First we go part way.

Lemma 20.6. *The inequality*

$$\sum_{i,j} |u_{ij}(x) - u_{ij}(y)| \le K[1 + |\ln x - y|]^{-1}, \quad x, y \in \overline{\Omega}, \tag{20.148}$$

holds.

Proof. We may suppose $x = 0$ and that $|y|$ is small. Furthermore we may make a linear transformation of variables (which is under control since we have excellent bounds for $H\{u_{ij}\}$) so that

$$u_{ij}(0) = \delta_{ij}.$$

The function ψ is then multiplied by a factor which we ignore. We know that $|u_{ij}| \le K$ and that, by (20.126) for any pure second derivative $T^2 u$,

$$LT^2 u \ge -C.$$

If ν is the exterior unit normal to $\partial\Omega$ at 0 consider the function (here y is fixed, and $\delta = |y|^{1/3}$ is small)

$$h(x) = T^2 u(0) + \frac{M}{|\log \delta|} - \frac{M}{\delta^2} x \cdot \nu - Cg$$

where g is a fixed smooth convex function in $\overline{\Omega}$, vanishing on $\partial\Omega$ and satisfying $Lg \geq 1$. For $x \in \partial\Omega$, $|x| < \delta$ we have by (20.113) for points on $\partial\Omega$,

$$|T^2u(x) - T^2u(0)| \leq \frac{C}{|\log \delta|}.$$

Consequently we may choose M large (under control) so that h is an upper barrier for T^2u in $\Omega \cap \{|x| < \delta\}$. It follows in particular that

$$T^2u(y) - T^2u(0) \leq \frac{C}{|\log |y|\,|}. \tag{20.149}$$

We now use the equation to obtain a lower bound. After our change of coordinates we have $\psi(0) = 1$ and so

$$\psi(y) \geq 1 - C|y|.$$

Here $\psi(y) = \lambda_1 \dots \lambda_n$, the product of the eigenvalues of H. By (20.149) it follows that

$$\lambda_i(y) \leq 1 + \frac{C}{|\log |y|\,|} \qquad \forall i.$$

We infer that

$$\lambda_i(y) \geq 1 - \frac{C}{|\log |y|\,|} \qquad \forall i,$$

and hence for every T as above

$$T^2u(y) \geq 1 - \frac{C}{|\log |y||}$$

in particular this holds for $u_{jj}(y)$. Taking $T = \frac{1}{\sqrt{2}}(\partial_i \pm \partial_j)$ we conclude that

$$|u_{ij}(y)| \leq \frac{C}{|\log |y|\,|} \quad \text{for} \quad i \neq j.$$

Thus we have proved (20.148), i.e. (20.113) for $x \in \partial\Omega$, $y \in \overline{\Omega}$.

Using this result it is now easy to give the

Conclusion of the Proof (20.113). Consider $x, y \in \overline{\Omega}$ with $d = d(x) \leq d(y) = \bar{d}$ and $\delta = |x - y|$; we may assume d, $\bar{d}$ and δ are all small. Distinguish two cases.

Case (i). $d \geq \delta^{1/2}$.
In this case by the theorem of the mean we have from (20.130″)

$$\begin{aligned} |u_{ij}(x) - u_{ij}(y)| &\leq \frac{2C}{d}\delta \\ &\leq 2C\delta^{1/2} \\ &\leq \frac{K}{|\log \delta|}. \end{aligned}$$

Case (ii). $d < \delta^{1/2}$.
In this case $d + \delta \leq 2\delta^{1/2}$. Using (20.148) we see easily that

$$\begin{aligned} |u_{ij}(x) - u_{ij}(y)| &\leq \frac{C}{|\log(d+\delta)|} \\ &\leq \frac{C}{|\log(2 \cdot \delta^{1/2})|} \\ &\leq \frac{K}{|\log \delta|}. \end{aligned}$$

We have established (20.113) and therefore Theorem 20.8 is proved.

20.11 More General Monge–Ampere Equations

In this section we take up an extension to more general Monge–Ampere equations — always in a bounded domain Ω with strictly convex C^∞ boundary $\partial\Omega$. We seek a strictly convex solution $u \in C^\infty(\overline{\Omega})$ of an equation of the form

$$\begin{aligned} \det(u_{ij}) &= \psi(x, u, \nabla u) \quad \text{in} \quad \Omega \\ u &= \phi \in C^\infty \quad \text{on} \quad \partial\Omega \end{aligned} \tag{20.150}$$

where $\psi(x, u, p)$ is a positive C^∞ function for $x \in \overline{\Omega}$, $u \leq \max \phi$, $p \in R^n$.

We assume there exists a convex subsolution $\underline{u} \in C^2(\overline{\Omega})$ which equals ϕ on $\partial\Omega$ and satisfies

$$\det(\underline{u}_{ij}) \geq \psi(x, \underline{u}, \nabla \underline{u}) \quad \text{in} \quad \Omega. \tag{20.151}$$

As always we suppose ϕ is a C^∞ strictly convex function in all of $\overline{\Omega}$.

Theorem 20.9. *Under condition (20.151) there exists a strictly convex solution $u \in C^\infty(\overline{\Omega})$ of (20.150) with $u \geq \underline{u}$. If $\psi_u \geq 0$, this solution is unique.*

This interesting theorem has been proved by Caffarelli, Nirenberg, Spruck [1].

Part III. Geometric Methods in Elliptic Equations of Second Order. Applications to Calculus of Variations, Differential Geometry and Applied Mathematics.

In Part III we present a priori bounds and geometric maximum principles for generalized and smooth solutions of the Dirichlet problem for quasilinear elliptic equations and various applications of these bounds and principles to the calcus of variations, differential geometry, and applied mathematics. We are also concerned with the uniqueness and stability of solutions of the Dirichlet problem for elliptic equations.

Chapter 7. Geometric Concepts and Methods in Nonlinear Elliptic Euler–Lagrange Equations

This chapter is concerned with global connections between the integrand $F(x, u, p)$ of n-multiple integrals

$$I(u) = \int_B F(x, u(x), Du(x))dx$$

and a priori estimates for solutions of the corresponding Euler–Lagrange equations, whose gradients satisfy some prescribed limitations. Such problems arise in the calculus of variations, differential geometry and continuum mechanics. Typical examples of these problems are presented in §§ 23, 24. Some of these estimates are called geometric maximum principles for variational problems.

Let B be a bounded domain in Euclidean space E^n, ∂B a closed continuous hypersurface in E^n and $\overline{B} = B \cup \partial B$. We denote by D an open domain in E^n with closure $\overline{D} \subset B$. We shall be concerned with C^0-estimates for solutions $u(x)$ belonging to $W_2^n(B) \cap C(\overline{B})$ of the Euler–Lagrange equation, that is, whose first and second (Sobolev) generalized derivatives are summable with power n in every such domain D. We are also concerned with the minimal assumptions of smoothness on the integrand $F(x, u, p)$ which provide C^0-estimates for solutions of the corresponding Euler–Lagrange equations. It turns out that these assumptions are conveniently stated in terms of the convexity of suitable hypersurfaces, constructed according to properties of the integrand $F(x, u, p)$ and in terms of the composition of generalized tangential mappings of solutions and convex hypersurfaces.

The main conclusions will be obtained in the form of necessary and sufficient conditions, which either interlock or coincide. They will be expressed by geometric inequalities between fundamental invariants of the equation. The estimates obtained in this chapter have various applications to the calculus of variations, differential geometry, continuum mechanics and the stability theory of solutions of the Dirichlet problem for elliptic equations.

§21. Geometric Constructions. Two-Sided C^0-Estimates of Functions with Prescribed Dirichlet Data

21.1 Geometric Constructions

Let E^n, P^n, Q^n be three n-dimensional Euclidean spaces and let $x = (x_1, x_2, \dots, x_n)$, $p = (p_1, p_2, \dots, p_n)$, $q = (q_1, q_2, \dots, q_n)$ be points in E^n, P^n, Q^n respectively. We also introduce two $(n+1)$-Euclidean spaces $E^{n+1} = E^n \times R = \{(x, z)\}$ and $P^{n+1} = P^n \times R = \{(p, w)\}$. Let O, O', O'' be the origins of E^n, P^n, Q^n. We assume that B is a bounded open domain in E^n. Let G be an open domain in P^n and let O' be an interior point of G. The only possibilities are either $\partial G \neq \emptyset$ or $\partial G = \emptyset$. In the first case G is a bounded or unbounded domain in P^n ,and $G \neq P^n$; but in the second case G must coincide with the space P^n. Let $r(G) = \text{dist}(O', \partial G)$.

Clearly $r(G) < +\infty$ if and only if $\partial G \neq \emptyset$. Hence $r(G) = +\infty$ only in the case when $G = P^n$.

We denote by $U(\rho)$ the open n-ball: $|p| < \rho$ in P^n. Many important applications are related to convex n-domains in G. In particular, if G is the n-ball: $|p| < a$, then $r(G) = a$ and $U(r(G)) = G$.

Now let $\phi(p)$ be a convex function in G. We introduce the set function

$$\omega(\phi, e') = |\chi_\phi(e')|_Q^{*)} \tag{21.1}$$

where χ_ϕ: $G \to P^n$ is the normal mapping of the convex function $\phi(p)$. The set function $\omega(\phi, e')$ is nonnegative and absolutely additive on the family of Borel subsets e' of G. If $\chi_\phi(G)$ is an unbounded set then $\omega(\phi, e')$ can take value $+\infty$ for some Borel subset $e' \subset G$. But

$$\omega(\phi, e') < +\infty \tag{21.2}$$

for every Borel subset $e' \subset G$ such that $\text{dist}(e', \partial G) > 0$.

Now we introduce the function

$$c(\rho) = \omega(\phi, U(\rho)) \tag{21.3}$$

for $0 \leq \rho < r(G)$ and denote by $C(\phi)$ the number

$$C(\phi) = \lim_{\rho \to r(G)} c(\rho). \tag{21.4}$$

Clearly

$$C(\phi) = \omega(\phi, U(r(G)) \quad \text{and} \quad 0 \leq C(\phi) \leq +\infty. \tag{21.5}$$

---

*) We denote by $|e|_E, |e'|_P, |e''|_Q$ the Lebesgue measures for the sets $e \subset E^n$, $e' \subset P^n$, $e'' \subset Q^n$.

The case $C(\phi) = 0$ can be realized if ϕ is a linear function in G, and more generally when the graph of ϕ is a convex cylinder. The case $C(\phi) = +\infty$ can be realized if meas $\chi_\phi(G) = +\infty$. Simple examples are as follows:

1) $G = P^n$; $\phi(p) = |p|^2$
2) G: $|p| < 1$, $\phi(p) = -(1 - |p|^2)^{1/2}$.

Clearly $C(\phi)$ coincides with the total area of the normal mapping χ_ϕ if $G = U(r(G))$.

Assume that the function $\phi(p)$ satisfies the following assumptions.

Assumption 1. The function $\omega(\phi, e')$ is absolutely continuous on the ring of Borel subsets of G, i.e.

$$\omega(\phi, e') = \int_{e'} \det\left(\frac{\partial^2\phi(p)}{\partial p_i \partial p_j}\right) dp. \tag{21.6}$$

Assumption 2. Let W be the subset of G such that

$$\det\left(\frac{\partial^2\phi(p)}{\partial p_i \partial p_j}\right) > 0 \tag{21.7}$$

everywhere. Then the n-volume of the set $W \cap A(\rho_1, \rho_2)$ is strictly positive for all numbers ρ_1, ρ_2 such that $0 < \rho_1 < \rho_2 < r(G)$, where $A(\rho_1, \rho_2)$ is the annulus $0 < \rho_1 < |p| < \rho_2 < r(G)$.

If a convex function $\phi(p)$ satisfies Assumptions 1 and 2, then the function $c(\rho)$ constructed above is nonnegative, continuous and strictly increasing in $[0, r(G))$; clearly $c(\rho) > 0$ for $\rho > 0$. Let $\rho = b(t)$ be the inverse for the function $t = c(\rho)$. Then $b(t)$ is also nonnegative, continuous and strictly increasing in $[0, C(\phi))$ and $b(t) > 0$ for $t > 0$.

If $r(G) < +\infty$, then the inverse $\rho = b(t)$ can be extended to $[0, C(\phi)]$ as a nonnegative, continuous, strictly increasing function. Clearly

$$b(C(\phi)) = r(G) < +\infty. \tag{21.8}$$

We will use (21.8) in our main estimates.

21.2 Convex and Concave Supports of Functions $u(x) \in W_2^n(B) \cap C(\overline{B})$

Let B be a bounded domain in E^n and let ∂B be a closed continuous hypersurface in E^n. We denote by $W_2^n(B) \cap C(\overline{B})$ the set of functions, which are continuous in $\overline{B}$ and whose first and second Sobolev generalized derivatives exist and are summable with degree n in every compact subdomain D of B, i.e. $\operatorname{dist}(D, \partial B) > 0$.

Let $u(x)$ be any function from $W_2^n(B) \cap C(\overline{B})$. We set

$$m = \inf u(x), \qquad M = \sup u(x) \tag{21.9}$$

for all $x \in \partial B$. The two-sided C^0-estimates for $u(x)$ in $\overline{B}$ are non-trivial only if

$$u_0 = \inf_{\overline{B}} u(x) < m \tag{21.10}$$

and

$$u_1 = \sup_{\overline{B}} u(x) > M. \tag{21.11}$$

It is sufficient only to establish the estimate of $u(x)$ from below; the estimate from above can be obtained in the similar way.

Let $\delta > 0$ be any real number satisfying the condition

$$m - \delta > u_0.$$

We will use the notation $m_\delta = m - \delta$. Let S_{m_δ} be the part of the graph of $u(x)$ located under the hyperplane

$$\gamma_{m_\delta}\colon\ z = m_\delta \tag{21.12}$$

in the space $E^{n+1} = E^n \times R$. Let $\overline{B}_{m_\delta}$ be a set in γ_{m_δ} whose projection on E^n coincides with $\overline{B}$ and let Γ_{m_δ} and Γ be the corresponding boundaries of $\overline{H}_{m_\delta}$ and $\overline{H}$, where $\overline{H}_m$ and $\overline{H}$ are the closed convex hulls of $\overline{B}_{m_\delta}$ and $\overline{B}$.

We denote by C_{m_δ} the closed convex hull of the set $B_{m_\delta} \cup S_{m_\delta}$. Then

$$\partial C_{m_\delta} = \overline{H}_{m_\delta} \cup S_{v_\delta}, \tag{21.13}$$

where S_{v_δ} is the graph of a convex function $v_\delta(x) \in C(\overline{H})$ such that

$$v_\delta(x) = m_\delta$$

for all $x \in \partial H$, and

$$v_\delta(x) \le u(x) \tag{21.14}$$

for all $x \in \overline{B}$. The convex function $v_\delta(x)$ is called the *convex support of the function* $u(x)$ (according to the number $\delta > 0$). If $S_{M_{\delta'}}$ is the part of the graph $u(x)$ located over the hyperplane $\gamma_{M_{\delta'}}\colon\ z = M_{\delta'}$ ($\delta' > 0$ is any number such that $M + \delta' < u_1$), then a similar geometric construction leads to the *concave support* $w_{\delta'}(x) \in C(\overline{H})$ *of the function* $u(x)$. Clearly $w_{\delta'}(x) = M_{\delta'}$ for all $x \in \partial H$ and $w_{\delta'}(x) \ge u(x)$ for all $x \in \overline{B}$.

21.3 Two-sided C^0-Estimates for Functions $u(x) \in W_2^n(B) \cap C(\overline{B})$

We assume as before that B is a bounded n-domain in E^n and ∂B is a closed continuous hypersurface in E^n. We also assume that

$$u(x) \in W_2^n(B) \cap C(\overline{B}) \tag{21.15}$$

and

$$\chi_u(B) \subset G, \tag{21.16}$$

where G is a prescribed n-domain in P^n, and the origin O' of P^n is an interior point of G. The concepts of the tangential mapping $\chi_u\colon B \to P^n$ and inclusion (21.16) for functions $u(x) \in W_2^n(B) \cap C(\overline{B})$ are explained below in this subsection.

Let $u_0 = \inf_{\overline{B}} u(x)$, $u_1 = \sup_{\overline{B}} u(x)$. Clearly only the cases

$$u_0 < m \quad \text{and} \quad u_1 > M \tag{21.17}$$

are interesting for consideration. Actually we have the inequalities $m \le u(x)$ and $u(x) \le M$ for all other cases.

It is sufficient to investigate the estimate for $u(x)$ from below, because the estimate for $u(x)$ from above can be obtained in a similar way. Let $\delta > 0$ satisfy the inequality

$$m - \delta > u_0 \tag{21.18}$$

and let $v_\delta(x)$ be the convex support of the function $u(x)$. We denote by x_0 an interior point of B such that

$$u_0 = \inf_{\overline{B}} u(x) = u(x_0). \tag{21.19}$$

Clearly the point $M_0(x_0, u_0)$ belongs simultaneously to both the graph of the function $u(x)$ and the graph of its convex support $v_\delta(x)$. According to our notation $m_\delta = m - \delta$. Therefore from (21.18) it follows that

$$m_\delta > u_0 \quad \text{or} \quad m_\delta - u_0 > 0. \tag{21.20}$$

Now consider two convex cones K_0 and K_1 with a common vertex at the point M_0. The cone K_0 has Γ_{m_δ} as its base and the cone K_1 is a convex cone of revolution, whose base is an $(n-1)$ sphere $\sum_{m_\delta} \subset \gamma_{m_\delta}$. The center of $\sum_{m_\delta}$ is a point (x_0, m_δ) in the hyperplane γ_{m_δ} and the radius of $\sum_{m_\delta}$ is equal to diam B. We denote by T_{m_δ} an open ball in γ_{m_δ}, whose boundary is $\sum_{m_\delta}$. Let T be the projection of T_{m_δ} on the hyperplane E^n: $z = 0$. Then clearly

$$\chi_{K_1}(T) \subset \chi_{K_0}(\text{int } H) \subset \chi_{v_\delta}(\text{int } H), \tag{21.21}$$

where H is the closed convex hull of the set $\overline{B}_m$.

Since K_1 is the cone of revolution, then $\chi_{K_1}(T)$ is a closed n-ball in P^n. The inequality

$$|p| \le \frac{h_\delta}{\text{diam } B} \tag{21.22}$$

describes the n-ball $\chi_{K_1}(T)$, where

$$h_\delta = m_\delta - u_0 > \tag{21.23}$$

is the height of the convex cone K_1.

Now we want to obtain the estimate

$$h_\delta \leq A, \tag{21.24}$$

where $A = \text{const} < +\infty$ depends only on some geometric invariants of the graph of the function $u(x)$ in the space E^{n+1}. These invariants will be constructed by means of prescribed convex function $\phi(p)$, which is defined in a given domain G in P^n and which satisfies Assumptions 1 and 2. Then from (21.18), (21.20), (21.23) and (25.24) it follows that

$$\inf_{\overline{B}} u(x) = u_0 = m_\delta - h_\delta \geq m - \delta - A, \tag{21.25}$$

where, as we outlined above, the constant A does not depend on δ. Finally we will prove that δ can be omitted in (21.25). Note that we consider estimates for functions $u(x) \in W_2^n(B) \cap C(\overline{B})$.

Clearly the inclusion

$$\chi_u(B) \subset G \tag{21.16}$$

is the natural assumption, which appears in the proof of estimate (21.24). But inclusion (21.16) is convenient to use only for functions $u(x) \in W_2^n(B) \cap C^1(\overline{B})$; this inclusion is not convenient for functions $u(x) \in W_2^n(B) \cap C(\overline{B})$. We will see below that some modification of Assumption 21.16) is convenient in the proof of estimate (21.24) for functions $u(x) \in W_2^n(B) \cap C(\overline{B})$. This modified assumption is based on the concept of global convex and global concave points of the graph of functions $u(x) \in C(\overline{B})$. A point $M(x, u(x))$ of the graph of $u(x)$, where $x \in B$ is said to be *globally convex (globally concave)*, if there exists at least one supporting hyperplane α passing through M and lying under (over) the graph of $u(x)$. If such a supporting hyperplane α has the equation $z = p_1^0 x_1 + \cdots + p_n^0 x_n + b^0$, then the point

$$p_0 \overset{\text{def}}{=} \chi(\alpha) = (p_1^0, p_2^0, \ldots, p_n^0) \in P^n$$

is called the normal image of α. We denote by

$$\nu_u^+(B) = \bigcup_{\alpha+} \chi(\alpha^+), \tag{21.26}$$

$$\nu_u^-(B) = \bigcup_{\alpha^-} \chi(\alpha^-), \tag{21.27}$$

$$\nu_u(B) = \nu_u^+(B) \cup \nu_u^-(B), \tag{21.28}$$

where α^+ (or α^-) is a supporting hyperplane passing through a global convex (or global concave) point of the graph of $u(x)$. All such supporting hyperplanes α^+ (or α^-) at all global convex (or global concave) points are considered in (21.26) (or in (21.27)).

Assumption 3. Let a convex function $\phi(p)$ be defined in a domain G in P^n. Then we consider only functions $u(x) \in W_2^n(B) \cap C(\overline{B})$ for which the inclusion

$$\nu_u(B) \subset G \tag{21.29}$$

holds.

Let

$$D_{m_\delta} = S_{v_\delta} \cap S_{m_\delta} \tag{21.30}$$

(see § 21.2 for notations in (21.30)), and let D be the projection of D_{m_δ} on the hyperplane E^n. Clearly D_δ is a closed subset of B, and $\operatorname{dist}(D_\delta, \partial B) > 0$. From the definition of a convex hull it follows that every supporting hyperplane α of the convex hypersurface S_{v_δ} is also a supporting hyperplane of the hypersurface S_{m_δ}: $z = u(x) \in W_2^n(B) \cap C(\overline{B})$ at some global convex point. Clearly $\alpha \cap D_{m_\delta} \neq \emptyset$, if $\alpha \cap S_{v_\delta}$ contains a point $(x, v_\delta(x))$, where $x \in \text{int } H$. Now we assume that the function $u(x)$ satisfies Assumption 3. Then

$$\chi_{v_\delta}(\text{int } H) \subset \nu_u^+(B) \subset \nu_u(B) \subset G. \tag{21.31}$$

Thus from (21.21) and (21.31) it follows that

$$\chi_{K_1}(T) \subset \chi_{v_\delta}(\text{int } H) \subset \nu_u^+(B) \subset \nu_u(B) \subset G. \tag{21.32}$$

Clearly

$$\chi_{v_\delta}(D_\delta) = \chi_{v_\delta}(B) = \chi_{v_\delta}(\text{int } H). \tag{21.33}$$

Thus we obtain the equalities

$$\begin{aligned} |(\chi_\phi \circ \chi_{v_\delta})(D_\delta)|_Q &= |(\chi_\phi \circ \chi_{v_\delta})(B)|_Q \qquad \text{(21.33a)} \\ &= |(\chi_\phi \circ \chi_{v_\delta})(\text{int } H)|_Q, \end{aligned}$$

where $(\chi_\phi \circ \chi_{v_\delta})(e) = \chi_\phi(\chi_v(e))$ for all Borel subsets e of the set int H. Now we can see that for every convex function $\phi(p)$ satisfying Assumptions 1 and 2 and for every function $u(x) \in W_2^n(B) \cap C(\overline{B})$ satisfying Assumption 3, there exists the number

$$\omega_+(\phi, u, \delta) = |(\chi_\phi \circ \chi_{v_\delta})(\text{int } H)|_Q. \tag{21.34}$$

Clearly

$$\omega_+(\phi, u, \delta) = |(\chi_\phi \circ \chi_{v_\delta})(B)|_Q = |(\chi_\phi \circ \chi_{v_\delta})(D_\delta)|_Q \tag{21.34a}$$

if $\delta \in (0, \delta_u)$, where $\delta_u = m - u_0 > 0$, $m = \inf\limits_{B} u(x)$ and $u_0 = \inf\limits_{\overline{B}} u(x)$. Clearly $\omega_+(\phi, u, \delta)$ is the area of the normal mapping χ_ϕ: $G \to Q^n$ computed on the set $\chi_{v_\delta}(\text{int } H)$. From (21.33) it follows that int H can be replaced either by D_δ or by B in the last statement.

Lemma 21.1. *Let $\partial G \neq \emptyset$ (i.e. $r(G) < +\infty$), $\phi(p)$ satisfy Assumptions 1 and 2, and the function $u(x) \in W_2^n(B) \cap C(\overline{B})$ satisfy Assumption 3. Let also $\delta_u = m - u_0 > 0$. Finally let the inequality*

$$\omega_+(\phi, u, \delta) \leq C(\phi) \tag{21.35}$$

hold for any $\delta \in (0, \delta_u)$, where $u_0 = \inf_{\overline{B}} u(x)$. Then the inequalities

$$0 < h_\delta \leq b(\omega_+(\phi, u, \delta)) \text{ diam } B \tag{21.36}$$

hold. Moreover, if $\omega_+(\phi, u, \delta) = C(\phi)$, then $b(\omega_+(\phi, u, \delta)) = \delta(G)$ in (21.36).

Proof. Since

$$0 < \delta < \delta_u = m - u_0 \tag{21.37}$$

then from (21.37) we obtain

$$h_\delta = m_\delta - u_0 = (m - \delta) - u_0 = \delta_u - \delta > 0.$$

The positive number h as we have seen above is the height of the convex cone of revolution K_1. According to our constructions

$$|(\chi_\phi \circ \chi_{K_1})(T)|_Q \leq |(\chi_\phi \circ \chi_{v_\delta})(\text{int } H)|_Q = \omega_+(\phi, u, \delta). \tag{21.38}$$

Now from (21.35) and the properties of the function $\rho = b(t)$, we obtain the inequality

$$b(\omega_+(\phi, u, \delta)) \leq b(C(\phi)) = r(G) < +\infty, \tag{21.39}$$

where the equality corresponds to the case

$$\omega_+(\phi, u, \delta) = C(\phi). \tag{21.40}$$

Since $\chi_{K_1}(T)$ is the n-ball

$$|p| \leq \frac{h_\delta}{\text{diam } B} \tag{21.41}$$

in the space P^n, then from (21.38–39) and (21.41) it follows that

$$\chi_{K_1}(T) \subset U(r(G)) \cap \chi_{v_\delta}(\text{int } H)$$

and

$$\frac{h_\delta}{\text{diam } B} \leq b(\omega_+(\phi, u, \delta)) \leq r(G) < +\infty \tag{21.42}$$

consecutively. Now the desired estimates (21.36) follow directly from inequalities (21.37a) and (21.42). The proof of Lemma 21.2 is completed. □

Lemma 21.2. *Let $\partial G = \emptyset$ (i.e. $r(G) = +\infty$), let the convex function $\phi(p)$ satisfy the conditions of Lemma 1 and let $u(x) \in W_2^n(G) \cap C(\overline{B})$. Let also $\delta_u = m - u_0 > 0$. Finally let the inequality*

$$\omega_+(\phi, u, \delta) < C(\phi) \tag{21.43}$$

hold for any $\delta \in (0, \delta_u)$. Then the inequalities

$$0 < h_\delta \leq b(\omega_+(\phi, u, \delta)) \text{ diam } B \tag{21.44}$$

hold for any $x \in \overline{B}$.

Since $r(G) = +\infty$ and $U(r(G)) = P^n$, then clearly

$$\chi_{K_1}(T) \subset \chi_v(\text{int } H) \subset P^n = U(r(G)).$$

Therefore Lemma 21.2 is a simplification of Lemma 21.1 both in the statement and in the proof.

Now we consider the properties of $\omega_+(\phi, u, \delta)$ which depend on the number $\delta \in (0, \delta_u)$, where $\delta_u = m - u_0 > 0$. If $0 < \delta_2 \leq \delta_1 < \delta_u$, then for every supporting hyperplane of the graph of function $v_{\delta_1}(x)$, $x \in \text{int } H$ there exists a parallel supporting hyperplane of the graph of function $v_{\delta_2}(x)$, $x \in \text{int } H$. This statement follows directly from the definition of a supporting hyperplane to the set and the constructions of the functions $v_{\delta_1}(x)$ and $v_{\delta_2}(x)$ for $x \in \text{int } H$. Thus

$$\omega_+(\phi, u, \delta_1) \leq \omega_+(\phi, u, \delta_2).$$

Hence the limit (finite or infinite) $\omega_+(\phi, u) = \lim_{\delta \to 0+} \omega_+(\phi, u, \delta)$ exists and is positive. The case $\omega_+(\phi, u) = +\infty$ is not excluded from our considerations.

If $\delta_u = m - u_0 = 0$, then $u(x) \geq m$ for all $x \in \overline{B}$. Thus $\omega_+(\phi, u) = 0$ in this case. Conversely, if $\omega_+(\phi, u) = 0$, then clearly $\delta_u = m - u_0 = 0$. The number $\omega_+(\phi, u)$ is called the ***ϕ-total area of convex support*** for a function $u(x) \in W_2^n(B) \cap C(\overline{B})$, subject to Assumption 3. The ***ϕ-total area*** $\omega_-(\Phi, u)$ ***of concave support*** for the same function $u(x)$ can be constructed in a similar way.

Theorem 21.1. *Let $\partial G \neq \emptyset$ (i.e. $r(G) < +\infty$) and a convex function $\phi(p)$, defined in $G \subset P^n$, satisfy Assumptions 1 and 2. Let the function $u(x) \in W_2^n(B) \cap C(\overline{B})$ satisfy Assumption 3. Then inequalities*

$$m - b(\omega_+(\phi, u))\text{diam } B \leq u(x) \leq M + b(\omega_-(\phi, u))\text{diam } B \tag{21.45}$$

hold for all $x \in \overline{B}$, if

$$\omega_\pm(\Phi, u) \leq C(\phi), \tag{21.46}$$

where, as usual, $m = \inf_{\partial B} u(x)$, $M = \sup_{\partial B} u(x)$.

Proof. The cases $\omega_+(\phi, u) = 0$ or $\omega_-(\phi, u) = 0$ lead to the respective estimates $u(x) \geq m$ or $u(x) \leq M$ for all $x \in \overline{B}$. Therefore only the cases

$$u_0 = \inf_{\overline{B}} u(x) < m \quad \text{and} \quad u_1 = \sup_{\overline{B}} u(x) > M$$

are interesting. Let $\delta > 0$ be an arbitrary number, subject to the condition $\delta < m - u_0$. Clearly

$$\omega_+(\phi, u, \delta) \leq \omega_+(\phi, u) \leq C(\phi).$$

Since

$$m_\delta - h_\delta \leq u_0 \leq u(x) \tag{21.47}$$

for all $x \in \overline{B}$, then from (21.47) and Lemma 21.1 it follows that

$$m - \delta - B(\omega_+(\phi, u, \delta)) \operatorname{diam} B \leq u_0 \leq u(x) \tag{21.48}$$

for all $x \in \overline{B}$. The function $\rho = b(t)$ is continuous and strictly increasing in $[0, C(\phi)]$ and

$$\lim_{\delta \to 0+} \omega_+(\phi, u, \delta) = \omega_+(\phi, u).$$

Thus (21.48) becomes

$$m - b(\omega_+(\phi, u)) \operatorname{diam} B \leq u(x)$$

for all $x \in \overline{B}$, if $\delta \to 0^+$ in inequalities (21.48).

If we apply the same considerations to the function $-u(x)$, then we obtain the inequalities

$$-u(x) \geq -M - b(\omega_-(\phi, u)) \operatorname{diam} B \tag{21.49}$$

for all $x \in \overline{B}$. From (21.48) and (21.49) we obtain the right inequality in (21.45) in $\overline{B}$. The proof of Theorem 21.1 is completed.

Remark. Since $r(G) < +\infty$, then $\omega_+(\phi, u)$ or $\omega_-(\phi, u)$ can be replaced by $C(\phi)$ in (21.45).

Theorem 21.2. *Let $\partial G = \emptyset$ (i.e. $r(G) = +\infty$) and let a convex function $\phi(p)$ defined in $G = P^n$ satisfy Assumptions 1 and 2. Let $u(x) \in W_2^n(B) \cap C(\overline{B})$.*

Then inequalities

$$m - b(\omega_+(\phi, u)) \operatorname{diam} B \leq u(x) \leq M + b(\omega_-(\phi, u)) \operatorname{diam} B, \tag{21.50}$$

hold for all $x \in \overline{B}$, if

$$\omega_\pm(\phi, u) < C(\phi). \tag{21.51}$$

The proof of this theorem can be obtained by Lemma 21.2 in the same way as the proof of Theorem 21.1 by Lemma 21.1.

Remark. Since $b(C(\phi)) = +\infty$, then inequalities (21.51) must be strict. In the opposite case, inequalities (21.50) are not increasing.

Now consider the computation of the number $C(\phi) = |(U(r(G)))|_Q$, which is one of the main invariants in the statements of Theorems 21.2 and 21.2. The total area $\sigma(\phi)$ of the normal mapping χ_ϕ is expressed by the formula

$$\sigma(\phi) = |\chi_\phi(G)|_Q = \int_Q \det\left[\frac{\partial^2\phi(p)}{\partial p_i \partial p_j}\right] dp. \tag{21.52}$$

Clearly $C(\phi) \leq \sigma(\phi)$. Below we assume that G is either the n-ball: $|p| < a$ or the entire space P^n. Then $U(r(G)) = G$ and therefore $C(\phi) = \sigma(\phi)$. Now we additionally assume that the graph of $\phi(p)$ in the space P^{n+1} is a complete convex hypersurface. Let K_ϕ be the asymptotic cone of this complete convex hypersurface $w = \phi(p)$. Since $\chi_\phi(G) = \chi_{K_\phi}(G)$, then $\sigma(\phi) = |\chi_{K_\phi}(G)|_Q$. If G is any convex bounded domain, then K_ϕ is a ray orthogonal to P^n. Hence $\chi_{K_\phi}(G) = Q^n$ and $\sigma(\phi) = +\infty$. Thus if G is the n-ball: $|p| < a$ ($a = \text{const} < +\infty$) and if the graph of $\phi(p)$ is a complete convex hypersurface, then

$$C(\phi) = \sigma(\phi) = +\infty. \tag{21.53}$$

Hence the crucial inequalities (21.46) become

$$\omega_\pm(\phi, u) \leq +\infty \tag{21.54}$$

(see Theorem 21.1). Now let K_ϕ be projected one-to-one onto P^n. Then K_ϕ is a non-degenerate convex cone. Therefore $\chi_{K_\phi}(G)$ is a bounded closed set in Q^n. Clearly $\sigma(\phi) = |\chi_{K_\phi}(G)|_Q < +\infty$ and $G = P^n$. Thus the crucial inequalities (21.51) become

$$\omega_\pm(\phi, u) < \sigma(\phi), \tag{21.55}$$

because $C(\phi) = \sigma(\phi)$ in this case.

§22. Applications to the Dirichlet Problem for Euler–Lagrange Equations

As mentioned above,

$$\sum_{i,k=1}^{n} F_{p_i p_k}(D(u(x))u_{ik} = n f_u(x, u(x)) \tag{22.0}$$

is the Euler–Lagrange equation for the functional

$$J(u) = \int_B |F(Du) + nf(x, u)]dx. \tag{22.1}$$

We assume that B is a bounded domain in E^n and ∂B is a closed continuous hypersurface in E^n, $u(x) \in W_2^n(B) \cap C(\overline{B})$ is a solution of equation (22.0) satisfying prescribed continuous Dirichlet data on ∂B and also satisfying Assumption 3:

$$\nu_u(B) \subset G \tag{22.2}$$

(see § 21), where G is prescribed bounded domain in P^n, for which $r(G) = \operatorname{dist}(O', \partial G) > 0$. Remember that $G = P^n$ if and only if $r(G) = +\infty$. We also assume that $F(p)$ is a convex function, defined in G, and that $F(p)$ satisfies Assumptions 1 and 2 (see § 21). Now we formulate Assumptions 4 and 5 concerning the properties of the function $f(x, u)$.

Assumption 4. The function $f(x, u) \in C(\overline{B} \times R)$ is convex with respect to u for every fixed $x \in \overline{B}$.

Therefore the derivative $f_u(x, u)$ is an increasing function of u for every fixed $x \in \overline{B}$. We denote by $f_u^+(x, u)$ and $f_u^-(x, u)$ the positive and negative parts of $f_u(x, u)$.

Assumption 5. The functions $f_u^+(x, k)$ and $f_u^-(x, k)$ are locally summable with degree n in B, where $-\infty < k < +\infty$ is any constant.

Theorem 22.1. *Let $u(x) \in W_2^n(B) \cap C(\overline{B})$ be a solution of the Euler–Lagrange equation for the functional $J(u)$ with prescribed continuous Dirichlet data on ∂B. If Assumptions 1–5 are satisfied, then the inequalities*

$$\omega_+(F, u) \leq \int_B [f_u^+(x, m)]^n dx \tag{22.3}$$

and

$$\omega_-(F, u) \leq \int_B [f_u^-(x, M)]^n dx \tag{22.4}$$

hold, where $m = \inf_{\partial B} u(x)$, $M = \sup_{\partial B} u(x)$ and the numbers $\omega_\pm(F, u)$ were defined in Subsection 21.3.

The integrals in (22.3) and (22.4) can take the value $+\infty$.

Proof. It is sufficient to prove inequality (22.3), because inequality (22.4) can be proved in the same way. If $\omega_+(F, u) = 0$, then (22.3) is trivial. We can assume

$$\omega_+(F, u) > 0. \tag{22.5}$$

Then from the definition of $\omega_+(F, u)$ it follows that

$$m > u_0 = \inf_{\overline{B}} u(x). \tag{22.6}$$

Let δ be any number from the open interval $(0, \delta_u)$, where $\delta_u = m - u_0 > 0$ and let $v_\delta(x)$ be the corresponding convex support of the function $u(x)$ (see Subsection 21.2). Then from (21.33) and (21.34) we obtained

$$\omega_+(F, u, \delta) = |\chi_F(\chi_{v_\delta}(D_\delta))|_Q \tag{22.7}$$

where D_δ is the set, on which $u(x) = v_\delta(x)$. We remind the reader that D is a closed subset of B and

$$\operatorname{dist}(D_\delta, \partial B) > 0. \tag{22.8}$$

It is well known that $\chi_{v_\delta}(D_\delta)$ is a closed subset of P^N.* From the definition of the convex function $v_\delta(x)$ it follows that every supporting hyperplane α of the hypersurface $z = v_\delta(x)$ is also a supporting hyperplane of the hypersurface $z = u(x)$. Moreover there exists a point $M_0(x_0, v(x_0)) \in \alpha$ such that $x_0 \in D_\delta$. Thus either

$$\chi_{v_\delta}(D_\delta) \subset \nu_u^+(B) \subset G \tag{22.9}$$

if $r(G) < +\infty$,** or

$$\chi_{v_\delta}(D_\delta) \subset U(a) \subset \nu_u^+(B) \subset G = P^n, \tag{22.10}$$

if $r(G) = +\infty$, where $U(\alpha)$ is the n-ball $|p| \leq a$ and the number a depends only on the number δ. Thus $\chi_{v_\delta}(D_\delta)$ is a compact subset of G. Hence

$$|\chi_{v_\delta}(D_\delta)|_P < +\infty. \tag{22.11}$$

Since $u(x) \in W_2^n(B) \cap C(\overline{B})$, from (22.8) and (22.10) it follows that

$$\delta(u_{ij}) \geq 0 \quad \text{and} \quad |\chi_{v_\delta}(D_\delta)|_P = \int_{D_\delta} \det(u_{ij}) dx. \tag{22.12}$$

We also use the information that every point $(x, u(x))$ of the graph of $u(x)$ is convex for $x \in D_\delta$, which follows directly from the definition of the set D_δ.

The same considerations lead to the proof that the set $\chi_F(\chi_{v_\delta}(D_\delta))$ is compact in the space Q^n. Hence

$$|\chi_F(\chi_{v_\delta}(D_\delta))|_Q < +\infty. \tag{22.13}$$

According to Assumption 1 for the function $F(p)$ we obtain the formula

$$|\chi_F(\chi_{v_\delta}(D_\delta))|_Q = \int_{\chi_{v_\delta}(D_\delta)} \det(F_{p_i p_j}(p)) dp. \tag{22.14}$$

Now from the theory of multiple integrals (see Schwartz [1], Chapter IV, § 10) and the facts presented above it follows that

$$0 < |\chi_F(\chi_{v_\delta}(D_\delta))|_Q = \int_{D_\delta} \det(F_{p_i p_j}(Du(x)) \det(u_{ij}(x)) dx. \tag{22.15}$$

Thus $|D_\delta|_E > 0$. The quadratic form $\sum_{i,k=1}^{n} u_{ik}(x)\xi_i\xi_k$ is nonnegative almost everywhere in D_δ, because all points $(x, u(x))$, $x \in D_\delta$ of the graph of $u(x)$ are

---

* An elementary proof of this fact can be found in Bakelman, Verner, and Kantor [1], Chapter 1, § 7.

** Recall that $r(G) = \operatorname{dist}(0', \partial G)$ and G is an open domain in P^n.

convex and because $u(x)$ has first and second Sobolev generalized derivatives almost everywhere in B. Since the function $F(p)$ is convex,

$$\sum_{i,j=1}^{n} F_{p_i p_j}(p)\xi_i\xi_j \tag{22.16}$$

is defined almost everywhere in G. Since $u(x) \in W_2^n(B)\cap C(\overline{B})$, the composite nonnegative quadratic form

$$\sum_{i,j=1}^{n} F_{p_i p_j}(D(u(x)))\xi_i\xi_j \tag{22.17}$$

is defined almost everywhere in B.

Thus the final part of the proof of Theorem 22.1 is reduced to the upper estimate of the integral in the right side of equality (22.15). Now we denote by D'_δ the subset of D_δ consisting of all points $x \in D_\delta$, where all generalized derivatives of the first and second orders of the function $u(x)$ take finite values, and where quadratic forms

$$\sum_{i,j=1}^{n} u_{ij}(x)\xi_i\xi_j \tag{22.18a}$$

and

$$\sum_{i,j=1}^{n} F_{p_i p_j}(D(u(x)))\xi_i\xi_j \tag{22.18b}$$

are defined and nonnegative. Clearly

$$|D'_\delta|_E = |D_\delta|_E > 0. \tag{22.19}$$

It is also clear that

$$\det(u_{ij}(x)) \geq 0 \quad \text{and} \quad \det(F_{p_i p_j}(Du(x))) \geq 0$$

for all $x \in D'_\delta$. Thus inequality (22.15) becomes

$$0 < \omega_+(F,u,\delta) = \int_{D'_\delta} \det(F_{p_i p_j}(D(u(x)))) \det(u_{ij}(x))dx. \tag{22.20}$$

Now we prove the inequality

$$[\det(F_{p_i p_j}(Du(x)))]^{1/n} \cdot [\det(u_{ij}(x))]^{1/n} \leq \frac{1}{n}\sum_{i,j=1}^{n} F_{p_i p_j}(Du(x))u_{ij}(x) \tag{22.21}$$

for all $x \in D'_\delta$.

Both determinants and the sum in inequality (22.21) are invariants of orthogonal transformations of the Euclidean space $R^n = \{\xi = (\xi_1, \xi_2, \ldots, \xi_n)\}$. If we fix any point $x \in D'_\delta$ and bring quadratic forms (22.18a and b) to canonical form, then (22.21) becomes the well known Cauchy inequality between the arithmetic and geometric means. Thus the proof of inequality (22.11) is completed.

Since $u(x)$ is a solution of the Dirichlet problem

$$\sum_{i,j=1}^{n} F_{p_i p_j}(Du(x))u_{ij}(x) = nf_u(x,u), \quad u|_{\partial B} = h(x) \in C(\partial B),$$

we have

$$0 \le \frac{1}{n}\sum_{i,j=1}^{n} F_{p_i p_j}(Du(x))u_{ij}(x) = f_u(x,u(x)) \le f_u^+(x,u(x)) \tag{22.22}$$

for all $x \in D'_\delta$. From (22.21), (22.22) and Assumptions 4 and 5 we obtain

$$0 < \omega_+(F,u,\delta) \le \int_{D'_\delta} [f_u^+(x,m)]^n dx,$$

where $m = \inf_{\partial B}\{h(x)\}$ as usual. Since

$$\int_{D'_\delta} [f_u^+(x,m)]^n dx \le \int_B [f_u^+(x,m)]^n dx,$$

we obtain the inequalities

$$0 < \omega_+(F,u,\delta) \le \int_B [f_u^+(x,m)]^n dx. \tag{22.23}$$

Since $\omega_+(F,u,\delta)$ is a non-increasing function of $\delta > 0$, from (22.23) and the definition of the number $\omega_+(F,u)$ it follows that

$$0 < \omega_+(F,u) = \lim_{\delta\to 0+} \omega_+(F,u,\delta) \le \int_B [f_u(x,m)]^n dx.$$

The inequality (22.3) is proved. The inequality (22.4) can be established in the same way. The proof of Theorem 22.1 is now complete. □

Theorem 22.2 (The Main Theorem of Estimates for Solutions of Euler–Lagrange Equation (22.0) in the Case $\partial G \neq \emptyset$). *Let $\partial G \neq \emptyset$ (i.e. $r(G) < \infty$) and let $u(x) \in W_2^n(B) \cap C(\overline{B})$ be a solution of the Euler–Lagrange equation (22.0). If Assumptions 1–5 are valid and if the inequalities*

$$\Omega_m^+ \le C(F) \tag{22.24}$$

and

$$\Omega_M^- \leq C(F) \quad , \tag{22.25}$$

hold, then the estimates

$$m - b(\Omega_m^+) \text{ diam } B \leq u(x) \leq M + b(\Omega_M^-) \text{ diam } B \tag{22.26}$$

hold for all $x \in \overline{B}$, *where*

$$C(F) = |\chi_F(U(r(G)))|_Q \tag{22.27}$$

$$= \int_{U(r(G))} \det(F_{p_i p_j}(p))dp,$$

$$\Omega_m^+ = \|f_u^+(x,m)\|_{L^n}^n, \tag{22.28}$$

$$\Omega_M^- = \|f_u^-(x,M)\|_{L^n}^n, \tag{22.29}$$

$m = \inf_{\partial G} u(x)$, $M = \sup_{\partial B} u(x)$. *The function* $\rho = b(t)$ *was introduced above (see Section 21.1).*

The proof of Theorem 22.2 follows directly from Theorems 21.1 and 22.1. Inequalities (22.24) and (22.25) are sharp. The corresponding examples will be considered in § 23. The number n in Assumptions 3: $u(x) \in W_2^n(B) \cap C(\overline{B})$ is not interchangeable with $n' < n$.

Theorem 22.3 (Main Theorem of Rstimates for Euler–Lagrange Equation (22.0) in the Case $\partial G = \emptyset$). *Let $\partial G = \emptyset$ (i.e. $r(G) = \infty$ or $G = P^n$) and let $u(x) \in W_2^n(B) \cap C(\overline{B})$ be a solution of the Euler–Lagrange equation (22.0). If Assumptions 1–5 are valid and the strict inequalities*

$$\Omega_m^+ < C(F) \tag{22.30}$$

$$\Omega_M^- < C(F) \tag{22.31}$$

hold, then estimates (22.26) hold for all $x \in \overline{B}$.

The notation is explained above (see Theorem 22.2). The proof of Theorem 22.3 follows directly from Theorems 21.2 and 22.1. Inequalities (22.30) and (22.31) are sharp. The corresponding examples will be considered in § 23.

It is possible to extend Theorems 22.2 and 22.3 to a few wide classes of nonlinear elliptic Euler–Lagrange equations, which correspond to multiple integrals

$$I(u) = \int_B F(x, u(x), Du(x))dx. \tag{22.32}$$

This extension is also for solutions $u(x) \in W_2^n(B) \cap C(\overline{B})$ of these equations. We also assume that the function $F(x,u,p)$ is convex with respect to $p \in G$ for any fixed $x \in B$, $u \in R$ and $F(x,u,p) \in W_2^n(B \times R \times G)$. These results will be considered in § 24.

§23. Applications to Calculus of Variations, Differential Geometry and Continuum Mechanics

23.1 Applications to Calculus of Variations

Theorem 23.1 (S.N Bernstein [4]). *Let $u(x,y) \in C^2(\overline{B})$ be a solution of the Euler–Lagrange equation for the following two-dimensional functional:*

$$J(u) = \int_B [F(u_x, u_y) + f(x,y,u)]dxdy. \tag{23.1}$$

Let the following conditions be fulfilled:

1) *B is a domain in a two-dimensional Euclidean plane $E^2 = \{(x,y)\}$ and ∂B is a closed curve in E^2;*
2) *$F(p,q) \in C^2(P^2)$, where $P^2 = \{(p,q)\}$ is a second two-dimensional Euclidean space;*
3) *Let*
$$|F(p,q)| \geq N(p^2+q^2)^{a/2} \tag{23.2}$$
for $p^2+q^2 \geq r^2$, where $N =$ const >0, $r =$ const >1 and $a =$ const >1;
4)
$$F_{pp}F_{qq} - F_{pq}^2 \geq F_0 = \text{const} > 0 \tag{23.3}$$
for all $(p,q) \in P^2$.
5) *$f(x,y,u)$ is a convex C^2-function in $\overline{B} \times R$ with respect to u.*

Then for $u(x,y)$ the estimate

$$|u(x,y)| \leq U_0, \quad (x,y) \in \overline{B}, \tag{23.4}$$

can be obtained, where the constant U_0 depends only on the properties of the functions $F(p,q)$, $f(x,y,u)$ and their derivatives up to second order, constants of conditions 3 and 4, and the numbers $m = \inf_{\partial B} u(x,y)$, $M = \sup_{\partial B} u(x,y)$.

First of all Bernstein estimated the integral

$$\int\int_B \{|u_{xx}| + u_{yy}| + |u_{xy}|\}dxdy \tag{23.5}$$

in the terms of data mentioned in his theorem and then he obtained the desired estimate (23.4). His technique is essentially two-dimensional.

The conditions and the proof of Theorem 22.3 are based on ideas and a technique different from the considerations of Bernstein. This permits us to omit the overly strong Bernstein's condition 3 (see inequality (23.2)) and also to consider generalized solutions $u(x) \in W_2^n(B) \cap C(\overline{B})$ of Euler-Lagrange equations instead of classical ones in the Bernstein's Theorem. Moreover, the convex functions $F(p,q)$ and $f(x,y,u)$ can be sufficiently non-smooth, and all considerations can be made for functions depending on n variables, where $n > 2$.

Thus the Bernstein Theorem can be significantly developed in many directions.

Now we present the statement and the proof of our theorem.

Theorem 23.2. *Let $F(p)$ be a convex function defined over the entire space P^n and satisfying Assumption 1. We also assume that the inequality*

$$\det(F_{p_i p_j}(p)) \geq F_0 > 0 \tag{23.6}$$

holds almost everywhere in P^n, where F_0 is any constant. Let $f(x,u)$ satisfy Assumption 4 and let the numbers*

$$\Omega_k^{\pm} = \|f_u^{\pm}(x,k)\|_{L^n(B)} < +\infty \tag{23.7}$$

*for any constant $k \in (-\infty, +\infty)$.***

Then the inequalities

$$m - \left[\frac{\Omega_m^+}{\mu_n F_0}\right]^{1/n} \cdot \operatorname{diam} B \leq u(x) \leq M + \left[\frac{\Omega_M^-}{\mu_n F_0}\right]^{1/n} \cdot \operatorname{diam} B \tag{23.8}$$

hold for all $x \in \overline{B}$ for any solution $u(x)$ of the Dirichlet problem

$$\sum_{i,j=1}^{n} F_{p_i p_j}(Du(x)) u_{ij}(x) = n f_u(x, u(x)), \tag{23.9}$$

$$u|_{\partial B} = h(x) \in C(\partial B), \tag{23.10}$$

which belongs to $W_2^n(B) \cap C(\overline{B})$, where as usual $m = \inf_{\partial B} h(x)$, $M = \sup_{\partial B} h(x)$.

Proof. We consider two convex functions $\phi_1(p) = \frac{1}{2}\sum_{i=1}^{n} p_i^2$ and $\phi_2(p) = F(p)$ in the entire space P^n. According to the definition of the functions $c_1(\rho)$, $c_2(\rho)$ (see Subsection 21.1, formula (21.3)) we obtain

$$c_1(\rho) = \int_{U(\rho)} \det(\phi_{1,ij}(p))dp = \int_{U(\rho)} 1dp = \mu_n \rho^n, \tag{23.11}$$

$$c_2(\rho) = \int_{U(\rho)} \det(F_{ij}(p))dp, \tag{23.12}$$

---

*Clearly this restriction on $F(p)$ is somewhat stronger than the restriction in Assumption 2 (see Subsection 21.1).

** The restriction (23.7) on $f_u^{\pm}(x,u)$ is somewhat stronger than the restriction in Assumption 5.

where as usual $U(\rho)$ is the n-ball $|p| < \rho$ in P^n and μ_n is the volume of $U(1)$. Inequality (23.6) gives the inequality

$$c_1(\rho) \leq \frac{1}{F_0} c_2(\rho), \tag{23.13}$$

which holds for all $\rho \in [0, +\infty)$. Clearly $c_1(\rho)$ and $c_2(\rho)$ are continuous and strictly increasing in $[0, +\infty)$. Since $C(\phi_1) = \lim_{\rho \to +\infty} c_1(\rho) = +\infty$, it follows from (23.13) that $C(\phi_2) = \lim_{\rho \to \infty} c_2(\rho) = +\infty$. Therefore for any $\rho \in [0, +\infty)$ there exists only one number $\rho^* \geq \rho$ such that

$$c_1(\rho^*) = \frac{1}{F_0} c_2(\rho). \tag{23.14}$$

Let $\rho = b_1(t)$ and $\rho = b_2(t)$ be inverses of $c_1(\rho)$ and $c_2(\rho)$. Then they are defined in $[0, +\infty)$, and are strictly increasing and continuous in $[0, +\infty)$. Thus

$$\rho^* = b_1 \left(\frac{1}{F_0} c_2(\rho) \right) \tag{23.15}$$

where $0 \leq \rho < \rho^* < +\infty$ are numbers considered in (23.14). Since the nonnegative numbers Ω_m^+ and Ω_M^- are finite,

$$\Omega_m^+ < C(\phi_2) = +\infty \tag{23.16}$$

and

$$\Omega_M^- < C(\phi_2) = +\infty. \tag{23.17}$$

Thus we can use Theorem 22.3 and obtain the estimates

$$m - b_2(\Omega_m^+) \operatorname{diam} B \leq u(x) \leq M - b_2(\Omega_M^-) \operatorname{diam} B \tag{23.18}$$

for all $x \in \overline{B}$. According to inequality (23.16), the nonnegative number $\rho = b_2(\Omega_m^+)$ is finite. Hence

$$\Omega_m^+ = c_2(\rho). \tag{23.19}$$

From (23.15) and (23.19) it follows that there exists only one number ρ^* such that

$$\rho \leq \rho^* < +\infty \tag{23.20}$$

and

$$\rho^* = b_1 \left(\frac{1}{F_0} c_2(\rho) \right) = b_1 \left(\frac{\Omega_m^+}{F_0} \right). \tag{23.21}$$

Inequalities (23.18), (23.20) and identity (23.21) now give the inequality

$$m - b_1 \left(\frac{\Omega_m^+}{F_0} \right) \operatorname{diam} B \leq u(x). \tag{23.22}$$

Similar considerations lead to the inequality

$$u(x) \leq M + b_1\left(\frac{\Omega_M^-}{F_0}\right) \text{ diam } B. \tag{23.23}$$

Since $t = c_1(\rho) = \mu_n \rho^n$, we have

$$b_1(t) = \left(\frac{t}{\mu_n}\right)^{1/n}. \tag{23.24}$$

Thus inequalities (23.22–23) and the formula (23.24) give the desired inequalities (23.8). The proof of Theorem 23.2 is complete. □

23.2 Applications to Differential Geometry

In this subsection we consider two-sided C^0-estimates of solutions of the Dirichlet problem for the mean curvature equation in Euclidean and Minkowski $n+1$-dimensional spaces.

a) Hypersurfaces with Prescribed Mean Curvature in Euclidean Space E^{n+1}.

Let a hypersurface S with prescribed mean curvature H be a graph of a function $u(x) \in W_2^n(B) \cap C(\overline{B})$. We assume that $H(x) \in L^n(B)$ and $u|_{\partial B} = h(x) \in C(\partial B)$. Clearly $u(x)$ is a solution of the Euler–Lagrange equation for the functional

$$I(u) = \int_B \left[\sqrt{1 + D(u(x))^2} + nH(x)u(x)\right] dx, \tag{23.25}$$

satisfying the Dirichlet boundary condition $u(x) = h(x)$ for all $x \in \partial B$.

The convex function $F(p) = \sqrt{1+|p|^2} \in C^\infty(P^n)$. Hence $G = P^n$ and $r(G) = +\infty$. The asymptotic cone K_F with the vertex at O' has the equation $w = 1 + |p|$, $p \in P^n$. Therefore $\chi_{K_F}(P^n)$ is the unit n-ball $|q| \leq 1$ in the space Q^n. According to (21.26)

$$C(F) = |\chi_{K_F}(P^n)|_Q = \mu_n, \tag{23.26}$$

where μ_n is the volume of the n-unit Euclidean ball.

Clearly $f_u^+(x,u) = H^+(x)$ nd $f_u^-(x,u) = H^-(x)$, where $H^+(x) \geq 0$ and $H^-(x) \geq 0$ are the positive and negative parts of $H(x)$. If all conditions of Theorem 22.3 are fulfilled then the crucial inequalities (22.30–31) become

$$\|H^+(x)\|_{L^n(B)} < \mu_n \quad \text{and} \quad \|H^-(x)\|_{L^n(B)} < \mu_n. \tag{23.27}$$

They provide the estimates (22.26) for the function $u(x)$, where the function

$$b(t) = \left[\frac{\mu_n^{2/n}}{\mu_n^{2/n} - t^{2/n}}\right]^{1/2}$$

is defined for $t \in [0, \mu_n)$.

The inequalities (23.27) are sharp. Actually the Dirichlet problem

$$\sum_{i=1}^{n} \frac{\partial}{\partial x_i} \left\{ \frac{u_i}{[1+(Du)^2]^{1/2}} \right\} = nH_0$$

$$u|_{\partial B} = 0$$

where $H_0 = \text{const} > 0$ and B is the n-ball $|x| < \frac{1}{H_0}$ is the corresponding example of this assertion.

The existence theorems of the Dirichlet problem for mean curvature equation by the interlocked necessary and sufficient conditions were established by Serrin [4], [5] and Bakelman [10], and Bakelman, Verner, and Kantor [1]. Two-sided C^0-estimates of solutions are very important in these investigations. The estimates considered in this subsection were established in Bakelman [10].

b) Spacelike Hypersurfaces with Prescribed Mean Curvature in the Minkowski Space M^{n+1}.

The space $R^{n+1} = \{(x,t)\} = \{(x_1, x_2, \ldots, x_n, t)\}$ with metric

$$ds^2 = \sum_{i=1}^{n} dx_i^2 - dz^2 \tag{23.28}$$

is called Minkowski space and its denoted by M^{n+1}. Let S be a hypersurface such that ds^2 is restricted to a positive form on S. Such an S is called spacelike. If S is the graph of a function $z = u(x)$, then S is spacelike if and only if $|Du(x)| < 1$ for any $x \in B$, where B is the domain of the function $u(x)$. The spacelike hypersurfaces were studied by Calabi [4] and Cheng and Yau [2] in connection with the Bernstein conjecture in M^{n+1}. The spacelike solutions $u(x) \in W_2^n(B) \cap C(\overline{B})$ of the Euler-Lagrange equation for

$$M(u) = \int_B [-[1-(Du)^2]^{1/2} + nH(x)u]dx \tag{23.29}$$

have prescribed mean curvature $H(x)$ in M^{n+1}. Clearly $H(x)$ is locally summable with degree n in the open domain B. According to our general considerations we conclude that the convex function $F(p) = -(1-|p|^2)^{1/2}$ is defined in the open ball $|p| < 1$ in P^n. Thus G is the ball $|p| < 1$ and $r(G) = 1$. Clearly

$$\chi_F(G) = Q^n. \tag{23.30}$$

Hence $C(F) = +\infty$. The crucial inequalities (22.24) and (22.25) in Theorem 22.2 become

$$\int_B H_\pm^n(x)dx \leq +\infty \tag{23.31}$$

for the Dirichlet problem for mean curvature equation in the space M^{n+1}. Since all other conditions of Theorem 22.2 are fulfilled, we can apply estimates (22.26) if and only if the functions $H^+(x)$ and $H^-(x)$ are locally summable with degree n in B.

Thus there is an essential difference between the solutions of the Dirichlet problem for the mean curvature equation in Euclidean and Minkowski spaces.

23.3 Applications to Continuum Mechanics

a) The Problem of Torsion of Hardening Rods.

Let P be a prismatic rod represented by the cylinder with the base $\overline{B} = B \cup \partial B$ and generators parallel to z-axis, where B is a bounded domain in the xy-plane. Let the base of P be clamped and let the rod P twist under the action of a moment M. We denote by w the torsion per unit length of the rod. Let $u(x, y)$ be the stress function of the rod P. Then $T = (u_x^2 + u_y^2)^{1/2}$ is the intensity of the tangential stress tensor and

$$\frac{\partial}{\partial x}[g(T^2)u_x] + \frac{\partial}{\partial y}g(T^2)u_y] = -2w \tag{23.32}$$

is *the equation of torsion of hardening rods*. The function $g(T^2)$ is called *the modulus of plasticity of the rod P*. It describes the dependence between the intensity Γ of the shear strain tensor and the intensity T of the tangential stress tensor by the formula $\Gamma = g(T^2)T$. The experimental law

$$\frac{d\Gamma}{dT} > 0 \tag{23.33}$$

is a necessary and sufficient condition of ellipticity for equation (23.32). The problem of hardening rods can be reduced to the Dirichlet problem for equation (23.32) with zero boundary data.

Now consider the Dirichlet problem

$$\sum_{i=1}^{n}[g(|Du|^2)u_{x_i}]_{x_i} = nH(x, u), \quad u|_{\partial B} = 0 \tag{23.34}$$

for equations which somewhat generalize equation (23.32) to n dimensions. Clearly (23.34) is the Euler–Lagrange equation for the functional

$$J(u) = \int_B \frac{1}{2}\left\{\int_0^{|Du(x)|^2} g(s)ds\right\} ds + n\int_B \left\{\int_0^{u(x)} [H(x, s)]dx\right\} dx. \tag{23.35}$$

It is appropriate to consider $g(|p|^2)$ either in the n-balls $U(a)$: $|p| < a$ or in the entire space P^n, where $0 < a = \text{const} < +\infty$. This makes it possible to consider $g(T^2)$ either as a function of a single variable T^2 or as a composite function of a single variable T. Now we introduce the function $\Gamma(T)g(T^2)T$, which is defined in the same domain as the function $g(T^2)$, and formulate assumptions for functions $g(T^2)$, $\Gamma(T)$ and $H(x,u)$, allowing Theorems 22.2; 22.3 to be applied for two-sided C^0-estimates of solutions of equations (23.34).

Below we assume that a positive constant a also takes the value $+\infty$. Thus both finite intervals $[0,a)$ and the ray $[0,+\infty)$ can be represented by $[0,a)$.

Let $s = T^2$ and $g(s) = g(T^2)$. Clearly $g(s)$ is defined on $[0,a^2)$.

Assumption 6. The function $g(s)$ is positive and absolutely continuous on $[0,a^2)$, i.e. $\frac{dg(s)}{ds}$ exists everywhere on $[0,a^2)$ and

$$g(s) = g(0) + \int_0^s \frac{dg(s)}{ds} ds \tag{23.36}$$

for all $s \in [0,a^2)$.

Assumption 7. Let $g(s)$ satisfy Assumption 6. Then

$$\frac{d\Gamma}{dT} > 0 \tag{23.37}$$

on a set of positive measure in any interval (a', a''), $0 < a' < a'' < a$, where $\Gamma(T) = g(T^2) \cdot T$ and $s = T^2$.

Assumption 8. The function $H(x,u)$ is increasing with respect to u for every fixed $x \in B$ and the functions $H^+(x,k)$ and $H^-(x,k)$ are locally summable with degree n in B for all values of the constant $k \in (-\infty,+\infty)$.

Below we consider functions $g(s)$ and $H(x,u)$ which satisfy Assumptions 6, 7, 8. From (23.35) it follows that the graph S_F of the function

$$F(p) = \frac{1}{2}\int_0^{|p|^2} \{g(s)\} ds \tag{23.38}$$

is a hypersurface of revolution. The meridian of this hypersurface has equation

$$w = F(T)$$

where $p = (T,0,0,\ldots,0) \in P^n$ and $0 \leq T < a$. Clearly

$$\frac{dw}{dT} = g(T^2)T = \Gamma(T)$$

for all $T \in [0,a)$. From Assumption 6 it follows that $\frac{dw}{dT}$ is an absolutely continuous function of T. Since $\Gamma(0) = 0$,

$$\frac{dw}{dT} = \int_0^T \frac{d\Gamma(\xi)}{dT} d\xi \tag{23.39}$$

for all $T \in [0,a)$. Now from (23.39) and Assumption 7 it follows that the function $w = F(T)$ is convex and the derivative $\frac{dF(T)}{dT}$ is strictly increasing absolutely continuous on $[0,a)$. Hence the area of the normal image $\omega(S_F, e')$ of the convex hypersurface of revolution S_F is absolutely continuous and

$$(S_F, e') = \int_{e'} \det(F_{p_i p_j}(p))dp. \tag{23.40}$$

The formula

$$\det(F_{p_i p_j}(p)) = g^{n-1}(|p|^2)\frac{d\Gamma(|p|)}{d|p|} \tag{23.41}$$

holds almost everywhere in the ball $U(a)$: $|p| < a$. Now from Assumption 6 and 7 and equalities (23.39–41) it follows that

$$\begin{aligned}\omega(S_F, U(\rho)) &= \mu_n \int_0^\rho g^{n-1}(T)\frac{d\Gamma(T)}{dT}T^{n-1}dT \\ &= \mu_n \int_0^\rho \Gamma^{n-1}(T)d\Gamma(T) = \mu_n\Gamma^n(\rho),\end{aligned}$$

where μ_n is the volume of the n-unit ball in E^n and $U(\rho)$ is the n-ball $|p| < \rho$ in P^n for $0 \le \rho < a$.

Thus from Assumptions 6 and 7 it follows that the function $F(p)$ introduced by (23.38) satisfies Assumptions 1 and 2 (see Subsection 21.1). Clearly Assumption 8 is equivalent to Assumptions 4 and 5 (see Section 22). Thus the following theorems can be obtained from Theorems 22.2 and 22.3.

Theorem 23.3. *Let G be the n-ball $|p| < a$, where $a < +\infty$ and let $u(x) \in W_2^n(B) \cap C(\overline{B})$ be a solution of the Dirichlet problem (23.34). If Assumptions 3, 6, 7, 8 are valid and if the inequalities*

$$\Omega_0^\pm \le \mu_n\Gamma^n(a) \tag{23.42}$$

hold, then the estimates

$$-b(\Omega_0^+)\text{diam } B \le u(x) \le b(\Omega_0^-)\text{diam } B \tag{23.43}$$

hold for all $x \in \overline{B}$, where

$$\Omega_0^\pm = \int_B [H^\pm(x,0)]^n dx, \tag{23.44}$$

and $\rho = b(t)$ is the inverse of the strictly increasing absolutely continuous function $\mu_n\Gamma^n(\rho)$, $0 \le \rho \le a$.

Theorem 23.4. *Let $G = P^n$ and let $u(x) \in W_2^n(B) \cap C(\overline{B})$ be a solution of the Dirichlet problem (23.34). If Assumptions 6, 7, 8 are valid and if the strict inequalities*

$$\Omega_0^\pm < \mu_n\Gamma^n(+\infty) \tag{23.45}$$

hold, then estimates (23.43) hold for all $x \in \overline{B}$.

All notations are explained in Theorem 23.3.

b) Equations Relating to Gas Dynamics.

It is well known that the stationary irrational flow of an ideal compressible fluid can be described by the following equation of continuity

$$\operatorname{div}(\sigma \cdot Du) = 0, \tag{23.46}$$

where the fluid density σ satisfies a density-speed relation $\sigma = \sigma(|Du|)$. For a perfect gas this relation is

$$\sigma = \left(1 - \frac{\gamma - 1}{2}|Du|^2\right)^{\frac{1}{\gamma-1}}, \tag{23.47}$$

where the constant γ is the ratio of specific heats of the gas and $\gamma > 1$.

In this subsection we consider non-homogeneous equations

$$\operatorname{div}(\sigma \cdot Du) = nH(x), \tag{23.48}$$

where σ is defined by (23.47). Clearly (23.48) is the Euler–Lagrange equation for the functional

$$S(u) = \int_B \frac{1}{2}\left\{\int_0^{T^2} \sigma(s)dx\right\} dx + \int_B \{nH(x)u\}dx, \tag{23.49}$$

where as usual, $T = |\text{grad } u(x)|$. Since $\gamma > 1$, the function

$$\sigma(|p|) = \left(1 - \frac{\gamma - 1}{2}|p|^2\right)^{\frac{1}{\gamma-1}}$$

and its derivatives are defined only in the ball: $|p| < \left(\frac{2}{\gamma-1}\right)^{1/2}$.

According to the previous subsection we should consider the function

$$F(T) = \frac{1}{2}\int_0^{T^2} \left(1 - \frac{\gamma - 1}{2}s^2\right)^{\frac{1}{\gamma-1}} ds$$

for $0 \le T < \left(\frac{2}{\gamma-1}\right)^{1/2}$. The function $F(T)$ is convex in $\left[0, \left(\frac{2}{\gamma+1}\right)^{1/2}\right)$ and it is concave in $\left(\left(\frac{2}{\gamma+1}\right)^{1/2}, \left(\frac{2}{\gamma-1}\right)^{1/2}\right)$. Thus the function

$$F(p_1, p_2, \dots, p_n) = \frac{1}{2}\int_0^{|p|^2} \left(1 - \frac{\gamma - 1}{2}s^2\right)^{\frac{1}{\gamma-1}} ds$$

is C^∞-strictly convex function only in the ball

$$G_\gamma\colon\ |p| < \left(\frac{2}{\gamma + 1}\right)^{1/2}. \tag{23.50}$$

Hence equation (23.49) is *elliptic* and the flow is *subsonic*, when $p = Du \in G_\gamma$.

Theorem 23.6. *Let $G = G_\gamma$ and let $u(x) \in W_2^n(B) \cap C(\overline{B})$ be a solution of the Dirichlet problem (23.34), satisfying Assumption 3 (see Subsection 21.3). If the inequalities*

$$\Omega^\pm < \mu_n \left(\frac{2}{\gamma+1}\right)^{\frac{n(\gamma+1)}{2(\gamma-1)}} \tag{23.51}$$

hold, then the estimates (23.43) hold.

We use the notation

$$\Omega^\pm = \int_B [H^\pm(x)]^n dx,$$

where $H^\pm(x)$ are the respective positive and negative parts of the function $H(x)$ and $\rho = b(t)$ is the inverse of the function $c(\rho) = \mu_n \left(1 - \frac{\gamma-1}{2}\rho^2\right)^{\frac{2}{\gamma-1}} \cdot \rho^n$, $0 \leq \rho \leq \left(\frac{2}{\gamma+1}\right)^{1/2}$.

Theorem 23.5 follows directly from Theorem 23.3. If we consider the boundary data $u|_{\partial B} = h(x) \in C(\partial B)$, then we replace inequalities (23.43) by inequalities

$$m - b(\Omega^+)\text{diam } B \leq u(x) \leq M + b(\Omega^-)\text{diam } B \tag{23.52}$$

for all $x \in \overline{B}$, where as usual $m = \inf_{\partial B} h(x)$, $M = \sup_{\partial B} h(x)$.

§24. C^2-Estimates for Solutions of General Euler–Lagrange Elliptic Equations

24.1 Introduction

In this section we are concerned with two-sided C^0-estimates for solutions of nonlinear elliptic Euler–Lagrange equations, relating to general functionals

$$I(u) = \int_B F(x, u(x), Du(x))dx. \tag{24.1}$$

We suppose as usual that B is a bounded domain in E^n, $n \geq 2$. In the case of general elliptic Euler–Lagrange equations the technique used in § 22 for equations (22.0) requires further development. For example the auxiliary convex hypersurfaces $w = F(p)$ will be constructed by means of solutions for the special auxiliary elliptic Monge–Ampere equations.

In § 24 we are also concerned with applications to calculus of variations and differential geometry.

24.2 Monge–Ampere Generators

Let G be a domain in P^n such that the origin O' of P^n is an interior point of G. Let $U(r(G)) \subset G$ be the n-ball introduced in § 21. Finally let $R(p)$ be a function in G such that:

a) $R(p)$ is nonnegative and locally summable in G;
b) $R(p) > 0$ on a subset of a positive measure in any annulus $A(\rho_1, \rho_2)$, where $0 < \rho_1 < \rho_2 < r(G)$;
c) the function
$$R^*(p) = \sup_{|p'|=|p|} R(p')$$
is summable in $U(\rho)$ for $0 < \rho < r(G)$, where $U(\rho)$ is the n-ball $|p| < \rho$ in P^n.

Functions $R(\rho)$ satisfying conditions a), b), c) are called *admissible*. A convex function $\phi(p)$ is called a *Monge–Ampere generator* if $\phi(p)$ is a convex generalized solution of the equation

$$\det\left(\frac{\partial^2 \phi(p)}{\partial p_i \partial p_j}\right) = R(p) \tag{24.2}$$

in $U(r(G))$. Thus every Monge–Ampere generator $\phi(p)$ satisfies the following conditions:

A) $\phi(p)$ is a convex function in $U(r(G))$ and $\phi(p)$ satisfies equation (24.2) almost everywhere in $U(r(G))$;

B) if χ_ϕ: $U(r(G)) \to Q^n$ is the normal mapping of the function $\phi(p)$, then

$$|\chi_\phi(e')|_Q = \int_{e'} R(p)dp, \tag{24.3}$$

where e' is any measurable subset of $U(r(G))$ such that either

$$\operatorname{dist}(e', \partial U(r(G))) > 0, \quad \text{if} \quad \partial G \neq \emptyset, \tag{24.4}$$

or e' is a bounded subset of P^n, if $\partial G = \emptyset$.

If $\phi(p)$ is a Monge–Ampere generator for a locally summable function $R(p) > 0$, then we call $R(p)$ the *density* of $\phi(p)$.

The existence of Monge–Ampere generators is equivalent to the solvability of equation (24.2) in the class of convex generalized solutions either in the balls: $|p| < a$ or in the entire space P^n. If a nonnegative function $R(p)$ satisfies the condition

$$\int_{|p|<r(G)} R(p)dp < +\infty, \tag{24.5}$$

then equation (24.2) has generalized convex solutions. These facts follow directly from §§ 11 and 17.

The limiting case

$$\int_{U(r(G))} R(p)dp = +\infty \tag{24.6}$$

can be solved for admissible densities $R(p)$ by the approximation of Monge–Ampere generators whose densities satisfy condition (24.5).

First of all we need to establish a few properties of the radial Monge–Ampere generators. The density $R(p)$ is called *radial*, if

$$R(p) = R(|p|) \tag{24.7}$$

for all $p \in U(r(G))$. Similarly a generalized convex solution $\phi(p)$ of the Monge–Ampere equation

$$\det\left(\frac{\partial^2\phi(p)}{\partial p_i \partial p_j}\right) = R(|p|) \tag{24.8}$$

is called *radial* or a radial Monge–Ampere generator, if $\phi(p)$ depends only on $|p|$, i.e.

$$\phi(p) = \phi(|p|). \tag{24.9}$$

Clearly the graph S_ϕ of the function $\phi(|p|)$ is a convex hypersurface of revolution around the w-axis in the space $P^{n+1} = P^n \times R = \{(p,w) = (p_1, p_2, \dots, p_n; w)\}$. Clearly $w = \phi(\rho)$, $0 \le \rho < r(G)$ is the equation of the meridian of S_ϕ. The left and right derivatives of a convex function $\phi(\rho)$ defined on an interval $[a,b)$*) exist except on a countable subset of $[a,b)$. Every such exceptional point is a corner point for the graph of $\phi(\rho)$. Since the convex hypersurface of revolution S_ϕ has the absolutely continuous area of its normal mapping χ_ϕ: $U(r(G)) \to Q^n$, the convex function $\phi(\rho)$ does not have any corner points. Hence $\phi'(\rho)$ exists at each point $\rho \in [0, r(G))$. From the absolutely continuity of the area of χ_ϕ it follows that $\phi'(\rho)$ is an absolutely continuous function of $\rho \in [0, r(G))$ **), $\phi'(0) = 0$ and

$$\begin{aligned} |\chi_\phi(U(\rho))|_Q &= \sigma_{n-1}\int_0^\rho (\phi'(\rho))^{n-1}\phi''(\rho)d\rho \\ &= \sigma_{n-1}\int_0^\rho [\phi'(\rho)]^{n-1} d\phi'(\rho), \end{aligned} \tag{24.10}$$

where $U(\rho)$ is the n-ball: $|p| < \rho$ in the space P^n and σ_{n-1} is the area of the unit $(n-1)$-sphere in P^n. Thus equation (24.8) for the meridian $w = \phi(\rho)$ becomes

$$[\phi'(\rho)]^{n-1}\phi''(\rho) = R(\rho)\rho^{n-1}. \tag{24.11}$$

*) We consider only the right derivative of $g(\rho)$ at the point $\rho = a$.

**) Since $\phi(\rho)$ is convex and $\phi'(0) = 0$, $\phi'(\rho) \ge 0$ in $[0, r(G))$. Clearly $\phi''(\rho) \ge 0$ in $[0, r(G))$.

Since

$$\int_{U(\rho)} R(p)dp = \sigma_{n-1} \int_0^\rho R(t)t^{n-1}dt, \tag{24.12}$$

the limiting condition (24.6) becomes

$$\int_0^{r(G)} R(t)t^{n-1}dt = +\infty. \tag{24.13}$$

Thus any solution $\phi(\rho)$, $0 \leq \rho < r(G)$ for the meridian of the desired radial solution of equation (24.8) satisfies the equation

$$(\phi'(\rho))^{n-1}\phi''(\rho) = R(\rho)\rho^{n-1} \tag{24.14}$$

and the following conditions

$$\phi'(0) = 0; \tag{24.15}$$

$$\int_0^\rho R(t)t^{n-1}dt < +\infty; \tag{24.16}$$

for any $\rho \in [0, r(G))$;

$$\lim_{\rho \to r(G)} \int_0^\rho R(t)t^{n-1}dt = +\infty. \tag{24.17}$$

Moreover we established above, that $\phi'(\rho)$ is absolutely continuous on $[0, r(G))$ and $\phi'(\rho) \geq 0$ for all these ρ. Since $\phi(\rho)$ is convex, then $\phi'(\rho)$ is increasing on $[0, r(G))$ and $\phi''(\rho)$ almost everywhere positive on the same interval.

From (24.14) and (24.15) we obtain

$$\phi'(\rho) = \left[n \int_0^\rho R(t)t^{n-1}dt\right]^{1/n} \tag{24.18}$$

and

$$\phi(\rho) = A_0 + \int_0^\rho \left[n \int_0^s R(t)t^{n-1}dt\right]^{1/n} ds \tag{24.19}$$

for any $\rho \in [0, r(G))$, where A_0 is any constant. We will prefer to choose $A_0 = 0$ in our further considerations.

From (24.18), (24.19) and assumption on the function $R(p)$ (see the beginning of Subsection 24.2) it follows that $\phi'(\rho)$ is strictly increasing in $[0, r(G))$ and therefore $\phi'(\rho) > 0$ in $(0, r(G))$. Thus the desired radial solution of equation (24.8) becomes

$$\phi(|p|) = A_0 + \int_0^{|p|} \left(n \int_0^s R(t)t^{n-1}dt\right)^{1/n} ds \tag{24.20}$$

for all $p \in U(r(G))$, where the constant A_0 will be chosen later.

Now we return to the existence of Monge–Ampere convex generators with prescribed densities $R(p)$ for which

$$\int_{U(r(G))} R(p)dp = +\infty. \tag{24.6}$$

We will assume that $R(p)$ is an admissible function (see the corresponding definition in the beginning of Subsection 24.2).

Below we will prove two existence theorems for Monge–Ampere generators. The first one is related to the case when $\partial G \neq \emptyset$, i.e. $r(G) < +\infty$, and the second one is related to the case $\partial G = \emptyset$, i.e. $r(G) = +\infty$ or $G = P^n$.

Theorem 24.1. *Let $\partial G \neq \emptyset$ and let $R(p)$ be an admissible function. Then the inequality*

$$\int_0^{r(G)} \left(\int_0^s R^*(t)t^{n-1}dt \right)^{1/n} ds < +\infty \tag{24.21}$$

provides the existence of Monge–Ampere generators for the function $R(p)$.

Proof. Let $\varepsilon > 0$ be any number less than $\frac{1}{4}r(G)$. We denote by $\varphi_\varepsilon(\rho)$ a C^1-function in $[0, r(G))$ such that

$$\varphi_\varepsilon(\rho) = \begin{cases} 1, & \text{if } \ 0 \le \rho \le r(G) - 2\varepsilon \\ \text{strictly decreasing function from 1 down to zero,} & \text{if } \ r(G) - 2\varepsilon < \rho < r(G) - \varepsilon \\ 0, & \text{if } \ r(G) - \varepsilon \le \rho < r(G). \end{cases}$$

Then the functions

$$R_\varepsilon(p) = R(p)\varphi_\varepsilon(|p|) \tag{24.22}$$

satisfy the inequalities

$$R_{\varepsilon_1}(p) \le R_{\varepsilon_2}(p) \le R(p) \tag{24.23}$$

for all $p \in U(r(G))$ and all $0 < \varepsilon_2 \le \varepsilon_1 \le \frac{1}{4}r(G)$. Since

$$\int_{U(r(G))} R_\varepsilon(p)dp \le \int_{|p| \le r(G) - \varepsilon} R(p)dp < +\infty$$

there exists a convex generalized solution $\psi_\varepsilon(p)$ of the Dirichlet problem

$$\det \left(\frac{\partial^2 \psi_\varepsilon(p)}{\partial p_i \partial p_j} \right) = R_\varepsilon(p), \tag{24.24}$$

$$\psi_\varepsilon(p)|_{\partial U(r(G))} = 0. \tag{24.25}$$

Now we use the comparison theorem for convex generalized solutions of the Dirichlet problem (24.24–25). Then the inequalities

$$\psi_{\varepsilon_1}(p) \geq \psi_{\varepsilon_2}(p) \geq \phi(p) \tag{24.26}$$

hold for all $\varepsilon \in \overline{U(r(G))}$ and for all $0 < \varepsilon_2 \leq \varepsilon_1 \leq \frac{1}{4}r(G)$, where $\phi(p) = \phi(|p|)$ is a radial convex generalized solution of the Dirichlet problem

$$\det\left(\frac{\partial^2\phi(p)}{\partial p_i \partial p_j}\right) = R^*(|p|), \tag{24.27}$$

$$\phi(p)|_{\partial U(r(G))} = 0. \tag{24.28}$$

Since $R(p)$ is admissible function and since $R^*(\rho)$ satisfies condition (24.21),

$$\phi^*(\rho) = \int_0^\rho \left(n\int_0^s R^*(t)t^{n-1}dt\right)^{\frac{1}{n}} ds \tag{24.29}$$

is a continuous nonnegative convex function in the closed interval $[0, r(G)]$ and $\phi^*(\rho)$ has the absolutely continuous increasing nonnegative derivative

$$\frac{d\phi^*(\rho)}{d\rho} = \left(n\int_0^\rho R^*(t)t^{n-1}dt\right)^{\frac{1}{n}} \tag{24.30}$$

in $[0, r(G))$. Clearly

$$\frac{d\phi^*(0)}{d\rho} = 0 \tag{24.31}$$

and

$$\lim_{\rho\to r(G)} \frac{d\phi^*(\rho)}{d\rho} = +\infty.^{*)} \tag{24.32}$$

Thus the number

$$\phi^*(r(G)) = \int_0^{r(G)} \left(n\int_0^s R^*(t)t^{n-1}dt\right)^{\frac{1}{n}} ds \tag{24.33}$$

is finite and positive.

From (24.20) it follows that the radial convex generalized solution $\phi(p)$ of the Dirichlet problem (24.27–28) can be expressed by the formula

$$\phi(p) = -\phi^*(r(G)) + \phi^*(|p|) = -\int_{|p|}^{r(G)} \left(n\int_0^s R^*(t)t^{n-1}dt\right)^{\frac{1}{n}} ds$$

---

*) We don't consider the case $\lim_{\rho\to r(G)} \phi^{*\prime}(p) < +\infty$ because Monge–Ampere generators for $R(p)$ exist in this case.

for all $p \in \overline{U(r(G))}$.

Since

$$|\phi(p)| \leq |\phi(0)| = |\phi^*(r(G))|$$

for all $p \in \overline{U(r(G))}$, then from (24.26) it follows that there exists

$$\psi(p) = \lim_{\varepsilon \to 0} \psi_\varepsilon(p),$$

which is a convex generalized solution of the Dirichlet problem

$$\det\left(\frac{\partial^2 \psi(p)}{\partial p_i \partial p_j}\right) = R(p)$$
$$\psi|_{|p|=r(G)} = 0.$$

The function $\psi(p)$ is the desired Monge–Ampere generator with prescribed admissible density $R(p)$. The proof is completed. □

Example 1. Let

$$R(p) = [1 - |p|^2]^{-\frac{n+2}{2}}. \tag{24.34}$$

The function $R(p)$ is defined in the n-ball G: $|p| < 1$. Clearly

$$G = U(r(G))$$

and

$$r(G) = 1.$$

Since $R(p) = R(|p|)$, then

$$R^*(|p|) = R(p). \tag{24.35}$$

We can find by elementary calculations that

$$\int_G R(p)dp = \int_G [1 - |p|^2]^{-\frac{n+2}{2}} = +\infty \tag{24.36}$$

and

$$\int_0^1 \left[\int_0^s R(t)t^{n-1}dt\right]^{1/n} ds < +\infty. \tag{24.37}$$

From (24.37) it follows that every Monge–Ampere generator with the density $R(p)$ is a convex bounded function in G: $|p| < 1$.

The convex function

$$\phi(p) = -\int_{|p|}^1 \left(n\int_0^s R(t)t^{n-1}dt\right)^{1/n} dt = -(1 - |p|^2)^{1/2} \tag{24.38}$$

is the Monge–Ampere generator for the function $R(p)$ satisfying the zero boundary data.

We use functions $\phi(p)$ and $R(p)$ for hypersurfaces with prescribed mean curvature in the Minkowski space M^{n+1} (see Subsection 23.2).

Example 2. Let p' and p'' be correspondingly the projections of any vector $p = (p_1, p_2, \dots, p_k, p_{k+1}, \dots, p_n)$ P^n on the k-plane, spanned on the first k coordinate axes, and on the $(n-k)$-plane, spanned on the last $(n-k)$ coordinate axes in P^n.

Then

$$p' = (p_1, p_2, \dots, p_k, 0, \dots, 0), \quad p'' = (0, 0, \dots, 0, p_{k+1}, \dots, p_n),$$
$$p = p' + p'';$$
$$|p'|^2 = \sum_{i=1}^{k} p_1^2; \quad |p''|^2 = \sum_{j=k+1}^{k} p_j^2 \quad |p|^2 = |p'|^2 + |p''|^2.$$

The function

$$R_{k,n-k}(p) = \frac{1}{3}[(1 - |p|^2)^{-\frac{n+2}{2}} + (1 - |p'|^2)^{-\frac{n+2}{2}} + (1 - |p''|^2)^{-\frac{n+2}{2}}]$$

is defined in the unit n-ball G; $|p| < 1$. Clearly $G = U(r(G))$ and $r(G) = 1$. The function $R_{k,n-k}(p)$ is admissible and

$$\int_{U(1)} R_{k,n-k}(p)dp = +\infty. \tag{24.39}$$

Since

$$R^*_{k,n-k}(p) \le R(p), \tag{24.40}$$

where $R(p)$ is the function investigated in Example 1. Thus all the conditions of Theorem 24.1 are fulfilled for the function $R^*_{n,k-k}(p)$ and therefore there exists at least one bounded convex Monge–Ampere generator for the original function $R_{k,n-k}(p)$.

Example 3. Let $k > 0$ be a constant. The function

$$R(p) = (1 - |p|^2)^{-\frac{k}{2}} \tag{24.41}$$

is defined in the ball G: $|p| < 1$. Clearly

$$G = U(r(G)) = U(1) \quad \text{and} \quad R^*(|p|) = R(p).$$

The inequality $k \ge 2$ is necessary and sufficient for the equality

$$\int_{|p| \le 1} (1 - |p|^2)^{-\frac{k}{2}} dp = +\infty. \tag{24.42}$$

Now the inequality $k \leq 2n+2$ is necessary and sufficient if the inequality

$$\int_0^1 \left[\int_0^s t^{n-1}(1-t^2)^{-\frac{k}{2}} dt\right]^{\frac{1}{n}} ds < +\infty$$

holds. Thus if the function $R(P)$ satisfies the condition (24.42), then $R(p)$ has a bounded radial Monge–Ampere generator if and only if

$$2 \leq k \leq 2n+2. \tag{24.43}$$

Now let $R(p)$ be a nonnegative locally summable function in the gradient space P^n and let

$$R^*(p) = \sup_{|p'|=|p|} R(p') \tag{24.44}$$

and

$$R^{**}(p) = \inf_{|p'|=|p|} R(p) \tag{24.45}$$

are radial functions. Therefore we will also use the notations

$$\widetilde{R}^*(|p|) = R^*(p)$$

and

$$\widetilde{R}^{**}(|p|) = R^{**}(p).$$

Thus we obtain two functions $R^*(\rho)$ and $R^{**}(\rho)$ of a single variable $\rho \in [0,+\infty)$. Clearly

$$0 \leq R^{**}(p) \leq R(p) \leq R^*(p) \tag{24.46}$$

for all $p \in P^n$.

Theorem 24.2. *Let the function $R(p)$ be defined in the entire space P^n and let the following condition hold:*

1) $R(p)$ is nonnegative and locally summable in P^n;
2)

$$\int_{P^n} R(p)dp = +\infty; \tag{24.47}$$

3) the functions $\widetilde{R}^*(\rho)$ and $\widetilde{R}^{**}(\rho)$ are locally summable in $[0,+\infty)$ and $\widetilde{R}^{**}(\rho)$ is strictly positive on a subset of a positive measure in any interval $(0,\rho)$, where $0 < \rho < +\infty$;
4)

$$\int_0^\rho \widetilde{R}^*(\rho)\rho^{n-1}d\rho < +\infty \tag{24.48}$$

for all $\rho \in [0,+\infty)$;

5)

$$\int_0^{+\infty} \widetilde{R}^{**}(\rho)\rho^{n-1}d\rho = +\infty. \tag{24.49}$$

Then there exists a convex Monge–Ampere generator $F(p)$ for the function $R(p)$, which is defined for all $p \in P^n$ and the asymptotic cone of which is a ray orthogonal to P^n in the space $P^{n+1} = P^n \times R$.

Proof. We consider the following radial convex Monge–Ampere generators

$$F^*(p) = 1 + \int_0^{|p|} \left\{ n \int_0^s R^*(t)t^{n-1}dt \right\}^{1/n} ds \tag{24.50}$$

and

$$F^{**}(p) = \int_0^{|p|} \left\{ n \int_0^s R^{**}(t)t^{n-1}dt \right\}^{1/n} ds \tag{24.51}$$

of the radial functions $R^*(p)$ and $R^{**}(p)$ for all $p \in P^n$. From (24.46) it follows that

$$F^*(p) \geq F^{**}(p) + 1 \tag{24.52}$$

for all $p \in P^n$. Clearly $F^*(p)$ and $F^{**}(p)$ are radial solutions of the Monge–Ampere equations

$$\det(F^*_{ij}(p)) = R^*(p) \tag{24.53}$$

and

$$\det(F^{**}_{ij}(p)) = R^{**}(p) \tag{24.54}$$

and the asymptotic cones of their graphs are rays orthogonal to P^n in the space $P^{n+1} = P^n \times R$.

Every convex function of one variable has everywhere the first derivative except maybe a countable set of points. Therefore there exists a sequence of points

$$0 < \rho_1 < \rho_2 < \cdots < \rho_m < \cdots$$

such that $\lim_{m\to\infty} \rho_m = +\infty$ and the derivatives $F^*_\rho(\rho_m)$ and $F^{**}_\rho(\rho_m)$ exist for all $m = 1, 2, \ldots$. Clearly

$$0 < F^{**}_\rho(\rho_m) \leq F^*_\rho(\rho_m) \tag{24.55}$$

where $m = 1, 2, \ldots$.

Now we introduce the following convex functions

$$\widetilde{F}^*_m(\rho) = \begin{cases} \widetilde{F}^*(\rho) & \text{if} \quad 0 \leq \rho < \rho_m; \\ \widetilde{F}^*_\rho(\rho_m)(\rho - \rho_m) + \widetilde{F}^*(\rho_m) & \text{if} \quad \rho \geq \rho_m \end{cases} \tag{24.56}$$

and

$$\widetilde{F}^{**}_m(\rho) = \begin{cases} \widetilde{F}^{**}(\rho) & \text{if} \quad 0 \le \rho < \rho_m \\ \\ \widetilde{F}^{**}_\rho(\rho_m)(\rho - \rho_m) + F^{**}(\rho_m) & \text{if} \quad \rho \ge \rho_m. \end{cases} \tag{24.57}$$

Let $F^*_m(p)$ and $F^{**}_m(p)$ be radial convex functions in P^n, whose graphs S^* and S^{**} are convex hypersurfaces of revolution with meridians $w = \widetilde{F}^*_m(\rho)$ and $w = \widetilde{F}^{**}_m(\rho)$. Then the normal images of the asymptotic cones K^*_m and K^{**}_m for S^*_m and S^{**}_m are the following balls

$$V^*_m\colon\ |q| \le F^*_\rho(\rho_m) \tag{24.58}$$

and

$$V^{**}_m\colon\ |q| \le F^{**}_\rho(\rho_m) \tag{24.59}$$

in the space Q^n.

From (24.57) it follows that

$$V^{**}_m \subset V^*_m.$$

Clearly

$$|V^*_m|_Q = \int_{|p|<\rho_m} R^*(p)dp \tag{24.60}$$

and

$$|V^{**}_m|_Q = \int_{|p|<\rho_m} R^{**}(p)dp. \tag{24.61}$$

Let V_m be a ball in Q^n whose equation is

$$|q| \le q_m$$

and whose volume is equal to

$$|V_m|_Q = \int_{|p|\le\rho_m} R(p)dp. \tag{24.62}$$

Since

$$|V^{**}_m|_Q \le |V_m|_Q \le |V^*_m|_Q$$

then

$$0 < \widetilde{F}^{**}_{m,\rho}(\rho_m) \le q_m = \left(\frac{|V_m|_Q}{\mu_n}\right)^{\frac{1}{n}} \le F^*_{m,\rho}(\rho_m) < +\infty. \tag{24.63}$$

We denote by K_m a convex cone of revolution whose normal image coincides with the ball V_m. According to Theorem 17.1, there exists only one generalized solution $F_m(p)$ of the Monge–Ampere equation

$$\det(F_{m,ij}(p)) = R_m(p), \tag{24.64}$$

whose asymptotic cone is K_m and which satisfies the initial condition $F_m(0) = \frac{1}{2}$. In (24.64) the function $R_m(p)$ is defined by the formula

$$R_m(p) = \begin{cases} R(p) & \text{if} \quad |p| < \rho_m; \\ 0 & \text{if} \quad |p| \geq \rho_m. \end{cases}$$

Moreover

$$F_m^{**}(p) \leq F_m(p) \leq F_m^{**}(p) \tag{24.65}$$

for all $p \in P^n$. The last assertion follows from the inclusions

$$V_m^{**} \subset V_m \subset V_m^*$$

for the normal images of asymptotic cones K_m^{**}, K_m, K_m^*; the inequalities

$$0 = F_m^{**}(0) < \frac{1}{2} = F_m(0) < 1 = F_m^*(0);$$

and the Comparison Theorem of solutions for the Dirichlet problem for Monge–Ampere equations

$$\det(u_{ij}) = f(x) \cdot g(Du)$$

(see § 11).

If $m \to +\infty$, then convex functions $F_m^{**}(p)$ and $F_m^*(p)$ converge correspondingly to convex functions $F^{**}(p)$ and $F^*(p)$. Clearly we can extract from the functions $F_m(p)$ a subsequence $F_{m_j}(p)$ which converges to some convex function $F(p)$ for all $p \in P^n$ and satisfying the initial condition $F(0) = \frac{1}{2}$. From (24.65) it follows that

$$F^{**}(p) \leq F(p) \leq F^*(p).$$

From the weak convergence of R-curvatures of convex functions (see Theorem 9.1) we obtain that $F(p)$ is a generalized solution of the original equation

$$\det(F_{ij}(p)) = R(p).$$

From the construction of the function $F(p)$ it follows that the asymptotic cone of the graph of $F(p)$ is a ray orthogonal to the hyperplane P^n.

The proof of Theorem 24.2 is complete. □

24.3 Assumptions Related to General Euler–Lagrange Equations

Let

$$\sum_{i=1}^{n} \frac{d}{dx_i}(F_{p_i}(x, u, Du)) - F_u(x, u, Du) = 0 \tag{24.66}$$

be the Euler–Lagrange equation for the functional

$$I(u) = \int_B F(x, u, Du)dx. \tag{24.67}$$

In our preliminary considerations we suppose that $F(x,u,p)$ $C^2(\overline{B} \times R \times G)$ solutions of equation (24.66) are $C^2(B)$ $C(\overline{B})$-functions. Then equation (24.66) becomes

$$\sum_{i,k=1}^{n} F_{p_i p_j}(x, u(x), Du(x))u_{ij}(x) = H(x, u(x), Du(x)), \tag{24.68}$$

where

$$H(x,u,p) = F_u(x,u,p) - \sum_{i=1}^{n} F_{p_i u}(x,u,p)p_i - \sum_{i=1}^{n} F_{p_i x_i}(x,u,p). \tag{24.69}$$

The equation (24.66) is elliptic if and only if the quadratic form

$$\sum_{i,k=1}^{n} F_{p_i p_k}(x,u,p)\xi_i\xi_k \tag{24.70}$$

is positive (negative) definite for all $(x,u,p) \in B \times R \times G$. If G is different from P^n, then we have the ***variational problem with limitations on the gradient of solutions***. We now formulate the assumptions that provide the two-sided C^0-estimates for solutions of Euler–Lagrange equations (24.68). There will be several series of such assumptions. The first series is related to general conditions for the function $H(x,u,p)$ and quadratic form (24.70). The special conditions for every term at the right side of (24.69) will not be considered in these assumptions. Other conditions reflecting special properties of the different groups of terms at the right side of (24.69) will be considered later.

Assumptions A.1. Let $F(x,u,p)$ be a $W_n^{(2)}$-function in $B \times R \times G$ and let

$$H(x,u,p) \leq \frac{g_1(x,u)}{R_1(p)} \tag{24.71}$$

and

$$-H(x,u,p) \leq \frac{g_2(x,u)}{R_2(p)} \tag{24.72}$$

correspondingly hold almost everywhere in $B \times R \times G$ if $D(x,u,p) \geq 0$ or $D(x,u,p) \leq 0$, where $g_1(x,u)$, $g_2(x,u)$ are nonnegative in $B \times R$ and $R_1(p)$, $R_2(p)$ are positive in G.

Assumption A.2. The quadratic form $\sum_{i,k=1}^{n} F_{p_ip_j}(x,u,p)\xi_i\xi_k$ is positive definite in $B \times R \times G$ and the inequalities

$$\ell_k(x,u)T_k(p) \leq R_k^n(p)\det(F_{p_ip_j}(x,u,p)) \tag{24.73}$$

$(k = 1,2)$ hold for all $(x,u,p) \in B \times R \times G$, where $\ell_k(x,u)$ are positive functions in $B \times R$ and $T_K(p)$ are nonnegative and locally summable in G. We also assume that

$$T_k(p) > 0$$

on the set of a positive measure in any annulus

$$A(\rho_1,\rho_2)\colon\ 0 < \rho_1 < |p| < \rho_2 < r(G). \tag{24.74}$$

Assumption A.3. If at least one of the integrals

$$\int_{U(r(G))} T_k(p)dp = +\infty, \qquad k = 1,2 \tag{24.75}$$

then either the conditions of Theorem 24.1 hold if $r(G) < +\infty$ or the conditions of Theorem 24.2 hold if $r(G) = +\infty$.

Assumption A.4. The functions

$$\psi_k(x,u) = \left[\frac{g_k(x,u)}{\ell_k(x,u)}\right]^n, \tag{24.76}$$

$k = 1,2$, are locally summable in B for every fixed $u_0 \in R$ and $\psi_k(x_0,u)$ are non-decreasing with respect to u for every fixed $x_0 \in B$.

Examples.

1) Let

$$F(x,u,p) = F(p) + nH(x)u \tag{24.77}$$

for all $(x,u,p) \in B \times R \times G$. Then the Euler–Lagrange equation for functional (24.77) is

$$\sum_{i,k=1}^{n} F_{p_ip_j}(Du)u_{ij} = nH(x). \tag{24.78}$$

Clearly

$$g_1(x,u) = nH_+(x), \quad g_2(x,u) = nH_-(x),$$
$$R_1(p) = R_2(p) = 1, \quad T_1(p) = T_2(p) = F(p).$$

Thus Assumptions A.1, A.2, A.3, A.4 coincide completely with the contents of Assumptions 1, 2, 3, 4 (see Subsection 21.1) for the case, when $F(x,u,p)$ is given by (24.77).

2) Let

$$F(x,u,p) = (1+|x|^2)^{1/2}(1+|p|^2)^{1/2} + nH(x)u$$

for all $(x,u,p) \in B \times R \times G$, where $|x|^2 = x_1^2 + \cdots + x_n^2$, $|p|^2 = p_1^2 + \cdots + p_n^2$. Then the Euler–Lagrange equation for the functional

$$I(u) = \int_B \{(1+|x|^2)^{1/2}(1+|p|^2)^{1/2} + nH(x)u\}dx$$

is

$$(1+|x|^2)^{1/2} \frac{(1+|Du|^2)\Delta u - \sum_{i=1}^{n} u_i u_j u_{ij}}{(1+|Du|^2)^{3/2}} = -\frac{\sum_{i=1}^{n} u_i x_i}{(1+|Du|^2)^{1/2}(1+|x|^2)^{1/2}} + nH(x).$$

Here

$$\frac{\partial^2 F}{\partial p_i \partial p_j} = \begin{cases} (1+|Du|^2 - u_i^2)(1+|x|^2)^{1/2}(1+|Du|^2)^{-\frac{3}{2}} & i=j=1,\ldots,n \\ -(1+|x|^2)^{1/2}(1+|Du|^2)^{-\frac{3}{2}} u_i u_j & i \neq j \end{cases}$$

and

$$H(x,u,Du) = -\frac{\sum_{i=1}^{n} u_i x_i}{(1+|Du|^2)^{1/2}(1+|x|^2)^{1/2}} + nH(x).$$

According to Assumptions A.1, A.2, A.3 and A.4 we can choose

$$g_1(x,u) = nH_+(x) + \frac{|x|}{(1+|x|^2)^{1/2}},$$
$$g_2(x,u) = nH_-(x) + \frac{|x|}{(1+|x|^2)^{1/2}},$$
$$R_1(p) = R_2(p) = 1.$$

This choice can be done, because

$$\frac{\left|\sum_{i=1}^{n} u_i x_i\right|}{(1+|Du|^2)^{1/2}(1+|x|^2)^{1/2}} \leq \frac{|Du| \cdot |x|}{(1+|Du|^2)^{1/2}(1+|x|^2)^{1/2}} \leq \frac{|x|}{(1+|x|^2)^{1/2}}$$

for all $(x, u, p) \in B \times R \times P^n$. Thus

$$H(x,u,p) \le nH_+(x) + \frac{|x|}{(1+|x|2)^{1/2}}$$

and

$$-H(x,u,p) \le nH_-(x) + \frac{|x|}{(1+|x|^2)^{1/2}}$$

for all $(x, u, p) \in B \times R \times P^n$.

Since

$$\det(F_{p_ip_j}) = (1+|x|^2)^{n/2}(1+|Du|^2)^{-\frac{n+2}{2}},$$

then inequalities (24.73) will be satisfied by the functions

$$\ell_1(x,u) = \ell_2,(x,u) = \ell(x) \equiv (1+|x|^2)^{\frac{n}{2}}, \quad x \in B,$$

and

$$T_1(p) = T_2(p) = T(p) \equiv (1+|p|^2)^{-\frac{n+2}{2}}, \quad p \in P^n.$$

We do not need to consider Assumption A.3, because

$$\int_{P^n} T(p)dp = \mu_n < +\infty$$

where μ_n is the volume of the n-unit ball in E^n.

Finally

$$\begin{aligned}\psi_1(x,u) = \psi(x) &\equiv \left(\frac{g_1(x,u)}{\ell(x)}\right)^n \\ &= (1+|x|^2)^{-\frac{n^2}{2}}\left[nH_+(x) + \frac{|x|}{(1+|x|^2)^{1/2}}\right]^n\end{aligned}$$

and

$$\begin{aligned}\psi_2(x,u) = \psi(x) &\equiv \left(\frac{g_2(x,u)}{\ell(x)}\right)^n \\ &= (1+|x|^2)^{-\frac{n^2}{2}}\left[nH_-(x) + \frac{|x|}{(1+|x|)^2)^{1/2}}\right]^n.\end{aligned}$$

Clearly Assumptions A.1, A.2 and A.4 are fulfilled.

We now find a Monge–Ampere generator $\phi(p)$ for the function $T(p)$. Since $T(p)$ is defined for all $p \in P^n$ and depends only on $|p|$, then we can find a radial Monge–Ampere generator for this function. Such generator is the special convex solution of the equation

$$\det\left(\frac{\partial^2\phi(p)}{\partial p_i \partial p_j}\right) = (1+|p|^2)^{-\frac{n+2}{2}},$$

whose asymptotic cone K_ϕ is a cone of revolution and the normal image of K_ϕ is the n-unit ball: $|q| \leq 1$ in the Euclidean Q^n. According to Theorem 17.1, this problem has radial convex generalized solutions, which are defined to within an additive constant. The analytic expression for these solutions is as follows

$$\phi(|p|) = \sqrt{1+|p|^2} + \text{const}$$

for $0 \leq |p| < +\infty$.

24.4 Two-sided Estimates for Solutions of Nonlinear Elliptic Euler–Lagrange Equations

We consider solutions $u(x) \in W_2^n(B) \cap C(\overline{B})$ of Euler–Lagrange equation (24.66) for functional (24.67). Let integrand $F(x,u,p)$ of functional (24.67) be defined in $B \times R \times G$, where G is a domain in P^n and the origin O' of P^n is an interior point of G.

If $G = P^n$, i.e. $\partial G = \emptyset$, then we do not consider any restriction for the values of $Du(x)$ for any solution $u(x) \in W_2^n(B) \cap C(\overline{B})$ of equation (24.66).

Thus we should consider carefully the case $\partial G \neq \emptyset$. In this case

$$Du(x) \in G \tag{24.79}$$

for all $x \in B$, where $u(x)$ is any solution of equation (24.66). But inclusion (24.79) is convenient to use only for functions $u(x) \in W_2^n(B) \cap C^1(\overline{B})$. Therefore we introduced some modification of condition (24.79) for $u(x) \in W_2^n(B) \cap C(\overline{B})$ (see Subsection 21.3). We now repeat this modification once more.

A point $M(x, u(x))$ of the graph of a function $u(x) \in C(\overline{B})$, where $x \in B$ is said to be *global convex (global concave)*, if there exists at least one supporting hyperplane α passing through M and lying under (over) the graph of $u(x)$. If such a supporting hyperplane α has the equation

$$z = p_1^0 x_1 + \cdots + p_n^0 x_n + b,$$

then the point

$$p_0 \stackrel{\text{def}}{=} \chi(\alpha) = (p_1^0 p_2^0, \ldots, p_n^0) \in P^n$$

is called the normal image of α. We denote by

$$\nu_u^+(B) = \bigcup_{\alpha^+} \chi(\alpha^+), \tag{24.80}$$

$$\nu_u^-(B) = \bigcup_{\alpha^-} (\alpha^-), \tag{24.81}$$

$$\nu_u(B) = \nu_u^+(B) \cup \nu_u^-(B), \tag{24.82}$$

where α^+ (or α^-) is a supporting hyperplane passing through a global convex (or global concave) point of the graph of $u(x)$. All such supporting hyperplanes

α^+ (or α^-) at all global convex (or global concave) points are considered in (24.80) (or in (24.81)).

Assumption 5. Let a function $F(x, u, p)$ be defined in $B \times R \times G$, where G is a domain in P^n. Then we consider only functions $u(x) \in W_2^n(B) \cap C(\overline{B})$ for which the inclusion

$$\nu_u(B) \subset G \tag{24.83}$$

holds.

Theorem 24.3. *Let $u(x) \in W_2^n(B) \cap C(\overline{B})$ be a solution of Euler–Lagrange equation (24.66), satisfying the condition*

$$\nu_u(B) \subset G \quad \text{and} \quad \partial G \neq \emptyset. \tag{24.84}$$

Let Assumptions A.1, A.2 and A.4 be fulfilled (see Subsection 24.3). If either one of functions $T_k(p)$, $k = 1, 2$, or both of these functions satisfy condition (24.75), then for such functions $T_k(p)$ Assumption A.3 holds. Finally let the inequalities

$$\Omega_1^+(m) \leq C(\phi_1), \tag{24.85}$$

$$\Omega_2^-(M) \leq C(\phi_2) \tag{24.86}$$

hold, where $m = \inf_{\partial B} u(x)$, $M = \sup_{\partial B} u(x)$,

$$\Omega_1^+(m) = \frac{1}{n^n} \int_B \psi_1(x, m)dx, \tag{24.87}$$

$$\Omega_2^-(M) = \frac{1}{n^n} \int_B \psi_2(x, M)dx, \tag{24.88}$$

$\phi_k(p)$ is a convex Monge–Ampere estimator constructed by means of the function $T_k(p)$, $k = 1, 2$, and

$$C(\phi_k) = |\chi_{\phi_k}(U(r(G)))|_Q = \int_{U(r(G))} T_k(p)dp. \tag{24.89}$$

Then the inequalities

$$m - b_1(\Omega_1^+(m))\text{diam } B \leq u(x) \leq M + b_2(\Omega_2^-(M))\text{diam } B \tag{24.90}$$

hold for all $x \in \overline{B}$, where $\rho = b_k(t)$, $k = 1, 2$, are inverses for the strictly increasing continuous functions

$$t = c_k(\rho) = \chi_{\phi_k}(U(\rho))_Q = \int_{|p|<\rho} \det(\phi_{k,ij}(p))dp \tag{24.91}$$

$$= \int_{|p|<\rho} T_k(p)dp.$$

If $\partial G = \emptyset$, then $G = P^n$ and inclusion (24.84) is trivial and always holds. If $G = P^n$ then (24.85), (24.86) have to be replaced by strict inequalities $\Omega_1^+(m) < C(\phi_1)$, $\Omega_2^-(M) < C(\phi_2)$.

Proof. From the statement of Theorem 24.3 it follows that there exist the convex Monge–Ampere estimators $\phi_1(p)$, $\phi_2(p)$ for functions $T_1(p)$, $T_2(p)$. Thus we can derive the desired estimates (24.90) from Theorem 22.2 (if $\partial G \neq \emptyset$) and from Theorem 22.3 (if $\partial G = \emptyset$), if the inequalities

$$|\chi_{\phi_1}(\nu_u^+(B))|_Q < \Omega_1^+(m) \tag{24.92}$$

and

$$|\chi_{\phi_2}(\nu_u^-(B))|_Q < \Omega_2^-(M), \tag{24.93}$$

are already established An analysis of the proof of Theorem 22.1 shows that inequalities (24.92–93) can be obtained from this proof, if we combine inequality (24.73) with inequality (22.21) and add to the proof of Theorem 22.1 a few straightforward considerations.

24.5 The Second Type of C^0-Estimates for Solutions for General Elliptic Euler–Lagrange Equations

We consider solutions $u(x) \in W_2^n(B) \cap C(\overline{B})$ of the Dirichlet problem

$$\sum_{i,k=1}^{n} F_{p_i p_j}(x,u,Du)u_{ik} = H(x,u,Du) \tag{24.94}$$

$$u|_{\partial B} = h(x), \tag{24.95}$$

where $H(x,u,Du)$ is defined by (24.69), B is a bounded domain in E^n, ∂B is a closed C^0-hypersurface in E^n and $h(x) \in C(\partial B)$.

We now formulate the second group of assumption, relating to functions $F(x,u,p)$ and $H(x,u,p)$. They will be different from Assumptions A.1–5 introduced in Subsections 24.3 and 24.4.

Assumption A.6. The function $F(x,u,p)$ is continuous in $\overline{B} \times R \times P^n$ and $F(x,u,p)$ has generalized first and second Sobolev partial derivatives of the class C^n on any compact set $\overline{B} \times R \times P^n$.

Assumption A.7. The form

$$\sum_{i,k=1}^{n} F_{p_i p_j}(x,u,p)\xi_i\xi_j$$

is positive definite at any fixed point $(x,u,p) \in B \times R \times P^n$.

Assumption A.8. The function

$$\frac{H(x,u,p)}{[\det(F_{p_i p_j}(x,u,p)]^{1/n}}$$

increases with u for any fixed $x \in \overline{B}$ and $p \in P^n$.

Assumption A.9. The inequalities

$$-d_L^-(x) - C_L^-(x)|p| \leq H(x, L, p)[\det(F_{p_i p_j}(x, L, p)]^{-1/n} \\ \leq d_L^+(x) + C_L^+(x)|p| \tag{24.96}$$

hold for any real number L, where $C_L^+(x)$, $C_L^-(x)$, $d_L^+(x)$, $d_L^-(x)$ are nonnegative functions in B, whose norms in $L^n(B)$ are finite. Clearly these functions depend on a number L.

Theorem 24.4. *Let $u(x) \in W_2^n(B) \cap C(\overline{B})$ be a solution of the Dirichlet problem (24.94–95). Let Assumptions A.6, A.7, A.8, A.9 be fulfilled for equation (24.94). Then the inequalities*

$$m - \tau_n \|d_m^+(x)\|_{L^n} \cdot \exp(\tau_n' \|C_m^+(x)\|_{L^n}) \text{diam } B \leq u(x) \\ \leq M + \tau_n \|d_M^-(x)\|_{L^n} \exp(\tau_n' \|C_M^-(x)\|_{L^n}) \text{diam } B, \tag{24.97}$$

where

$$m = \inf_{\partial B} h(x), \quad M = \sup_{\partial B} h(x)$$

$$\tau_n = \frac{2}{n} e^{1/n} \mu_n^{-1/n}, \quad \tau_n' = 2^n n^{-(n+1)} \mu_n^{-1},$$

and μ_n is the volume of the unit n-ball in E^n.

First of all we recall a few concepts and notations introduced in Subsection 21.2. Let $\delta > 0$ be any real number satisfying the condition

$$m - \delta > u_0.$$

We will use the notation $m_\delta = m - \delta$. Let S_{m_δ} be the part of the graph of $u(x)$ located under hyperplane

$$\gamma_{m_\delta}\colon\ z = m_\delta \tag{24.98}$$

in the space $E^{n+1} = E^n \times R$. Let $\overline{B}_{m_\delta}$ be a set in γ_{m_δ} whose projection on E^n coincides with $\overline{B}$ and let Γ_{m_δ} and Γ be the boundaries of $\overline{H}_{m_\delta}$ and $\overline{H}$ respectively, where $\overline{H}_{m_\delta}$ and $\overline{H}$ are the closed convex hulls of $\overline{B}_{m_\delta}$ and $\overline{B}$.

We denote by C_{m_δ} the closed convex hull of the set $\overline{B}_{m_\delta} \cup S_{m_\delta}$. Then

$$\partial C_{m_\delta} = \overline{H}_{m_\delta} \cup S_{v_\delta}, \tag{24.99}$$

where S_{v_δ} is the graph of a convex function $v_\delta(x) \in C(\overline{H})$ such that

$$v_\delta(x) = m_\delta$$

for all $x \in \partial H$, and

$$v_\delta(x) \leq u(x) \tag{24.100}$$

for all $x \in \overline{B}$. The convex function $v_\delta(x)$ is called the *convex support of the function* $u(x)$ with respect to the number $\delta > 0$. A similar geometric construction leads to the *concave support* $w_{\delta'}(x)$ *of the function* $u(x)$ with respect to the number $\delta' > 0$.

Proof of Theorem 24.4. Let $v_\delta(x)$ be the convex support of $u(x)$ with respect to the number $\delta > 0$. We consider the functions $u(x)$ and $v_\delta(x)$ only on the set $\overline{H}$. For this case the inequalities

$$0 \leq [\det(u_{ik})]^{\frac{1}{n}} \leq \frac{H(x,u,Du)}{n[\det(F_{p_ip_j}(x,u,p)]^{1/n}} \tag{24.101}$$

hold for all $x \in \overline{H}$ for each sufficiently small $\delta > 0$ (see the proofs of Theorems 22.1–3).

The inequalities

$$0 \leq [\det(u_{ik})]^{\frac{1}{n}} \leq \frac{1}{n}[C_m^+(x)|Du| + d_m^+(x)] \tag{24.102}$$

holds for all $x \in \overline{H}$. This follows from the conditions of Theorems 24.4. From (24.102) it follows that the inequalities

$$\begin{aligned} 0 &\leq [\det(u_{ik})]^{\frac{1}{n}} \\ &\leq \frac{1}{n}[(C_m^+(x))^n + \varepsilon^n(d_m^+(x))^n]^{\frac{1}{n}}(|Du|^{\frac{n}{n-1}} + \varepsilon^{-\frac{n}{n-1}})^{\frac{n-1}{n}} \end{aligned}$$

hold for all $x \in \overline{H}$, where 0 is any sufficiently small number.

Let $p = (p_1, p_2, \ldots, p_n)$ be an arbitrary vector in P^n. We consider the function

$$R(p) = [|p|^n + \varepsilon^{-n}]^{-1}$$

on P^n. Then

$$\begin{aligned} n &= \inf_{P^n} \frac{\left[|p|^{\frac{n}{n-1}} + \varepsilon^{-\frac{n}{n-1}}\right]^{-(n-1)}}{R(p)} \\ &= \inf_{[0,+\infty)} \frac{\xi^n + \varepsilon^{-n}}{\left(\xi^{\frac{n}{n-1}} + \varepsilon^{-\frac{n}{n-1}}\right)^{n-1}} > 2^{-n} \end{aligned}$$

because the function

$$\lambda(\xi) = \frac{\xi^n + \varepsilon^{-n}}{\left(\xi^{\frac{n}{n-1}} + \varepsilon^{-\frac{n}{n-1}}\right)^{n-1}}$$

satisfies the conditions:

(a) $\lambda(0) = \lambda(\infty) = 1$;
(b) $\lambda(\xi) < 1$ for $\xi \in (0, +\infty)$;
(c) the equation $\lambda'(\xi) = 0$ has only one root $\xi_0 = \frac{1}{\delta}$ and

$$\inf_{[0,+\infty)} \lambda(\xi) = \lambda(\xi_0) = \lambda\left(\frac{1}{\delta}\right) > 2^{-n}.$$

Therefore the inequality

$$R(Du)\det(u_{ij}) \le \left(\frac{2}{n}\right)^n [(C_m^+(x))^n + \varepsilon^n (d_m^+(x))^n]$$

holds everywhere on $\overline{H}$ for any sufficiently small $\varepsilon > 0$.

From the last inequality it follows that

$$\omega(R, v_\delta, G) \le \left(\frac{2}{n}\right)^n [\|C_m^+(x)\|^n_{L^n} + \varepsilon^n \|d_m^+(x)\|^n_{L^n}], \tag{24.103}$$

where $\omega(R, v_\delta, G)$ is the R-curvature of the convex function $v_\delta(x)$ for the convex domain G, where the function v_δ is defined (see §§ 21, 22). Only the case

$$v_\delta(x_0) = \inf_B v_\delta(x) < \inf_{\partial G} v_\delta(x) = m,$$

$x_0 \in B$ is non-trivial. Clearly

$$v_\delta(x_0) = u(x_0)$$

and

$$m - u(x_0) > 0.$$

From the results of Subsection 21.3 we obtain

$$\omega(R, v_\delta, G) \ge \int_{|p| \le \rho_m} R(p)dp, \tag{24.104}$$

where

$$\rho_m = \frac{m - v_\delta(x_0)}{\operatorname{diam} B}.$$

We also present the proof of inequality (24.104) at the end of the proof of Theorem 24.4.

From inequality (24.104) it follows that

$$\begin{aligned}
\omega(R, v_\delta, G) &\ge \int_{|p| \le \rho_m} \frac{dp}{|p|^n + \varepsilon^{-n}} \\
&= \sigma_{n-1} \int_0^{\frac{m-u(x_0)}{\operatorname{diam} B}} p^{n-1}(p^n + \varepsilon^{-n})dp \\
&= \mu_n \ln\left[1 + \left(\frac{m - u(x_0)}{\operatorname{diam} B}\right)^n \cdot \varepsilon^n\right],
\end{aligned} \tag{24.105}$$

where μ_n is the volume of the unit n-ball and σ_{n-1} is the area of the unit $(n-1)$-sphere.

We now obtain the inequality

$$\mu_n \ln\left[1 + \left(\frac{m - u(x_0)}{\text{diam } B}\right)^n \varepsilon^n\right] - \varepsilon^n \left(\frac{2}{n}\right)^n \|d_m^+(x)\|_{L^n}^n \\ \leq \left(\frac{2}{n}\right)^n \|C_m^+(x)\|_{L^n}. \tag{24.106}$$

There are now two possibilities:

(a) $\frac{m-u(x_0)}{\text{diam } B} \leq \frac{2}{n(\mu_n)^{1/n}} \|d_m^-(x)\|_{L^n}$;

(b) $\frac{m-u(x_0)}{\text{diam } B} > \frac{2}{n(\mu_n)^{1/n}} \|d_m^+(x)\|_{L^n}$.

Since

$$\exp\left(\frac{2}{n}\right)^n \frac{1}{\mu_n} \|C_m^+(x)\|_{L^n}^n \geq 1$$

we obtain the inequality

$$\frac{m - u(x_0)}{\text{diam } B} \leq \frac{2}{n(\mu_n)^{1/n}} \|d_m^+(x)\|_{L^n} \exp\left[\left(\frac{2}{n}\right)^n \frac{1}{\mu_n} \|C_m^+(x)\|_{L^n}^n\right]$$

in the case (a). Therefore

$$m - \frac{2 \text{ diam } B}{n(\mu_n)^{1/n}} \|d_m^+(x)\|_{L^n} \exp\left[\left(\frac{2}{n}\right)^n \frac{1}{\mu_n} \|C_m^+(x)\|_{L^n}^n\right] \\ \leq u(x_0) = \inf_B u(x)$$

everywhere in B. Thus the lower estimate for $u(x)$ is proved in case (a).

Now consider case (b). We have the inequality

$$\frac{m - u(x_0)}{\text{diam } B} > \lambda_n I_n$$

in the case (b), where $m - u(x_0) > 0$, $\lambda_n = \frac{2}{n(\mu_n)^{1/n}}$ and $I_n = \|d_m^+(x)\|_{L^n}$. Hence there exists a real number $q > 0$ such that

$$\frac{m - u(x_0)}{\text{diam } B} = \lambda_n [q + I_n]. \tag{24.107}$$

Clearly the inequality

$$[\sigma q + I_n] > 0 \tag{24.108}$$

holds for any $\sigma \in (0, 1]$. This follows directly from (24.107). We now fix any $\sigma \in (0, 1/2]$, then

$$\frac{m - u(x_0)}{\text{diam } B} > \lambda_n [\sigma q + I_n]. \tag{24.109}$$

We set

$$\varepsilon = \left(\lambda_n^{-n}(\sigma q + I_n)^{-n} - \left[\frac{m - u(x_0)}{\text{diam } B}\right]^{-n} \right)^{\frac{1}{n}}.$$

From (24.109) it follows that $\varepsilon > 0$. Direct calculations lead to the identity

$$1 + \left[\frac{m - u(x_0)}{\text{diam } B}\right]^n \varepsilon^n = \frac{[m - u(x_0)]^n}{\lambda_n^n(\sigma q + I_n)^n(\text{diam } B)^n}. \tag{24.110}$$

Notice that

$$\frac{[m - u(x_0)]^n}{\lambda_n^n[(\sigma q + I_n) \text{ diam } B]^n} > 1. \tag{24.111}$$

Thus inequality (24.106) becomes

$$n \ln \left[\frac{m - u(x_0)}{\lambda_n(\sigma q + I_n)\text{diam } B}\right] - 1 + \left[\frac{mu(x_0)}{\text{diam } B}\right]^{-n} \lambda_n^n(\sigma q + I_n)^n$$
$$\leq \lambda_n^n \|C_m^+(x)\|_{L^n}. \tag{24.112}$$

From inequalities (24.111) and (24.112) it follows that

$$0 < \left[\frac{m - u(x_0)}{\text{diam } B}\right]^{-n} (\sigma q + I_n)^n \lambda_n^n < 1.$$

Thus we obtain

$$\ln \frac{[m - u(x_0)]}{\lambda_n \cdot (\sigma q + 1)\text{diam } B} \leq \frac{1}{n} + \frac{1}{n}\lambda_n^n \|C_m^+(x)\|_{L^n}^n.$$

Therefore

$$0 \leq m - u(x_0) \leq \lambda_n(\text{diam } B)(\sigma q + I_n) \exp\left[\frac{1}{n} + \frac{1}{n}\lambda_n^n \|C_m^+(x)\|_{L^n}^n\right].$$

Finally we obtain

$$\begin{aligned} u(x) \geq u(x_0) = \inf_{\overline{B}} u(x) &\geq m - \lambda_n(\sigma q + \|d_m(x)\|_{L^n}) e^{\frac{1}{n}}(\text{diam } B) \\ &\times \exp\left[\frac{1}{n}\lambda_n^n \|C_m^+(x)\|_{L^n}^n\right] \\ &= \frac{2}{n(\mu_n)^{1/n}} e^{1/n}(\text{diam } B)(\sigma q + \|d_m^+(x)\|_{L^n}) \\ &\exp\left[\frac{2^n}{n^{n+1}\mu_n} \|C_m^+(x)\|_{L^n}^n\right]. \end{aligned}$$

Now letting $\sigma \to 0$ and using the notations introduced in the statement of the present theorem, we obtain the final estimate $u(x)$ from below:

$$m - \tau_n \|d_m^+(x)\|_{L^n} \cdot \exp[\tau_n' \|C_m^+(x)\|_{L^n}) \text{diam } B \leq u(x)$$

for all $x \in \overline{B}$.

The desired estimate for $u(x)$ from above (see the statement of Theorem 24.4) has the similar proof.

This completes the proof of Theorem 24.4. □

We now establish inequality (24.104). This inequality is related to convex functions. Let G be a bounded open convex domain in E^n and let $u(x) \in C(\overline{G})$ be an arbitrary convex function satisfying the boundary condition

$$u|_{\partial G} = m = \text{const.} \tag{24.113}$$

We consider the convex cone K with vertex $(x_0, u(x_0))$ and base $L = [(x, m), x \in \partial G]$, where x_0 is an inner point of G.

Let $\chi_u(G)$ and $\chi_K(G)$ be the normal images of G with respect to $u(x)$ and the convex cone K. Then

$$\chi_u(G) \supset \chi_K(G).$$

Hence

$$\omega(R, K, G) = \int_{\chi_K(G)} R(p)dp \leq \int_{\chi_u(G)} R(p)dp = \omega(R, u, G).$$

We consider the n-ball $Q \subset E^n$ with the center at the point x_0 and the radius diam G. We denote by Q_m the n-ball $[(x, m), x \in Q]$. Let K_0 be the convex cone of revolution with the vertex $(x_0, u(x_0))$ and with the base ∂Q_m. Clearly

$$\chi_{K_0}(Q) \subset \chi_K(G).$$

Therefore

$$\omega(R, K_0, Q) < \omega(R, K, G).$$

The set $\chi_{K_0}(Q)$ is the n-ball in R^n with the center $(0, 0, \ldots, 0)$ and the radius

$$\rho_m = \frac{m - u(x_0)}{\text{diam } G}.$$

Thus

$$\omega(R, u, G) \geq \omega(R, K, G) \geq \omega(R, K_0, Q) = \int_{|p| \leq \rho_m} R(p)dp.$$

Inequality (24.104) is proved.

Comments to Chapter 7.

In Chapter 7 we investigated geometric two-sided C^0-estimates for solutions of the Dirichlet problem for nonlinear elliptic Euler–Lagrange equations. For this class of quasilinear elliptic equations the connections between the geometric form of solutions, the integrand of the functional, and the theory of convex functions and bodies are the most profound. These connections were studied in Sections 21, 22 of Chapter 7. These connections lead to the nontrivial sufficiently simple necessary and sufficient conditions for obtaining the desired C^0-estimates for solutions of elliptic Euler–Lagrange equations.

We explain these conditions in the case of the simplest functional

$$I(u) = \int_B [F(Du) + nf(x)u]dx,$$

where B is a bounded domain in the Euclidean space $E^n = \{x = (x_1, \ldots, x_n0\}$, $F(p)$ is a C^2-function in the n-dimensional ball $U(a)$: $|p| < a$ in the Euclidean space $P^n = \{p = (p_1, \ldots, p_n)\}$ and $f(x) \in L^n(B)$.

The Euler–Lagrange equation for the functional $I(u)$ is elliptic if and only if the function $F(p)$ is strictly convex, i.e. all eigenvalues of the matrix

$$\begin{Vmatrix} F_{11} & F_{12} & \ldots & F_{1n} \\ F_{21} & F_{22} & \ldots & F_{2n} \\ \cdots & \cdots & \cdots & \cdots \\ \cdots & \cdots & \cdots & \cdots \\ F_{n1} & F_{n2} & \ldots & F_{nn} \end{Vmatrix}$$

are positive for all $p \in U(a)$. Let χ_F: $U(a) \to Q^n$ be the tangential mapping of the hypersurface $w = F(p)$ in the third Euclidean n-dimensional space Q^n.

Now we introduce a few concepts and notations:

a)

$$m = \inf_{\partial B} u(x), \quad M = \sup_{\partial B} u(x),$$

where $u(x) \in W_2^n(B) \cap C(\overline{B})$;

b)

$$C(F) = \operatorname*{meas}_{Q^n} \chi_F(U(a))$$

c) The formula

$$t = c_F(\rho) = \int_{|p|<\rho} \det(F_{p_i p_j}(p))dp$$

defines a continuous strictly increasing function. This function is defined on $[0, a]$ if $a < +\infty$, and the same function is defined in $[0, +\infty)$, if $a = +\infty$. In both cases

$$C(F) = \lim_{\rho \to a} c_F(\rho).$$

The inverse function

$$\rho = b_F(t)$$

is defined either on $[0, C(F)]$ if $a < +\infty$, or in $[0, C(F))$ if $a = +\infty$.

d)

$$\Omega^+(f) = \int_B [f^+(x)]^n dx,$$

$$\Omega^-(f) = \int_B [f^-(x)]^n dx.$$

It turns out that necessary and sufficient conditions for obtaining the desired C^0-estimates for solutions of the Euler–Lagrange equation

$$\sum_{i=1}^{n} [F_{p_i}(Du)]_{x_i} = nf(x) \tag{$*$}$$

of the functional $I(u)$ are given by the inequalities

$$\Omega^+(f) \le C(F) \quad \text{and} \quad \Omega^-(f) \le C(F) \tag{$**$}$$

if $0 < a < +\infty$.

If $a = +\infty$, i.e. $U(a) = P^n$, then inequalities $(**)$ are the necessary conditions for obtaining the desired C^0-estimates for solutions of equation $(*)$. The strict inequalities

$$\Omega^+(f) < C(F) \quad \text{and} \quad \Omega^-(f) < C(F) \tag{29.11}$$

are the sufficient conditions for obtaining in C^0-estimates for solutions of the same equation $(*)$.

The desired estimates for solutions $u(x) \in W_2^n(B) \cap C(\overline{B})$ are as follows

$$m - b_F(\Omega^+(f)) \text{ diam } B \le u(x) \le M + b_F(\Omega^-(f)) \text{ diam } B. \tag{$***$}$$

In Sections 22 and 23 these estimate were extended to a wider class of Euler–Lagrange equations

$$\sum_{i=1}^{n} [F_{p_i}(Du)]_{x_i} = nf_u(x, u).$$

In § 23 these estimates were applied to various Dirichlet problems, from calculus of variations, global differential geometry and continuum mechanics. In § 24 we introduced the concept of Monge–Ampere generators, which are constructed by the coefficients of given elliptic Euler–Lagrange equation and established the existence of such generators with different boundary data. By means of the Monge–Ampere generators estimates (∗ ∗ ∗) were established for the Euler–Lagrange equations, relating to general functionals

$$\int_B F(x, u, Du)dx.$$

The results of §§ 21, 22, 23 were first published in the author's papers [10], [18]. The results of § 24 are published here for the first time.

Chapter 8. The Geometric Maximum Principle for General Non-Divergent Quasilinear Elliptic Equations

In this chapter we present the geometric maximum principles for solutions $u(x) \in W_2^n(B) \cap C(\overline{B})$ of the elliptic Dirichlet problem

$$\sum_{i,j=1}^{n} a_{ij}(x, u, Du)u_{ij} = b(x, u, Du), \tag{VIII.1}$$

$$u|_{\partial B} = h(x), \tag{VIII.2}$$

where B is a bounded domain in the Euclidean space E^n and ∂B is a closed continuous hypersurface in E^n. We assume that $h(x) \in C(\partial B)$.

The following structure conditions will be imposed on the functions $a_{ij}(x, u, p)$ and $b(x, u, p)$:

1) The functions $a_{ij}(x, u, p)$ and $b(x, u, p)$ are defined in $B \times R \times P^n$ and the quadratic form $a_{ij}(x, u, p)\xi_i\xi_j$ is positive definite in $B \times R \times P^n$. We also assume that the functions $a_{ij}(x, u(x), Du(x))$, $b(x, u(x), Du(x))$ and

$$\frac{|b(x, u(x), Du(x))|^n}{[\det(a_{ij}(x, u(x), Du(x))]}$$

are summable in B.

Since $a_{ij}(x,u,p)\xi_i\xi_j$ is a positive definite form, $\det[(a_{ij}(x,u,p)]$ is positive almost everywhere in $B \times R \times P^n$.

2) Let $m = \min[0, \inf_{\partial B} h(x)]$, $M = \max[0, \sup_{\partial B} h(x)]$, $R_m^- = (-\infty, m]$, $R_M^+ = [M, +\infty)$.

Then inequality

$$\frac{b(x,u,p)}{[\det(a_{ij}(x,u,p))]^{1/n}} \leq \frac{\varphi(x,u) - c(x)u + |d(x)|\,|p|}{[R(|p|)]^{1/n}} \tag{VIII.3}$$

holds for all $(x,u,p) \in B \times R_m^- \times P^n$, where $b(x,u,p) \geq 0$; while the inequality

$$\frac{b(x,u,p)}{[\det(a_{ij}(x,u,p)]^{1/n}} \geq \frac{\varphi(x,u) - c(x)u - |d(x)| \cdot |p|}{[R(|p|)]^{1/n}} \tag{VIII.4}$$

holds for all $(x,u,p) \in B \times R_M^+ \times P^n$, where $b(x,u,p) \leq 0$, and the functions $\varphi(x,u)$, $c(x)$, $d(x)$ and $R(|p|)$ satisfy the following conditions:

a) $\varphi(x,u)$ is non-decreasing with respect to $u \in (-\infty, +\infty)$ for all fixed $x \in B$ and $\|\varphi(x,k)\|_{L^n(B)} < +\infty$, where $k \in (-\infty, +\infty)$ is any constant;
b) $\|c^+(x)\|_{L^n(B)} < +\infty$, where $c^+(x)$ is the positive part of the function $c(x)$;
c) $\|d(x)\|_{L^n(B)} < +\infty$;
d) if $\rho = |p|, p \in P^n$, then the function $R(\rho)$, $\rho \in [0, +\infty)$, is strictly positive, non-increasing and locally summable in P^n.

The main interest is related to the asymptotic behavior of the function $R(\rho)$ as $\rho \to +\infty$. Therefore without loss of generality one can assume that $R(\rho) = R_0 = \text{const} > 0$ for all $\rho \in [0, z]$, where the number z satisfies the inequalities $0 \leq z < \frac{1}{2 \operatorname{diam} B}$.

The geometric maximum principles can be briefly expressed by inequalities

$$\begin{aligned} &m - \phi_R(\|\varphi(x,m)\|_{L^n}, \|c^+(x)\|_{L^n}, \|d(x)\|_{L^n}, CoB) \\ &\quad \leq u(x) \leq M + \phi_R(\|\varphi(x,M)\|_{L^n}, \|c^+(x)\|_{L^n}, \|d(x)\|_{L^n}, CoB), \end{aligned} \tag{VIII.5}$$

which hold for all $x \in \overline{B}$. The analytic structure of the function ϕ_R is determined by the function $R(|p|)$. One of the main properties of the function ϕ_R is that

$$\phi_R(0,0,0,CoB) = 0, \tag{VIII.6}$$

which provides the validity of the classical maximum principle for solutions $u(x) \in W_2^n(B) \cap C(\overline{B})$ of the Dirichlet problem (VIII.1)—(VIII.2).

Three important special cases of the geometric maximum principles were established earlier. They can be described by inequalities (VIII.3) and (VIII.4), if functions $\varphi(x,u)$, $c(x)$, $d(x)$ and $R(|p|)$ satisfy additional conditions.

The first case is related to functions $c(x) = 0$, $d(x) = 0$ and an arbitrary function $R(p)$. (It is not necessary that R depends on $|p|$.) This case was investigated by Bakelman [1], [4], [6], [7], and [9] in connection with estimates of the eigenvalues of elliptic problems and the Dirichlet problem for mean curvature equation in Euclidean, Hyperbolic and Riemannian spaces.

The second one is related to linear elliptic equations in non-divergent form. In this case $\varphi(x,u) = f(x)$ and $R(|p|) = 1$. The main results were obtained by Alexandrov [6], [7], Bakelman [6], [7], [9], and Pucci [1], [2], [3] in the sixties.

The third case is related to nonlinear Euler–Lagrange equations. In this case $d(x) = 0$ and $R(|p|) \geq R_0 = \text{Const} > 0$. The main results are obtained by the author in 1986–88. This case has interesting applications to Calculus of Variations (developments of S.N. Bernstein results [4] to several variables) and Continuum Mechanics, see Bakelman [17], [18], [19], [20] and Chapter 7 of this monograph.

In all these investigations the geometric ideas and techniques play the crucial role.

The most difficult case for deriving the geometric maximum principles are related to the general case of inequalities (VIII.3) and (VIII.4), i.e. to the assumptions $\|c^+(x)\|_{L^n} \neq 0$, $\|d(x)\|_{L^n} \neq 0$ and $\lim\limits_{|p|\to+\infty} R(|p|) = 0$. There are many interesting applications of this case to Calculus of Variations, Differential Geometry and Continuum Mechanics. In Spring and Summer of 1990 Bakelman and Perry [1] obtained the geometric maximum principles under these general assumptions. They are sharp and cover all maximum principles mentioned above. My recent investigations are based on new geometric and analytic ideas and techniques.

In this chapter the presentation is mainly related to quasilinear elliptic Dirichlet problems. For the special case of linear elliptic Dirichlet problems the geometric maximum principles can be obtained as sufficiently simple corollaries. Therefore we do not systematically present maximum principles for linear elliptic Dirichlet problems.

§25. The First Geometric Maximum Principle for Solutions of the Dirichlet Problem for General Quasilinear Equations

25.1 The First Geometric Maximum Principle for General Quasilinear Elliptic Equations and Linear Elliptic Equations of the Form

$$\sum_{i,k=1}^{n} a_{ik}(x)u_{ik} + c(x)u = f(x). \tag{25.1}$$

We consider the Dirichlet problem

$$\sum_{i,k=1}^{n} a_{ik}(x,u,Du)u_{ik} = b(x,u,Du), \tag{25.2}$$

$$u|_{\partial B} = h(x), \tag{25.3}$$

where B is a bounded open domain in the Euclidean space E^n, ∂B is a closed continuous hypersurface in E^n and $h(x) \in C(\partial B)$. We assume that:

A.1. The functions $a_{ik}(x,u,p), b(x,u,p)$ are defined in $B \times R \times G$ and at each point $(x,u,p) \in B \times R \times G$ the form

$$\sum_{i,k=1}^{n} a_{ik}(x,u,p)\xi_i\xi_j \tag{25.4}$$

is positive definite, where G is a domain in the n-dimensional Euclidean space $P^n = \{p = (p_1, p_2, \dots, p_n)\}$ and $O' = (0,0,\dots,0)$ is an point of G. The case $G = P^n$ is not excluded.

A.2. In each point $(x,u,p) \in B \times R \times G$ the inequality

$$\frac{b^n(x,u,p)}{n^n \det(a_{ik}(x,u,p))} \le \frac{\varphi_1(x)}{R_1(p)} \tag{25.5}$$

holds if $b(x,u,p) \ge 0$, while the inequality

$$\frac{|b^n(x,u,p)|}{n^n \det(a_{ik}(x,u,p))} \le \frac{\varphi_2(x)}{R_2(p)} \tag{25.6}$$

holds if $b(x,u,p) \le 0$.

A.3. In inequalities (25.5) and (25.6) the functions $\varphi_1(x), \varphi_2(x)$ are nonnegative and locally summable in B and the functions $R_1(p)$, $R_2(p)$ are positive and locally summable in G.

Let $r(G) = \operatorname{dist}\{O', \partial G\}$. Below we consider two different cases

$$r(G) < +\infty \tag{25.7}$$

and

$$r(G) = +\infty \tag{25.8}$$

(see § 21). Clearly (25.8) is equivalent to $G = P^n$.

We will consider solutions $u(x) \in W_2^n(B) \cap C(\overline{B})$ of the Dirichlet problem (25.2)–(25.3). If $r(G) = +\infty$, i.e. $G = P^n$, then we do not impose any restriction on the values of $Du(x)$ for any solution $u(x) \in W_2^n(B) \cap C(\overline{B})$ of equation (25.2).

Now we consider carefully the case (25.7), i.e. $G \subset P^n$ and $G \neq P^n$. In this case

$$Du(x) \subset G \tag{25.9}$$

for all $x \in B$, where $u(x)$ is any solution of equation (24.66). But inclusion (25.9) is only convenient for functions $u(x) \in C(B) \cap C^1(\overline{B})$. Therefore we introduced some modification of condition (25.9) for $u(x) \in W_2^n(B) \cap C(\overline{B})$ (see Subsection 21.3). We now repeat this modification once more.

A point $M(x, u(x))$ of the graph of a function $u(x) \in C(\overline{B})$, where $x \in B$, is said to be *global convex* (global concave), if there exists at least one supporting hyperplane α passing through M and lying under (over) the graph of $u(x)$. If such a supporting hyperplane α has the equation

$$z = p_1^0 x_1 + \cdots + p_n^0 x_n + b,$$

then the point

$$p_0 \stackrel{\text{def}}{=} \chi(\alpha) = (p_1^0, \ldots, p_n^0) \in P^n$$

is called the normal image of α. We denote by

$$\nu_u^+(B) = \bigcup_{\alpha+} (\alpha^+), \tag{25.10}$$

$$\nu_u^-(B) = \bigcup_{\alpha^-} (\alpha^-), \tag{25.11}$$

$$\nu_u(B) = \nu_u^+(B) \cup \nu_u^-(B), \tag{25.12}$$

where α^+ (or α^-) is a supporting hyperplane passing through a global convex (or global concave) point of the graph of $u(x)$. All such supporting hyperplanes α^+ (or α^-) at all global convex (or global concave) points are taken into account in (25.10) (or in (25.11)).

We now formulate the last assumption of this subsection.

A.4. If the functions $a_{ik}(x, u, p)$ and $b(x, u, p)$ are defined in $B \times R \times G$, where G is a domain in P^n and $\partial G \neq \emptyset$, then we consider only functions $u(x) \in W_2^n(B) \cap C(\overline{B})$ for which the inclusion

$$\nu_u(B) \subset G \tag{25.13}$$

holds.

Theorem 25.1. *Let $r(G) = dist\{O', \partial G\} < +\infty$,* i.e. *$\partial G \neq \emptyset$, and let $u(x) \in W_2^n(B) \cap C(\overline{B})$ be a solution of the Dirichlet problem (25.2–3). If Assumptions A.1–A.4 are valid and if the inequalities*

$$\int_B \varphi_1(x)dx \leq \int_{U(r(G))} R_1(p)dp \tag{25.14}$$

and

$$\int_B \varphi_2(x)dx \leq \int_{U(r(G))} R_2(p)dp \tag{25.15}$$

hold, then the estimates

$$m - b_1(\Omega^+) \text{ diam } B \leq u(x) \leq M + b_2(\Omega^-) \text{ diam } B \tag{25.16}$$

hold for all $x \in \overline{B}$, where

$$m = \inf_{\partial B} h(x), \quad M = \sup_{\partial B} h(x); \tag{25.17}$$

$$\Omega^+ = \int_B \varphi_1(x)dx, \quad \Omega^- = \int_B \varphi_2(x)dx \tag{25.18}$$

and $\rho = b_i(t)$ is the inverse for the functions

$$t = \int_{U(\rho)} R_i(p)dp, \qquad i = 1, 2. \tag{25.19}$$

Remarks on the Statement of Theorem 25.1.

1) In (25.19), $U(\rho)$ is the n-dimensional ball $|p| < \rho$ in P^n, where $0 < \rho < +\infty$. Let

$$C(R_i) = \int_{U(r(G))} R_i(p)dp, \qquad i = 1, 2. \tag{25.20}$$

Then the inverse functions $\rho = b_i(t)$, $i = 1, 2$ are defined for all $t \in [0, C(R_i)]$. The cases $C(R_i) = +\infty$ are not excluded.

2) Inequalities (25.14) and (25.15) can be rewritten in the form

$$\Omega^+ \leq C(R_1) \tag{25.14'}$$

and

$$\Omega^- \leq C(R_2). \tag{25.15'}$$

Therefore the cases $\Omega^+ = +\infty$ or $\Omega^- = +\infty$ are not excluded.

Proof. The proof of Theorem 25.1 can be obtained from Theorem 24.3 by means of the Monge–Ampere generators $F_1(p)$ and $F_2(p)$, which correspond to any convex generalized solutions of Monge–Ampere equations

$$\det \left[\frac{\partial^2 F_1(p)}{\partial p_i \partial p_j} \right] = R_1(p)$$

and

$$\det \left[\frac{\partial^2 F_2(p)}{\partial p_i \partial p_j} \right] = R_2(p)$$

in the ball $U(r(G)) \subset P^n$.

Nevertheless the direct proof of Theorem 25.1 deserves to be considered. It can be done on the basis of the geometric constructions presented in the first half of Subsection 21.3. In this proof Monge–Ampere generators and some specific details relating to elliptic Euler–Lagrange equations will not be used.

It is sufficient to prove the inequality

$$m - b_1(\Omega^+) \operatorname{diam} B \le u(x), \qquad x \in \overline{B}, \tag{25.21}$$

because the second inequality in (25.16) can be proved in a similar way.

Only the case

$$m > u_0 = \inf_{\overline{B}} u(x) \tag{25.22}$$

is worth considering.

Subsection 1. The Main Geometric Constructions. Let δ be any number from the open interval $(0, \delta_u)$, where $\delta_u = m - u_0 > 0$ and let $v_\delta(x)$ be the convex support of the function $u(x)$ (see Subsection 21.2). We denote by x_0 an interior point of B such that

$$u_0 = \inf_{\overline{B}} u(x) = u(x_0). \tag{25.23}$$

Clearly the point $M_0(x_0, u_0)$ belongs simultaneously to both the graph of the function $u(x)$ and the graph of its convex support $v_\delta(x)$. Let $m_\delta = m - \delta$. Then

$$m_\delta >_0 \quad \text{or} \quad m_\delta - u_0 > 0. \tag{25.24}$$

We now consider two convex cones K_0 and K_1 with a common vertex at the point M_0. The cone K_0 has Γ_{m_δ} as its base (see Subsection 21.2) and the cone K_1 is a convex cone of revolution, whose base is an $(n-1)$-sphere $\sum_{m_\delta} \subset \gamma_{m_\delta}$ (see Subsection 21.2). The center of $\sum_{m_\delta}$ is a point (x_0, m_δ) in the hyperplane γ_{m_δ}: $z = m_\delta$, and the radius of $\sum_{m_\delta}$ is equal to diam B. We denote by T_{m_δ} an open ball in γ_{m_δ}, whose boundary is $\sum_{m_\delta}$. Let T be the projection of T_{m_δ} on the hyperplane E^n: $z = 0$. Then clearly

$$\chi_{K_1}(T) \subset \chi_{K_0}(\operatorname{int} H) \subset \chi_{v_\delta}(\operatorname{int} H), \tag{25.25}$$

where H is the closed convex hull of the set $\overline{B}_m$. Since K_1 is the cone of revolution, then $\chi_{K_1}(T)$ is a closed n-ball in P^n. Clearly the inequality

$$|p| \le \frac{h_\delta}{\operatorname{diam} B} \tag{25.26}$$

describes the n-ball $\chi_{K_1}(T)$, where

$$h_\delta = m_\delta - u_0 > 0$$

is the height of the convex cone K_1.

Let S_m be the part of the graph of $u(x)$ located under the hyperplane γ_{m_δ} in $E^{n+1} = E^n \times R$ and let

$$D_{m_\delta} = S_{v_\delta} \cap S_{m_\delta}. \tag{25.27}$$

We denote by D_δ the projection of D_{m_δ} on E^n. Clearly D_δ is a closed subset of B, and

$$\operatorname{dist}(D_\delta, \partial B) > 0.$$

From the definition of a convex hull it follows that every supporting hyperplane α of the convex hypersurface S_{v_δ} is also a supporting hyperplane of the hypersurface S_{m_δ} at some global convex point. Clearly $\alpha \cap D_{m_\delta} \neq \emptyset$, if $\alpha \cap S_{v_\delta}$ contains a point $(x, v_\delta(x))$, where $x \in \text{int } H$.

Since $u(x)$ satisfies Assumption A.4,

$$\chi_{v_\delta}(\text{int } H) \subset \nu_u^+(B) \subset \nu_u(B) \subset G. \tag{25.28}$$

Now from (25.25) and (25.28) it follows that

$$\chi_{K_1}(T) \subset \chi_{v_\delta}(\text{int } H) \subset \nu_u^+(B) \subset \nu_u(B) \subset G. \tag{25.29}$$

Clearly

$$\chi_{v_\delta}(D_\delta) = \chi_{v_\delta}(B) = \chi_{v_\delta}(\text{int } H). \tag{25.30}$$

Thus we obtain the inequalities

$$\begin{aligned}\int_{\chi_{v_\delta}(D_\delta)} R_1(p)dp &= \int_{\chi_{v_\delta}(D_\delta)} R_1(p)dp \\ &= \int_{\chi_{v_\delta}(\text{int } H)} R_1(p)dp.\end{aligned} \tag{25.31}$$

For every positive function $R_1(p)$ satisfying Assumption A.3 and for every function $u(x) \in W_2^n(B) \cap C(\overline{B})$ satisfying Assumption A.4 the number

$$\omega_+(R_1, u, \delta) = \int_{\chi_{v_\delta}(\text{int } H)} R_1(p)dp \tag{25.32}$$

exists. From (25.32) it follows that

$$\omega_+(R_1, u, \delta) = \int_{\chi_{v_\delta}(B)} R_1(p)dp = \int_{\chi_{v_\delta}(\gamma_\delta)} R_1(p)dp, \tag{25.33}$$

if $\delta \in (0, \delta_u)$ and $\delta_u = m - u_0 > 0$.

Subsection 2. Estimates for h_δ.

Lemma 25.1. *Let $r(G) < +\infty$ and a positive function $R_1(p)$ satisfy Assumption A.3. Let a function $u(x)$ satisfy Assumption A.4 and $\delta_u = m - u_0 > 0$. Finally let the inequality*

$$\omega_+(R_1, u, \delta) \le C(R_1) \tag{25.34}$$

hold for any $\delta \in (0, \delta_u)$. Then the inequalities

$$0 < h_\delta \le b_1[\omega_+(r_1, u, \delta)] \text{ diam } B \tag{25.35}$$

hold. Moreover if $\omega_+(R_1, u, \delta) = C(R_1)$, then $b_1[\omega_+(R_1, u, \delta)] = r(G) < +\infty$ in (25.35).

Proof. Since

$$0 < \delta < \delta_u = m - u_0, \tag{25.36}$$

then from (25.36) we obtain

$$h_\delta = m_\delta - u_0 = (m - \delta) - u_0 = \delta_u - \delta > 0. \tag{25.37}$$

The number $h_\delta > 0$ is the height of the convex cone of revolution K_1. According to our constructions

$$\int_{\chi(K_1(T))} R_1(p)dp = \int_{\chi_{v_\delta}(\text{int } H)} R_1(p)dp = \omega_+(R_1, u, \delta). \tag{25.38}$$

Now from (25.34) and the properties of the function $\rho = b_1(t)$ we obtain the inequality

$$b_1(\omega_+(R_1, u, \delta)) \le b_1(C(R_1)) = r(G) < +\infty, \tag{25.39}$$

where the equality corresponds to the case

$$\omega_+(R_1, u, \delta) = C(R_1). \tag{25.40}$$

Since $\chi_{K_1}(T)$ is the n-ball

$$|p| \le \frac{h_\delta}{\text{diam } B} \tag{25.41}$$

in the space P^n, then from (25.38–39) and (25.41) it follows that

$$\chi_{K_1}(T) \subset U(r(G)) \cap \chi_{v_\delta}(\text{int } H)$$

and

$$\frac{h_\delta}{\text{diam } B} \le b_1(\omega_+(R_1, u, \delta)) \le r(G) < +\infty \tag{25.42}$$

consecutively. Now the desired estimates (25.35) follow directly from inequalities (25.37) and (25.42). The proof of Lemma 25.1 is completed. □

Subsection 3. Two-Sided Estimates for Functions $u(x) \in W_2^n(B) \cap C(\overline{B})$ Independent on the Number δ. First of all we consider the properties of $\omega_+(R_1, u, \delta)$, where $\delta \in (0, \delta_u)$ and $\delta_u = m - u_0 > 0$. If $0 < \delta_2 < \delta_1 < \delta_u$, then for every supporting hyperplane of the graph of the function $v_{\delta_1}(x)$, $x \in \operatorname{int} H$ there exists a parallel supporting hyperplane of the graph of function $v_{\delta_2}(x)$, $x \in \operatorname{int} H$. This statement follows directly from the definition of a supporting hyperplane to the set and the constructions of the functions $v_{\delta_1}(x)$ and $v_{\delta_2}(x)$ for $x \in \operatorname{int} H$. Thus

$$\omega_+(R_1, u, \delta_1) \leq \omega_+(R_1, u, \delta_2).$$

Hence the limit (finite or infinite)

$$\omega_+(R_1, u) = \lim_{\delta \to 0+} \{\omega_+(R_1, u, \delta)\}$$

exists and is positive. The case $\omega_+(R_1, u) = +\infty$ is not excluded from our considerations.

If $\delta_u = m - u_0 = 0$, then $u(x) \geq m$ for all $x \in \overline{B}$. Thus $\omega_+(R_1, u) = 0$ in this case. Conversely, if $\omega + (R_1, u) = 0$, then $\delta_u = m - u_0 = 0$.

The number

$$\omega_+(R_1, u) \text{ is called } R_1\textit{-total area of convex support}$$

for a function $u(x) \in W_2^n(B) \cap C(\overline{B})$. The *$R_2$-total area $\omega_-(R_2, u)$ of concave support* for the same function $u(x)$ can be constructed in a similar way.

Lemma 25.2. *Let $r(G) < +\infty$ and let positive functions $R_1(p)$, $R_2(p)$ satisfy Assumption A.3. Let a function $u(x) \in W_2^n(B) \cap C(\overline{B})$ satisfy Assumption A.4. Then inequalities*

$$m - b_1[\omega_+(R_1, u)] \operatorname{diam} B \leq u(x) \leq M + b_2[\omega_-(R_2, u)] \operatorname{diam} B \tag{25.43}$$

hold for all $x \in \overline{B}$, if

$$\omega_+(R_1, u) \leq C(R_1) \quad \text{and} \quad \omega_-(R_2, u) \leq C(R_2), \tag{25.44}$$

where as usual $m = \inf_{\partial B} u(x)$, $M = \sup_{\partial B} u(x)$.

Proof. The cases $\omega_+(R_1, u) = 0$ and $\omega_-(R_2, u) = 0$ lead to the respective estimates $u(x) \geq m$ or $u(x) \leq M$ for all $x \in \overline{B}$. Therefore only the cases

$$u_0 = \inf_{\overline{B}} u(x) < m \quad \text{and} \quad u_1 = \sup_{\overline{B}} u(x) > M$$

deserve consideration. Let $\delta > 0$ be an arbitrary number satisfying the condition $\delta < m - u_0$. Clearly

$$\omega_+(R_1, u) \leq \omega_+(R_1, u) \leq C(R_1).$$

Since

$$m_\delta - h_\delta \leq u_0 \leq u(x) \tag{25.45}$$

for all $x \in \overline{B}$, then from (25.45) and Lemma 25.1 it follows that

$$m - \delta - b_1[\omega_+(R_1, u, \delta)]\text{diam } B \leq u_0 \leq u(x) \tag{25.46}$$

for all $x \in \overline{B}$. The function $\rho = b_1(t)$ is continuous and strictly increasing in $[0, C(R_1)]$ and

$$\lim_{\delta \to 0+} \omega_+(R_1, u, \delta) = \omega_+(R_1, u).$$

Thus (25.46) becomes

$$m - b_1[\omega_+(R_1, u)] \text{ diam B} \leq u(x)$$

for all $x \in \overline{B}$, if $\delta \to 0^+$ in inequalities (25.46).

If we apply the same consideration to the function $-u(x)$ and replace $R_1(p)$ by $R_2(p)$, we obtain the inequality

$$-u(x) \geq -M - b_2[\omega_-(R_2, u)] \text{ diam B} \tag{25.47}$$

for all $x \in \overline{B}$. From (25.46) and (25.47) it follows that inequalities (25.43) hold in $\overline{B}$. The proof of Lemma 25.2 is completed.

Remark. Since $r(G) < +\infty$, then for the limiting case $\omega_+(R_1, u)$ or $\omega_-(R_2, u)$ can be replaced respectively by $C(R_1)$ or $C(R_2)$.

Subsection 4. The Final Part of the Proof of Theorem 25.1.

Lemma 25.3. *Let $u(x) \in W_2^n(B) \cap C(\overline{B})$ be a solution of the Dirichlet problem (25.2–3) and let Assumptions A.1, A.2, A.3 and A.4 are fulfilled. Then the inequalities*

$$\omega_+(R_1, u) \leq \int_B \varphi_1(x)dx \tag{25.48}$$

and

$$\omega_-(R_2, u) \leq \int_B \varphi_2(x)dx \tag{25.49}$$

hold.

The integrals in (25.48) and (25.49) can optionally take the value $+\infty$.

Proof. It is sufficient to prove inequality (25.48), because inequality (25.49) can be proved in the same way. If $\omega_+(R_1, u) = 0$, then (25.48) is trivial. Let

$$\omega_+(R_1, u) > 0, \tag{25.50}$$

then from the definition of $\omega_+(F, u)$ it follows that

$$m > u_0 = \inf_{\overline{B}} u(x). \tag{25.51}$$

Let δ be any number from the open interval $(0, \delta_u)$, where $\delta_u = m - u_0 > 0$ and let $v_\delta(x)$ be the corresponding convex support of the function $u(x)$. Then according to (25.32–33) we have

$$\omega_+(R_1, u, \delta) = \int_{\chi_{v_\delta}(D_\delta)} R_1(p)dp, \tag{25.32}$$

where D_δ is the set, on which $u(x) = v_\delta(x)$. We must note again that D_δ is a closed subset of B and

$$\text{dist}(D_\delta, \partial B) > 0. \tag{25.53}$$

It is well known and mentioned above that $\chi_{v_\delta}(D_\delta)$ is a closed subset of P^n. From the definition of the convex function $v_\delta(x)$ it follows that every supporting hyperplane α of the hypersurface $z = v_\delta(x)$ is also a supporting hyperplane of the hypersurface $z = u(x)$. Moreover there exists a point $M_0(x_0, v(x_0)) \in \alpha$ such that $x_0 \in D_\alpha$. Thus

$$\chi_{v_\delta}(D_\delta) \subset \nu_u^+(B) \subset G, \tag{25.54}$$

if $r(G) < +\infty$. Thus $\chi_{v_\delta}(D_\delta)$ is a compact subset of G. Hence

$$\text{meas}\ \chi_{v_\delta}(D_\delta) < +\infty. \tag{25.55}$$

Since $u(x) \in W_2^n(B) \cap C(\overline{B})$, then from (25.53), (25.54) and (25.55) it follows

$$\det(u_{ij}) \geq 0 \quad \text{and} \quad \text{meas}\ \chi_{v_\delta}(D_\delta) = \int_{D_\delta} \det(u_{ij})dx. \tag{25.56}$$

We also use the information that every point $(x, u(x))$ of the graph of $u(x)$ is convex for $x \in D_\delta$, which follows directly from the definition of the set D_δ. According to Assumption A.3 and A.4 we obtain the formula

$$\omega_+(R_1, u, \delta) = \int_{\chi_{v_\delta}(D_\delta)} R_1(p)dp = \int_{D_\delta} R_1(Du)\det(u_{ij})dx. \tag{25.57}$$

From (25.52) it follows that

$$0 < \omega_+(R_1, u, \delta) = \int_{\gamma_\delta} R_1(Du)\det(u_{ij})dx. \tag{25.58}$$

Thus

$$\text{meas}\ D_\delta > 0. \tag{25.59}$$

The quadratic form

$$\sum_{i,k=1}^{n} u_{ik}(x)\xi_i\xi_k \tag{25.60}$$

is nonnegative almost everywhere in D_δ, because all point $(x, u(x))$, $x \in D_\delta$ of the graph of $u(x)$ are convex and because $u(x)$ has first and second Sobolev generalized derivatives almost everywhere in B. Since $u(x) \in W_2^n(B) \cap C(\overline{B})$, then the composite positive quadratic form

$$\sum_{i,k=1}^{n} a_{ik}(x, u(x), Du(x))\xi_i\xi_k \tag{25.61}$$

is defined almost everywhere in B.

Now we denote by D'_δ the subset of D_δ consisting of all points $x \in D_\delta$, where all generalized derivatives of the first and second orders of the function $u(x)$ take finite values, and where quadratic forms

$$\sum_{i,k=1}^{n} u_{ij}(x)\xi_i\xi_j \tag{25.62}$$

and

$$\sum_{i,k=1}^{n} a_{ik}(x, u(x), Du(x))\xi_i\xi_k \tag{25.63}$$

are defined and nonnegative. Clearly

$$\text{meas } D'_\delta = \text{means } D_\delta > 0. \tag{25.64}$$

It is also clear that

$$\det(u_{ik}(x)) \geq 0 \quad \text{and} \quad \det(a_{ik}(x, u(x), Du(x)) > 0.$$

Thus inequality (25.58) becomes

$$0 < \omega_+(R_1, u, \delta) = \int_{D'_\delta} R_1(Du(x)) \det(u_{ik}(x))dx. \tag{25.65}$$

We now prove the inequality

$$\begin{aligned}(\det[a_{ik}(x, u(x), Du(x)])^{1/n} \cdot [\det(u_{ik}(x)]^{1/n} \\ \leq \frac{1}{n}\sum_{i,k=1}^{n} a_{ik}(x, u(x), Du(x))u_{ik}(x)\end{aligned} \tag{25.66}$$

for all $x \in D'_\delta$.

Both determinants and the sum in inequality (25.66) are invariants of orthogonal transformations of the Euclidean space $R^n = \{\xi = (\xi_1, \xi_2, \ldots, \xi_n)\}$. If we fix any point $x \in D'_\delta$ and bring quadratic forms (25.62) and (25.63) to canonical form, then (25.66) becomes the well known Cauchy inequality relating to the arithmetic and geometric means. Thus the proof of inequality (25.66) is completed.

Since $u(x)$ is a solution of the Dirichlet problem (25.2–3), then

$$0 \le \frac{1}{n} \sum_{i,k=1}^{n} a_{ik}(x, u(x), Du(x)) u_{ik}(x) \qquad (25.67)$$
$$= \frac{1}{n} b(x, u(x), Du(x))$$

for all $x \in D'_\delta$. From (25.66), (25.67) and Assumptions A.1–4 we obtain

$$0 < \omega_+(R_1, u, \delta) \le \int_{D'_\delta} \varphi_1(x) dx.$$

Since

$$\int_{D'_\delta} \varphi_1(x) dx \le \int_B \varphi_1(x) dx = \Omega^+$$

we obtain the inequalities

$$0 < \omega_+(R_1, u, \delta) \le \Omega^+. \qquad (25.68)$$

Since $\omega_+(R_1, u, \delta)$ is a non-decreasing function of $\delta > 0$, then from (25.68) and the definition of the number $\omega_+(R_1, u)$ it follows that

$$0 < \omega_+(R_1, u) = \lim_{\delta \to 0+} \omega_+(R_1, u, \delta) \le \int_B \varphi_1(x) dx.$$

Thus the inequality (25.48) is proved. Inequality (25.49) can be established in the same way. The proof of Lemma 25.3 is completed.

The proof of Theorem 25.1 follows directly from Lemmas 25.2 and 25.3.

Inequalities (25.14) and (25.15) are sharp. The number n in Assumption A.4 is not interchangeable with $n' < n$. The corresponding example will be considered below in this section.

Theorem 25.2. *Let $r(G) = +\infty$, i.e. $G = P^n$, and let $u(x) \in W_2^n(B) \cap C(\overline{B})$ be a solution of the Dirichlet problem (25.2–3). If Assumptions A.1, A.2, A.3 are fulfilled and if the strict inequalities*

$$\Omega^+ < C(R_1) \quad \text{and} \quad \Omega^- < C(R_2)$$

hold, then estimates (25.16) hold for all $x \in \overline{B}$.

The simple analysis shows that Theorem 25.2 and the corresponding lemmas are simplifications of Theorem 25.1 and Lemmas 25.1, 2, 3 both in the statements and the proofs.

We now consider the *applications of Theorems 25.1 and 25.2 to the Dirichlet problem for important classes of linear and quasilinear elliptic equations*. In these investigations we also use the techniques and results developed in Chapter 7.

1. The Dirichlet Problem for Linear Elliptic Equations $\sum_{i,k=1}^{n} a_{ik}u_{ik} + cu = f$.

Notations and Assumptions. Let B be a bounded domain in E^n and ∂B be a closed continuous hypersurface in E^n. Let $a_{ik}(x)$, $c(x)$ and $f(x)$ be measurable functions in B satisfying the following properties:

a) the quadratic form $\sum_{i,k=1}^{n} a_{ik}(x)\xi_i\xi_k$ is positive definite for all $x \in B$;

b) the functions

$$\varphi^{\pm}(x) = \frac{f^{\pm}(x)}{[\det(a_{ik}(x))]^{1/n}}, \tag{25.70}$$

$$\gamma^{\pm}(x) = \frac{c^{\pm}(x)}{[\det(a_{ik}(x))]^{1/n}} \tag{25.71}$$

are summable with degree n in B, where $f^{\pm}(x)$, $c^{\pm}(x)$ are positive and negative parts of the functions $f(x)$ and $c(x)$.

Theorem 25.3. *Let $u(x) \in W_2^n(B) \cap C(\overline{B})$ be a solution of the Dirichlet problem*

$$\sum_{i,k=1}^{n} a_{ik}(x)u_{ik}(x) = f(x) \tag{25.72}$$

$$u|_{\partial B} = h(x) \in C(\partial B). \tag{25.73}$$

Then the estimates

$$m - \frac{1}{n(\mu_n)^{1/n}}\|\varphi^+(x)\|_{L^n(B)} \text{ diam } B \le u(x) \tag{25.74}$$

$$\le M + \frac{1}{n(\mu_n)^{1/n}}\|\varphi^-(x)\|_{L^n(B)} \text{ diam } B$$

hold for all $x \in \overline{B}$, where $m = \inf_{\partial B} h(x)$, $M = \sup_{\partial B} h(x)$ and μ_n is the volume of the unit n-ball in E^n.

Proof. Let $R_1(p) = R_2(p) = 1$ for all $p \in P^n$. Then the function

$$t = \int_{U(\rho)} 1 dp = \mu_n \rho^n$$

is defined on $[0,+\infty)$ and its inverse

$$\rho = b_1(t) = b_2(t) = \left(\frac{t}{\mu_n}\right)^{1/n}$$

is also defined on $[0,+\infty)$.

According to notation (25.70) and Assumptions A.1, A.2, A.3 (see p. 414) we can apply Theorem 25.2 and obtain the desired estimates (25.74).

The proof of Theorem 25.3 is completed.

Theorem 25.3 is called geometric maximum principle for the Dirichlet problem (25.72–73). From estimates (25.74) it follows that the Dirichlet problem (25.72–73) has not more than one solution in the functional class $W_2^n(B)\cap C(\overline{B})$.

If we consider solutions $u(x) \in W_2^{n'}(B) \cap C(\overline{B})$ of the Dirichlet problem (25.72–73) with $n' < n$, then Theorem 25.3 is incorrect. This assertion follows from the following example constructed by Gilbarg and Serrin [1].

The Dirichlet problem

$$\sum_{i,k=1}^{n} a_{ik}(x)u_{ik}(x) = 0, \tag{25.75}$$

$$u|_{\partial K_R} = R \tag{25.76}$$

has two solutions in the n-ball K_R: $|x| \le R$, where

$$\lambda < 1;$$

$$a_{ij}(x) = \delta_i^j + b\frac{x_i x_j}{r^2}, \qquad i,j = 1,2,\ldots,n;$$

$$b = -1 + \frac{n-1}{1-\lambda};$$

δ_i^j is the Kronecker symbol, and $r = \left(\sum_{i=1}^{n} x_i^2\right)^{1/2}$.

The first solution of the problem (25.75–76) is the constant R^λ and the second one is the function $u(x) = r^\lambda$. If $b > -1$, i.e. $\lambda < 1$, then equation (25.75) is elliptic and

$$u_{x_i} = \lambda r^{\lambda-2}x_i \in L^q(K_R),$$
$$u_{x_i x_j} = \lambda r^{\lambda-4}[(\lambda-2)x_i x_j + \delta_i^j r^2] \in L^p(K_R),$$

where $p < \frac{n}{2-\lambda}$ and $q < \frac{n}{1-\lambda}$. The number p approaches n and the number q grows infinitely, if λ approaches 1.

We now consider the Dirichlet problem

$$\sum_{i,k=1}^{n} a_{ik}(x)u_{ik} + c(x)u(x) = f(x) \tag{25.77}$$

$$u|_{\partial B} = h(x) \in C(\partial B). \tag{25.78}$$

Let all notations and assumptions immediately preceding Theorem 25.3 be fulfilled. We are concerned with the estimate of the C^0-norm of solutions $u(x) \in W_2^n(B) \cap C(\overline{B})$ for the Dirichlet problem (25.77–78).

Let $m = \inf_{\partial B} h(x)$ and $M = \sup_{\partial B} h(x)$. We now set

$$m^* = \max\{0; m\} \quad \text{and} \quad M^* = \max\{0; M\}. \tag{25.79}$$

Clearly $m^* \leq 0$ and $M^* \geq 0$. It is sufficient to establish the estimate for $u(x)$ from below. First of all we consider the case $m \leq 0$. In this case $m = m^* = -|m|$.

Let $u(x) \in W_2^n(B) \cap C(\overline{B})$ be a solution of the Dirichlet problem (25.77–78). We apply a few geometric constructions used in the proof of Lemma 25.3 (see Subsection 4 of the proof of Theorem 25.1) to the graph of the function $u(x)$. Let D_δ be the subset of B on which $u(x) = v_\delta(x)$, where $v_\delta(x)$ is the support of $u(x)$ for a sufficiently small $\delta > 0$ (see p. 424). Clearly $u(x) < 0$ on D_δ. Since $u(x)$ is a solution of equation (25.77), then the following line of equalities and inequalities

$$\begin{aligned}\sum_{i,k=1}^{n} a_{ik}(x)u_{ik}(x) &= f(x) - c(x)u(x) \\ &= f(x) - c^+(x)u(x) \\ &\quad + c^-(x)u(x) \leq f^+(x) - c^+(x)u(x)\end{aligned} \tag{25.80}$$

holds for all $x \in D_\delta$. According to (25.59) meas $D_\delta > 0$. Let D'_δ be the subset of D_δ consisting of all points $x \in D_\delta$, where all generalized derivatives of the first and second orders of $u(x)$ take finite values and where quadratic forms (25.62) and (25.63) exist and are nonnegative. Clearly meas $D'_\delta =$ meas D_δ (see (25.64)). From (25.66) it follows that

$$[\det(a_{ik}(x))]^{1/n}[\det(u_{ik}(x))]^{1/n} \leq \sum_{i,k=1}^{n} a_{ik}(x)u_{ik}(x). \tag{25.81}$$

From (25.80) and (25.81) it follows that

$$0 \leq [\det(u_{ik}(x))]^{1/n} \leq \frac{f^+(x) - c^+(x)u(x)}{n[\det(a_{ik}(x)]^{1/n}}.$$

If we use notations (25.70) and (25.71), then the last inequality becomes

$$0 \le [\det(u_{ik}(x))]^{1/n} \le \frac{1}{n}[\varphi^+(x) - \gamma^+(x)u(x)] \tag{25.82}$$

for all $x \in D'_\delta$. We now set $R(p) = 1$, $p \in P^n$, and according to (25.65) obtain

$$\begin{aligned} 0 < \omega_+(1, u, \delta) &= \int_{D'_\delta} \det(u_{ij}(x))dx \\ &\le \frac{1}{n^n} \int_{D'_\delta} [\varphi^+(x) - \gamma^+(x)u(x)]^n dx. \end{aligned} \tag{25.83}$$

We have only inequality $u(x) \le m$ for all $x \in D'_\delta$ and $u(x) \in C(\overline{B})$, then according to (25.82) we obtain the inequalities

$$0 \le \varphi^+(x) - \gamma^+(x)u(x) \le \varphi^+(x) + \gamma^+(x)\|u(x)\|_{C(\overline{B})}. \tag{25.84}$$

Now from (25.83) and (25.84) we obtain the following estimates

$$\begin{aligned} 0 < \omega_+(1, u, \delta) &\le \frac{1}{n^n} \int_{D'_\delta} [\varphi^+(x) + \gamma^+(x)\|u(x)\|_{C(\overline{B})}]^n dx \\ &\le \frac{1}{n^n} \int_B [\varphi^+(x) + \gamma^+(x)\|u(x)\|_{C(\overline{B})}]^n dx. \end{aligned}$$

Thus finally

$$\begin{aligned} 0 < \omega_+(1, u) &= \lim_{\delta \to 0+} \omega_+(1, u, \delta) \\ &\le \frac{1}{n^n} \int_B [\varphi^+(x) + \gamma^+(x)\|u\|_{C(\overline{B})}]^n dx. \end{aligned}$$

Since $C(R_1) = C(1) = +\infty$, then from Theorem 25.2 we obtain the inequality

$$m = \frac{\operatorname{diam} B}{n(\mu_n)^{1/n}} \|\varphi^-(x) + \gamma^+(x)\|u\|_{C(\overline{B})}\|_{L^n(B)} \le u(x) \tag{25.85}$$

for all $x \in \overline{B}$.

Hence

$$m - \frac{\operatorname{diam} B}{n(\mu_n)^{1/n}} \|\varphi^+(x) + \gamma^+(x)\|u(x)\|_{C(\overline{B})}\|_{L^n(B)} \le \inf_{\overline{B}} u(x). \tag{25.86}$$

The inequality

$$-\frac{\operatorname{diam} B}{n(\mu_n)^{1/n}} \|\varphi^-(x) + \gamma^+(x)\|u(x)\|_{C(\overline{B})}\|_{L^n(B)} \le \inf_{\overline{B}} u(x) \tag{25.87}$$

can be established in the similar way if $m > 0$. In this case we consider only the part of the graph of $u(x)$ lying under the hyperplane $z = 0$.

Inequalities (25.86) and (25.87) can be combined by the following inequality

$$m^* - \frac{\operatorname{diam} B}{n(\mu_n)^{1/n}} \|\varphi^+(x) + \gamma^+(x)\|u(x)\|_{C(\overline{B})}\|_{L^n(B)} \leq \inf_{\overline{B}} u(x). \tag{25.88}$$

The inequality

$$M^* + \frac{\operatorname{diam} B}{n(\mu_n)^{1/n}} \|\varphi^-(x) + \gamma^+(x)\|u(x)\|_{C(\overline{B})}\|_{L^n(B)} \geq \sup_{\overline{B}} u(x) \tag{25.89}$$

can be obtained in the same way.

We now obtain the estimate for $\|u(x)\|_{C(\overline{B})}$. We want to have a sufficiently simple formula for this estimate. As we know $m^* \leq 0$ and $M^* \geq 0$. Therefore the number $N = \max\{|m^*|, M^*\}$ is nonnegative. Clearly

$$\|\varphi^{\pm}(x)\|_{L^n(B)} \leq \|\varphi(x)\|_{L^n(B)}, \tag{25.90}$$

where $\varphi(x) = \varphi^+(x) + \varphi^-(x)$. Thus from (25.88), (25.89) and (25.90) it follows that

$$\|u(x)\|_{C(\overline{B})} \leq N + \frac{\operatorname{diam} B}{n(\mu_n)^{1/n}} [\|\varphi(x)\|_{L^n(B)} + \|\gamma^+(x)\|_{L^n(B)} \|u(x)\|_{C(\overline{B})}].$$

From the last inequality we obtain the desired estimates for $\|u(x)\|_{C(\overline{B})}$, if we assume that the inequality

$$\|\gamma^+(x)\|_{L^n(B)} < n(\mu_n)^{1/n} [\operatorname{diam} B]^{-1} \tag{25.91}$$

holds. This estimate is as follows

$$\|u(x)\|_{C(\overline{B})} \leq \frac{N n(\mu_n)^{1/n} [\operatorname{diam} B]^{-1} + \|\varphi(x)\|_{L^n(B)}}{n(\mu_n)^{1/n} [\operatorname{diam} B]^{-1} - \|\gamma^+(x)\|_{L^n(B)}}. \tag{25.92}$$

Thus we established the following theorem.

Theorem 25.4. *Let $u(x) \in W_2^n(B) \cap C(\overline{B})$ be a solution of the Dirichlet problem (25.77–78), where B is a bounded domain in E^n and ∂B is a closed continuous hypersurface in E^n. Let $a_{ik}(x)$, $c(x)$ and $f(x)$ be measurable functions in B satisfying the following conditions:*

a) the quadratic form $\sum_{i,k=1}^{n} a_{ik}(x)\xi_i\xi_k$ is positive definite for all $x \in B$;

b) the functions

$$\varphi^{\pm}(x) = \frac{f^{\pm}(x)}{[\det(a_{ij}(x)]^{1/n}} \tag{25.93}$$

and

$$\gamma^+(x) = \frac{c^+(x)}{[\det(a_{ij}(x)]^{1/n}} \tag{25.94}$$

are summable in B with degree n;

c)

$$\|\gamma^+(x)\|_{L^n(B)} < \frac{n(\mu_n)^{1/n}}{\operatorname{diam} B}. \tag{25.95}$$

Then the estimate (25.92) for $u(x)$ holds.

Corollary. *Estimate (25.92) becomes*

$$\|u(x)\|_{C(\overline{B})} \leq \frac{\|\varphi(x)\|_{L^n(B)}}{n(\mu_n)^{1/n}[\operatorname{diam} B]^{-1} - \|\gamma^+(x)\|_{L^n(B)}} \tag{25.96}$$

if $u(x)|_{\partial B} = 0$.

2. The Dirichlet Problem for Elliptic Nonlinear Euler–Lagrange Equations.

In §§ 22 and 23 we derived by geometric methods two-sided C^0-estimates for solutions $u(x) \in W_2^n(B) \cap C(\overline{B})$ of the Dirichlet problem for elliptic Euler–Lagrange equations

$$\sum_{i=1}^n \frac{\partial}{\partial x_i}\left[\frac{\partial F(Du)}{\partial p_i}\right] = n f_u(x,u). \tag{25.97}$$

As we know equation (25.97) corresponds to the functional

$$I(u) = \int_B [F(Du) + nf(x,u)]dx. \tag{25.98}$$

The ellipticity of equation (25.97) is equivalent to convexity of the function $F(p)$ in the gradient space P^n. Therefore the basic assumptions with respect to ellipticity of equation (25.97) were stated in terms of geometric invariants of the convex functions $F(p)$.

In our previous investigations (see §§ 22, 23) we also supposed that the second function $f(x,u)$ is convex with respect to u for every fixed $x \in \overline{B}$. This assumption leads to the establishment of the geometric maximum principal considered in § 22 and its various applications considered in § 23.

Now we want to derive the geometric maximum principle for the Dirichlet problem in the class of solutions $u(x) \in W_2^n(B) \cap C(\overline{B})$ without the assumption of convexity of the function $f(x,u)$ with respect to u.

We study this problem for Euler–Lagrange equation, which corresponds to the functional

$$J(u) = \int_B \left\{F(Du) + n\left[f(x,u) - \frac{1}{2}c(x)u^2\right]\right\} dx \tag{25.99}$$

in the class of functions $u(x) \in W_2^n(B) \cap C(\overline{B})$, where ∂B is a closed continuous hypersurface in E^n and B is a bounded open domain in E^n.

We will use notations and concepts introduced in §§ 21, 22, 23.

Let $F(p)$ be a convex function defined in the entire space P^n and χ_F: $P^n \to Q^n$ be its normal mapping, where $Q^n = \{q = (q_1, q_2, \ldots, q_n)\}$ is a new copy of n-dimensional Euclidean space.

We introduce the set function

$$\omega(F, e') = \text{meas}_{Q^n} \chi_F(e'). \tag{25.100}$$

The set function $\omega(F, e')$ is nonnegative and absolutely additive on the family of Borel subsets $e' \subset P^n$. If $\chi_F(P^n)$ is an unbounded set in Q^n, then $\omega(F, e')$ can take value $+\infty$ for some Borel subsets $e' \subset P^n$. But

$$\omega(F, e') < +\infty$$

for every Borel subset e' of P^n with the compact closure. Now we formulate the basic assumptions concerning the functions $F(p)$, $f(x, u)$ and $c(x)$.

Assumption I. The function $\omega(F, e')$ is absolutely continuous on the family of Borel subsets e' of P^n, i.e.

$$\omega(F, e') = \int_{e'} \det(F_{p_i p_j}(p)) dp. \tag{25.101}$$

Assumption II. Let $F(p)$ be a convex function defined on the entire space P^n and satisfying Assumption I. Then the inequality

$$\det(F_{p_i p_j}(p)) \geq F_0 > 0 \tag{25.102}$$

holds almost everywhere in P^n, where $F_0 = \text{const} > 0$.

Assumption III. A function $f(x, u) \in C(\overline{B} \times R)$ is convex with respect to u for every fixed $x \in \overline{B}$.

Therefore the derivative $f_u(x, u)$ is nondecreasing with respect to u for any fixed $x \in \overline{B}$. We denote by $f_u^+(x, u)$ and $f_u^-(x, u)$ the positive and negative parts of $f_u(x, u)$.

Assumption IV. The functions $f_u^+(x, k)$ and $f_u^-(x, k)$ are summable with degree n in B, where $-\infty < k < +\infty$.

Assumption V. The measurable function $c(x)$ in B satisfies the inequality

$$\int_B [c^+(x)]^n dx < +\infty \tag{25.103}$$

where $c^+(x)$ is a positive part of $c(x)$.

Theorem 25.5. *Let $u(x) \in W_2^n(B) \cap C(\overline{B})$ be a solution of the Dirichlet problem*

$$\sum_{i=1}^{n} \frac{\partial}{\partial x_i}\left(\frac{\partial F(Du(x))}{\partial p_i}\right) = n f_u(x, u(x)) - nc(x)u(x), \tag{25.104}$$

$$u|_{\partial B} = h(x) \in C(\partial B). \tag{25.105}$$

Let all Assumptions I, II, III, IV, V be fulfilled. Then the following inequalities

$$m^* - \frac{\operatorname{diam} B}{(\mu_n F_0)^{1/n}}[\|f_u^+(x, m^*)\|_{L^n(B)} + \|c^+(x)\|_{L^n(B)}\|u(x)\|_{C(\overline{B})}] \le u(x) \tag{25.106}$$

and

$$M^* + \frac{\operatorname{diam} B}{(\mu_n F_0)^{1/n}}[\|f_u^-(x, M^*)\|_{L^n(B)} + \|c^+(x)\|_{L^n(B)}\|u(x)\|_{C(\overline{B})}] \ge u(x) \tag{25.107}$$

hold for all $x \in \overline{B}$, where $m = \inf_{\partial B} h(x)$, $M = \sup_{\partial B} h(x)$, $m^* = \min\{0; m\} \le 0$ and $M^* = \max\{0; M\} \ge 0$.

Proof. According to our assumptions equation (25.104) becomes

$$\sum_{i,j=1}^{n} F_{p_i p_j}(Du(x))u_{ij}(x) = n f_u(x, u(x)) - nc(x)u(x). \tag{25.108}$$

It is sufficient to establish inequality (25.106). First of all consider the case $m \le 0$. In this case $m = m^* = -|m|$.

Let $u(x) \in W_2^n(B) \cap C(\overline{B})$ be a solution of the Dirichlet problem (25.104–105). We now apply the same geometric constructions as in the proof of Theorem 25.4. Let D_δ be the subset of B on which $u(x) = v_\delta(x)$, where $v_\delta(x)$ is the convex support of $u(x)$ for a sufficiently small $\delta > 0$ (see p. 424). Clearly $u(x) < 0$ on D_δ. Since $u(x)$ is a solution of equation (25.108), then the following line of equalities and inequalities

$$\begin{aligned}\sum_{i,j=1}^{n} F_{p_i p_j}(Du(x))u_{ij} &= n f_u(x, u(x)) - n[c^+(x) - c^-(x)]u(x) \\ &\le n f_u(x, m) - nc^+(x)u(x) \\ &\le n f_u^+(x, m) - nc^+(x)u(x)\end{aligned} \tag{25.109}$$

holds for all $x \in D_\delta$.

If we apply the techniques developed in the proofs of Theorems 22.1, 22.3, 23.2, 25.3 and 25.4, then we obtain the inequality

$$m - \frac{\operatorname{diam} B}{(\mu_n F_0)^{1/n}} [\|f_u^+(x,m)\|_{L^n(B)} + \|c^+(x)\|_{L^n(B)} \|u(x)\|_{C(\overline{B})}] \le u(x) \quad (25.110)$$

for all $x \in \overline{B}$. It should be noted that $m \le 0$ in inequality (25.110).

If $m > 0$, then inequality

$$-\frac{\operatorname{diam} B}{(\mu_n F_0)^{1/n}} [\|f_u^+(x,0)\|_{L^n(B)} + \|c^+(x)\|_{L^n(B)} \|u(x)\|_{C(\overline{B})}] \le u(x) \quad (25.111)$$

hold for all $x \in \overline{B}$. This can be proved in the similar way, because it is sufficient to use only the part of the graph of $u(x)$ lying under the hyperplane $z = 0$. Inequalities (25.110) and (25.111) can be combined in the following way

$$m^* - \frac{\operatorname{diam} B}{(\mu_n F_0)^{1/n}} [\|f_u^+(x,m^*)\|_{L^n(B)} + \|c^+(x)\|_{L^n(B)} \|u(x)\|_{C(\overline{B})}] \\ \le u(x)$$

for all $x \in \overline{B}$.

Finally the inequality

$$M^* + \frac{\operatorname{diam} B}{(\mu_n F_0)^{1/n}} [\|f_u^-(x,M^*)\|_{L^n(B)} + \|c^+(x)\|_{L^n(B)} \|u(x)\|_{C(\overline{B})}] \\ \ge u(x)$$

for all $x \in \overline{B}$ can be established in the same way as above.

The proof of Theorem 25.5 is completed.

We are now concerned with the suitable estimate for $\|u(x)\|_{C(\overline{B})}$. First of all we introduce two nonnegative numbers N and $A(m^*, M^*)$, depending on the boundary data $u|_{\partial B} = h(x)$. The definitions of these numbers are as follows

$$N = \max\{|m^*|, M^*\},$$
$$A(m^*, M^*) = \max\{\|f_u^\pm(x,m^*)\|_{L^n(B)}, \quad \|f^\pm(u(x,M^*)\|_{L^n(B)}\}.$$

Clearly $N \ge 0$ and $A(m^*, M^*) \ge 0$.

Theorem 25.6. *Let $u(x) \in W_2^n(B) \cap C(\overline{B})$ be a solution of the Dirichlet problem (25.104–105). Let Assumptions I, II, III, IV, V and the inequality*

$$\|c^+(x)\|_{L^n(B)} < (\mu_n F_0)^{1/n} [\operatorname{diam} B]^{-1} \quad (25.112)$$

be fulfilled. Then the following estimate

$$\|u(x)\|_{C(\overline{B})} \leq \frac{N[\mu_n F_0]^{1/n}(\text{diam } B)^{-1} + A(m^*, M^*)}{(\mu_n \cdot F_0)^{1/n}[\text{diam } B]^{-1} - \|c^+(x)\|_{L^n(B)}} \tag{25.113}$$

holds.

Proof. From Theorem 25.5 it follows

$$\begin{aligned} &-N - \frac{\text{diam } B}{(\mu_n F_0)^{1/n}}[A(m^*, M^*) + \|c^+(x)\|_{L^n(B)}\|u(x)\|_{C(\overline{B})} \\ &\leq \inf_{\overline{B}} u(x) \leq \sup_{\overline{B}} u(x) \\ &\leq N + \frac{\text{diam } B}{(\mu_n F_0)^{1/n}}[A(m^*, M^*) + \|c^+(x)\|_{L^n(B)}\|u(x)\|_{C(\overline{B})}]. \end{aligned} \tag{25.114}$$

Now the desired estimate (25.113) follows from inequalities (25.114) and (25.112).
The proof of Theorem 25.6 is completed.

25.2 The Improvement of Estimates (25.16) for Solutions of General Quasilinear Elliptic Equations Depending on Properties of the Functions $\det(a_{ik}(x, u, p))$ and $b(x, u, p)$

In this subsection we are concerned with the replacement of diam B by $(\text{vol } CoB)^{1/n}$ in estimates (25.16), where CoB is the convex hull of a bounded domain in E^n. We are also concerned with the improvement of inequalities (25.16) permitting to estimate the possible value of solutions of the Dirichlet problem (25.2–3) at a given point of the domain B. Below we establish the positive answer for these problems in the terms of the estimators for the function $b^n(x, u, p)[\det(a_{ik}(x, u, p))]^{-1}$. We are concerned with the development of (25.5–6). More precisely we suppose that the following Assumptions A.2-a and A.3-a will be fulfilled.

A.2-a. In addition to Assumption A.1 (see p. 414) the functions $a_{ik}(x, u, p)$, $b(x, u, p)$ are defined in $B \times R \times P^n$ and in each point $(x, u, p) \in B \times R^- \times P^n$ the inequality

$$\frac{b(x, u, p)}{n[\det(a_{ik}(x, u, p))]^{1/n}} \leq \frac{\varphi_1(x) - c(x)u}{[R_1(p)]^{1/n}} \tag{25.115}$$

holds if $b(x, u, p) \geq 0$, while in each point $(x, u, p) \in B \times R^+ \times P^n$ the inequality

$$\frac{|b(x, u, p)|}{n[\det(a_{ik}(x, u, p))]^{1/n}} \leq \frac{\varphi_2(x) + c(x)u}{[R_2(p)]^{1/n}} \tag{25.116}$$

holds if $b(x, u, p) \leq 0$, where $R^- = (-\infty, 0]$ and $R^+ = [0, +\infty)$.

A.3-a. In inequalities (25.115) and (25.116) the functions $\varphi_1(x)$, $\varphi_2(x)$ are nonnegative and belong to $L^n(B)$, the functions $R_1(p)$, $R_2(p)$ are defined in P^n and the inequalities

$$R_1(p) \geq C_1 = \text{const } > 0, \tag{25.117}$$

$$R_2(p) \geq C_2 = \text{const } > 0 \tag{25.118}$$

hold for all $p \in P^n$. Finally $\|c^+(x)\|_{L^n(B)} < +\infty$, where $c^+(x)$ is the positive part of $c(x)$.

We are concerned with solutions $u(x) \in W_2^n(B) \cap C(\overline{B})$ of the Dirichlet problem

$$a_{ik}(x,u,Du)u_{ik} = b(x,u,Du), \tag{25.119}$$

$$u|_{\partial B} = 0. \tag{25.120}$$

We now recall a few concepts and facts associated with solutions of the Dirichlet problem (25.119–120).

Let $v(x)$ be the largest of the convex functions $\tilde{v}(x)$ such that $\tilde{v}(x) \leq u(x)$ everywhere in B. It is defined in CoB, where CoB is the convex hull of a bounded domain B. We say that $v(x)$ is the *convex support* of $u(x)$. The point $x_0 \in B$ is called a convexity point of the function $u(x)$ if $u(x_0) = v(x_0)$. The set of all convexity points of $u(x)$ is called the *convexity contact set* of $u(x)$. We denote this set by M_u^+. The concepts of the concave support $w(x)$, concavity points and the concavity contact set M_u^- of the function $u(x)$ can be introduced in the similar way. Clearly $M_u^\pm$ is at least a Borel subset of B.

Since $u(x) \in W_2^n(B) \cap C(\overline{B})$, then as we know, the formulas

$$\omega(R,v,CoB) = \int_{M_u^+} R(Du)\det(u_{ik})dx \tag{25.121}$$

and

$$\omega(R,w,CoB) = \int_{M_u^-} R(Du)|\det(u_{ik})|dx \tag{25.122}$$

hold for any locally summable positive function $R\colon P^n \to (-\infty,+\infty)$. If in addition Assumptions A.1, A.2-a and A.3-a are fulfilled, then the inequalities

$$\Omega^+ < +\infty \quad \text{and} \quad \Omega^- < +\infty \tag{25.123}$$

hold, where the notations

$$\Omega^+ = \int_B [\varphi_1(x) + c^+(x)|u(x)|]^n dx \tag{25.124}$$

and

$$\Omega^- = \int_B [\varphi_2(x) + c^+(x)|u(x)|]^n dx \tag{25.125}$$

are introduced for the brevity. The considerations made in the proof of Lemma 25.3 lead to the following inequalities

$$\omega(R_1, v, CoB) = \int_{M_u^+} R_1(Du)\det(u_{ij})dx \tag{25.126}$$
$$\leq \int_{M_u^+} \frac{R_1(Du)b^n(x,u,Du)}{n^n \det(a_{ik}(x,u,Du))}dx$$
$$\leq \int_{M_u^+} [\varphi_1(x) + c^+(x)|u(x)|]^n dx \leq \Omega^+ < +\infty$$

and

$$\omega(R_2, w, CoB) = \int_{M_u^-} R_2(Du)|\det(u_{ij})|dx \leq \Omega^- < +\infty. \tag{25.127}$$

We now want to obtain the desired estimates for any solution $u(x) \in W_2^n(B) \cap C(\overline{B})$ of the Dirichlet problem (25.119–120) basing on Assumptions A.2-a and A.3-a, and on inequalities (25.123) and (25.126–127). As we know it is sufficient to establish lower estimates, since upper estimates are obtained from those by changing the sign of the solution with the appropriate changes in Assumptions A.2-a and A.3-a.

Let $v(x)$ be the convex support for a solution $u(x)$ of the Dirichlet problem (25.119–120). Below we frequently use the following convex functions and hypersurfaces. Let x_0 be an arbitrary inner point of CoB. We denote by K_{x_0} and $K_{x_0}(v)$ the lateral surfaces of two convex cones with the common base $Co(B \cup \partial B)$, whose vertices are corresponding the points $(x_0, -1)$ and $(x_0, v(x_0))$. Let $K_{x_0}(x)$ and $K_{x_0,v}(x)$ be convex functions, whose graphs are respectively K_{x_0} and $K_{x_0}(v)$. Clearly

$$K_{x_0,v}(x) = |v(x_0)|K_{x_0}(x) \tag{25.128}$$

for all $x \in Co(B \cup \partial B)$.

We denote by $B^*(v, x_0)$ and $B^*(x_0)$ the normal images of the set CoB with respect to convex functions $K_{x_0,v}(x)$ and $K_{x_0}(x)$. Then the following lemma is valid.

Lemma 25.4. *The identity*

$$B^*(v, x_0) = |v(x_0)|B^*(x_0) \tag{25.129}$$

holds for all inner points x_0 of CoB, i.e. $B^*(v, x_0)$ is homothetic to $B^*(x_0)$ with respect to the origin $0'$ of the space P^n and $|v(x_0)|$ is the coefficient of this homothety.

Proof. If $\inf_B u(x) = 0$, then $v(x) = 0$ for all $x \in \overline{B}$ and formula (25.129) is trivial. Thus only the case $\inf_B u(x) < 0$ is interesting. In this case $v(x) < 0$ for any inner point x of CoB.

Since

$$B^*(v, x_0) = \chi K_{x_0,v}(x_0)$$

and

$$B^*(x_0) = \chi K_{x_0}(x_0),$$

then both $B^*(v, x_0)$ and $B^*(x_0)$ are closed convex n-domains in P^n.

Every supporting hyperplane γ of the hypersurface K_{x_0} passes through the point $(x_0, 1)$*). Therefore γ has the following explicit equation

$$z = (p^0, x - x_0) - 1, \tag{25.130}$$

where $p^0 \in B^*(x_0)$. When p^0 passes through the whole set $B^*(x_0)$, then all supporting hyperplanes of K_{x_0} will be taken into account. The $(n-1)$-plane

$$\pi\colon\ (p^0, x - x_0) - 1 = 0,$$

is the intersection of γ and the hyperplane E^n: $z = 0$. Clearly π lies outside the interior of CoB. Hence π and the point $(x_0, v(x_0))$ define the unique supporting hyperplane γ_v of $K_{x_0}(v)$, passing through π and the point $(x_0, v(x_0))$. The elementary calculation shows that

$$z = (|v(x_0)|p^0, x - x_0) - |v(x_0)| \tag{25.131}$$

is the equation of γ_v. Hence $|v(x_0)|p^0$ is a point of $B^*(x_0, v)$. It is evident that the point $|v(x_0)|p^0$ passes through the whole set $|v(x_0)|B^*(x_0)$, if p^0 passes through the whole set $B^*(x_0)$. Thus we obtain the inclusion

$$|v(x_0)|B^*(x_0) \subset B^*(x_0, v). \tag{25.132}$$

The opposite inclusion

$$|v(x_0)|B^*(x_0) \supset B^*(x_0, v) \tag{25.133}$$

can be proved by the replacement of the set $|v(x_0)|B^*(x_0)$ by the set $B^*(x_0, v)$. From (25.132) and (25.133) it follows that identity (25.129) is valid.

The proof of Lemma 25.4 is completed.

Lemma 25.5. *Let $u(x) \in W_2^n(B) \cap C(\overline{B})$ be a solution of the Dirichlet problem (25.119–120) and let Assumptions A.1, A.2-a, A.3-a (see p. 414 and p. 440) be fulfilled. Then the inequalities*

$$-\frac{\|\varphi_1(x)\|_{L^n(B)} + \|c^+(x)u(x)\|_{L^n(B)}}{C_1^{1/n}[\mathrm{vol}(B^*(x_0))]^{1/n}} \leq v(x_0) \leq 0 \tag{25.134}$$

---

*) We recall that the graphs of all functions and hypersurfaces are considered in the Euclidean $(n+1)$-space $E^{n+1} = E^n \times R$.

hold for any inner point x_0 of CoB, where $v(x)$ is the convex support of $u(x)$ and vol($B^*(x_0)$) is the Lebesgue measure of the set $B^*(x_0) \subset P^n$.

Remark. If $v(x_0) = 0$ at least at one inner point x_0 of CoB, then $v(x_0) = 0$ in $\overline{CoB}$ and inequalities (25.134) are trivial.

Proof. Thus only the case $v(x_0) < 0$ for any inner point x_0 of CoB is interesting. From inequality (25.126) it follows that

$$\omega(R_1, v, CoB) \leq \Omega^+, \tag{25.135}$$

where

$$\Omega^+ = \int_B [\varphi_1(x) + c^+(x)|u(x)|]^n dx. \tag{25.136}$$

On the other hand from Assumption A.3-a and Lemma 25.4 it follows that

$$\begin{aligned}\omega(R_1, v, CoB) &= \int_{B^*(v,x_0)} R_1(p)dp \geq C_1 \ \mathrm{vol}[B^*(v, x_0)] \qquad (25.137)\\ &= C_1 \ \mathrm{vol}[|v(x_0)|B^*(v, x_0)| = C_1^n |v(x_0)|^n \ \mathrm{vol}[B^*(x_0)].\end{aligned}$$

Now from (25.135–137) we obtain the inequality

$$|v(x_0)| \leq \frac{(\Omega^+)^{1/n}}{C_1^{1/n}[\mathrm{vol}(B^*(x_0))]^{1/n}}. \tag{25.138}$$

Since $v(x_0) = -|v(x_0)|$, then (25.138) implies the desired estimates (25.134).

The proof of Lemma 25.5 is completed. □

Lemma 25.6. *Let T be a bounded n-dimensional convex body and let S be the centroid of T. We denote by E the solid ellipsoid of the minimal volume, which has the center at S and contains T. Then the solid ellipsoid E_1 homothetic to E with respect to the center S and the coefficient $\left(\frac{1}{n}\right)^{3/2}$ is contained inside T**).

Proof. Let γ_1 and γ_2 be two parallel supporting hyperplanes of T with the opposite outer unit normals. Let h_1 and h_2 be the distances from S to γ_1 and γ_2 respectively. Then

$$\frac{1}{n} \leq \frac{h_1}{h_2} \leq n.$$

The proof of this auxiliary assertion is as follows. Let the hyperplane γ pass through the point S and be parallel to γ_1 and γ_2. We consider the convex cone K with the vertex $A_1 \in \partial T \cap \gamma_1$, which projects the set $T \cap \gamma$ on the hyperplane

---

*) Lemma 25.6 is a well known geometric fact. We present a proof of this lemma due to Pogorelov [7], pp. 91, 92.

γ_2. Note that the base of K lies in the hyperplane γ_2. Clearly the centroid S' of K is nearer to γ_2 than the centroid S of T. But

$$\frac{h_1'}{h_2'} = n$$

for the centroid S' of K. Hence

$$\frac{h_1}{h_2} \leq n. \tag{25.139}$$

The inequality

$$\frac{h_2}{h_1} \leq n$$

can be proved in the similar way. Thus

$$\frac{h_1}{h_2} \geq \frac{1}{n}. \tag{25.140}$$

From (25.139) and (25.140) it follows that

$$\frac{1}{n} \leq \frac{h_1}{h_2} \leq n. \tag{25.141}$$

Without loss of generality we can assume that the ellipsoid E is a n-ball, since the general case can be reduced to this one by a suitable affine transformation. Let r be the radius of E. Let the point Q be the nearest one to S. If Lemma 25.6 is incorrect, then

$$\operatorname{dist}(Q, S)\frac{r}{n^{3/2}}. \tag{25.142}$$

Let γ_1 be a supporting hyperplane of T passing through the point Q and let γ_2 be a supporting hyperplane of T parallel to γ_1. Clearly γ_1 and γ_2 have the opposite unit outer normals. From inequalities (25.141) it follows that

$$b = \operatorname{dist}(S, \gamma_2) < \frac{r}{n^{1/2}}. \tag{25.143}$$

We now introduce Cartesian coordinates $x_1, x_2, \ldots, x_n$ such that S is their origin and the axis x_1 is perpendicular to the hyperplane γ_1 and γ_2. Since T is contained inside the ball E and it is also contained between the hyperplanes γ_1 and γ_2, then

$$|x_1| < b, \qquad x_1^2 + x_2^2 + \cdots + x_n^2 \leq r^2$$

for all $x = (x_1, x_2, \ldots, x_n) \in T$.

Let E' be the following solid ellipsoid

$$\frac{x_1^2}{\alpha^2} + \frac{1}{\beta^2}\sum_{k=2}^{n} x_k^2 = 1,$$

where $\alpha = b\sqrt{n}$, $\beta = \left[\frac{n(r^2-b^2)}{n-1}\right]^{1/2}$. Below two assertions will be proved:

$$\text{a)}\quad T \subset E', \quad \text{b)}\quad \text{vol } E' < \text{vol } E.$$

Since $b < \frac{r}{\sqrt{n}}$, then α, β. Indeed

$$\alpha < \frac{r}{\sqrt{n}}\sqrt{n} = r = \sqrt{\frac{n}{n-1}} \cdot \left(r^2 - \frac{r^2}{n}\right)^{1/2} < \sqrt{\frac{n}{n-1}} \cdot (r^2 - b^2)^{1/2}$$
$$= \beta.$$

If $x = (x_1, x_2, \ldots, x_n) \in T$, then

$$\frac{x_1^2}{\alpha^2} + \frac{1}{\beta^2} \sum_{k=2}^{n} x_k^2 < b^2 \left(\frac{1}{\alpha^2} - \frac{1}{\beta^2}\right)^{1/2} = \frac{r^2}{\beta^2} = 1.$$

Thus $x \in E'$. Hence $T \subset E'$.

From the equalities

$$\frac{\text{vol } E'}{\text{vol } E} = \frac{\alpha \beta^{n-1}}{r^n} = \sqrt{n}\, \frac{b}{r} \left[\frac{n\left(1 - \frac{b^2}{r^2}\right)}{n-1}\right]^{n-1}$$

and the inequality

$$\frac{b}{r} < \frac{1}{\sqrt{n}}$$

it follows that $\frac{\text{vol } E'}{\text{vol } E} < 1$.

But this is impossible, because according to the statement of Lemma 25.6 E has the smallest volume among all solid ellipsoids containing T and having centers at the point S.

The proof of Lemma 25.6 is completed.

Comments to Lemmas 25.5 and 25.6

1) Let $w(x)$ be the concave support of any solution $u(x) \in W_2^n(B) \cap C(\overline{B})$ of the Dirichlet problem (25.119–120). Then the inequalities

$$0 \le w(x_0) \le \frac{\|\varphi_2(x)\|_{L^n(B)} + \|c^+(x)u(x)\|_{L^n(B)}}{C_2^{1/n}[\text{vol } B^*(x_0)]^{1/n}} \tag{25.144}$$

hold for any inner point x_0 of CoB.

The proof of this statement can be obtained from inequalities (25.134) by changing the sign of the solution $u(x)$ with the appropriate changes in Assumptions A.2-a and A.3-a.

2) Inequalities (25.134) and (25.144) are valid without the assumption that the Dirichlet problem (25.119–120) has not more than one solution $u(x) \in W_2^n(B) \cap C(\overline{B})$.

3) Inequalities (25.134) and (25.144) permit to estimate the possible values of the convex and concave supports at every given interior point for any solution $u(x) \in W_2^n(B) \cap C(\overline{B})$ of the Dirichlet problem (25.119–120).

4) The joint application of Lemma 25.6 and inequalities (25.134) and (25.144) permit to obtain the effective and sufficiently simple non-trivial estimates of the C^0-norms for all solutions $u(x)$ of the Dirichlet problem (25.119–120). Below this topic will be studied in Theorems 25.7–10 together with the topic of the boundary values for the convex and concave supports of all solutions $u(x)$ of the Dirichlet problem (25.119–120).

5) The uniqueness theorem for the Dirichlet problem (25.119–120) can be established in the terms of $\|c^+(x)\|_{L^n(B)}$. Below we shall also study interlocking necessary and sufficient conditions of this uniqueness theorem for linear elliptic equations (25.77) and elliptic Euler–Lagrange equations (25.97).

Theorem 25.7. *Let $u(x) \in W_2^n(B) \cap C(\overline{B})$ be a solution of the Dirichlet problem (25.119–120) and let Assumptions A.1, A.2-a, A.3-a be fulfilled. Then the inequalities*

$$\begin{aligned} &-\frac{1}{C_1^{1/n}[\text{vol } B^*(x_0)]^{1/n}}[\|\varphi_1(x)\|_{L^n(B)} \\ &+ \|c^+(x)u(x)\|_{L^n(B)}] \le u(x_0) \\ &\le \frac{1}{C^{1/n}[\text{vol } B^*(x_0)]^{1/n}}[\|\varphi_2(x)\|_{L^n(B)} + \|c^+(x)u(x)\|_{L^n(B)}] \end{aligned} \tag{25.145}$$

hold for any $x_0 \in \overline{B}$, where the set $B^*(x_0)$ is introduced above (see p. 443).

Proof. Let $v(x)$ and $w(x)$ be respectively convex and concave supports of a solution $u(x)$ introduced in Theorem 25.7. Since

$$v(x) \le u(x) \le w(x)$$

for all $x \in \overline{B}$, then from (25.134) and (25.144) it follows that the desired estimates (25.145) are valid.

Theorem 25.7 is proved.

Supplements and Effective Estimates for vol $B^*(x_0)$

Supplement 1. The General Theorem for the Estimate of $\|u(x)\|_{C(\overline{B})}$

Below we use a somewhat more special form of Assumptions A.2-a and A.3-a for obtaining estimates for C^0-norms of solutions of the Dirichlet problem (25.119–120).

A.2-b. In addition to Assumption A.1 (see p. 414) the functions $a_{ik}(x,u,p)$, $b(x,u,p)$ are defined in $B\times R\times P^n$ and in each point $(x,u,p)\in B\times R^-\times P^n$ the inequality

$$\frac{b(x,u,p)}{n[\det(a_{ik}(x,u,p))]^{1/n}}\leq\frac{|\varphi(x)|-c(x)u}{[R(p)]^{1/n}} \tag{25.146}$$

holds if $b(x,u,p)\geq 0$, while in each point $(x,u,p)\in B\times R^+\times P^n$ the inequality

$$\frac{|b(x,u,p)|}{n[\det(a_{ik}(x,u,p))]^{1/n}}\leq\frac{|\varphi(x)|+c(x)u}{[R(p)]^{1/n}} \tag{25.147}$$

holds if $b(x,u,p)\leq 0$, where $R^-=(-\infty,0]$ and $R^+=[0,+\infty)$.

A.3-b. In inequalities (25.146–147) $\varphi(x)\in L^n(B)$ and $R(p)$ is locally summable in P^n and the inequality

$$R(p)\geq C=\text{const }>0 \tag{25.148}$$

holds for all $p\in P^n$. Finally $\|c^+(x)\|_{L^n(B)}<+\infty$, where $c^+(x)$ is the positive part of $c(x)$.

If we choose $\varphi_1(x)$ and $\varphi_2(x)$ as the positive and negative parts of $\varphi(x)$ and set $R_1(x)=R_2(p)=R(p)$ for all $p\in P^n$, then Assumptions A.2-b and A.3-b are satisfied as corollaries of Assumptions A.2-a and A.3-a.

Theorem 25.8. *Let $u(x)\in W_2^n(B)\cap C(\overline{B})$ be a solution of the Dirichlet problem (25.119–120) and let Assumptions A.1, A.2-b, A.3-b be fulfilled. Then the estimate*

$$|u(x_0)|\leq\frac{\|\varphi\|_{L^n(B)}+\|c^+u\|_{L^n(B)}}{C^{1/n}[\text{vol }^*(x_0)]^{1/n}} \tag{25.149}$$

holds for any $x_0\in\overline{B}$.

The proof of this theorem follows directly from (25.145).

Supplement 2. Effective Estimates for vol $B^*(x_0)$

Lemma 25.7. *Let α be a supporting hyperplane of the convex cone K_{x_0} (see p. 443) with the largest slope among all supporting hyperplanes of K_{x_0}. Let $p(\alpha)$ be the slope of α and let $\beta=\alpha\cap E^n$ be a $(n-1)$-dimensional supporting plane of $\partial(CoB)\subset E^n$. Then*

$$\text{vol }B^*(x_0)\geq\frac{\mu_n}{[\text{diam }B]^n}\left[1+\frac{\mu_{n-1}}{n\mu_n}\left(\frac{1-\rho(x_0)}{1+\rho(x_0)}\right)^{\frac{n-1}{2}}\frac{1-\rho(x_0)}{\rho(x_0)}\right], \tag{25.150}$$

where x_0 is any inner point of the set CoB, μ_n is the volume of the n-unit ball and $\text{dist}(x_0,\beta)=\rho(x_0)$ diam B. (Clearly $0<\rho(x_0)<1$ for any inner point x_0 of the set CoB.)

Remark. From inequality (25.150) it follows that vol $B^*(x_0) \to \infty$, if x_0 approaches $\partial(CoB)$.

Proof. Let $T(x_0)$ be the closed n-ball with the center x_0 and the radius equal to diam B. We denote by $K_{T(x_0)}$ the lateral surface of the solid cone of revolution with the base $T(x_0)$ and with the vertex at the point $(x_0, -1)$. Clearly the normal image $K^*_{T(x_0)}$ of $K_{T(x_0)}$ is the n-ball

$$|p| \leq \frac{1}{\text{diam } B} \tag{25.151}$$

in the gradient space P^n.

Let α be a supporting hyperplane of the convex cone K_{x_0} mentioned in the statement of Lemma 25.7 and let $\beta = \alpha \cap E^n$ be a $(n-1)$-dimensional supporting plane of $\partial(CoB)$. Clearly

$$p(\alpha) = \frac{1}{\text{dist}(x_0, \beta)} \tag{25.152}$$

where $p(\alpha)$ is the slope of the hyperplane α. Since

$$\text{dist}(x_0, \beta) \leq \text{dist}(x_0, \beta \cap (CoB)),$$

then

$$\lim p(\alpha) = +\infty, \tag{25.153}$$

if the point $x_0 \in B$ approaches $\partial(CoB)$. Let the point p^* be the normal image of the hyperplane α. Then

$$B^*(x_0) \supset Co\{K^*_{T(x_0)} \cup \{p^*\}\}. \tag{25.154}$$

Clearly p^* is not an inner point of the n-ball $K^*_{T(x_0)}$. We now set

$$\text{dist}(x_0, \beta) = \rho(x_0) \text{ diam } B. \tag{25.155}$$

Clearly

$$0 < \rho(x_0) < 1$$

for all inner points $x_0 \in CoB$ and $\rho(x_0) \to 0$ if $x_0)$ approaches $\partial(CoB)$. Since

$$p^* = p(\alpha) = \frac{1}{\rho(x_0) \text{ diam } B}, \tag{25.156}$$

then from (25.154) and (25.156) it follows that

$$\text{vol } B^*(x_0) \geq \frac{\mu_n}{(\text{diam } B)^n} + \frac{1}{n}\mu_{n-1} q_1^{n-1} q, \tag{25.157}$$

where

$$q = \frac{1}{\operatorname{diam} B}\left[\frac{1}{\rho(x_0)} - 1\right] = \frac{1-\rho(x_0)}{\rho(x_0)\operatorname{diam} B} \tag{25.158}$$

and

$$\begin{aligned} q_1 &= \frac{q}{\operatorname{diam} B}\left(|p^*|^2 - \frac{1}{\operatorname{diam} B^2}\right)^{1/2} \\ &= \left[\frac{1-\rho(x_0)}{1+\rho(x_0)}\right]^{1/2}\frac{1}{\operatorname{diam} B}. \end{aligned} \tag{25.159}$$

Finally from (25.157–159) it follows that the desired estimate (25.150) is valid.
Lemma 25.7 is proved.

Corollary. *Let all conditions of Theorem 25.8 be fulfilled. Then estimate (25.149) leads to more effective estimate*

$$|u(x_0)| \le \frac{\operatorname{diam} B(\|\varphi\|_{L^n(B)} + \|c^*\|_{L^n(B)})\rho^{\frac{1}{n}}(x_0)}{C^{1/n}\mu_n^{1/n}\left[\rho(x_0) + \frac{\mu_n}{\mu_{n-1}}\left(\frac{1-\rho(x_0)}{1+\rho(x_0)}\right)^{n-1/2}(1-\rho(x_0))\right]^{1/n}} \tag{25.160}$$

for any $x_0 \in \overline{B}$.

We are now concerned with other lower estimates for vol $B^*(x_0)$ by means of various invariants for convex domains.

Let x_0 be an arbitrary inner point of CoB. Let the origin of E^n be chosen at the point x_0. We denote by S^{n-1} the $(n-1)$-unit sphere in E^n. Let $\nu \in S^{n-1}$ be a unit vector, i.e. $\nu = \overrightarrow{x_0 s}$, where s is a point of S^{n-1}. We denote by $h(\nu)$ the supporting function of the closed convex surface $\partial(CoB)$. Then $h(\nu) > 0$ for all $\nu \in S^{n-1}$ and

$$(X,\nu) = h(\nu) \tag{25.161}$$

is the equation of the $(n-1)$-dimensional supporting plane $\beta(\nu)$ of $\partial(CoB)$ with the outer unit normal ν. Clearly $(X,\nu) < h(\nu)$ for any inner point X of CoB. Let $\gamma^*(\nu)$ be the supporting hyperplane of the convex cone K_{x_0} passing through the $(n-1)$-plane $\beta(\nu)$. Then the equation for the hyperplane $\gamma(\nu)$ is as follows

$$\alpha z + [(X,\nu) - h(\nu)] = 0, \tag{25.162}$$

where the value of α can be found from the condition $z(x_0) = -1$. Thus from (25.162) it follows that

$$-\alpha - h(\nu) = 0. \tag{25.163}$$

Hence equation (25.162) becomes

$$\gamma(\nu)\colon\ z = \frac{1}{h(\nu)}(X,\nu) = 1. \tag{25.164}$$

Let $\gamma^*(\nu)$ be the normal image of the hyperplane $\gamma(\nu)$. Then

$$\gamma^*(\nu) = \left(\frac{\nu_1}{h(\nu)}, \frac{\nu_2}{h(\nu)}, \ldots, \frac{\nu_n}{h(\nu)}\right) = \frac{1}{h(\nu)}\nu. \tag{25.165}$$

Therefore

$$\text{vol } B^*(x_0) = \frac{1}{n}\int_{S^{n-1}} \frac{d\sigma}{h^n(\nu)}. \tag{25.166}$$

In (25.164–166) the origin in E^n was chosen at the point x_0. If we choose the origin at any point $Q \in \text{int } CoB$, then formulas (25.164–166) become

$$\gamma(\nu) = \frac{(X,\nu)}{h(\nu) - (\overrightarrow{Qx_0},\nu)} - \frac{h(\nu)}{h(\nu) - (\overrightarrow{Qx_0},\nu)} \tag{25.164-a}$$

$$\gamma^*(\nu) = \left(\frac{\nu_1}{h(\nu) - (\overrightarrow{Qx_0},\nu)}, \ldots, \frac{\nu_n}{h(\nu) - (\overrightarrow{Qx_0},\nu)}\right)$$

$$= \frac{1}{h(\nu) - (\overrightarrow{Qx_0},\nu)} \tag{25.165-a}$$

$$\text{vol } B^*(x_0) = \frac{1}{n}\int_{S^{n-1}} [h(\nu) - (\overrightarrow{Qx_0},\nu)]^{-n} d\sigma \tag{25.166-b}$$

where S^{n-1} is the unit sphere with the center Q. In (25.165-a–166-a) x_0 and Q are two arbitrary points of CoB and $h(\nu)$ is the supporting function of $\partial(CoB)$ with respect to the point Q, which we consider as the origin of E^n.

Below we choose Q as the centroid of the convex set CoB.

Example 1. Let B be the n-ball: $|x| \le R$ in E^n. We choose the origin of E^n at the point $0 = (0, 0, \ldots, 0)$. Then $h(\nu) = R$ with respect to 0. Let x_0 be any inner point of CoB, then

$$\text{vol } B^*(x_0) = \frac{1}{n}\int_{S^{n-1}} [R - |x_0|\cos(\overrightarrow{0x_0},\nu)]^{-n} d\sigma \tag{25.167}$$

where $\overrightarrow{0x_0} = x_0$ and S^{n-1} is the unit sphere with the center 0.

Example 2. Let B be the solid ellipsoid

$$\frac{x_1^2}{a_1^2} + \frac{x_2^2}{a_2^2} + \cdots + \frac{x_n^2}{a_n^2} = 1 \tag{25.168}$$

in E^n and let x_0 be any inner point of CoB. Then

$$\text{vol } B^*(x_0) = \frac{1}{n}\int_{S^n} \frac{[\tau(y)]^n d\sigma}{[1 - \langle x_0, y\rangle]^n}, \tag{25.169}$$

where

$$\langle x, y\rangle = \sum_{i=1}^{n} \frac{x_i y_i}{a_i^2}, \quad \|x\|^2 = \langle x, x\rangle, \tag{25.170}$$

and

$$\tau(y) = \left[\frac{y_1^2}{a_1^4} + \cdots + \frac{y_2^2}{a_n^4}\right]^{1/2}. \tag{25.171}$$

In (25.169) the point $y \in \partial B$.

Remark. If $a_1 = a_2 = \cdots = a_n = R$ then (25.174) turns into (25.173).

Proof of the Formula (25.169). Let $y = (y_1, y_2, \ldots, y_n)$ be any point of ∂B, then

$$\|y\|^2 = 1 \tag{25.172}$$

and the tangent $(n-1)$-plane $\beta(y)$ of ∂B at the point y has the equation

$$\frac{X_1 y_1}{a_1^2} + \frac{X_2 y_2}{a_2^2} + \cdots + \frac{X_n y_n}{a_n^2} - 1 = 0. \tag{25.173}$$

Hence the outer unit normal $\nu(y)$ of ∂B at the point y is as follows

$$\nu(y) = \frac{1}{\tau(y)} \left\{\frac{y_1}{a_1^2} \frac{y_2}{a_2^2}, \ldots, \frac{y_2}{a_n^2}\right\}, \tag{25.174}$$

where $\nu(y)$ is defined by (25.171). From (25.173–174) it follows that the supporting function of ∂B with respect to the origin 0 has the following form

$$h(\nu(y)) = [\tau(y)]^{-1}. \tag{25.175}$$

Thus we obtain

$$\operatorname{vol} B^*(x_0) = \frac{1}{n} \int_{S^{n-1}} [(\tau(y))^{-1} - (\overrightarrow{0x_0}, \nu(y))]^{-n} d\sigma \tag{25.176}$$

$$= \frac{1}{n} \int_{S^{n-1}} \frac{(\tau(y))^n d\sigma}{(1 - \langle x_0, y\rangle)^n}. \tag{25.177}$$

Thus the desired formula (25.169) is established.

We now establish the final lower estimate for vol $B^*(x_0)$, where B is a bounded domain in E^n. Let $\overline{CoB}$ be the closed convex hull of a bounded domain in E^n and let S be the centroid of $\overline{CoB}$. We denote by $\overline{Q}$ the closed solid ellipsoid of the minimal volume, which contains (CoB) and has the center at the point S. Let $K_{x_0}(\overline{CoB})$ and $K_{x_0}(\overline{Q})$ be two convex cones with the common vertex $(x_0, -1)$ and bases $\overline{CoB}$ and $\overline{Q}$. Clearly $K_{x_0}(\overline{CoB})$ lies inside $K_{x_0}(\overline{Q})$. We denote by $B^*(x_0)$ and $B^{**}(x_0)$ the normal images of $K_{x_0}(\overline{CoB})$ and $K_{x_0}(\overline{Q})$. Then the following theorem is correct.

Theorem 25.9.

$$\text{vol } B^*(x_0) \geq \frac{1}{n} \int_{S^{n-1}} \frac{[\tau(y)]^n d\sigma}{(1 - \langle x_0, y \rangle)^{1/n}}, \tag{25.178}$$

where all notations made in (25.169–171).

Proof. From our constructions it follows that

$$B^*(x_0) \supset B^{**}(x_0).$$

Then from this inclusion and (25.177) we obtain the desired inequality (25.178).

Supplement 3. Pointwise Estimates of the Possible Values of Solutions for the Dirichlet Problem (2.119–120)

We now want to eliminate the term $\|c^+(x)u(x)\|_{L^n(B)}$ in inequalities (25.145) and (25.149). The positive solution of this problem depends on the mutual integral properties of the functions $c^+(x)$ and vol $B^*(x)$. It is convenient to introduce the following numerical characteristic

$$g(c^+, B) = \|c^+(x)(\text{vol } B^*(x))^{-\frac{1}{n}}\|_{L^n(B)} \tag{25.179}$$

of these mutual properties.

Theorem 25.10. *Let $u(x) \in W_2^n(B) \cap C(\overline{B})$ be a solution of the Dirichlet problem (25.119–120) and let Assumptions A.1, A.2-a, A.3-a and inequality*

$$g(c^+, B) < C^{1/n} \tag{25.180}$$

be fulfilled, where the constant C is defined by inequality (25.148). Then the estimate

$$|u(x_0)| \leq \frac{\|\varphi\|_{L^n(B)}}{(\text{vol } B^*(x_0))^{1/n}(C^{1/n} - g(c^+, B))} \tag{25.181}$$

holds for all $x_0 \in \overline{B}$.

Proof. According to Assumption A.3-a $\|c^+(x)\|_{L^n(B)} < +\infty$. Since $u(x) \in W_2^n(B) \cap C(\overline{B})$, then $\|c^+(x)u(x)\|_{L^n(B)} < +\infty$. Now from (25.149) it follows that

$$\|c^+u\|_{L^n(B)} \leq \frac{\|\varphi\|_{L^n(B)} + \|c^+u\|_{L^n(B)}}{C^{1/n}} g(c^+, B),$$

where $g(c^+, B)$ is defined by (25.179). Hence

$$\|c^+u\|_{L^n(B)} \leq [1 - C^{-1/n} g(c^+, B)] \leq C^{-1/n} \|\varphi\|_{L^n(B)} g(c^+, B).$$

Now from (25.180) it follows that

$$\|c^+u\|_{L^n(B)} \leq \frac{\|\varphi\|_{L^n(B)} g(c^+, B)}{C^{1/n} - g(c^+, B)}. \tag{25.182}$$

Thus we obtain the desired inequality (25.181) directly from (25.149) and (25.182). The proof of Theorem 25.10 is completed.

Supplement 4. Estimates of C^2-norms for Solutions of the Dirichlet Problem (25.119–120)

In this supplement we assume that all the conditions of Theorem 25.10 are fulfilled. Then from estimate (25.181) it follows that

$$\|u(x)\|_{C(\overline{B})} \leq \frac{\|\varphi\|_{L^n(B)}}{[C^{1/n} - g(c^+, B)]} \sup_{\overline{B}} \frac{1}{[\text{vol } B^*(x_0)]^{1/n}}. \tag{25.183}$$

Estimate (25.183) can be replaced by a rather rough one, which will be more effective. Let $\overline{Q}$ be the closed solid ellipsoid of the minimal volume, which contains CoB and has the center at the point S, where S is the centroid of CoB (see p. 455). Let $K_{x_0}(\overline{CoB})$ and $K_{x_0}(\overline{Q})$ be two convex cones with the common vertex $(x_0, -1)$ and bases $\overline{CoB}$ and $\overline{Q}$. We denote by $B^*(x_0)$ and $B^{**}(x_0)$ the normal images of $K_{x_0}(\overline{CoB})$ and $K_{x_0}(\overline{Q})$. Since

$$B^*(x_0) \supset B^{**}(x_0)$$

(see p. 456), then

$$\text{vol } B^*(x_0) \geq \text{vol } B^{**}(x_0)$$

for all $x_0 \in CoB$. Hence we can replace vol $B^*(x_0)$ by vol $B^{**}(x_0)$ in inequality (25.181). Then we obtain the inequality

$$\|u(x)\|_{C(B)} \leq \frac{\|\varphi\|_{L^n(B)}}{[C^{1/n} - g(c^+, B)]} (\text{vol } \overline{Q})^{1/n} \left(\frac{1}{\mu_n}\right)^{\frac{2}{n}} \tag{25.184}$$

since

$$\text{vol } \overline{Q} = \inf_{CoB} \frac{\mu_n^2}{(\text{vol } B^{**}(x_0))}.$$

Finally if $\overline{Q}'$ is the n-dimensional solid ellipsoid homothetic to $\overline{Q}$ with respect to the point S, whose homothety coefficient is $n^{-3/2}$, then $\overline{Q}' \subset CoB$ (see Lemma 25.6). Since

$$\text{vol}(\overline{Q}') = n^{-3n/2} \text{ vol}(\overline{Q})$$

then (25.184) can be rewritten in the following wy

$$\|u(x)\|_{C(\overline{B})} \leq \frac{n^{3/2} (\text{vol } \overline{Q}')^{1/n}}{C^{1/n} - g(c^+, B)} \|\varphi\|_{L^n(B)} \left(\frac{1}{\mu_n}\right)^{\frac{2}{n}}. \tag{25.185}$$

Clearly the factor $(\text{vol } \overline{Q}')^{1/n}$ can be replaced by $(\text{vol}(CoB))^{1/n}$.

25.3 The Improvement of Estimate (25.106–107) and (25.113) for Solutions $u(x) \in W_2^n(B) \cap C(\overline{B})$ of the Dirichlet Problem for Euler–Lagrange Equations

In Subsection 25.1 the estimates (25.106–107) have been obtained as illustrations of the geometric maximum principle for the most general class of quasilinear elliptic equations. But Euler–Lagrange equations, investigated in Subsection 25.1, satisfy Assumptions I, II, III, IV, V. They form somewhat narrower class of quasilinear elliptic equations. Therefore the more powerful technique, developed in Subsection 25.2, can be applied to this class of Euler–Lagrange equations.

In this subsection we are concerned with the improvement of estimates (25.106–107) and (25.113) by means of the technique developed in Subsection 25.2.

We shall study the Dirichlet problem for Euler–Lagrange equation associated with the functional

$$J(u) = \int_B \left[F(Du) + n\left(f(x,u) - \frac{1}{2}c(x)u^2\right)\right] dx \tag{25.99}$$

in the class of functions $u(x) \in W_2^n(B) \cap C(\overline{B})$, where ∂B is a closed continuous hypersurface in E^n and B is a bounded open domain in E^n.

We shall use notations and concepts introduced in §§ 21, 22, 23 and in 25.1. We suppose that Assumptions I, II, III, IV, V are fulfilled.

More precisely we shall study estimates of solutions $u(x) \in W_2^n(B) \cap C(\overline{B})$ of the Dirichlet problem

$$\sum_{i,j=1}^{n} F_{p_i p_j}(Du(x)) u_{ij}(x) = n f_u(x, u(x)) - nc(x)u(x), \tag{25.186}$$

$$u|_{\partial B} = 0. \tag{25.187}$$

We now check up the validity of Assumptions A.1, A.2-b, A.3-b.

1) The convex function $F(p)$ is defined in the entire space P^n and has the first and second differentials almost everywhere in P^n. According to Assumption 1 (see p. 436) the area of the normal mapping $\omega(F, e')$ of the function $F(p)$ is absolutely continuous on the family of Borel subsets e' of P^n, i.e.

$$\omega(F, e') = \int_{e'} \det(F_{p_i p_j}(p)) dp. \tag{25.188}$$

According to Assumption II (see p. 436)

$$\det(F_{p_i p_j}(p)) \geq F_0 = \text{const} > 0 \tag{25.189}$$

almost everywhere in P^n. Hence the form

$$\sum_{i,j=1}^{n} F_{p_ip_j}(p)\xi_i\xi_j$$

is positive definite.

Clearly the function $b(x,u,p) = n(f_u(x,u) - c(x)u)$ is defined in $B \times R$. Thus the Assumption A.1 is fulfilled.

2) The identity

$$\frac{b(x,u,p)}{n(\det(a_{ij}(x,u,p))^{1/n}} = \frac{f_u(x,u) - c(x)u}{(\det F_{p_ip_j}(p))^{1/n}} \tag{25.190}$$

holds in $B \times R \times P^n$. According to Assumption III (see p. 436) the function $f_u(x,u)$ is non-decreasing with respect to u for every fixed $x \in B$. Hence

$$f_u(x,0) \geq f_u(x,u), \tag{25.191}$$

if $u \leq 0$, i.e. $u \in R^- = (-\infty, 0]$, and

$$f_u(x,0) \leq f_u(x,u), \tag{25.192}$$

if $u \geq 0$, i.e. $u \in R^+[0,+\infty)$.

Let (x,u,p) be any point of $B \times R^- \times P^n$, where $b(x,u,p) \geq 0$. Then

$$\frac{b(x,u,p)}{n(\det(a_{ij}(x,u,p)))^{1/n}} = \frac{f_u(x,u) - c(x)u}{(\det(F_{p_ip_j}(p)))^{1/n}} \leq \frac{f_u(x,0) - c(x)u}{(\det(F_{p_ip_j}(p)))^{1/n}}$$
$$\leq \frac{f_u(x,0) - c(x)u}{(\det(F_{p_ip_j}(p)))^{1/n}}. \tag{25.193}$$

Similarly let (x,u,p) be any point of $B \times R^+ \times P^n$, where $b(x,u,p) \leq 0$. Then

$$\frac{b(x,u,p)}{n(\det(a_{ij}(x,u,p)))^{1/n}} = \frac{-f_u(x,u) + c(x)u}{(\det(F_{p_ip_j}(p)))^{1/n}} \leq \frac{-f_u(x,0) + c(x)u}{(\det(F_{p_ip_j}(p)))^{1/n}}$$
$$\leq \frac{|f_u(x,0)| + c(x)u}{(\det(F_{p_ip_j}(p)))^{1/n}}. \tag{25.195}$$

Thus Assumption A.2-b is fulfilled, if we set

$$\varphi(x) = f_u(x,0) \tag{25.195}$$

and

$$R(p) = \det(F_{p_jp_j}(p)) \tag{25.196}$$

in inequalities (25.193) and (25.194).

3) From Assumptions I, II, IV, V it follows that $\varphi(x) = f_u(x,0) \in L^n(B)$, $R(p)$ is locally summable in P^n and

$$R(p) = \det(F_{p_ip_j}(p)) \geq F_0 = \text{const} > 0 \tag{25.197}$$

for all $p \in P^n$.

Thus Assumption A.3-b is fulfilled.

We now consider the improvements of estimates (25.106), (25.107) and (25.113) for solutions of the Dirichlet problem (25.186–187).

Let us give these improvements by the following theorems.

Theorem 30.11. *Let $u(x) \in W_2^n(B) \cap C(\overline{B})$ be a solution of the Dirichlet problem (25.186–187) and let Assumptions I, II, III, IV, V be fulfilled. Then the estimate*

$$|u(x_0)| \leq \frac{\|f_u(x,0)\|_{L^n(B)} + \|c^+(x)u(x)\|_{L^n(B)}}{F_0^{1/n}(\text{vol } B^*(x_0))^{1/n}} \tag{25.198}$$

holds for all $x_0 \in \overline{B}$.

The proof follows directly from (25.189), (25.193), (25.194) and Theorem 25.8.

The effective estimates for vol $B^*(x_0)$ can be taken from Supplement 2, Subsection 25.2.

Theorem 25.12. *Let $u(x) \in W_2^n(B) \cap C(\overline{B})$ be a solution of the Dirichlet problem (25.186–187) and let Assumptions I, II, III, IV, V and the inequality*

$$g(c^+, B) < F_0^{1/n} \tag{25.199}$$

be fulfilled, where $g(c^+, B)$ and the constant F_0 are defined by (25.179) and (25.189). Then the estimate

$$|u(x_0)| \leq \frac{\|f(x,0)\|_{L^n(B)}}{(\text{vol } B^*(x_0))^{1/n}(F_0^{1/n} - g(c^+, B))} \tag{25.200}$$

holds for all $x_0 \in B$.

Theorem 25.13. *Let all conditions of Theorem 25.12 be fulfilled. Then the following estimates*

$$\|u(x)\|_{C(\overline{B})} \leq \frac{\|f(x,0)\|_{L^n(B)}}{F_0^{1/n} - g(c^+, B)} \tag{25.201}$$

$$\sup_{\overline{B}} \frac{1}{(\text{vol } B^*(x_0))^{1/n}},$$

$$\|u(x)\|_{C(\overline{B})} \leq \frac{\|f(x,0)\|_{L^n(B)}}{F_0^{1/n} - g(c^+, B)} (\text{vol } \overline{Q})^{1/n} \left(\frac{1}{\mu_n}\right)^{\frac{2}{n}} \tag{25.202}$$

$$\|u(x)\|_{C(\overline{B})} \leq \frac{n^{\frac{3}{2}}\|f(x,0)\|_{L^n(B)}}{F_0^{1/n} - g(c^+, B)} (\text{fol } \overline{Q}')^{1/n} \left(\frac{1}{\mu_n}\right)^{\frac{2}{n}} \tag{25.203}$$

hold, where the ellipsoids $\overline{Q}$ and $\overline{Q}'$ have been introduced in Supplement 4, Subsection 25.2. Clearly the factor $(\text{vol } \overline{Q}')^{1/n}$ can be replaced by $(\text{vol}(Co(B))^{1/n}$.

The proofs of Theorems 25.12 and 25.13 follow from Theorem 25.11 in the similar way as the proofs as Theorem 25.12 and 25.13 follow from Theorem 25.11 in the similar way as the proofs of Theorem 25.10 and the corresponding assertions in Supplement 4 from Theorem 25.8.

25.4 Final Remarks Relating to Subsections 25.2 and 25.3

1) If $c^+(x) = 0$, then (25.149) and (25.181) coincide (see Theorems 25.8 and 25.10). Analogously (25.198) and (25.200) coincide if $c^+(x) = 0$.
2) In inequalities (25.149) and (25.198) we do not assume that the Dirichlet problems (25.119–120) and (25.186–187) have an unique solution. These inequalities can be applied to non-zero solutions of these problems. If $\|\varphi\|_{L^n(B)} = 0$, then from (25.149) (or (25.198)) it follows that the maximum of the product $u(x)(\text{vol } B^*(x))^{1/n}$ can not be too sharp. The sharpness of this maximum depends on the properties of the function $c^+(x)$.
3) The right parts of (25.149), (25.181), (25.198), (25.200) are functions majorizing solutions of the Dirichlet problems (25.119–120) and (25.186–187). These majorants vanish on ∂B for any convex bounded domain B in E^n.

25.5 Polar Reciprocal Convex Bodies. Estimates and Majorants for Solutions of the Dirichlet Problems (25.119–120) and (25.186–187) Depending on vol(CoB)

First of all we present the necessary concepts and facts relating to polar reciprocal convex compact sets in E^n. Let Q be a convex compact set in E^n and H: $E^n \to R$ be its supporting function. Then

$$H(u) = \max_{x \in Q}(x, u). \tag{25.204}$$

As we know (see § 6) $H(u)$ satisfies two important properties

$$H(\lambda u) = \lambda H(u) \tag{25.205}$$

for all $\lambda \geq 0$ and

$$H(u + v) \leq H(u) + H(v) \tag{25.206}$$

for all $u, v \in E^n$.

Conversely any function H: $E^n \to R$ satisfying (25.205) and (25.206) is the supporting function of a convex set; this set is the intersection of the half-spaces $(x, u) \leq H(u)$. The restriction of H to the unit $(n-1)$-sphere S^{n-1}:

$$h(u) = H(u), \quad u \in S^{n-1} \tag{25.207}$$

is also called the supporting function of Q. Let Q be a convex compact set containing the origin 0, then $h(u)$ is the distance from 0 to the supporting $(n-1)$-plane with the outer normal $u \in S^{n-1}$; if $0 \notin Q$, the numbers $h(u)$ may differ from this distance in sign.

Below we shall consider only convex compact domains Q, containing the origin 0 as an inner point of Q, i.e. some neighborhood of 0 in E^n belongs to int Q. The function

$$r(u) = \max\{\lambda \geq 0, \lambda u \in Q\} \tag{25.208}$$

is called the *radial function* of Q. The function

$$D(u) = \inf\{\lambda \geq 0; u \in \lambda Q\} = \begin{cases} r^{-1}(u) & \text{for} \quad u \neq 0, \\ 0 & \text{for} \quad u = 0 \end{cases} \qquad (25.209)$$

is called the *distance function* of Q.

The distance function has been introduced by Minkowski. He proved that the distance function is convex. Hence the distance function of a convex compact domain Q turns out to be the supporting function of a certain convex set Q^* defined by

$$Q^* = \{v \in E^n; (u,v) \leq 1 \quad \text{for all} \quad u \in Q\}. \qquad (25.210)$$

Q^* is called the *polar* of Q with respect to the unit sphere S^{n-1} centred at origin 0, where 0 is an inner point of Q. The sets Q and Q^* are also called *polar reciprocal* to each other with respect to S^{n-1} centred at the origin 0.

Clearly $(Q^*)^* = Q$ for the polar reciprocal convex compact domain Q containing the origin as an inner point. Finally the distance function of Q^* coincides with the supporting function of Q and conversely.

Let Q be a centrally symmetric convex compact domain in E^n with its centre at the origin 0, then it is known that

$$\left(\frac{1}{n^n}\right)^{1/n} \mu_n^2 \leq \text{vol}(Q)\ \text{vol}(Q^*) \leq \mu_n^2. \qquad (25.211)$$

The left-hand inequality is due to Bambah [1], while that on the right-hand side, which is clearly best possible, is due to Santalo [1]. Recently Bourgain and Milman [1] improved the left-hand inequality in (25.211). They established the inequality

$$k\mu_n^2 \leq \text{vol}(Q)\ \text{vol}(Q^*) \qquad (25.212)$$

where k is a constant depending on n.

If Q does not have the origin 0 as centre, the product $\text{vol}(Q)\ \text{vol}(Q^*)$ can be a suitable choice of Q be made arbitrarily small or arbitrarily large.

In this subsection Lemmas 25.8, 25.9 and 25.10 will be useful. They are due to John, Macbeath and Bambah respectively.

Lemma 25.8. *Let Q be a centrally symmetric n-dimensional convex compact body with centre at the origin 0. Let E be the ellipsoid of the minimal volume centred at 0 that contains Q, then*

$$\text{vol}(E) \leq n^{\frac{n}{2}}\ \text{vol}(Q). \qquad (25.213)$$

Lemma 25.9. *Let Q be a n-dimensional convex compact body (not necessarily with a centre of symmetry) and let P be the parallelepiped of the minimal volume that contains Q. Then*

$$\text{vol}(P) \leq n!\ \text{vol}(Q). \qquad (25.214)$$

Lemma 25.10. *Let P be a parallelepiped that contains the origin 0 and P be a closed set in E^n. Let P^* be the generalized octahedron polar reciprocal to P. Then*

$$\mathrm{vol}(P)\ \mathrm{vol}(P^*) \geq \frac{4^n}{n!}. \tag{25.215}$$

Since the proof of Lemma 25.10 is sufficiently simple and elementary we present it below.

Proof. Let the faces of P be denoted by $F_1, F_2, \dots, F_n$, $F_1', F_2', \dots, F_n'$, where F_i' is parallel to F_i ($i = 1, 2, \dots, n$). By a suitable choice of orthogonal axes we can assume that the foot of the perpendicular from 0 to F_i has coordinates $(a_{i1}, a_{i2}, \dots, a_{ii}, 0, 0, \dots, 0)$. If the origin 0 is an inner point of P, the coordinates of the foot of the perpendicular from 0 to F_i' are $(-k_i a_{i1}, -k_i a_{i2}, \dots, -k_i a_{ii}, 0, \dots, 0)$ where k_i is a positive real number. With this choice of axes the equations of F_i and F_i' take the form

$$a_{i1}x_1 + \cdots + a_{ii}x_i = a_{i1}^2 + \cdots + a_{ii}^2,$$
$$a_{i1}x_1 + \cdots + a_{ii}x_i = -k_i(a_{i1}^2 + \cdots + a_{ii}^2).$$

Clearly

$$\mathrm{vol}(P) = (1 + k_1)\dots(1 + k_n)\prod_{i=1}^{n}(a_{i1}^2 + \cdots + a_{ii}^2)|a_{11}\dots a_{nn}|^{-1}.$$

The poles P_i, P_i' of F_i, F_i' with respect to S^{n-1} have coordinates

$$\left(\frac{a_{i1}}{A_i}, \dots, \frac{a_{ii}}{A_i}, 0, \dots, 0\right)$$

and

$$\left(-\frac{a_{i1}}{k_i A_i}, \dots, -\frac{a_{ii}}{k_i A_i}, 0, \dots, 0\right)$$

respectively, where $A_i = a_{i1}^2 + \cdots + a_{ii}^2$. Since P^* is the convex hull of $P_1, \dots, P_n$, $P_1', \dots, P_n'$, then

$$\mathrm{vol}(P^*) = \frac{1}{n!}|a_{11}\dots a_{nn}|\prod_{i=1}^{n}\left(1 + \frac{1}{k_i}\right)\left(\prod_{i=1}^{n} A_i\right)^{-1}.$$

Thus

$$\mathrm{vol}(P)\ \mathrm{vol}(P^*) = \frac{1}{n!}(1 + k_1)^2 \dots (1 + k_n)^2 (k_1 \dots k_n)^{-1} \geq 4^n (n!)^{-1}$$

if 0 is an inner point of P. Clearly this inequality is correct if $0 \in \partial P$.

Now we consider the proof of the left-hand inequality in (25.211). Let E be the ellipsoid of the minimal volume centred at 0 that contains Q. Then E^*, the ellipsoid polar reciprocal to E, is contained in Q^*. Thus using Lemma 9.8 we have

$$\begin{aligned}\operatorname{vol}(Q)\operatorname{vol}(Q^*) &\geq \operatorname{vol}(Q)\operatorname{vol}(E')\\ &\geq \left(\frac{1}{n^n}\right)^{1/2}\operatorname{vol}(E)\operatorname{vol}(E') = \left(\frac{1}{n^n}\right)^{1/2}\mu_n^2.\end{aligned}$$

Now let K be a n-dimensional convex compact domain that contains the origin 0 then Bambah has proved the inequality

$$\operatorname{vol}(K)\operatorname{vol}(K^*) \geq \frac{4^n}{(n!)^2} \tag{25.216}$$

in the following way: Let P be the parallelepiped of the minimal volume that contains K. Then P contains the origin 0. Clearly $P^* \subset K^*$. Therefore using Lemmas 25.9 and 25.10 we obtain

$$\operatorname{vol}(K)\operatorname{vol}(K^*) \geq \operatorname{vol}(K)\operatorname{vol}(P^*) \geq \frac{1}{n}\operatorname{vol}(P)\operatorname{vol}(P^*) \geq \frac{4^n}{(n!)^2}.$$

Thus inequality (25.216) is proved.

Now we consider the applications of inequalities (25.211–216) to estimate norms for solutions of the Dirichlet problems (25.119–120) and (25.186–187).

First of all we recall the notation used above in Subsection 25.2. We consider the Dirichlet problem (25.119–120) in a bounded domain B in E^n for which ∂B is a continuous closed hypersurface.

Let $\varphi(x)$, $c(x)$ and $R(p)$ be functions appearing in Assumptions A.2-b and A.3-b. Then

$$R(p) \geq C = \text{const } > 0 \tag{25.148}$$

and

$$g(c^+, B) = \|c^+(x)[\operatorname{vol} B^*(x)]^{-1}\|_{L^n(B)}. \tag{25.179}$$

Theorem 25.14. *Let $u(x) \in W_2^n(B) \cap C(\overline{B})$ be a solution of the Dirichlet problem (25.119–120) and let Assumptions A.1, A.2-b, A.3-b and inequality*

$$g(c^+, B) < C^{1/n} \tag{25.180}$$

be fulfilled. Then the estimate

$$\|u(x)\|_{C(\overline{B})} \leq \frac{(n!)^{\frac{2}{n}}}{4}\frac{\|\varphi\|_{L^n(B)}(vol\ Co(B))^{\frac{1}{n}}}{C^{1/n} - g(c^+, B)} \tag{25.217}$$

holds.

Proof. From (25.216) it follows that

$$\operatorname{vol}(\overline{CoB})\operatorname{vol}(B^*(x)) \geq \frac{4^n}{(n!)^2}$$

for all $x \in \overline{CoB}$. Hence

$$\frac{1}{\operatorname{vol} B^*(x)} \leq \frac{(n!)^2}{4^n}\operatorname{vol}(Co(B)) \tag{25.218}$$

for all $x \in \overline{CoB}$. Thus from (25.181) and (25.218) we obtain the desired estimate (25.217).

The proof of Theorem 25.14 is completed.

Remark. Estimate (25.217) is established for any bounded domain B in E^n, the boundary of which is a continuous closed hypersurface. Therefore the closed convex hull $\overline{CoB}$ of B is any convex compact n-dimensional domain in E^n. It is important to note that appearance of the factor $(\operatorname{vol}(\overline{CoB}))^{1/n}$ in estimate (25.217). This factor appears instead of either diam B or the volume of auxiliary convex bodies that contain $\overline{CoB}$, which is taken in degree $\frac{1}{n}$.

If we use only inequalities (25.211–216), then the best estimate for $\|u(x)\|_{C(\overline{B})}$ is inequality (25.217). Inequality (25.217) can be improved if we consider only centrally symmetric domains B. For such domains B the set $\overline{CoB}$ is also centrally symmetric and the centres of B and $\overline{CoB}$ coincide. From (25.183) and (25.212) it follows that the inequality

$$\|u(x)\|_{C(\overline{B})} \leq \frac{1}{k\mu_2^2}\frac{\|\varphi\|_{L^n(B)}}{C^{1/n} - g(c^+, B)}(\operatorname{vol}(\overline{CoB}))^{1/n} \tag{25.219}$$

holds for any bounded centrally symmetric domain B.

§26. The Geometric Maximum Principle for General Quasilinear Elliptic Equations (Continuation and Development)

In this section we develop the geometric maximum principle for solutions $u(x) \in W_2^n(B) \cap C(\overline{B})$ of the Dirichlet problem

$$\sum_{i,k=1}^{n} a_{ik}(x,u,Du)u_{ik} = b(x,u,Du), \tag{26.1}$$

$$u|_{\partial B} = 0 \tag{26.2}$$

for wider classes of quasilinear elliptic equations than those which have been studied in Section 25. We also consider applications of this principle to linear, semilinear and nonlinear Euler–Lagrange elliptic equations.

26.1 The Main Assumptions

We suppose that the following conditions will be fulfilled:

Condition C.1. B is a bounded domain in E^n and ∂B is a closed continuous hypersurface in E^n.

Condition C.2. The quadratic form

$$\sum_{i,k=1}^{n} a_{ik}(x,u,p)\xi_i\xi_k$$

is positive definite in $B \times R \times P^n$. In each point $(x,u,p) \in B \times R^- \times P^n$ the inequality

$$\frac{b(x,u,p)}{n[\det(a_{ik}(x,u,p))]^{1/n}} \leq \frac{\varphi(x,u) - c(x)u + |d(x)|\,|p|}{[R(|p|)]^{1/n}} \tag{26.3}$$

holds if $b(x,u,p) \geq 0$, while in each point $(x,u,p) \in B \times R^+ \times P^n$ the inequality

$$\frac{b(x,u,p)}{n[\det(a_{ik}(x,u,p))]^{1/n}} \geq \frac{\varphi(x,u) - c(x)u - |d(x)|\,|p|}{[R(|p|)]^{1/n}} \tag{26.4}$$

holds if $b(x,u,p) \leq 0$, where $R^- = (-\infty, 0]$, $R^+ = [0, +\infty)$ and the functions $\varphi(x,u)$, $R(|p|)$, $c(x)$ and $d(x)$ satisfy the following conditions:

a) the function $\varphi(x,u)$ is non-decreasing with respect to u for all fixed $x \in B$ and

$$\|\varphi(x,0)\|_{L^n(B)} < +\infty; \tag{26.5}$$

b)

$$\|c^+(x)\|_{L^n(B)} < +\infty; \tag{26.6}$$

where $c^+(x)$ is the positive part of the function $c(x)$;

c)

$$\|d(x)\|_{L^n(B)} < +\infty; \tag{26.7}$$

d) If we set $\rho = |p|, p \in P^n$, then the function $R(|p|)$, participating in inequalities (26.3) and (26.4), becomes $R(\rho)$ of a single variable $\rho \in [0, +\infty)$. We suppose that $R(\rho)$ is strictly positive, locally summable and non-increasing in $[0, +\infty)$. We also suppose that

$$R(0) \leq 1. \tag{26.8}$$

If $\rho \to +\infty$, then two possibilities can occur:

$$\text{a)} \qquad \lim_{\rho \to +\infty} R(\rho) = R(+\infty) > 0 \tag{26.9 – a}$$

and

$$\text{b)} \qquad \lim_{\rho\to+\infty} R(\rho) = R(+\infty) = 0. \qquad (26.9-\text{b})$$

In the case a) we do not need to impose any additional condition on the function $R(p)$. The case b) is significantly more complicated. Its study requires to impose additional conditions on $R(p)$ and use more profound techniques (see Subsection 27.4).

26.2 Concepts and Notations Related to Solutions of the Dirichlet Problem (26.1–2)

Let $u(x) \in W_2^n(B) \cap C(\overline{B})$ be a solution of the Dirichlet problem (26.1–2). We denote by $B^-(u)$ the subset of B, where $u(x) < 0$, and by $B^+(u)$ the subset of B, where $u(x) > 0$. Clearly $B^-(u)$ and $B^+(u)$ are open subsets of B. Let $v(x)$ be the *convex support* of $u(x)$ and M_u^+ be the *convexity set* of $u(x)$ (see Subsection 25.2). As we know M_u^+ is the collection of points $x_0 \in B$ such that $u(x_0) = v(x_0)$. Since $u(x)$ vanishes on ∂B, then $u(x) < 0$ for all $x \in M_u^+$. If the set $B^-(u)$ is not empty, then

$$M_u^+ \subset B^-(u) \qquad (26.10)$$

and

$$\omega(Q, v, CoB) = \int_{M_u^+} Q(Du)\det(u_{ij})dx \qquad (26.11)$$

for any positive locally summable function $Q(p)$ in P^n. Let $K(u, x_0)$ be the convex cone with the base $\partial(CoB)$ and the vertex $(x_0, u(x_0))$. If $x_0 \in M_u^+$, then

$$0 < \int_{K^*(u,x_0)} Q(p)dp \leq \int_{\chi_v(CoB)} Q(p)dp \qquad (26.12)$$
$$= \omega(Q, v, CoB) = \int_{M_u^+} Q(Du)\det(u_{ij})dx,$$

where $K^*(u, x_0)$ and $\chi_v(CoB)$ are the normal images of the cone $K(u, x_0)$ and the convex function $v(x)$. From (26.12) it follows that

$$\text{meas} B^-(u) \geq \text{meas } M_u^+ > 0 \qquad (26.13)$$

if the set $B^-(u)$ is not empty.

In the similar way we can introduce the *concave support* $w(x)$ and the *concavity set* M_u^- for the function $u(x)$. If the set $B^+(u)$ is not empty, then $M_u^- \subset B^+(u)$ and all facts (26.11–13) are correct for the concave support $w(x)$.

We now obtain the basic inequalities leading to the lower and upper estimates for solutions $u(x) \in W_2^n(B) \cap C(\overline{B})$ of the Dirichlet problem (26.1–2). It is sufficient to consider only the lower estimate for these solutions. For such solutions the inequality

$$u(x) < 0 \qquad (26.14)$$

holds for all $x \in M_u^+$ and the inequality

$$0 \le (\det(u_{ij}))^{1/n} \le \frac{b(x,u,Du)}{n[\det(a_{ij}(x,u,Du)]^{1/n}} \tag{26.15}$$

holds almost everywhere in M_u^+. From inequality (26.3) it follows that inequality (26.15) becomes

$$\begin{aligned} 0 \le [\det(u_{ij})]^{1/n} &\le \frac{(\varphi(x,0) + c^+(x)|u(x)|) + |d(x)|\;|Du(x)|}{[R(|Du(x)|)]^{1/n}} \\ &\le \frac{(\varphi^+(x,0) + c^+(x)|u(x)|) + |d(x)|\;|Du(x)|}{[R(|Du(x)|)]^{1/n}} \end{aligned} \tag{26.16}$$

almost everywhere in M_u^+. We now use the Hölder inequality. Then (26.16) becomes

$$\begin{aligned} 0 \le [\det(u_{ij})]^{1/n} &\le [\alpha^n(\varphi^+(x,0) + c^+(x)|u(x)|)^n + |d(x)|^n]^{1/n} \\ &\cdot [\alpha^{-\frac{n}{n-1}} + |Du(x)|^{\frac{n}{n-1}}]^{\frac{n-1}{n}} [R(|Du(x)|)]^{-\frac{1}{n}} \end{aligned} \tag{26.17}$$

almost everywhere in M_u^+, where α is any real positive number.

Thus from (26.11), (26.15) and (26.17) we obtain the following inequality

$$\begin{aligned} \omega(Q_\alpha, v, CoB) &= \int_{M_u^+} Q_\alpha(|Du|)\det(u_{ij})dx \\ &\le \alpha^n[\|\varphi^+(x,0)\|_{L^n(B^-(u))} + \|c^+(x)u(x)\|_{L^n(B^-(u))}]^n \\ &+ \|d(x)\|^n_{L^n(B^-(u))}, \end{aligned} \tag{26.18}$$

where $Q_\alpha(|p|) = R(|p|)[\alpha^{-(n/n-1)} + |p|^{(n/n-1)}]^{-(n-1)}$.

We now introduce a few characteristics related to the solution $u(x) \in W_2^n(B) \cap C(\overline{B})$ of the Dirichlet problem (26.1–2), which we studied in this section. The quantities

$$\sigma^+(u) = \|c^+(x)u(x)\|_{L^n(B^-(u))}, \tag{26.19}$$

$$\sigma^-(u) + \|c^+(x)u(x)\|_{L^n(B^+(u))}, \tag{26.20}$$

$$\sigma(u) = \|c^+(x)u(x)\|_{L^n(B)} \tag{26.21}$$

we called the positive, negative and absolute spectral characteristics of the solution $u(x)$. Clearly

$$\sigma^+(u) \ge 0, \quad \sigma^-(u) \ge 0 \quad \text{and} \quad \sigma(u) \ge \sigma^+(u) + \sigma^-(u). \tag{26.22}$$

We also introduce the following three quantities

$$\phi^+(u) = \|\varphi^+(x,0)\|_{L^n(B^-(u))}, \quad \phi^-(u) = \|\varphi^-(x,0)\|_{L^n(B^+(u))} \tag{26.23}$$

and

$$\phi = \|\varphi(x,0)\|_{L^n(B)}. \tag{26.24}$$

We call ϕ^+, ϕ^- and ϕ the positive, negative and absolute integral curvatures of $u(x)$. Clearly

$$\phi^+(u) \geq 0, \phi^-(u) \geq 0 \quad \text{and} \quad \phi \geq \phi^+(u) + \phi^-(u). \tag{26.25}$$

The functionals

$$k^+(u) = (\mu_n)^{-\frac{1}{n}}[\phi^+(u) + \sigma^+(u)] \tag{26.26}$$

and

$$k^-(u) = (\mu_n)^{-\frac{1}{n}}[\phi^-(u) + \sigma^-(u)] \tag{26.27}$$

are important for deriving the lower and upper estimates of $u(x)$ at any point $x \in B^-(u)$ and $x \in B^+(u)$ respectively, where μ_n is the volume of the n-unit ball in E^n.

26.3 The Development of Techniques Related to Functions $Q_\alpha(|p|)$ and $R(|p|)$

Let $x_0 \in B^-(u)$, then

$$v(x_0) \leq u(x_0) < 0, \tag{26.28}$$

where $u(x) \in W_2^n(B) \cap C(\overline{B})$ is a solution of the Dirichlet problem (26.1–2) and $v(x)$ is the convex support for $u(x)$. The solution $u(x)$ has studied in Subsection 26.2.

Let $K(u, x_0)$ and $K(v, x_0)$ be convex cones with the common base $\partial(CoB)$ and vertices $(x_0, u(x_0))$ and $(x_0, v(x_0))$ respectively. As we know

$$K^*(u, x_0) \subset K^*(V, x_0) \subset \chi_v(CoB),$$

where $K^*(u, x_0), K^*(v, x_0)$ and $\chi_v(CoB)$ are the normal images of convex cones $K(u, x_0)$, $K(v, x_0)$ and convex function $v(x)$. Therefore

$$\omega(Q_\alpha, K^*(u, x_0)) \leq \omega(Q_\alpha, K^*(v, x_0)) \leq \omega(Q_\alpha, v, CoB). \tag{26.29}$$

As we proved in Section 25 the closed ball

$$|p| \leq \frac{|u(x_0)|}{\text{diam } B} \tag{26.30}$$

is contained in $K^*(u, x_0)$. Therefore from (26.18–19), (26.23), (26.26), (26.29–30) it follows that

$$\int_{|p| \leq \frac{|u(x_0)|}{\text{diam } B}} Q_\alpha(|p|)dp \leq \omega(Q_\alpha, v, CoB) \tag{26.31}$$
$$\leq \alpha^n \mu_n [k^*(u)]^n + \|d(x)\|^n_{L^n(B^-(u))}.$$

The elementary calculations lead to the identity

$$\begin{aligned}
&\int_{|p|\le\frac{|u(x_0)|}{\operatorname{diam} B}} Q_\alpha(|p|)dp \\
&= n\mu_n \int_0^{|u(x_0)|/\operatorname{diam} B} \alpha^n \rho^{n-1}[1+(\alpha\rho)^{n/n-1}]^{-(n-1)} R(\rho)d\rho \\
&= (n-1)\mu_n \cdot \int_1^{1+\left(\frac{\alpha|u(x_0)|}{\operatorname{diam} B}\right)^{\frac{n}{n-1}}} \frac{(q-1)^{n-2} R\left(\frac{(q-1)^{\frac{n-1}{n}}}{\alpha}\right)\cdot dq}{q^{n-1}},
\end{aligned}$$

where $q = 1+(\alpha\rho)^{\frac{n}{n-1}}$ and α is any real positive number.

Inequality (26.31) and identity (26.32) are the development of the basic inequality (26.18). The further progress is related to the study of the integral in the right side of identity (26.32). We denote this integral by $I_n(\alpha)$. First of all we decompose integral $I_n(\alpha)$ in the following way:

$$\begin{aligned}
I_n(\alpha) &= (n-1)\mu_n \int^{1+\left(\frac{\alpha|u(x_0)|}{\operatorname{diam} B}\right)^{\frac{n}{n-1}}} \frac{R\left(\frac{(q-1)^{\frac{n}{n-1}}}{q}\right)}{q} dq \\
&\quad - (n-1)\mu_n \int_1^{1+\left(\frac{\alpha|u(x_0)|}{\operatorname{diam} B}\right)^{\frac{n}{n-1}}} \\
&\qquad R\left(\frac{(q-1)^{\frac{n}{n-1}}}{\alpha}\right)\left[\frac{q^{n-2}-(q-1)^{n-2}}{q^{n-1}}\right] dq.
\end{aligned} \tag{26.33}$$

The function

$$n_\alpha(\tau) = \int_1^\tau \left[\frac{q^{n-2}-(q-1)^{n-2}}{q^{n-1}}\right] R\left(\frac{(q-1)^{\frac{n}{n-1}}}{\alpha}\right) dq \tag{26.34}$$

is strictly increasing and continuous in $[1,+\infty)$. Since $n_\alpha(1)=0$, then

$$n_\alpha(\tau) > n_\alpha(1) = 0 \tag{26.35}$$

for all $\tau \in (1,+\infty)$ and all real positive $\alpha > 0$.

The function $R(\rho)$ is non-increasing for $0 \le \rho < +\infty$. Therefore

$$R\left(\frac{(q_1-1)^{n-n-1}}{\alpha}\right) \ge R\left(\frac{(q_2-1)^{n/n-1}}{\alpha}\right) \tag{26.36}$$

for $1 \le q_1 < q_2$, where $q_1 = 1+(\alpha\rho_1)^{\frac{n}{n-1}}$ and $q_2 = 1+(\alpha\rho_2)^{\frac{n}{n-1}}$. From (26.34–36) it follows that

$$\begin{aligned}
0 < n_\alpha(\tau) &\le \sum_{k=1}^{n-2}\binom{n-2}{k}\int_1^\tau \frac{R\left(\frac{(q-1)^{\frac{n}{n-1}}}{\alpha}\right)}{q^{k+1}} dq \\
&\le R(0)\sum_{k=1}^{n-2}\frac{1}{k}\binom{n-2}{k}\left[1-\frac{1}{\tau^k}\right] \le R(0)\sum_{k=1}^{n-2}\frac{1}{k}\binom{n-2}{k}.
\end{aligned}$$

Thus $\lim_{\tau\to+\infty} n_\alpha(\tau) = n_\alpha(+\infty)$ exists and the inequalities

$$0 < n_\alpha(+\infty) < R(0)\sum_{k=1}^{n-2}\frac{1}{k}\binom{n-2}{k}$$

hold for any positive number α.

Finally the inequalities

$$\begin{aligned} 0 &< n_\alpha(\tau) < n_\alpha(+\infty) \\ &= \int_1^{+\infty} R\left(\frac{(q-1)^{\frac{n}{n-1}}}{\alpha}\right)\left[\frac{q^{n-2}-(q-1)^{n-2}}{q^{n-1}}\right]dp \\ &< R(0)\sum_{k=1}^{n-2}\frac{1}{k}\binom{n-2}{k} \le \sum_{k=1}^{n-2}\frac{1}{k}\binom{n-2}{k} \end{aligned}$$

hold for any $\tau \in (1,+\infty)$ and any fixed $\alpha > 0$, because $R(0) \le 1$ according to (26.8).

Remark. From (26.34) it follows that $n_\alpha(\tau) \equiv 0$ for the two-dimensional Dirichlet problem (26.1–2). Therefore estimate (26.37) will only be used for the n-dimensional Dirichlet problem (26.1–2) with $n \ge 3$.

We now introduce the numbers

$$A_n^{\pm} = \begin{cases} \frac{1}{2} + \frac{1}{2\pi}\|d(x)\|^2_{L^2(B^{\mp}(u))}, & n = 2; \\ \frac{n-1}{n}\sum_{s=1}^{n-2}\frac{1}{s}\binom{n-2}{s} + \frac{1}{n} + \frac{1}{n\mu_n}\|d(x)\|^n_{L^n(B^{\mp}(u))}, & n \ge 3. \end{cases} \tag{26.38}$$

Clearly $A_n^{\pm} \le \frac{1}{2}$ for all integers $n \ge 2$.

In (26.32) we replace the variable $\rho \in [0,+\infty)$ by the variable $q \in [1,+\infty)$, where

$$q = 1 + (\alpha\rho)^{\frac{n}{n-1}} \tag{26.39}$$

and α is any positive number. It is evident that

$$\rho = \frac{1}{\alpha}(q-1)^{\frac{n-1}{n}}, \qquad q \in [1,+\infty). \tag{26.40}$$

The choice of the number α will be done below. This choice depends on the properties of the function $R(\rho)$.

26.4 The Main Estimates for Solutions of Problem (26.1–2) if $R(p)$ Satisfies (26.9-a)

We now estimate from below the value of the expression

$$J_n(\alpha) = (n-1)\mu_n \int_1^{1+(\alpha|u(x_0)|[\text{diam } B]^{-1})^{n/n-1}} R[\rho(q)]q^{-1}dq \tag{26.41}$$

in the relation (26.33).

First of all we study the case

$$\lim_{\rho\to+\infty} R(\rho) = R(+\infty) > 0. \tag{26.9-a}$$

Since $R(\rho)$ is non-increasing strictly positive function of ρ, then

$$J_n(\alpha) \geq (n-1)\mu_n R(+\infty)\ln[1 + (\alpha|u(x_0)|[\text{diam } B]^{-1})^{\frac{n}{n-1}}]. \tag{26.42}$$

The number $\alpha > 0$ was arbitrary up to now. We now show that α can be chosen in the way that the estimate

$$u(x_0) \geq -\frac{\text{diam } B}{\mu_n^{\frac{1}{n}}}(\|\varphi^+(x,0)\|_{L^n(B^-(u))} + \|c^+(x)u(x)\|_{L^n(B^-(u))})\exp\left(\frac{A_n^+}{R(+\infty)}\right) \tag{26.43}$$

holds for all $x_0 \in \overline{B}$.

If $B^-(u) = \emptyset$, then $u(x) \geq 0$ in $\overline{B}$. In such a case inequality (26.43) is valid. Therefore it is sufficient to derive inequality (26.43) by the assumption that $B^-(u)$ is not empty and $x_0 \in B^-(u)$.

Since $R(+\infty) \leq R(0) \leq 1$ and $A_n^+ \geq \frac{1}{2}$, then

$$\frac{A_n^+}{R(+\infty)} \geq \frac{1}{2}. \tag{26.44}$$

First we study the general case

$$k^+(u) = \frac{1}{\mu_n^{\frac{1}{n}}}[\|\varphi^+(x,0)\|_{L^n(B^-(u))} + \|c(x)u(x)\|_{n(B^-(u))}] > 0. \tag{26.45}$$

The case $k^+(u) = 0$ will be considered at the end of the proof of inequality (26.43).

From the inequality

$$\frac{|u(x_0)|}{\text{diam } B} \leq k^+(u)e^{1/3} \tag{26.46}$$

it follows that inequality (26.43) is valid at the point x_0. This fact follows from (26.44) and (26.45). Thus only the case

$$\frac{|u(x_0)|}{\text{diam } B} > k^+(u)e^{1/3} \tag{26.47}$$

is interesting for considerations. We now set

$$\alpha = \frac{\text{diam } B}{|u(x_0)|}\cdot\left[\left(\frac{|u(x_0)|}{k^+(u)\text{ diam } B}\right)^{\frac{n}{n-1}} - 1\right]^{\frac{n-1}{n}}. \tag{26.48}$$

From (26.47) and (26.48) it follows that

$$\alpha \geq \frac{1}{k^+(u)}(1 - e^{-\frac{n}{3(n-1)}})^{\frac{n-1}{n}}. \tag{26.49}$$

Relations (26.31), (26.32), (26.37), (26.42), (26.47), (26.48) and the identity

$$1 + \left(\frac{\alpha|u(x_0)|}{\text{diam } B}\right)^{\frac{n}{n-1}} = \left(\frac{|u(x_0)|}{k^+(u) \text{ diam } B}\right)^{\frac{n}{n-1}} \tag{26.50}$$

lead to the inequality

$$\ln \frac{|u(x_0)|}{k^+(u) \text{ diam } B} \leq \frac{n-1}{n} \frac{1}{R(+\infty)} \tag{26.51}$$

$$\left[\sum_{s=1}^{n-2} \frac{1}{s}\binom{n-2}{s} + \frac{1}{n-1}\alpha^n[k^+(u)]^n + \frac{1}{(n-1)\mu_n}\|d(x)\|^n_{L^n(B^-(u))}\right].$$

Since

$$\alpha^n[k^+(u)]^n = \left[1 - \left(\frac{k^+(u) \text{ diam } B}{|u(x_0)|}\right)^{\frac{n}{n-1}}\right]^{n-1} \leq 1, \tag{26.52}$$

then (26.51) becomes

$$\ln\left(\frac{|u(x_0)|}{k^+(u) \text{ diam } B}\right) \leq \frac{1}{R(+\infty)} A_n^+.$$

Thus

$$|u(x_0)| \leq k^+(u)(\text{diam } B) \exp\left(\frac{A_n^+}{R(+\infty)}\right). \tag{26.53}$$

Since

$$u(x_0) < 0,$$

then from (26.53) we obtain the desired inequality (26.43). We now study the case

$$k^+(u) = \frac{1}{\mu_n^{1/n}}[\|\varphi^+(x,0)\|_{L^n(B^-(u))} + \|c^+(x)u(x)\|_{L^n(B^-(u))}] = 0. \tag{26.54}$$

Let $\varepsilon > 0$ be an arbitrary number. Then inequalities (26.17) can be strengthen in the following way

$$0 \leq [\det(u_{ij})]^{1/n} \leq \frac{[\alpha^n(\varphi^+(x,0) + c^+(x)|u(x)| + \varepsilon)^n + |d(x)|^n]^{1/n}}{[\alpha^{-\frac{n}{n-1}} + |Du(x)|^{\frac{n}{n-1}}]^{\frac{n-1}{n}}[R(|Du(x)|)]^{1/n}} \tag{26.55}$$

almost everywhere in M_u^+, where α is any real positive number. From (26.11), (26.15), (26.31), (26.54) and (26.55) it follows that

$$\int_{|p|\le|u(x_0)|(\text{diam } B)^{-1}} Q_\alpha(|p|)dp \le \omega(Q_\alpha, v, CoB) \tag{26.56}$$
$$\le \alpha^n\varepsilon^n \text{ meas } B^-(u) + \|d(x)\|^n_{L^n(B^-(u))},$$

where $Q_\alpha(|p|)$ is defined above by (26.18). We set

$$k(\varepsilon) = \varepsilon\left(\frac{\text{meas } B}{\mu_n}\right)^{1/n}. \tag{26.57}$$

Clearly $k(\varepsilon) > 0$. We now replace $k^+(u)$ by $k(\varepsilon)$ in all formulas and relations (26.40–53). Then we obtain the inequality

$$u(x_0) \ge -k(\varepsilon)[\text{diam } B]\exp\left(\frac{A_n^+}{R(+\infty)}\right) \tag{26.58}$$

for all $x_0 \in B^-(u)$, where $\varepsilon > 0$ is an arbitrary number. Letting $\varepsilon \to 0$ in (26.58) we establish the inequality

$$u(x_0) \ge 0 \tag{26.59}$$

for all $x_0 \in B^-(u)$. Hence the set $B^-(u)$ is empty and $u(x) \ge 0$ for all $x \in \overline{B}$. Thus inequality (26.43) is valid if $k^+(u) = 0$.

The inequality

$$u(x_0) \le \frac{\text{diam } B}{\mu_n^{1/n}}(\|\varphi^-(x,0)\|_{L^n(B^+(u)} + \|c^+(x)u(x)\|_{L^n(B^+(u))})\exp\left(\frac{A_n^-}{R(+\infty)}\right)$$

holds for all $x_0 \in \overline{B}$. The proof of the last inequality can be obtained in the same way as the proof of inequality (26.43).

Thus the following theorem is established.

Theorem 26.1. *Let $u(x) \in W_2^n(B) \cap C(\overline{B})$ be a solution of the Dirichlet problem (26.1–2). Let Conditions C.1, C.2 and relation (26.9-a) be fulfilled. Then the following estimates*

$$-\frac{\text{diam } B}{\mu_n^{1/n}}(\|\varphi^+(x,0)\|_{L^n(B^-(u))} + \|c^+(x)u(x)\|_{L^n(B^-(u))})\exp\left(\frac{A_n^+}{R(+\infty)}\right) \tag{26.60}$$
$$\le u(x_0) \le \frac{\text{diam } B}{\mu_n^{1/n}}(\|\varphi^-(x,0)\|_{L^n(B^+(u))} + \|c^+(x)u(x)\|_{L^n(B^+(u))})\exp\frac{A_n^-}{R(+\infty)}$$

hold for all $x_0 \in \overline{B}$.

Theorem 26.2. *Let all conditions of Theorem 26.1 be fulfilled, then the inequality*

$$|u(x_0)| \leq \frac{\operatorname{diam} B}{\mu_n^{1/n}}[\|\varphi(x,0)\|_{L^n(B)} + \|c^+(x)u(x)\|_{L^n(B)}]\exp\left(\frac{A_n}{R(+\infty)}\right) \tag{26.61}$$

holds, where

$$A_n = \begin{cases} \frac{1}{2} + \frac{1}{2\pi}\|d(x)\|^2_{L^n(B)} & \text{for} \quad n=2; \\ \frac{n-1}{n}\sum\limits_{k=1}^{n-2}\frac{1}{k}\binom{n-2}{k} + \frac{1}{n} + \frac{1}{n\mu_n}\|d(x)\|^n_{L^n(B)} & \text{for} \quad n \geq 3. \end{cases} \tag{26.62}$$

The proof of this theorem follows directly from estimate (26.60) and the following evident inequalities

$$\|\varphi^{\pm}(x,0)\|_{L^n(B^{\mp}(u))} \leq \|\varphi(x,0)\|_{L^n(B)}, \tag{26.63}$$

$$A_n^{\pm} \leq A_n. \tag{26.64}$$

Theorem 26.3. *Let all conditions of Theorem 26.1 be fulfilled and let the inequality*

$$\|c^+(x)\|_{L^n(B)} < \frac{\mu_n^{1/n}}{\exp\frac{A_n}{R(+\infty)}\operatorname{diam} B} \tag{26.65}$$

hold. Then the following estimate

$$|u(x_0)| \leq \frac{\|\varphi(x,0)\|_{L^n(B)}[\operatorname{diam} B]\exp\left(\frac{A_n}{R(+\infty)}\right)}{\mu_n^{1/n} - \|c^+(x)\|_{L^n(B)}[\operatorname{diam} B]\exp\left(\frac{A_n}{R(+\infty)}\right)} \tag{26.66}$$

holds for all $x \in \overline{B}$.

Proof. From inequality (26.61) it follows that

$$\|c^+(x)u(x)\|_{L^n(B)} \leq \frac{\operatorname{diam} B}{\mu_n^{1/n}}(\|\varphi(x,0)\|_{L^n(B)} + \|c^+(x)u(x)\|_{L^n(B)})$$
$$\exp\left(\frac{A_n}{R(+\infty)}\right)\|c^+(x)\|_{L^n(B)}.$$

Thus the following inequality

$$\|c^+(x)u(x)\|_{L^n(B)} \left(1 - \frac{\operatorname{diam} B}{\mu_n^{1/n}}\exp\left(\frac{A_n}{R(+\infty)}\right)\|c^+(x)\|_{L^n(B)}\right)$$
$$\leq \left(\frac{\operatorname{diam} B}{\mu_n^{1/n}}\right)\|\varphi(x,0)\|_{L^n(B)}\exp\left(\frac{A_n}{R(+\infty)}\right)\cdot\|c^+(x)\|_{L^n(B)}. \tag{26.67}$$

We now use condition (26.65) and obtain from inequalities (26.61) and (26.67) the desired estimate (26.66). The proof of Theorem 26.3 is complete. □

26.5 Uniform Estimates for Solutions of the Dirichlet Problem (26.1–2) (Continuation and Development of Subsection (26.4))

In this subsection we derive two-sided estimates for solutions $u(x) \in W_2^n(B) \cap C(\overline{B})$ of the Dirichlet problem (26.1–2) if the function $R(\rho)$ satisfies more general conditions than in Subsection 26.4. In particular every function $R(p)$, which satisfies either (26.9-a) or (26.9-b), also satisfies these conditions. Therefore we need to modify the Assumption d) in Condition C.2 in the following way:

Assumption d′:

d_1': $R(\rho)$ is a positive continuous non-increasing function in $[0, +\infty)$;

d_2':

$$R(0) \leq 1 \tag{26.73}$$

d_3': the function

$$F(T,k) = (T-1)\int_0^1 \frac{R(k \cdot (sT)^{\frac{n-1}{n}})ds}{1+(T-1)s} \tag{26.74}$$

is defined for all $T \in [1, +\infty)$ and for all $k \in [0, +\infty)$; we also assume that the function $s = F(T, k)$ as function of $T \in [1, +\infty)$ is positive, continuous, strictly increasing and

$$\lim_{T \to +\infty} F(T,k) = +\infty \tag{26.75}$$

for any fixed $k \in [0, +\infty)$.

Remarks. 1) From property d_1') it follows that

$$F(T, k_1) \geq F(T, k_2)$$

if $k_1 \leq k_2$. Thus if $\lim_{T\to+\infty} F(T, \bar{k}) = +\infty$, then $\lim_{T\to+\infty} F(T,k) = +\infty$ for all $k \in [0, \bar{k}]$.

2) Since $s = F(T, k)$ is a positive, continuous, strictly increasing function of $T \in [1, +\infty)$ for any fixed $k \in [0, +\infty)$, there exists the inverse

$$T = g_k(s), \quad s \in [0, +\infty) \tag{26.76}$$

for any fixed $k \in [0, +\infty)$. The functions $g_k(s)$ are positive, continuous and strictly increasing

$$g_k(s) \geq g_k(0) = 1 \tag{26.77}$$

for all $k \in [0, +\infty)$.

Theorem 26.4. *Let $u(x) \in W_2^n(B) \cap C(\overline{B})$ be a solution of the Dirichlet problem (26.1–2). Let Condition C.1 and modified Condition C.2 be fulfilled. Then the following inequalities*

$$
\begin{aligned}
&-\frac{\operatorname{diam} B}{\mu_n^{1/n}}(\|\varphi^+(x,0)\|_{L^n(B^-(u))} + \|c^+(x)u(x)\|_{L^n(B^-(u))}) \\
&\cdot g_{k^+(u)}^{\frac{n-1}{n}}\left(\Phi(+\infty) + \frac{\|d\|_{L^n(B^-(u))}}{(n-1)\mu_n}\right) \le u(x_0) \qquad (26.78)\\
&\le \frac{\operatorname{diam} B}{\mu_n^{1/n}}(\|\varphi^-(x,0)\|_{L^n(B^+(u))} + \|c^+(x)u(x)\|_{L^n(L^-(u))}) \\
&\cdot g_{k^-(u)}^{\frac{n-1}{n}}\left(\phi(+\infty) + \frac{\|d\|_{L^n(B^-(u))}}{(n-1)\mu_n}\right)
\end{aligned}
$$

hold for all $x_0 \in \overline{B}$, where

$$
\begin{aligned}
\phi(T) = R(0)\sum_{k=0}^{n-3}\frac{1}{n-k-2}\binom{n-2}{k}\left(1-\frac{1}{T^{n-k-2}}\right) \qquad (26.79)\\
+\left(1-\frac{1}{T}\right)^{n-1}\frac{1}{n-1}
\end{aligned}
$$

for $1 \le T < +\infty$, and

$$
\begin{aligned}
k^+(u) &= \frac{1}{\mu_n^{1/n}}[\|\varphi^+(x,0)\|_{L^n(B^-(u))} \qquad (26.80)\\
&\|c^+(x)u(x)\|_{L^n(B^-(u))}],\\
k^-(u) &= \frac{1}{\mu_n^{1/n}}[\|\varphi^-(x,0)\|_{L^n(B^+(u))} \qquad (26.81)\\
&+\|c^+(x)u(x)\|_{L^n(B^+(u))}].
\end{aligned}
$$

Proof. It is sufficient to derive only the left inequality in (26.78). First we consider the case $k^+(u) > 0$. We start with the inequality

$$
\begin{aligned}
n\mu_n \cdot \int_0^{\frac{|u(x_0)|}{\operatorname{diam} B}} \frac{\alpha^n\rho^{n-1}R(\rho)d\rho}{[1+(\alpha\rho)^{n/n-1}]^{n-1}} \qquad (26.82)\\
\le \alpha^n\mu_n[k^+(u)]^n + \|d(x)\|^n_{L^n(B^-(u))},
\end{aligned}
$$

which follows directly from inequality (26.31) and identity (26.32). Here α is any positive number. There are only two possibilities for the number $u(x_0)$: either

$$|u(x_0)| \le k^+(u)\cdot \operatorname{diam} B \qquad (26.83)$$

or

$$|u(x_0)| > k^+(u) \cdot \text{diam } B. \tag{26.84}$$

From (26.77) and (26.83) it follows that the left inequality in (26.78) is valid. Therefore it is sufficient to study only the case (26.83). Thus below it is always assumed, that

$$|u(x_0)| > k^+(u) \text{ diam } B.$$

We now set

$$\alpha = \frac{\text{diam } B}{|u(x_0)|}\left[\left(\frac{|u(x_0)|}{k^+(u) \text{ diam } B}\right)^{\frac{n}{n-1}} - 1\right]^{\frac{n-1}{n}}. \tag{26.85}$$

According to (26.84) α is a positive number. We now simplify the integral in the left side of inequality (26.82). Let

$$T = \left(\frac{|u(x_0)|}{k^+(u) \text{ diam } B}\right)^{\frac{n}{n-1}}. \tag{26.86}$$

From Assumption (26.84) it follows that $T > 1$. We now introduce a new variable

$$s = \frac{(\alpha\rho)^{\frac{n}{n-1}}}{T-1}, \tag{26.87}$$

where

$$\alpha = \frac{(T-1)^{\frac{n-1}{n}}}{k^+(u)T^{n-1/n}}. \tag{26.88}$$

From (26.81), (26.84), (26.87–88) it follows that the variable s runs from 0 up to 1 and

$$\begin{aligned}
n\mu_n \cdot &\int_0^{\frac{|u(x_0)|}{\text{diam } B}} \frac{\alpha^n \rho^{n-1} R(\rho) d\rho}{[1+(\alpha\rho)^{n/n-1}]^{n-1}} \\
&= (n-1)(T-1)\mu_n \int_0^1 \frac{[(T-1)s]^{n-2} R(k^+(sT)^{\frac{n-1}{n}})ds}{[1+(T-1)s]^{n-1}} \\
&= (n-1)(T-1)\mu_n \left[\int_0^1 \frac{R(k^+(sT)^{\frac{n-1}{n}})ds}{1+(T-1)s}\right. \\
&\quad \left. - \int_0^1 \frac{[1+(T-1)s]^{n-2} - [(T-1)s]^{n-2}}{(1+(T-1)s)^{n-1}} R(k^+(sT)^{\frac{n-1}{n}})ds\right].
\end{aligned}$$

In the right side of the least identity both integrals are positive numbers. Thus inequality (26.81) becomes

$$\begin{aligned}
(T-1)\int_0^1 &\frac{R(k^+(sT)^{\frac{n-1}{n}})ds}{1+(T-1)s} \le (T-1) \\
&\int_0^1 \frac{[1+(T-1)s]^{n-2} - [(T-1)s]^{n-2}}{[1+(T-1)s]^{n-1}} R(k^+(sT)^{\frac{n-1}{n}} ds \\
&+ \frac{(T-1)^{n-1}}{(n-1)T^{n-1}} + \frac{1}{(n-1)\mu_n}\|d(x)\|_{L^n(B^-(u))}.
\end{aligned} \tag{26.90}$$

We now estimate from above the integral in the right side of inequality (26.90):

$$
\begin{aligned}
&(T-1)\int_0^1 \frac{[1+(T-1)s]^{n-2}-[(T-1)s]^{n-2}}{[1+(T-1)s]^{n-1}} R(k^+(sT)^{\frac{n-1}{n}})ds \\
=&(T-1)\sum_{k=0}^{n-3}\binom{n-2}{k}\int_0^1 \frac{(T-1)^k s^k}{[1+(T-1)s]^k}\cdot\frac{R(k^+(sT)^{\frac{n-1}{n}})ds}{[1+(T-1)s]^{n-k-1}} \\
\leq& R(0)\cdot(T-1)\sum_{k=0}^{n-3}\binom{n-2}{k}\int_0^1 \frac{ds}{[1+(T-1)s]^{n-k-1}} \qquad (26.91)\\
=&R(0)\sum_{k=0}^{n-3}\frac{1}{n-k-2}\binom{n-2}{k}\left(1-\frac{1}{T^{n-k-2}}\right).
\end{aligned}
$$

The function

$$\phi(T)=R(0)\sum_{k=0}^{n-3}\frac{1}{n-k-2}\binom{n-2}{k}\left(1-\frac{1}{T^{n-k-2}}\right)+\frac{(T-1)^{n-1}}{(n-1)T^{n-1}} \qquad (26.92)$$

is defined in $[1,+\infty)$. This function is positive, continuous and strictly increasing in $[1,+\infty)$. Clearly $\phi(1)=0$ and

$$0\leq\phi(T)\leq\phi(+\infty)=R(0)\sum_{k=0}^{n-3}\frac{1}{n-k-2}\binom{n-2}{k}+\frac{1}{n-1}<+\infty. \qquad (26.93)$$

Thus

$$\Phi(+\infty)+\frac{1}{(n-1)\mu_n}\|d(x)\|_{L^n(B^-(u))}<+\infty. \qquad (26.94)$$

Finally from (26.74) and (26.90–94) it follows that

$$F(T,k^+)\leq\phi(+\infty)+\frac{1}{(n-1)\mu_n}\|d(x)\|_{L^n(B^-(u))}. \qquad (26.95)$$

According to the properties of the function $F(T,k^+(u))$ and its inverse $g_{k^+(u)}(s)$ we obtain the following inequality

$$T\leq g_{k^+(u)}(\phi(+\infty)+\frac{1}{(n-1)\mu_n}\|d(x)\|_{L^n(B^-(u))}). \qquad (26.96)$$

Since

$$T=\left(\frac{|u(x_))|}{k^+(u)\ \text{diam}\ B}\right)^{\frac{n}{n-1}}$$

and

$$k^+(u)=\frac{1}{\mu_n^{1/n}}[\|\varphi^+(x,0)\|_{L^n(B^-(u))}+\|c^+(x)u(x)\|_{L^n(B^-(u))}],$$

then from (26.96) we obtain the left desired estimate

$$u(x_0) \geq -\frac{\text{diam } B}{\mu_n^{1/n}}(\|\varphi^+(x,0)\|_{L^n(B^-(u))} + \|c^+(x)u(x)\|_{L^n(B^-(u))})$$
$$\cdot \left[g_{k^+(u)}\phi(+\infty) + \frac{\|d\|_{L^n(B^-(u))}}{(n-1)\mu_n}\right]^{\frac{n-1}{n}}.$$

The right desired estimate in (26.78) can be proved in the similar way.

We now study the case

$$k^+(u) = \frac{1}{\mu_n^{1/n}}[\|\varphi^+(x,0)\|_{L^n(B^-(u))} + \|c^+(x)u(x)\|_{L^n(B^-(u))}] = 0. \quad (26.97)$$

Let $\varepsilon > 0$ be an arbitrary number. Then inequalities (26.17) can be strengthen in the following way

$$0 \leq [\det(u_{ij})]^{1/n} \leq \frac{[\alpha^n(\varphi^+(x,0)+c^+(x)\cdot|u(x)|+\varepsilon)^n + |d(x)|^n]^{1/n}}{[\alpha^{-\frac{n}{n-1}} + |Du(x)|^{\frac{n}{n-1}}]^{\frac{n-1}{n}}[R(|Du(x)|)]^{\frac{1}{n}}} \quad (26.98)$$

almost everywhere in M_u^+, where ε is any real positive number. From (26.11). (26.15), (26.31), (26.54) and (26.55) it follows that

$$\int_{|p|\leq\frac{|u(x_0)|}{\text{diam } B}} Q_\alpha(|p|)dp \leq \omega(Q_\alpha, v, CoB) \quad (26.99)$$
$$\leq \alpha^n\varepsilon^n \text{ meas } B^-(u) + \|d(x)\|^n_{L^n(B^-(u))},$$

where $Q_\alpha(|p|)$ is defined by (26.18). We set

$$k(\varepsilon) = \varepsilon\left(\frac{\text{meas } B}{\mu_n}\right)^{\frac{1}{n}}. \quad (26.100)$$

Clearly $k(\varepsilon) > 0$. We now replace $k^+(u)$ by $k(\varepsilon)$ in all formulas and relations (26.82–86). Then we obtain the inequality

$$u(x_0) \gtrsim k(\varepsilon)\cdot \text{diam } B \cdot \left[g_{k(\varepsilon)}\left(\phi(+\infty) + \frac{\|d(x)\|_{L^n(B^-(u))}}{(n-1)\mu_n}\right)\right]^{\frac{n-1}{n}}. \quad (26.101)$$

For any fixed $T \in [1, +\infty)$ the function

$$F(T, k(\varepsilon)) = (T-1)\int_0^1 \frac{R(k(\varepsilon)\cdot(sT)^{\frac{n-1}{n}})ds}{1+(T-1)s} \quad (26.102)$$

is a strictly non-increasing, positive, continuous function of $k(\varepsilon) \in (0, +\infty)$. Let $T(\varepsilon) \in [1, +\infty)$ be a number for which

$$F(T(\varepsilon), k(\varepsilon)) = \phi(+\infty) = \frac{\|d(x)\|_{L^n(B^-(u))}}{(n-1)\mu_n}. \quad (26.103)$$

Clearly there exists only one number $T(\varepsilon)$ for any $\varepsilon > 0$.

The properties of the numbers $T(\varepsilon)$.

Since

$$\phi(+\infty) + \frac{\|d(x)\|_{L^n(B^-(u))}}{(n-1)\mu_n} \geq \phi(+\infty) > \frac{1}{n-1} \quad (26.104)$$

then $T(\varepsilon) > 1$. Below we establish a more precise estimate for $T(\varepsilon)$.

Lemma 26.1. *For all $\varepsilon > 0$ the numbers $T(\varepsilon)$ satisfy the following properties:*

a) *For all $\varepsilon > 0$ the following inequalities*

$$T(\varepsilon) \geq e^{\phi(+\infty)} \geq e^{1/n-1} = const > 1 \tag{26.105}$$

hold.

b) *$T(\varepsilon)$ is an increasing function of $\varepsilon \in (0, +\infty)$.*

c) *For all $\varepsilon \in (0, +1]$ the following inequalities*

$$1 < e^{1/n-1} \leq T(\varepsilon) \leq T(1) = const < +\infty, \tag{26.106}$$

hold.

d) *$\lim_{\varepsilon \to 0+} T(\varepsilon)$ exists and $\lim_{\varepsilon \to 0+} T(\varepsilon) \geq e^{1/n-1} > 1$.*

Proof. a) According to the definition of $T(\varepsilon)$ we obtain

$$\frac{1}{n-1} < p(+\infty) + \frac{\|d(x)\|_{L^n(B^-(u))}}{(n-1)\mu_n} = (T(\varepsilon) - 1)$$
$$\int_0^1 \frac{R(k(\varepsilon) \cdot (sT)^{\frac{n-1}{n}} ds}{1 + (T(\varepsilon) - 1)s}.$$

Since $R(\rho)$ is a non-decreasing continuous function in $[0, +\infty)$, then

$$\frac{1}{n-1} < \Phi(+\infty) \leq [T(\varepsilon) - 1]R(0) \int_0^1 \frac{ds}{1 + (T(\varepsilon) - 1)s} = R(0) \cdot \ln T(\varepsilon).$$

According the Condition d) of the present theorem $0 < R(0) \leq 1$. Therefore the inequalities

$$1 < e^{1/n-1} < e^{\phi(+\infty)} \leq T(\varepsilon)$$

hold for all $\varepsilon > 0$.

The assertion a) is proved.

b) Let $0 < \varepsilon' < \varepsilon$, then from (26.100) it follows that

$$k(\varepsilon') < k(\varepsilon). \tag{26.107}$$

Since $R(\rho)$ is a non-increasing, continuous function in $[0, +\infty)$, then inequality (26.107) leads to the following inequality

$$F(T, k(\varepsilon')) = (T - 1) \int_0^1 \frac{R(k(\varepsilon') \cdot (sT)^{\frac{n-1}{n}})ds}{(1 + (T - 1)s)} \tag{26.108}$$
$$\geq (T - 1) \int_0^1 \frac{R(k(\varepsilon) \cdot (sT)^{\frac{n-1}{n}})ds}{(1 + (T - 1)s)} = F(T, k(\varepsilon)).$$

Thus $F(T, k(\varepsilon))$ *is a non-increasing function of* $\varepsilon > 0$ *for any fixed* $T \in (1, +\infty)$. From the definition of the numbers $T(\varepsilon)$ and $T(\varepsilon')$ it follows that

$$F(T(\varepsilon), k(\varepsilon)) = F(T(\varepsilon'), k(\varepsilon')) = \phi(+\infty) + \frac{\|d(x)\|_{L^n(B^-(u))}}{(n-1)\mu_n} = \text{const } < +\infty. \tag{26.109}$$

Since $\varepsilon' < \varepsilon$, then

$$F(T(\varepsilon), k(\varepsilon)) \leq F(T(\varepsilon), k(\varepsilon')). \tag{26.110}$$

As we now the function $F(T, k(\varepsilon'))$ is a strictly increasing function of T for any fixed value of $k(\varepsilon')$. Hence from (26.109–110) it follows that

$$T(\varepsilon') \leq T(\varepsilon)$$

if $0 < \varepsilon' < \varepsilon$.

The assertion b) is proved.

c) Let $\varepsilon = 1$, then from (26.103) it follows that $T(1)$ is a finite positive number. We can consider $T(1)$ as a constant. From assertions a) and b) of the present lemma we obtain the desired inequalities:

$$1 < e^{1/n+1} < e^{\phi(+\infty)} \leq T(\varepsilon) \leq T(1). \tag{26.111}$$

The assertion c) is proved.

d) The assertion d) follows directly from assertions b) and c).

According to the definition of $T(\varepsilon)$ we obtain

$$T(\varepsilon) = g_{k(\varepsilon)}\left[\phi(+\infty) + \frac{1}{(n-1)\mu_n}\|d(x)\|_{L^n(B^-(u))}\right]. \tag{26.112}$$

Thus inequality (26.101) becomes

$$u(x_0) \geq -k(\varepsilon)[\text{diam } B](T(\varepsilon))^{\frac{n-1}{n}}. \tag{26.113}$$

Letting $\varepsilon \to 0$ in (26.113) we establish the inequality

$$u(x_0) \geq 0 \tag{26.114}$$

for all $x_0 \in B^-(u)$. Hence the set $B^-(u)$ is empty and $u(x) \geq 0$ for all $x \in \overline{B}$. Thus in (26.78) the left inequality is correct for $k^+(u) = 0$.

If $k^-(u) = 0$, then the proof of the validity of the right inequality in (26.78) is similar. The proof of Theorem 26.4 is complete. □

Theorem 26.5. *Let $u(x) \in W_2^n(B) \cap C(\overline{B})$ be a solution of the Dirichlet problem (26.1–2). Let Condition C.1 and modified Condition C.2 be fulfilled. Then the inequality*

$$|u(x_0)| \leq \frac{\text{diam } B}{\mu_n^{1/n}}[\|\varphi(x,0)\|_{L^n(B)} + \|c^+(x)u(x)\|_{L^n(B)}]$$
$$\cdot \left[g_{k(u)}\left(\phi(+\infty) + \frac{\|d(x)\|_{L^n(B)}}{(n-1)\mu_n}\right)\right]^{\frac{n-1}{n}}$$

holds for all $x_0 \in \overline{B}$, where

$$k(u) = \frac{1}{\mu_n^{1/n}}[\|\varphi(x,0)\|_{L^n(B)} + \|c^+(x)u(x)\| + L^n(B)] > 0.$$

If $k(u) = 0$, then the Dirichlet problem (26.1–2) has only zero solution and inequalities (26.113) are fulfilled.

Theorem 26.6. *Let all conditions of Theorem 26.5 be fulfilled and let the inequality*

$$\|c^+(x)\|_{L^n(B)} < \frac{\mu_n^{1/n}}{\text{diam } B\left[g_{k(u)}\left(\phi(+\infty) + \frac{\|d(x)\|_{L^n(B)}}{(n-1)\mu_n}\right)\right]^{\frac{n-1}{n}}} \tag{26.115}$$

holds. Then the following estimate

$$|u(x_0)| \leq \tag{26.116}$$
$$\frac{\|\varphi(x,0)\|_{L^n(B)} \cdot \text{diam } B \cdot \left[g_{k(u)}\left(\phi(+\infty) + \frac{1}{(n-1)\mu_n}\|d(x)\|_{L^n(B)}\right)\right]^{\frac{n-1}{n}}}{\mu_n^{1/n} - \|c(x)\|_{L^n(B)} \cdot \text{diam } B \cdot g_{k(u)}^{\frac{n}{n-1}}\left(\phi(+\infty) + \frac{1}{(n-1)\mu_n}\|d(x)\|_{L^n(B)}\right)}$$

holds for all $x_0 \in \overline{B}$, where $k(u) > 0$.

If $k(u) = 0$, then $\|\varphi(x,0)\|_{L^n(B)} = 0$ and according to Theorems 26.4 and 26.5 $u(x) = 0$ everywhere in B. Thus only the case $k(u) > 0$ is interesting in Condition (26.115) and estimate (26.116).

26.6 Comments to the Modified Condition C.2

1) The modification of Condition C.2 only touches upon Assumption d). The modified Assumption d) is as follows (see Subsection 26.5):

Assumption *d′*:

d'_1: $R(\rho)$ is a positive continuous non-increasing function in $[0, +\infty)$;

d'_2:

$$R(0) \leq 1; \tag{26.73}$$

d_3': the function

$$F(T,k) = (T-1)\int_0^1 \frac{R(k(sT)^{\frac{n-1}{n}})ds}{1+(T-1)s} \tag{26.74}$$

is defined for all $T \in [1,+\infty)$ and we also assume that for all $k \in (0,+\infty)$ the function $v = F(T,k)$, as a function of T, is strictly increasing, positive, continuous and

$$\lim_{T\to+\infty} F(T,k) = +\infty \tag{26.75}$$

for any fixed $k \in (0,+\infty)$.

First of all we are concerned with examples of functions $R(\rho)$ which satisfy Assumption d$'$). One of the simplest examples of such kind is as follows:

$$R(\rho) = \frac{1}{\ln e\left[1+\left(\frac{\rho}{k}\right)^{\frac{n}{n-1}}\left(1-\frac{1}{T}\right)\right]}, \tag{26.117}$$

where $k > 0$ and $T > 1$ are fixed numbers. From equations (26.86–88) it follows that

$$\rho = k(sT)^{\frac{n-1}{n}} \tag{26.118}$$

and

$$(T-1)s = \left(\frac{\rho}{k}\right)^{\frac{n}{n-1}}\left(1-\frac{1}{T}\right), \tag{26.119}$$

where the variable s runs from 0 up to 1. Thus formula (26.117) becomes

$$R(k(sT)^{\frac{n-1}{n}}) = \frac{1}{\ln e[1+(T-1)s]}. \tag{26.120}$$

Hence the function $F(T,k)$ constructed by formula (26.74) is as follows:

$$F(T,k) = (T-1)\int_0^1 \frac{ds}{[1+(T-1)s]\ln[e[1+(T-1)s]]} = \ln\ln(eT), \tag{26.121}$$

where $1 \le T < +\infty$.

Clearly the function $F(T,k) = \ln\ln(eT)$ satisfies all conditions which are imposed on $F(T,k)$ in Assumption A$_3'$. We see that $F(T,k)$ is independent of the number k.

The inverse $g(v)$, $v \in [0,+\infty)$ for the function $F(T,k)$ is independent of k and has the following expression

$$g(v) = e^{e^v - 1}. \tag{26.122}$$

Therefore the estimate for $|u(x_0)|$ in Theorem 26.5 has the following form

$$|u(x_0)| \le \frac{\text{diam } B}{\mu_n^{1/n}} [\|\varphi(x,0)\|_{L^n(B)} + \|c^+(x)u(x)\|_{L^n(B)}] \cdot \left[\exp\left[\exp\left(\phi(+\infty) + \frac{\|d(x)\|_{L^n(B)}}{(n-1)\mu_n}\right) - 1\right]\right] \tag{26.123}$$

for all $x_0 \in \overline{B}$, if $R(\rho) = \frac{1}{\ln e\left[1+(\frac{\rho}{k})^{\frac{n}{n-1}}(1-\frac{1}{T})\right]}$,

$$k = \frac{1}{\mu_n^{1/n}}[\|\varphi(x,0)\|_{L^n(B)} + \|c^+(x)u(x)\|_{L^n(B)}] > 0 \quad \text{and}$$

$$T = \left(\frac{|u(x_0)|}{k \text{ diam } B}\right)^{\frac{n}{n-1}} > 1.$$

The similar expressions have estimates (26.78–79) and (26.116) in Theorems 26.4 and 26.6.

2) The development of example presented in the previous subsection can be done in a few ways. We consider one of them. Let

$$R(\rho) = \frac{1}{\prod\limits_{k=1}^{p} \underbrace{\ln e\left[\ln e\left[\ln e \ldots \left[\ln e\left(1 + \left(\frac{\rho}{k}\right)^{\frac{n}{n-1}}\left(1 - \frac{1}{T}\right)\right] \ldots\right]\right]}_{k-\text{times}}}.$$

It is clear that

$$R(k(sT)^{\frac{n-1}{n}}) = \frac{1}{\prod\limits_{k=1}^{p} \underbrace{\ln e[\ln e \ldots [\ln e[1+(T-1)s]] \ldots]}_{k-\text{times}}}.$$

Therefore

$$F(T,k) = \ln[\ln e[\ln e \ldots [\ln(eT)] \ldots]]. \tag{26.124}$$

The generalizations of formula (26.122) and inequality (26.123) can be directly obtained from (26.124).

3) Theorems 26.4–6 are also correct, if Assumption d_3' is replaced by a somewhat weaker Assumption d_3''.

Assumption d_3'': There exists a positive continuous strictly increasing function $F^*(T,k)$ such that

$$F^*(T,k) \le F(T,k) \tag{26.125}$$

for all $1 \le T < +\infty; 0 < k < +\infty$ and $\lim\limits_{T\to+\infty} F^*(T,k) = +\infty$ for any fixed $k > 0$.

§27. Pointwise Estimates for Solutions of the Dirichlet Problem for General Quasilinear Elliptic Equations

In this section we shall consider solutions $u(x) \in W_2^n(B) \cap C(\overline{B})$ of the Dirichlet problem

$$\sum_{i,k=1}^{n} a_{ik}(x, u, Du)u_{ik} = b(x, u, Du), \tag{27.1}$$

$$u|_{\partial B} = 0. \tag{27.2}$$

We assume that Condition C.1 and the modified Condition C.2 are fulfilled (see Subsections 26.1 and 26.6). Thus B is a bounded domain in an Euclidean n-dimensional space E^n and ∂B is a continuous hypersurface in E^n. We recall that the modified Condition C.2 can be obtained from Condition C.2 if we replace Subcondition d) by Subcondition d").

In Section 26 we established uniform two-sided estimates for solutions $u(x)$ of the Dirichlet problem (27.1–2), see Theorems 26.1–6, which are sharp in the terms of the norms $\|u(x)\|_{C(\overline{B})}$.

In the present section we obtain the more subtle two-sided pointwise estimates for the same solutions $u(x)$, which are sharp at any point $x \in B$. For obtaining these estimates it is required that we introduce a few new geometric concepts and study more deeply the geometric properties of solutions of the Dirichlet problem (27.1–2).

27.1 Integral $I(\lambda, \alpha, x_0)$

Let $E^{n+1} = E^n \times R = \{(x, z) = (x_1, x_2, \dots, x_n; z)\}$ be the $(n+1)$-dimensional space with the canonical norm

$$|(x, z)| = [|x|^2 + z^2]^{1/2} = (x_1^2 + x_2^2 + \dots + x_n^2 + z^2)^{1/2}.$$

Let CoB be the convex hull of B in E^n and let x_0 be any inner point of CoB. We denote by $\overline{CoB}$ the closure of CoB. Let $K(\lambda, x_0)$ be the convex cone with vertex (x_0, λ) and base $\partial(\overline{CoB})$, where λ is any real number. If $\lambda < 0$, then $K(\lambda, x_0)$ is the graph of a convex function; while $K(\lambda, x_0)$ is the graph of a concave function if $\lambda > 0$. We denote by $K^*(\lambda, x_0)$ the normal image of the convex cone $K(\lambda, x_0)$. As we know $K^*(\lambda, x_0)$ is a bounded convex closed domain in the gradient space P^n. $K^*(\lambda, x_0)$ degenerates into the point $p = 0$ if and only if $\lambda = 0$. Finally $K^*(\lambda, x_0)$ can be unbounded closed convex domain if $x_0 \in \partial(\overline{CoB})$.

It is sufficient to obtain only the desired lower estimate for solutions $u(x)$ of the Dirichlet problem (27.1–2), because the upper desired estimate for the same solution $u(x)$ can be established in the similar way. Therefore we consider below only convex cones $K(\lambda, x_0)$ for $\lambda < 0$. We now present a few evident properties of the normal images of convex cones $K(\lambda, x_0)$:

For any $\lambda \leq 0$ the following equation

$$K^*(\lambda, x_0) = |\lambda| K^*(-1, x_0) \tag{27.3}$$

holds. From (27.3) it follows that

$$K^*(\lambda_1, x_0) \subset K^*(\lambda_2, x_0) \tag{27.4}$$

and

$$K^*(\lambda_2, x_0) \backslash K^*(\lambda_1, x_0) \neq \emptyset \tag{27.5}$$

if

$$0 \geq \lambda_1 > \lambda_2 > -\infty.$$

Clearly

$$\bigcup_{\lambda \in (-\infty, 0]} K^*(\lambda, x_0) = P^n \tag{27.6}$$

or in other words

$$\lim_{\lambda \to -\infty} K^*(\lambda, x_0) = P^n.$$

We proved in Section 25, that $K^*(\lambda, x_0)$ contains the n-ball

$$U\left(\frac{|\lambda|}{\text{diam } B}\right): \ |p| \leq \frac{|\lambda|}{\text{diam } B}. \tag{27.7}$$

We fix an arbitrary system of spherical coordinates ρ, θ with the origin $p = 0$ in P^n. Then $\rho = |p|$ for any $p \in P^n$. Let S^{n-1} be the unit hypersphere $|p| = 1$ in P^n and let $p \neq 0$ be any element of P^n, then θ is the collection of the geographical coordinates of $\tilde{p} = \frac{p}{|p|} \in S^{n-1}$. Thus θ can be associated with a point of S^{n-1}. We denote by

$$\rho = \rho_{x_0}(\theta) \tag{27.8}$$

the equation of the closed convex hypersphere $\partial K^*(-1, x_0)$ in the spherical coordinates introduced above. Then for any $\lambda \leq 0$

$$K^*(\lambda, x_0) = |\lambda| K^*(-1, x_0). \tag{27.9}$$

The formula

$$\rho = |\lambda| \rho_{x_0}(\theta) \tag{27.10}$$

defines the equation of $\partial K^*(\lambda, x_0)$. Only the case $\lambda = 0$ corresponds to the degenerate closed convex hypersurface $\partial K^*(0, x_0)$, which is the point $p = 0$.

Finally we present the relationship between the function $\rho_{x_0}(\theta)$ and the support function $h(\theta)$ of the convex domain CoB with respect to an inner point

x_0 of CoB. This relationship is as follows: the convex domain $K^*(\lambda, x_0)$ can be described by the inequality

$$\rho \le \frac{|\lambda|}{h(\theta)}$$

i.e.

$$\rho_{x_0}(\theta) = \frac{1}{h(\theta)}.$$

Here we associate θ with the outer unit normal of the support hyperplane of $K^*(\lambda, x_0)$. The formulas presented above follows directly from the definition of the normal mapping.

Let now $R(|p|)$ be a function introduced in the modified Condition C.2 (see Subsections 26.1 and 26.5). The integral

$$I(\lambda, \alpha, x_0) = \int_{K^*(\lambda, x_0)} \frac{\alpha^n R(|p|)dp}{[1 + (\alpha|p|)^{n/n-1}]^{n-1}} \tag{27.11}$$

will be essentially used in our further considerations. In (27.11) α is any positive number and λ is any negative number. From the properties of the function $R(|p|)$ and inequalities (27.4–5) it follows that $I(\lambda, \alpha, x_0)$ is a strictly decreasing continuous function of $\lambda \in (-\infty, 0]$ for any fixed $\alpha > 0$ and any fixed inner point x_0 of CoB. The function $I(\lambda, \alpha, x_0)$ is positive for $\lambda < 0$ and vanishes at $\lambda = 0$.

In the spherical coordinates ρ, θ, introduced above, the function $I(\lambda, \alpha, x_0)$ has the following representation

$$I(\lambda, \alpha, x_0) = \int_{S^{n-1}} \left[\int_0^{|\lambda|\rho_{x_0}(\theta)} \frac{\alpha^n R(\rho)\rho^{n-1} d\rho}{[1 + (\alpha\rho)^{n/n-1}]^{n-1}} \right] d\sigma_\theta. \tag{27.12}$$

27.2 The Mapping's Mean

Let

$$\varphi\colon [a, +\infty) \to [b, +\infty)$$

be a homeomorphism.

Let S^{n-1} be the unit $(n-1)$-dimensional sphere in P^n with center at the origin $p = 0$ and let $f(\theta)$ be a continuous function on S^{n-1}. We introduce the number $(f)_\varphi$ by the equation

$$(f)_\varphi = \varphi^{-1} \left[\frac{1}{\sigma_n} \cdot \int_{S^{n-1}} \varphi(f(\theta)) d\sigma_\theta \right]. \tag{27.13}$$

The number $(f)_\varphi$ is called the *φ-mean of a function* $f(\theta)$. In (27.13) σ_n is the area of S^{n-1} and $d\sigma_\theta$ is the element of the area of S^{n-1}. Clearly

$$\varphi[(f)_\varphi] = \frac{1}{\sigma_n} \int_{S^{n-1}} \varphi(f(\theta)) d\sigma_\theta.$$

If $\eta = \psi \circ \varphi$ is the composition of homeomorphisms ψ and φ, then the following formula

$$(f)_\eta = (f)_{\psi\circ\varphi} = (\varphi^{-1} \circ \psi^{-1}) \left[\frac{1}{\sigma_n} \int_{S^{n-1}} (\psi \circ \varphi)[f(\theta)] d\sigma_\theta\right]$$
$$= \varphi^{-1}\left[\psi^{-1}\left[\frac{1}{\sigma_n}\int_{S^{n-1}} \psi(\varphi(f(\theta))d\sigma_\theta)\right]\right] \tag{27.14}$$

holds. Certainly in (27.13) and (27.14) the condition $f(\theta) \in [a, +\infty)$ holds for all $\theta \in S^{n-1}$. Below it will be assumed that this condition is fulfilled.

We now fix a homeomorphism φ: $[a, +\infty) \to [b, +\infty)$. We call the function F: $[b, +\infty) \to R$ convex with respect to the mapping $\varphi(f(\theta))$ if the inequality

$$F[t(\varphi(f(\theta_2)) + (1-t)(\varphi(f(\theta_1))] \le tF[\varphi(f(\theta_2))]$$
$$+ (1-t)F[\varphi(f(\theta_1))] \tag{27.15}$$

holds for all $t \in [0, 1]$, where θ_1 and θ_2 are any points of S^{n-1}. This definition can also be stated more obviously. Let $\xi = \varphi(f(\theta))$. Then a function F: $[b, +\infty) \to R$ is convex with respect to φ if and only if the inequality

$$F[(1-t)\xi_1 + t\xi_2] \le (1-t)F(\xi_1) + tF(\xi_2) \tag{27.16}$$

holds for all $t \in [0, 1]$.

If we replace inequality (27.15) by the opposite one, then we obtain the definition of concave functions F with respect to φ.

Lemma 27.1. *If F is a convex function with respect to a mapping φ, then*

$$\frac{1}{\sigma_n}\int_{S^{n-1}} F[\varphi(f(\theta))]d\sigma_\theta \ge F[\varphi(f)_\varphi]. \tag{27.17}$$

Proof. We decompose S^{n-1} by the system of its disjunctive Borel subsets $e_1, e_2, \ldots, e_{m_\delta}$, which satisfy the following properties:

a) $\operatorname{diam} e_s < \delta, \quad s = 1, 2, \ldots, m_\delta;$

b) $\operatorname{meas}_{S^{n-1}} e_s = \dfrac{\sigma_n}{m_\delta}, \quad s = 1, 2, \ldots, m_\delta.$

Hence

$$\frac{1}{\sigma_n}\int_{S^{n-1}} F[\varphi(f(\theta))]d\sigma_\theta = \lim_{\delta\to 0}\frac{1}{m_\delta}\sum_{i=1}^{m_\delta} F[\varphi(f(\theta_i))], \tag{27.18}$$

where $\theta_i \in e_i$, $i = 1, 2, \ldots, m_\delta$. Since F is a convex function with respect to φ, then from (27.15), (27.18) and (27.14) it follows that

$$\frac{1}{\sigma_n}\int_{S^{n-1}} F[\varphi(f(\theta))]d\sigma_\theta \ge \lim_{\delta\to 0} F\left[\frac{1}{\sigma_n}\sum_{i=1}^{m_\delta}\varphi(f(\theta_i))\frac{\sigma_n}{m_\delta}\right]$$
$$= F\left[\frac{1}{\sigma_n}\cdot\int_{S^{n-1}}\varphi(f(\theta))d\sigma_\theta\right] = F[\varphi(f)_\varphi]].$$

The proof of Lemma 27.2 is completed.

Corollary. *Let $\eta = \psi \circ \varphi$ be the homeomorphism considered above. Then inequality(27.17) becomes*

$$\frac{1}{\sigma_n}\int_{S^{n-1}} F[\psi(\varphi(f(\theta)))]d\sigma_\theta \geq F\left[\frac{1}{\sigma_n}\int_{S^{n-1}} \psi(\varphi(f(\theta)))d\sigma_\theta\right]. \qquad (27.19)$$

All concepts and facts presented in Subsection 27.2 can be extended to the compositions of three or more mappings.

27.3 The General Lemma of Convexity

Let x_0 be any inner point of a bounded convex domain CoB. The support function $h(\theta)$ of $\partial(CoB)$ with respect to x_0 satisfies the inequalities

$$0 < h_0 = \text{const} \leq h(\theta) \leq h_1 = \text{const} \ < \ \text{diam } B \qquad (27.20)$$

for all $\theta \in S^{n-1}$. In (27.20) the constant h_0 satisfies the inequality

$$h_0 \geq \text{dist}(x_0, \partial(CoB)) > 0. \qquad (27.21)$$

Thus the radial function

$$\rho = \rho_{x_0}(\theta), \qquad \theta \in S^{n-1}$$

of the closed convex hypersurface $\partial K^*(-1, x_0)$ satisfies the following inequalities

$$0 < \frac{1}{\text{diam } B} < \frac{1}{h_1} \leq \rho_{x_0}(\theta) \leq \frac{1}{h_0} < +\infty \qquad (27.22)$$

for all $\theta \in S^{n-1}$. Inequalities (27.22) follows from the formula

$$\rho_{x_0}(\theta) = \frac{1}{h(\theta)}$$

and inequalities (27.20). Let $a < A$ be two arbitrary numbers. The cases $a = -\infty$ and $A = +\infty$ are not excluded. Let f: $(0, +\infty) \to (a, A)$ be a C^2-function, which satisfies the following conditions:

$$\text{a)} \quad f((0, +\infty)) = (a, A); \qquad (27.23)$$

$$\text{b)} \quad f'(\rho) > 0 \quad \text{for all} \quad \rho \in (0, +\infty); \qquad (27.24)$$

$$\text{c)} \quad f''(\rho) < 0 \quad \text{for all} \quad \rho \in (0, +\infty). \qquad (27.25)$$

Thus $f(\rho)$ is a strictly increasing and strictly concave function in $(0, +\infty)$. Moreover $f'(\rho)$ is strictly decreasing and $\frac{1}{f'(\rho)}$ is strictly increasing in $(0, +\infty)$. Clearly $\zeta = f(\rho)$ has the C^2-inverse

$$\rho = g(\zeta), \qquad \zeta \in (a, A),$$

which satisfies the following conditions:

$$\frac{d\rho}{d\zeta} = \frac{1}{\frac{d\zeta}{d\rho}} = \frac{dg(\zeta)}{d\zeta} > 0 \quad \text{for all} \quad \zeta \in (a, A); \tag{27.26}$$

$$\frac{d^2\rho}{d\zeta^2} = \frac{d^2 g(\zeta)}{d\zeta^2} > 0 \quad \text{for all} \quad \zeta \in (a, A). \tag{27.27}$$

Thus $\rho = g(\zeta)$ is a strictly increasing and strictly convex function in (a, A). Moreover $g'(\zeta)$ is strictly increasing and $\frac{1}{g'(\zeta)}$ is strictly decreasing in (a, A).

Thus the function $\zeta = f(\rho)$ determines a C^2-diffeomorphism of the C^∞-manifold $(0, +\infty)$ onto C^∞-manifold (a, A).

Lemma 27.2 (The First Convexity Lemma). *Let $f(\rho)$ be a C^2-function in $(0, +\infty)$, which satisfies conditions (27.23–35), and let $\lambda < 0$ be any fixed number. Finally let the function*

$$J(\rho) = \frac{R(\rho)}{\rho \cdot f'(\rho)} \tag{27.28}$$

be non-decreasing in the interval $(0, +\infty)$. Then the function

$$H = \int_0^{|\lambda|\rho_{x_0}(\theta)} \frac{R(t)t^{n-1}\,dt}{(\alpha^{-\frac{n}{n-1}} + t^{\frac{n}{n-1}})^{n-1}} \tag{27.29}$$

is convex with respect to C^2-mapping $\zeta = f(|\lambda|\rho_{x_0}(\theta))$ for any fixed $\alpha > 0$ and any positive continuous function $R(p)$ in $(0, +\infty)$.

Proof. The function

$$\eta = f(|\lambda|\rho) \tag{27.30}$$

is continuous and strictly increasing in $(0, +\infty)$. Let $g(\eta) = f^{-1}(\eta)$ be the inverse of function (27.30). Clearly $\eta \in (a, A)$ and

$$|\lambda|\rho_{x_0}(\theta) = g(f(|\lambda|\rho_{x_0}(\theta))) \tag{27.31}$$

for all $\theta \in S^{n-1}$. We introduce the following function

$$Q(\eta) = \int_0^{g(\eta)} \frac{R(t)t^{n-1}\,dt}{(\alpha^{-\frac{n}{n-1}} + t^{\frac{n}{-1}})^{n-1}}, \quad \eta \in (a, A) \tag{27.32}$$

and establish its convexity. Then according to the definition of convexity for functions with respect to mappings (see Subsection 27.2) the proof of Lemma 27.3 will be completed. From (27.32) it follows that

$$\frac{dQ(\eta)}{d\eta} = \frac{[g(\eta)]^n}{(\alpha^{-\frac{n}{n-1}} + [g(\eta)]^{\frac{n}{n-1}})^{n-1}} \cdot \frac{R(g(\eta))}{g(\eta)} \cdot \frac{dg(\eta)}{d\eta}.$$

The function $g(\eta)$ is strictly increasing in (a, A). This fact follows from (27.26) because $\rho = g(\eta)$. Since

$$\eta = f(|\lambda|\rho)$$

is a C^2-diffeomorphism of the C^∞-manifold $(0, +\infty)$ onto the C^∞-manifold (a, A), then

$$\frac{R(g(\eta))}{g(\eta))} \cdot \frac{dg(\eta)}{d\eta} = \frac{R(|\lambda|\rho)}{|\lambda|\rho \cdot f'(|\lambda|\rho)}, \tag{27.33}$$

where $\lambda < 0$ is any fixed number. Since

$$J(\rho) = \frac{R(\rho)}{\rho \cdot f'(\rho)}$$

is a non-decreasing function of ρ (see conditions of Lemma 27.2), then from (27.33) it follows that $\frac{dQ(\eta)}{d\eta}$ is a non-decreasing function of η. Hence $Q(\eta)$ is a convex function in (a, A).

The proof of Lemma 27.2 is complete. □

We now consider two examples.

Example 1. Let $f(\rho) = \ln[(\text{diam } B)\rho]$, where $\rho \in (0, +\infty)$. Then $f(\rho) \in C^\infty(0, +\infty)$ and

$$f(0, +\infty) = (-\infty, +\infty),$$
$$f'(\rho) = \frac{1}{\rho} > 0 \quad \text{for all} \quad \rho \in (0, +\infty),$$
$$f''(\rho) = -\frac{1}{\rho^2} < 0 \quad \text{for all} \quad \rho \in (0, +\infty).$$

Thus $f(\rho)$ is a strictly increasing and strictly concave function in $(0, +\infty)$. Let $R(\rho)$ be a continuous positive function in $(0, +\infty)$. Then

$$J(\rho) = \frac{R(\rho)}{f'(\rho)} = R(\rho).$$

Thus all conditions of Lemma 27.2 will be fulfilled for Example 1, if $R(\rho)$ is a continuous strictly positive non-decreasing function. In particular $R(\rho)$ can be a positive constant. Moreover the inequality

$$R(\rho) \geq R(0) = \lim_{\rho \uparrow 0+} R(\rho) = \text{const} > 0 \tag{27.34}$$

holds for all $\rho \in (0, +\infty)$.

From Lemma 27.2 it follows that the function H is convex with respect to mapping $\ln[(\text{diam } B) \cdot \rho_{x_0}(\theta)]$ if and only if the function $R(\rho)$ satisfies inequality (27.34) for all $\rho \in (0, +\infty)$.

The statement and the proof of Lemma 27.2 show that we consider values of ρ, which are more or equal to $\frac{1}{\text{diam } B}$. In the next subsection of § 27 we shall use the numbers λ satisfying the condition $|\lambda| > 1$. Therefore it is sufficient to consider functions $f(\rho)$ only for $\rho \geq \frac{1}{\text{diam } B}$ and use numbers λ such that $|\lambda| > 1$.

Example 2. Let

$$f(\rho) = \ln\ln[(2e \text{ diam } B)\rho], \quad \rho > \frac{1}{2 \text{ diam } B}. \tag{27.35}$$

Then $f(\rho) \in C^\infty\left[\left(\frac{1}{2 \text{ diam } B}, +\infty\right)\right]$ and

$$f\left[\left(\frac{1}{2 \text{ diam } B}, +\infty\right)\right] = (-\infty, +\infty);$$

$$f'(\rho) = \frac{1}{[\ln(2e \text{ diam } B)\rho] \cdot \rho} > 0 \quad \text{for all} \quad \rho > \frac{1}{2 \text{ diam } B};$$

$$f''(\rho) = -\frac{1 + [\ln(2e \text{ diam } B)\rho]}{[\ln(2e \text{ diam } B)\rho^2]\rho^2} < 0 \quad \text{for all} \quad \rho > \frac{1}{2 \text{ diam } B}.$$

Thus the function $f(\rho)$ is strictly increasing and strictly concave in $\left(\frac{1}{2 \text{ diam } B}, +\infty\right)$.

Let $R(\rho)$ be a continuous positive function in $(0, +\infty)$. Then

$$J(\rho) = \frac{R(\rho)}{\rho \cdot f'(\rho)} = R(\rho)\ln[(2e \text{ diam } B) \cdot \rho] \quad \text{for} \quad \rho > \frac{1}{2 \text{ diam } B}.$$

If all conditions of Lemma 27.2 are fulfilled, then $J(\rho)$ is a non-decreasing function in $\left(\frac{1}{2 \text{ diam } B}, +\infty\right)$. Thus

$$R(\rho) = \frac{J(\rho)}{\ln[(2e \text{ diam } B)\rho]} \tag{27.36}$$

for all $\rho > \frac{1}{2 \text{ diam } B}$. Since

$$J(\rho) \geq J\left[\frac{1}{2 \text{ diam } B}\right] \lim_{\rho\uparrow\frac{1}{(2 \text{ diam } B)^*}} J[\rho] = \text{const} > 0,$$

then the function $R(\rho)$ satisfies the inequality

$$R(\rho) \geq \frac{J\left[\frac{1}{2 \text{ diam } B}\right]}{\ln[(2e \text{ diam } B) \cdot \rho]}. \tag{27.37}$$

Thus from Lemma 27.2 it follows that the function H is convex with respect to the mapping $\ln[\ln(2e \text{ diam } B) \cdot \rho_{x_0}(\theta)]$ if and only if the function $R(\rho)$ satisfies inequality (27.37).

27.4 The Pointwise Estimates for Solutions of the Dirichlet Problem (27.1–2)

In this subsection we derive the desired two-sided pointwise estimates for solutions $u(x) \in W_2^{(n)}(B) \cap C(\overline{B})$ of the Dirichlet problem (27.1–2), which were mentioned in the introduction to § 27. We now formulate all the conditions, which will be used for obtaining these estimates.

Condition C.1. B is a bounded domain in E^n and ∂B is a closed continuous hypersurface in E^n.

Condition C.2". The quadratic form

$$\sum_{i,k=1}^{n} a_{ik}(x,u,p)\xi_i\xi_k$$

is positive definite in $B \times R \times P^n$. In each point $(x,u,p) \in B \times R^- \times P^n$ the inequality

$$\frac{b(x,u,p)}{n[\det(a_{ik}(x,u,p))]^{1/n}} \leq \frac{\varphi(x,u) - c(x)u + |d(x)| \cdot |p|}{[R(|p|)]^{1/n}} \tag{27.38}$$

holds if $b(x,u,p) \geq 0$, while in each point $(x,u,p) \in B \times R^+ \times P^n$ the inequality

$$\frac{b(x,u,p)}{n[\det(a_{ik}(x,u,p))]^{1/n}} \geq \frac{\varphi(x,u) - c(x)u - |d(x)| \cdot |p|}{[R(|p|)]^{1/n}} \tag{27.39}$$

holds if $b(x,u,p) \leq 0$, where $R^- = (-\infty, 0]$, $R^+ = [0, +\infty)$ and the functions $\varphi(x,u)$, $R(|p|)$, $c(x)$ and $d(x)$ satisfy the following conditions:

a) the function $\varphi(x,u)$ is non-decreasing with respect to u for all fixed $x \in B$ and

$$\|\varphi(x,0)\|_{L^n(B)} < +\infty; \tag{27.40}$$

b)

$$\|c^+(x)\|_{L^n(B)} < +\infty, \tag{27.41}$$

where $c^+(x)$ is the positive part of the function $c(x)$;

c)

$$\|d(x)\|_{L^n(B)} < +\infty;$$

d") let $\rho = |p|, p \in P^n$, then the function $R(\rho)$, $\rho \in [0, +\infty)$ is strictly positive, non-increasing and locally summable in $(0, +\infty)$. The main interest is related to the asymptotic behavior of functions $R(\rho)$ as $\rho \to +\infty$. Therefore without loss of generality one can assume that $R(\rho) = R_0 = \text{const} > 0$ for all $\rho \in [0,1]$.

We now recall a few basic concepts and facts established in Sections 25 and 26. Let $u(x) \in W_2^n(B) \cap C(\overline{B})$ be a solution of the Dirichlet problem

(27.1–2). We denote by $B^-(u)$ the subset of B, where $u(x) < 0$, and by $B^+(u)$ the subset of B where $u(x) > 0$. $B^-(u)$ and $B^+(u)$ are open subsets of B. Let $v(x)$ be the *convex support* of $u(x)$ and M_u^+ be the *convexity set* of $u(x)$ (see Subsection 25.2). As we know M_u^+ is the collection of points $x_0 \in B$ such that $u(x_0) = v(x_0)$. Since $u(x)$ vanishes on ∂B, then $u(x) < 0$ for all $x \in M_u^+$. If the set $B^-(u)$ is not empty, then

$$M_u^+ \subset B^-(u) \tag{27.42}$$

and

$$\omega(Q, v, CoB) = \int_{M_u^+} Q(u) \det(u_{ij}) dx \tag{27.43}$$

for any positive locally summable function $Q(p)$ in P^n. Let $K(u, x_0)$ be the convex cone with base $\partial(CoB)$ and vertex $(x_0, u(x_0))$. If $x_0 \in M_u^+$, then

$$\begin{aligned} 0 < \int_{K^*(u,x_0)} Q(p)dp &\leq \int_{\chi_v(CoB)} Q(p)dp \\ &= \omega(Q, v, CoB) = \int_{M_u^+} Q(u) \det(u_{ij}) dx, \end{aligned} \tag{27.44}$$

where $K^*(u, x_0)$ and $\chi_v(CoB)$ are the normal images of the cone $K(u, x_0)$ and the convex function $v(x)$ respectively. From (27.44) it follows that

$$\text{meas } B^-(u) \geq \text{meas } M_u^+ > 0, \tag{27.45}$$

if the set $B^-(u)$ is not empty.

In a similar way we introduce the *concave support* $w(x)$ and the *concavity set* M_u^- for the function $u(x)$. If the set $B^+(u)$ is not empty, then $M_u^- \subset B^+(u)$ and all facts (27.43–45) are correct for the concave support $w(x)$.

In Section 26, the following basic inequalities were established:

$$\begin{aligned} 0 < \int_{K^*(u,x_0)} Q_\alpha(p)dp &\leq \omega(Q_\alpha, v, CoB) \\ &= \int_{M_u^+} Q_\alpha(|Du|) \det(u_{ij}) dx \\ &\leq \alpha^n [\|\varphi^+(x,0)\|_{L^n(B^-(u))} + \|c^+(x)u(x)\|_{L^n(B^-(u))}]^n \\ &\quad + \|d(x)\|^n_{L^n(B^-(u))}, \end{aligned} \tag{27.46}$$

and

$$\begin{aligned} \int_{K^*(u,x_0)} Q_\alpha(p)dp &= \omega(Q_\alpha, v, CoB) \\ &= \int_{M_u^-} Q_\alpha(|Du|) \det(u_{ij}) dx \\ &\leq \alpha^n [\|\varphi^-(x,0)\|_{L^n(B^+(u))} \\ &\quad + \|c^+(x)u(x)\|_{L^n(B^+(u))}]^n + \|d(x)\|^n_{L^n(B^+(u))}, \end{aligned} \tag{27.47}$$

if $x_0 \in M_u^+$ or $x_0 \in M_u^-$ respectively. In (27.46–47) α is any real positive number and

$$Q_\alpha(|p|) = \frac{R(|p|)}{(\alpha^{-\frac{n}{n-1}} + |p|^{\frac{n}{n-1}})^{n-1}}. \tag{27.48}$$

We now estimate the integral

$$\int_{K^*(u,x_0)} Q_\alpha(|p|)dp \tag{27.49}$$

from below. Without loss of generality we assume that $x_0 \in M_u^+$. Thus

$$u(x_0) < 0. \tag{27.50}$$

Then

$$\int_{K^*(u,x_0)} Q_\alpha(|p|)dp = I(u(x_0), \alpha, x_0) = \tag{27.51}$$
$$\int_{S^{n-1}} \left(\int_0^{|u(x_0)|\rho_{x_0}(\theta)} \frac{R(t)t^{n-1}dt}{(\alpha^{-\frac{n}{n-1}} + t^{\frac{n}{n-1}})^{n-1}} \right) d\sigma_\theta,$$

where

$$\rho = \rho_{x_0}(\theta) \tag{27.51}$$

is the equation of the closed convex hypersurface $\partial K^*(-1, x_0)$ in spherical coordinates in P^n. Let f: $(z, +\infty) \to (a, A)$*) be a C^2-function, which satisfies the following conditions:

$$\text{a)}\quad 0 < z \le \frac{1}{2 \text{ diam } B} \quad \text{and}$$
$$f((z, +\infty)) = (a, A); \tag{27.52}$$
$$\text{b)}\quad f'(\rho) > 0 \quad \text{for all} \quad \rho > z; \tag{27.53}$$
$$\text{c)}\quad f''(\rho) < 0 \quad \text{for all} \quad \rho > z. \tag{27.54}$$

The next condition describes mutual properties of functions $f(\rho)$ and $R(\rho)$.

Condition C.2. The function

$$J(\rho) = \frac{R(\rho)}{\rho f'(\rho)} \tag{27.55}$$

is non-decreasing for $\rho > z$, where $R(\rho)$ is a function satisfying Condition C.2" and $f(\rho)$ is a C^2-function satisfying conditions (27.52–54).

---

*) The cases $a = -\infty$ and $A = +\infty$ are not excluded.

Lemma 27.3. *Let $u(x) \in W_2^n(B) \cap C(\overline{B})$ be a solution of the Dirichlet problem (27.1–2). Let $x_0 \in M_u^+$ and $u(x_0) \leq -1$. Finally let Conditions C.1, C.2", C.3 be fulfilled. Then the inequality*

$$I(u(x_0), \alpha, x_0) \geq \sigma_n \int_0^{g[F]} \frac{R(t)t^{n-1}dt}{(\alpha^{-\frac{n}{n-1}} + t^{\frac{n}{n-1}})^{n-1}} \tag{27.56}$$

holds for any positive number α, where g is the inverse for the function $\zeta = f(|u(x_0)| \cdot \rho)$, $\rho > z$, σ_n is the area of S^{n-1} and

$$F = f[(|u(x_0)|\rho_{x_0}(\theta))_f] = \frac{1}{\sigma_n} \int_{S^{n-1}} f(|u(x_0)|\rho_{x_0}(\theta)) d\sigma_\theta \tag{27.57}$$

is the value of the function f at the f-mean of the function $|u(x_0)|\rho_{x_0}(\theta)$ (see (27.13), (27.14)).

Proof. From Lemma 27.2 and remarks, relating to this lemma, it follows that the function

$$H = \int_0^{|u(x_0)|\rho_{x_0}(\theta)} \frac{R(t)t^{n-1}dt}{(\alpha^{-\frac{n}{n-1}} + t^{\frac{n}{n-1}})^{n-1}} \tag{27.58}$$

is convex with respect to the C^2-function

$$f(|u(x_0)|\rho_{x_0}(\theta)) \tag{27.59}$$

for any fixed $\alpha > 0$. Since $|u(x_0)| \geq 1$, then $|u(x_0)|\rho > z$ for all $\rho > z$. Let g be the inverse for the function $\zeta = f(|u(x_0)|\rho)$, where $\rho > z$. Clearly

$$|u(x_0)|\rho = g(f[|u(x_0)|\rho]) \tag{27.60}$$

for all $\rho > z$. Therefore

$$|u(x_0)|\rho_{x_0}(\theta) = g(f[|u(x_0)|\rho_{x_0}(\theta)]). \tag{27.61}$$

From (27.56) and (27.61) we obtain that

$$I(u(x_0), \alpha, x_0) = \int_{S^{n-1}} \left(\int_0^{g(f[|u(x_0)|\rho_{x_0}(\theta)])} \frac{R(t)t^{n-1}dt}{(\alpha^{-\frac{n}{n-1}} + t^{\frac{n}{n-1}})^{n-1}} \right) d\sigma_\theta.$$

Since the function H is convex with respect to function (27.59), then from Lemma 27.1 it follows that

$$I(u(x_0), \alpha, x_0) \geq \sigma_n \int_0^{g(F)} \frac{R(t)t^{n-1}dt}{(\alpha^{-\frac{n}{n-1}} + t^{\frac{n}{n-1}})^{n-1}}, \tag{27.62}$$

where F is the value of the function f at the f-mean of the function $|u(x_0)|\rho_{x_0}(\theta)$ (see the statement of Lemma 27.3).

Lemma 27.3 is proved.

Remarks to Lemma 27.3

1) If $|u(x_0)| \geq 1$, then the inequality

$$g[F] \geq \frac{|u(x_0)|}{\text{diam } B} \tag{27.63}$$

holds.

Proof. From (27.22) it follows that

$$\rho_{x_0}(\theta) \geq \frac{1}{\text{diam } B}.$$

Since $|u(x_0)| \cdot \rho_{x_0}(\theta) \geq \frac{1}{\text{diam } B}$, then from (27.57) and (27.53) we obtain

$$F \geq \frac{1}{\sigma_n} f\left(|u(x_0)| \frac{1}{\text{diam } B}\right) \sigma_n = f\left(|u(x_0)| \frac{1}{\text{diam } B}\right).$$

Hence

$$g[F] \geq g\left[f\left(|u(x_0)| \frac{1}{\text{diam } B}\right)\right] = \frac{|u(x_0)|}{\text{diam } B}.$$

Thus inequality (27.63) is proved.

2) From (27.63) it follows that

$$\lim_{|u(x_0)| \to +\infty} g[F] = +\infty. \tag{27.64}$$

3) *A more subtle estimate for* $g[F]$. We now are concerned with a more subtle estimate for $g[F]$ than estimate (27.63). Such estimate should reflect the position of a point x_0 with respect to the boundary of CoB. We only present the two-dimensional case in order to avoid superfluous calculations. However the general n-dimensional case does not significantly differ from the two-dimensional one.

Let $\overline{CoB}$ be the closed convex hull of B. We fix a point $x_0 \in M_u^+ \subset \text{int}(\overline{CoB})$. Let $k(\theta)$ be the support function of $\partial(\overline{CoB})$ with respect to x_0. Clearly

$$h_0 = \inf_{\theta \in S^1} k(\theta), \tag{27.65}$$

where S^1 is the unit circle in E^2 with center x_0. It is well known that

$$\frac{h_0}{\text{diam } B} \leq \frac{1}{2}. \tag{27.66}$$

We denote by $K(-1,x_0)$ the convex cone with base $\partial(\overline{CoB})$ and vertex $(x_0,-1)$, which is located in the space $E^3 = E^2 \times R$. Let $K^*(-1,x_0)$ be the normal image of $K(-1,x_0)$. As we know $K^*(-1,x_0)$ is a closed bounded convex domain in P^2, which contains the following closed disk

$$U = \left\{|p| \leq \frac{1}{diam\ B}, p \in P^n\right\}. \tag{27.67}$$

Let α be a support line of $\partial(\overline{CoB})$ in E^2 such that

$$\operatorname{dist}(x_0,\alpha) = h_0. \tag{27.68}$$

We denote by β a support plane of the cone $K(-1,x_0)$ in E^3, which passes through the support line α. If $p(\beta)$ be the normal image of β, then

$$p(\beta) \in K^*(-1,x_0) \tag{27.69}$$

and

$$\operatorname{dist}(p(\beta),0') = \frac{1}{h_0}, \tag{27.70}$$

where $0'$ is the origin of P^2. Thus

$$Q = \overline{Co(p(\beta) \cup U)} \tag{27.71}$$

is a closed bounded convex set, which is contained in $K^*(-1,x_0)$. According to our notation

$$\rho = \rho_{x_0}(\theta) = \frac{1}{k(\theta)}$$

is the equation of $K^*(-1,x_0)$ in polar coordinates in P^2. Let

$$\rho = q(\theta) \tag{27.72}$$

be the equation of ∂Q in the same polar coordinates in P^2. Then (27.73)

$$0 < q(\theta) \leq \rho_{x_0}(\theta) = \frac{1}{k(\theta)}.$$

We now assume that the polar axis passes through the vector $\overrightarrow{0'p(\beta)}$. Then

$$\rho_{x_0}(\theta) = \begin{cases} \frac{1}{(\text{diam } B)\cos(\alpha-\theta)} & \text{if } 0 \leq \theta \leq \alpha; \\ \frac{1}{\text{diam } B} & \text{if } \alpha < \theta < 2\pi - \alpha; \\ \frac{1}{(\text{diam } B)\cos[(\theta+\alpha)-2]} & \text{if } 2\pi - \alpha \leq \theta \leq 2\pi, \end{cases} \tag{27.74}$$

where the angle α is defined by the formula

$$\alpha = \arccos \frac{h_0}{\text{diam } B}. \tag{27.75}$$

From (27.66) it follows that the angle α satisfies the following inequalities

$$\frac{\pi}{3} \le \alpha < \frac{\pi}{2}.$$

Thus

$$\begin{aligned} F &= \frac{1}{2\pi} \int_{S^1} f(|u(x_0)|\rho_{x_0}(\theta))d\theta \\ &= \frac{1}{2\pi} \int_0^{2\pi} f\left[\frac{|u(x_0)|}{\text{diam } B}\right] d\theta \\ &\quad + \frac{1}{\pi} \int_0^{\alpha} \left[f\left(\frac{|u(x_0)|}{\text{diam } B} \cdot \frac{1}{\cos(\alpha - \theta)}\right) - f\left(\frac{|u(x_0)|}{\text{diam } B}\right)\right] d\theta \\ &= f\left[\frac{|u(x_0)|}{\text{diam } B}\right] \\ &\quad + \frac{1}{\pi} \int_0^{\alpha} f'\left[\frac{|u(x_0)|}{\text{diam } B} \frac{1}{\cos(\alpha - \theta^*)}\right] \frac{|u(x_0)|}{\text{diam } B} \left[\frac{1}{\cos(\alpha - \theta)} - 1\right] d\theta; \end{aligned} \tag{27.76}$$

where $\zeta = |u(x_0)|\rho_{x_0}(\theta)$ and $0 < \theta^* < \theta$ for any $\theta \in (0, \alpha)$.

Since f'_ζ is a non-increasing fiction of ζ, then identity (27.76) leads to the following inequality

$$\begin{aligned} F &\ge f\left[\frac{|u(x_0)|}{\text{diam } B}\right] \\ &\quad + \frac{|u(x_0)|}{\text{diam } B} f'\left[\frac{|u(x_0)|}{\text{diam } B}\right] \int_0^{\alpha} \left[\frac{1}{\cos(\alpha - \theta)} - 1\right] d\theta. \end{aligned} \tag{27.77}$$

Clearly both terms in the right side of inequality (27.77) are positive.

Since

$$\int_0^{\alpha} \left[\frac{1}{\cos(\alpha - \theta)} - 1\right] d\theta = \ln \frac{1 + tn\frac{\alpha}{2}}{1 - tn\frac{\alpha}{2}} - \alpha$$

and

$$0 < \cos\alpha = \frac{h_0}{\text{diam } B} \le \frac{1}{2},$$

then

$$\int_0^{\alpha} \left(\frac{1}{\cos(\alpha - \theta)} - 1\right) d\theta = \ln \frac{1 + \sqrt{1 - (h_0/\text{diam } B)^2}}{(h_0/\text{diam } B)} - \arccos(h_0/\text{diam } B),$$

where

$$\frac{\pi}{3} \le \alpha = \arccos(h_0/\text{diam} B) < \frac{\pi}{2}.$$

Let $\eta = \frac{h_0}{\text{diam } B}$, then $\eta \in (0, 1/2]$. We introduce the function $\varphi(\eta)$ by the following formula

$$\psi(\eta) = \ln \frac{1 + \sqrt{1 - \eta^2}}{\eta} - \arccos \eta \tag{27.78}$$

for $\eta \in (0, 1/2]$. Since

$$\psi'(\eta) = -\frac{1}{\eta(1 - \eta^2)^{\frac{1}{2}}} < 0$$

for $0 < \eta \leq \frac{1}{2}$ and

$$\begin{aligned} \psi(1/2) &= \ln(2 + \sqrt{3}) - \frac{\pi}{3} = \ln\left(\frac{1 + \frac{\sqrt{3}}{3}}{1 - \frac{\sqrt{3}}{3}}\right) - \frac{\pi}{3} \\ &\geq 2 \cdot \frac{\sqrt{3}}{3}\left[1 + \frac{1}{3} \cdot \frac{1}{3} + \frac{1}{5} \cdot \frac{1}{9} + \frac{1}{7} \cdot \frac{1}{27} + \cdots\right] - 1.05 > 0.41, \end{aligned}$$

then $\psi(\eta)$ is a strictly positive decreasing function in $(0, 1/2]$, for which

$$\lim_{\eta \to 0+} \psi(\eta) = +\infty. \tag{27.79}$$

Thus the inequalities

$$+\infty > \psi(\eta) > 0.41 \tag{27.80}$$

hold for $0 < \eta = h_0/\text{diam } B \leq \frac{1}{2}$. Finally from (27.77–79) it follows that

$$F \geq f\left(\frac{|u(x_0)|}{\text{diam } B}\right) + \frac{|u(x_0)|}{\pi \cdot \text{diam } B} f'\left(\frac{|u(x_0)|}{\text{diam } B}\right) \psi\left(\frac{h_0}{\text{diam } B}\right), \tag{27.81}$$

where the function $\psi\left(\frac{h_0}{\text{diam } B}\right)$ satisfies inequalities (27.80) and

$$\lim_{h_0 \to 0} \psi\left(\frac{h_0}{\text{diam } B}\right) = +\infty. \tag{27.82}$$

Thus

$$g[F] \geq g\left\{f\left(\frac{|u(x_0)|}{\text{diam } B}\right) + \frac{|u(x_0)|}{\pi \cdot \text{diam } B} f'\left[\frac{|u(x_0)|}{\text{diam } B}\right] \psi\left(\frac{h_0}{\text{diam } B}\right)\right\}. \tag{27.83}$$

Inequalities (27.81) and (27.83) are the desired estimates for F and $g[F]$ respectively. Since both terms in the right side of (27.83) are positive, then our previous estimate (27.63) follows directly from (27.83).

As mentioned above we only present two-dimensional variant of estimates (27.81) and (27.83) in order to avoid superfluous calculations. However the general n-dimensional case does not significantly differ from estimates (27.81) and (27.83). In this case two-dimensional integrals in polar coordinates are replaced by n-dimensional integrals in spherical coordinates.

4) **Examples.** Here we apply estimates (27.81) and (27.83) to functions $f(\rho)$ introduced above. Let $x_0 \in M_u^+$ be any inner point of $\overline{CoB}$, $k(\theta)$ be the suppor function of $\partial(\overline{CoB})$ with respect to the point x_0 and

$$h_0 = \inf_{0\leq\theta\leq 2\pi} k(\theta) > 0. \tag{27.84}$$

We assume that

$$|u(x_0)| \geq 1. \tag{27.85}$$

We also recall that

$$0 < \frac{h_0}{\text{diam } B} \leq \frac{1}{2}. \tag{27.86}$$

Example 1. The function

$$\zeta = f(\rho) = \ln[(\text{diam } B)\rho] \tag{27.87}$$

is a C^∞-function in $(0, +\infty)$. Since

$$f'(\rho) = \frac{1}{\rho} \quad \text{for all} \quad \rho > 0,$$
$$f''(\rho) = -\frac{1}{\rho^2} \quad \text{for all} \quad \rho > 0,$$

then $f(\rho)$ is a strictly continuous and strictly concave function in $(0, +\infty)$. Clearly

$$\rho = g(\zeta) = \frac{1}{\text{diam } B} e^\zeta, \quad \zeta \in (-\infty, +\infty) \tag{27.88}$$

is the inverse for $f(\rho)$.

As we know

$$F = \frac{1}{2\pi}\int_0^{2\pi} \ln[(\text{diam } B)|u(x_0)|\rho_{x_0}(\theta)]d\theta \tag{27.89}$$
$$= \ln(|u(x_0)| \cdot \text{diam } B) + \frac{1}{2\pi}\int_0^{2\pi} \ln \rho_{x_0}(\theta)d\theta.$$

Hence

$$g[F] = |u(x_0)|e^{\frac{1}{2\pi}\int_0^{2\pi} \ln \rho_{x_0}(\theta)d\theta} = |u(x_0)|(\rho_{x_0}(\theta))_{\ln}, \tag{27.90}$$

where $(\rho_{x_0}(\theta))_{\ln}$ is the ln-mean of $\rho_{x_0}(\theta) = \frac{1}{k(\theta)}$.

Since

$$\rho_{x_0}(\theta) \geq \frac{1}{\text{diam } B},$$

then from (27.90) it follows that

$$g[F] \geq \frac{|u(x_0)|}{\text{diam } B}.$$

Thus we again obtain the first estimate (27.63) for $g[F]$.

The multiplier $(\rho_{x_0}(\theta))_{\ln}$ in (27.90) depends on x_0. But the exact formula for $(\rho_{x_0}(\theta))_{\ln}$ is sufficiently complicated in the terms of $\rho_{x_0}(\theta)$. Therefore we only present the following non-trivial estimate for $g[F]$:

$$g[F] \geq \frac{|u(x_0)|}{\text{diam } B} \left(\frac{\text{diam } B + [(\text{diam } B)^2 - h_0^2]^{1/2}}{h_0} \right)^{1/\pi}$$
$$e^{-\frac{1}{\pi} \arccos \frac{h_0}{\text{diam } B}},$$

which follows directly from (27.83). This estimate is precise. The multiplier

$$\left(\frac{\text{diam } B + [(\text{diam } B)^2 h_0^2]^{1/2}}{h_0} \right)^{1/\pi} \tag{27.91}$$

describes completely the properties of $g[F]$, which reflect the position of a point x_0 with respect to $\partial(\overline{CoB})$.

Since, as shown above,

$$\frac{\pi}{3} \leq \alpha = \arccos \frac{h_0}{\text{diam } B} < \frac{\pi}{2}$$

we have

$$e^{-\frac{1}{2}} \leq e^{-\frac{1}{\pi} \arccos \frac{h_0}{\text{diam } B}} \leq e^{-\frac{1}{3}} \tag{27.92}$$

Example 2. The function

$$\zeta = f(\rho) = \ln \ln[(2e \cdot \text{ diam } B)\rho], \quad \rho > \frac{1}{2 \text{ diam } B}, \tag{27.93}$$

is a C^∞-function in $\left(\frac{1}{2 \text{ diam } B}, +\infty\right)$. Since

$$f'(\rho) = \frac{1}{[\ln(2e \cdot \text{ diam } B)\rho]\rho} > 0 \text{ for all } \rho > \frac{1}{2 \text{ diam } B}$$
$$f''(\rho) = -\frac{1 + \ln(2e \cdot \text{ diam } B)\rho]}{[\ln(2e \cdot \text{ diam } B)\rho]^2 \rho^2} < 0 \text{ for all } \rho > \frac{1}{2 \text{ diam} B},$$

then $f(\rho)$ is a strictly continuous and strictly concave function in $\left(\frac{1}{2 \text{ diam } B}, +\infty\right)$.

Clearly

$$\rho = g(\zeta) = \frac{e^{e^\zeta}}{2e \cdot \text{ diam } B}, \quad \zeta \in (-\infty, +\infty), \tag{27.94}$$

is the inverse for $f(\rho)$. Clearly $g(\zeta)$ is a strictly increasing and strictly convex C^∞-function. As we know

$$F = \frac{1}{2\pi} \int_0^{2\pi} \ln[\ln[(2e \cdot \text{ diam } B) \cdot |u(x_0)| \cdot \rho_{x_0}(\theta)]] d\theta. \tag{27.95}$$

Hence

$$g(F) = \frac{1}{2e \cdot \text{ diam } B} e^{e^{\frac{1}{2\pi} \int_0^{2\pi} \ln[\ln[(2e \cdot \text{ diam } B) \cdot |u(x_0)| \rho_{x_0}(\theta)]] d\theta}}. \tag{27.96}$$

We now apply estimates (27.81) and (27.83) to integrals (27.95) and (27.96). Then we obtain

$$F \geq \ln[\ln 2e \cdot |u(x_0)|] + \frac{1}{\pi \ln[2e \cdot |u(x_0)|]} \psi \left(\frac{h_0}{\text{diam } B} \right) \tag{27.97}$$

and

$$g[F] \geq \frac{|u(x_0)|}{\text{diam } B} [2e \cdot |u(x_0)|]^{e^{A(|u(x_0)|, h_0)}} - 1, \tag{27.98}$$

where

$$A(|u(x_0)|, h_0) = \exp \psi \left\{ \frac{h_0}{\text{diam } B} \right\} \cdot [\pi \cdot \ln(2e|u(x_0)|]^{-1} - 1. \tag{27.99}$$

Since $x_0 \in B^-(u)$, then $u(x_0) < 0$. We now replace the convex set $K^*(u, x_0)$ by the n-ball

$$\cup \left(\frac{|u(x_0)|}{\text{diam } B} \right) : \ |p| \leq \frac{|u(x_0)|}{\text{diam } B}. \tag{27.100}$$

Then from inequality (27.46) one obtain the uniform estimate

$$|u(x_0)| \leq S_0 < +\infty \tag{27.101}$$

for all $x_0 \in B^-(u)$. The constant S_0 only depends on $\|\varphi^+(x, 0)\|_{L^n(B)}$, $\|c^+(x)\|_{L^n(B)}$, $\|d(x)\|_{L^n(B)}$ and diam B. Thus the following estimate

$$g[F] \geq \frac{|u(x_0)|}{\text{diam } B} [2e|u(x_0)|]^{[\exp \bar{A}(h_0)] - 1} \tag{27.101}$$

holds, where

$$\bar{A}(h_0) = \exp \psi \left\{ \frac{h_0}{\text{diam } B} \right\} \cdot [\pi \ln[2eS_0]]^{-1} - 1.$$

Clearly estimate (27.102) is more subtle than the estimate

$$g[F] \geq \frac{|u(x_0)|}{\text{diam } B}.$$

Below it is more convenient to write $F(|u(x_0)|)$ and $g(F(|u(x_0)|))$ instead of F and$g(F)$ respectively. Thus

$$F(|u(x_0)|) = \frac{1}{\sigma_n} \int_{S^{n-1}} f(|u(x_0)| \rho_{x_0}(\theta)) d\sigma_\theta, \tag{27.103}$$

where σ_n is the area of the unit hypersphere in P^n.

We now introduce the following characteristic of $\partial(CoB)$ with respect to any inner point x_0 of CoB:

$$T(f, x_0) = \inf_{s\in(0,+\infty)} \frac{1}{s} g\left(\frac{1}{\sigma_n}\int_{S^{n-1}} f(s\rho_{x_0}(\theta))d\sigma_\theta\right), \tag{27.104}$$

where $f(\rho)$ is any concave function, satisfying conditions (27.52–54). From inequality (27.63) it follows that

$$T(f; x_0) \geq \frac{1}{\text{diam } B} > 0. \tag{27.105}$$

The more subtle lower estimate for $T(f; x_0)$ follows from inequality (27.83) and its n-dimensional generalization. Clearly the estimate

$$g(F(|u(x_0)|) \geq |u(x_0)| \cdot T(f; x_0) \tag{27.106}$$

is correct for any $x_0 \in B^-(u)$, where $u(x) \in W_2^n(B) \cap C(\overline{B})$ is any solution of the Dirichlet problem (27.1–2).

In the final part of Section 27 we are concerned with the derivation of a lower estimate for solutions $u(x) \in W_2^n(B) \cap C(\overline{B})$ of the Dirichlet problem (27.1–2). Let Conditions C.1, C.2", C.3 be fulfilled and let x_0 be any point of $B^-(u)$. Clearly x_0 is an inner point of CoB. From inequalities (27.46), (27.56) and (27.106) it follows that

$$\begin{aligned} 0 < \sigma_n &\int_0^{|u(x_0)|T(f;x_0)} \frac{R(t)t^{n-1}dt}{(\alpha^{-\frac{n}{n-1}} + t^{\frac{n}{n-1}})^{n-1}} \\ \leq \sigma_n &\int_0^{g(F(|u(x_0)|))} \frac{R(t)t^{n-1}dt}{(\alpha^{-\frac{n}{n-1}} + t^{\frac{n}{n-1}})^{n-1}} \\ \leq \alpha^n &[\|\varphi^+(x,0)\|_{L^n(B^-(u))} + \|c^+(x)u(x)\|_{L^n(B^-(u))}]^n \\ &+ \|d(x)\|^n_{L^n(B^-(u))}. \end{aligned} \tag{27.107}$$

Let

$$\zeta = 1 + (\alpha \cdot t)^{\frac{n}{n-1}}. \tag{27.108}$$

Then

$$\begin{aligned} \sigma_n \cdot \int_0^{|u(x_0)|T(f;x_0)} \frac{R(t)t^{n-1}dt}{(\alpha^{-\frac{n}{n-1}} + t^{\frac{n}{n-1}})^{n-1}} &= \frac{n-1}{n}\sigma_n \\ \cdot \int_1^{1+[\alpha|u|T]^{\frac{n}{n-1}}} \frac{\overline{R}(\zeta;\alpha)(\zeta)^{n-2}d\zeta}{\zeta^{n-1}} \\ = \frac{n-1}{n}\sigma_n \int_1^{1+[\alpha|u|T]^{\frac{n}{n-1}}} \frac{\overline{R}(\zeta;\alpha)}{\zeta}d\zeta - \frac{n-1}{n} \\ \cdot \sigma_n \int_1^{1+[\alpha|u|T]^{\frac{n}{n-1}}} \frac{\overline{R}(\zeta;\alpha)[\zeta^{n-2} - (\zeta-1)^{n-2}]d\zeta}{\zeta^{n-1}}, \end{aligned}$$

where $u = u(x_0), T = T(f; x_0)$ and

$$\overline{R}(\zeta; \alpha) = R\frac{(\zeta - 1)^{\frac{n-1}{n}}}{\alpha}. \tag{27.109}$$

Below we used the functional

$$k^+(u) = (\mu_n)^{-\frac{1}{n}}[\|\varphi^+(x,0)\|_{L^n(B^-)} + \|c^+(x)u(x)\|_{L^n(B^-)}], \tag{27.110}$$

where μ_n is the volume of the n-unit ball. This functional has been introduced in Section 26.

First we study the general case $k^+(u) > 0$. The inequalities obtained by the condition $k^+(u) > 0$ will be presented in Theorems 27.1–5. If $k^+(u) = 0$ and $b(x,u,p) \geq 0$ at a point $(x,u,p) \in B \times R^- \times P^n$, then inequality (27.38) becomes

$$b(x,u,p)[\det(a_{ik}(x,u,p))]^{-\frac{1}{n}} \leq n|d(x)| \cdot |p| \cdot [R(|p|)]^{\frac{1}{n}}.$$

This case deserves the special consideration.

In the final part of Section 27 we establish that at any point $x_0 \in B^-(u)$, i.e. at any $x_0 \in B$, where $u(x_0) < 0$, the desired lower estimate for solutions $u(x) \in W_2^n(B) \cap C(\overline{B})$ of problem (27.1–2) is as follows

$$0 \geq u(x_0) \geq -\frac{k^+(u)}{T(f;x_0)} \cdot (1 + \varepsilon_0^+)^{\frac{n-1}{n}\,*)}. \tag{27.111}$$

The finite positive number ε_0^+ can be effectively determined by prescribed data of the Dirichlet problem (27.1–2). Below we describe the process of the determination for the positive number ε_0^+ in detail.

For the number $u(x_0)$ can only be two possibilities either

$$|u(x_0)|T(f,x_0) \leq k^+(u), \tag{27.112}$$

or

$$|u(x_0)| \cdot T(f;x_0) > k^+(u). \tag{27.113}$$

If for the number $u(x_0)$ inequality (27.112) is correct, then clearly the desired estimate (27.111) also is correct. Therefore we need to establish estimate (27.111) only in the case, when inequality (27.113) holds. We now define parameter $\alpha > 0$ by the following equation

$$\alpha^{\frac{n}{n-1}} = [k^+(u)]^{-\frac{n}{n-1}} - [|u(x_0)|T(f;x_0)]^{-\frac{n}{n-1}}. \tag{27.114}$$

---

*) We assume that $k^+(u) > 0$.

From (27.114) we obtain

$$1+[\alpha|u(x_0)|T(f;x_0)]^{\frac{n}{n-1}} = \left(\frac{|u(x_0)|T(f;x_0)}{k^+(u)}\right)^{\frac{n}{n-1}}. \tag{27.114-a}$$

We now derive a lower estimate for the integral

$$Q_n = \frac{n-1}{n}\sigma_n \int_1^{\left(\frac{|u(x_0)|T(f;x_0)|}{k^+(u)}\right)^{n/n-1}} \frac{\overline{R}(\zeta;\alpha)d\zeta}{\zeta}, \tag{27.115}$$

where α is chosen by the equation (27.114). The inequalities

$$0 < (\zeta-1)^{\frac{n-1}{n}} \leq \zeta^{\frac{n-1}{n}} \leq \zeta$$

hold for all numbers $\zeta > 1$. Since $R(t)$ is a non-increasing function of $t \in [0,+\infty)$, then the following inequalities

$$\overline{R}(\zeta;\alpha) = R\left(\frac{(\zeta-1)^{\frac{n-1}{n}}}{\alpha}\right) \geq R\left(\frac{\zeta^{\frac{n-1}{n}}}{\alpha}\right) \geq R\left(\frac{\zeta}{\alpha}\right) \tag{27.116}$$

hold for all $\zeta > 1$. Thus from (27.114), (27.115) and (27.116) it follows that

$$Q_n \geq \frac{n-1}{n}\sigma_n \int_1^{\left(\frac{|u(x_0)|T(f;x_0)}{k^+(u)}\right)^{\frac{n-1}{n}}} \frac{R(\zeta/\alpha)d\zeta}{\zeta}, \tag{27.117}$$

where α is defined by equation (27.114). Let

$$\eta = \frac{\zeta}{\alpha}, \tag{27.118}$$

then $\zeta = \alpha \cdot \eta$ and $d\zeta = \alpha \cdot d\eta$. Thus inequality (27.117) becomes

$$Q_n \geq \frac{n-1}{n}\sigma_n \int_{1/\alpha}^{1/\alpha\left(\frac{|u(x_0)|T(f;x_0)}{k^+(x_0)}\right)^{n/n-1}} \frac{R(\eta)}{\eta} d\eta. \tag{27.119}$$

According to inequality (27.113) we can set

$$\frac{|u(x_0)|T(f;x_0)}{k^+(u)} = (1+\varepsilon)^{\frac{n-1}{n}}, \tag{27.120}$$

where $\varepsilon > 0$ is some number. At the fixed point x_0 the numbers $T(f;x_0)$ and $k^+(u)$ independent on $|u(x_0)|$. Hence these numbers can be considered as constants at the point x_0. From (27.120) it follows that ε is a strictly increasing continuous function of $|u(x_0)|$ and $\varepsilon \to +\infty$ if $|u(x_0)| \to +\infty$. Conversely

$|u(x_0)|$ also is a strictly increasing continuous function of ε and $|u(x_0)| \to +\infty$ if $\varepsilon \to +\infty$. From equation (27.120) we obtain the following formulas:

$$\alpha = \frac{1}{k^+(u)} \left(\frac{\varepsilon}{1+\varepsilon}\right)^{\frac{n-1}{n}}, \tag{27.121}$$

$$\frac{1}{\alpha} = k^+(u) \left(\frac{1+\varepsilon}{\varepsilon}\right)^{\frac{n-1}{n}}, \tag{27.123}$$

$$\frac{1}{\alpha}\left(\frac{|u(x_0)| T(f;x_0)}{k^+(u)}\right)^{\frac{n-1}{n}} = \tag{27.124}$$

$$k^+(u) \left(\frac{1+\varepsilon}{\varepsilon}\right)^{\frac{n-1}{n}} (1+\varepsilon).$$

Thus inequality (27.117) becomes

$$Q_n \geq \frac{n-1}{n} \sigma_n \int_{k^+(u)\left(\frac{1+\varepsilon}{\varepsilon}\right)^{\frac{n-1}{n}}}^{k^+(u)\left(\frac{1+\varepsilon}{\varepsilon}\right)^{\frac{n-1}{n}}(1+\varepsilon)} \frac{R(\eta)}{\eta} d\eta, \tag{27.125}$$

where ε is any positive number.

Lemma 27.4. *Let*

$$\phi^+(\varepsilon) = \int_{k^+(u)\left(\frac{1+\varepsilon}{\varepsilon}\right)^{\frac{n-1}{n}}}^{k^+(u)\left(\frac{1+\varepsilon}{\varepsilon}\right)^{\frac{n-1}{n}}\cdot(1+\varepsilon)} \frac{R(\eta)}{\eta} d\eta. \tag{27.126}$$

Then $\phi^+(\varepsilon)$ is a strictly increasing positive C^1-function of $\varepsilon \in (0, +\infty)$, which satisfies the conditions

$$\lim_{\varepsilon \to 0} \phi^+(\varepsilon) = 0 \tag{27.127}$$

and

$$\lim_{\varepsilon \to +\infty} \phi^+(\varepsilon) = \int_{k^+(u)}^{+\infty} \frac{R(\eta)}{\eta} d\eta. \tag{27.128}$$

Proof. $R(\eta)$ is a bounded positive non-increasing continuous function in $(0, +\infty)$; $\varepsilon > 0$ and $k^+(u) > 0$. Hence

$$\phi^+(\varepsilon) \in C^1(0, +\infty).$$

Clearly

$$\frac{d\phi^+(\varepsilon)}{d\varepsilon} = \frac{R(k^+(u)\left(\frac{1+\varepsilon}{\varepsilon}\right)^{\frac{n-1}{n}} \cdot (1+\varepsilon))}{1+\varepsilon}$$
$$+ \frac{(n-1)\left[R\left(k^+(u)\left(\frac{1+\varepsilon}{\varepsilon}\right)^{\frac{n-1}{n}}\right) - R(k^+(u)\left(\frac{1+\varepsilon}{\varepsilon}\right)^{\frac{n-1}{n}}(1+\varepsilon))\right]}{n(1+\varepsilon)\varepsilon}.$$

According to the properties of the function $R(\eta)$ mentioned above it follows that

$$\frac{d\phi^+(\varepsilon)}{d\varepsilon} > 0.$$

Hence $\phi^+(\varepsilon)$ is a strictly increasing C^1-function in $(0, +\infty)$.

Since

$$0 < R(\eta) \leq R_0 = \text{const } < +\infty \tag{27.129}$$

for all $\varepsilon \in (0, +\infty)$, then

$$0 < \phi^+(\varepsilon) \leq R_0 \ln(1+\varepsilon)$$

for all $\varepsilon > 0$. Hence

$$\lim_{\varepsilon \to 0} \phi^+(\varepsilon) = 0.$$

Clearly

$$\lim_{\varepsilon \to +\infty} \phi^+(\varepsilon) = \int_{k^+(u)}^{+\infty} \frac{R(\eta)d\eta}{\eta}.$$

Lemma 27.4 is proved completely.

Lemma 27.5. *Let*

$$V_n = \int_1^{\left(\frac{|u(x_0)|T(f;x_0)}{k^+(u)}\right)^{\frac{n}{n-1}}} \frac{\overline{R}(\zeta;\alpha)[\zeta^{n-2} - (\zeta-1)^{n-2}]d\zeta}{\zeta^{n-1}}, \tag{27.130}$$

then

$$0 < V_n \leq R_0 \cdot D_n, \quad \textit{for} \quad n \geq 3, \tag{27.131}$$

where

$$0 < R_0 = \lim_{\eta \to 0+} R(\eta) < +\infty,$$

$$D_n = 1 + \frac{1}{2} + \frac{1}{3} + \cdots + \frac{1}{n-2};$$

and $V_n = 0$ *for* $n = 2$.

Proof. Let $n \geq 3$. Only this case is interesting to study. Then

$$0 < V_n \leq R_0 \int_1^{+\infty} \frac{\zeta^{n-2} - (\zeta-1)^{n-2}}{\zeta^{n-1}} d\zeta. \tag{27.132}$$

We now compute the integral in the right side of inequality (27.132). Let

$$D_n(x) = \int_1^{1+x} \frac{1}{\zeta}\left[1 - \left(1 - \frac{1}{\zeta}\right)^{n-2}\right] \zeta,$$

where x is any real positive number. Clearly

$$\lim_{x\to+\infty} D_n(x) = \int_1^{+\infty} \frac{\zeta^{n-2} - (\zeta-1)^{n-2}}{\zeta^{n-1}} d\zeta. \tag{27.133}$$

We set

$$\tau = 1 - \frac{1}{\zeta},$$

where $\zeta \in [1, +\infty)$ is the original variable. Then

$$\frac{1}{\zeta} = 1 - \tau \quad \text{and} \quad d\tau = \frac{d\zeta}{\zeta^2}.$$

Thus

$$\begin{aligned} D_n(x) &= \int_0^{\frac{x}{1+x}} (1 + \tau + \tau^2 + \cdots + \tau^{n-2}) d\tau \\ &= \frac{x}{1+x} + \frac{1}{2}\left(\frac{x}{1+x}\right)^2 + \frac{1}{3}\left(\frac{x}{1+x}\right)^3 + \cdots \\ &+ \frac{1}{n-2}\left(\frac{x}{1+x}\right)^{n-2} \end{aligned}$$

and

$$D_n = 1 + \frac{1}{2} + \frac{1}{3} + \cdots + \frac{1}{n-2}. \tag{27.134}$$

From (27.132–134) it follows the validity of Lemma 27.5 for $n \geq 3$. For $n = 2$ this lemma is evident.

Now inequalities (27.107), (27.119), equations (27.114), (27.121–124) and Lemmas 27.4, 27.5 lead to the following inequality

$$\begin{aligned} \frac{n-1}{n}\sigma_n \phi^+(\varepsilon) \leq \frac{n-1}{n}\sigma_n \cdot R_0 \cdot D_n + \mu_n \left(\frac{\varepsilon}{1+\varepsilon}\right)^{n-1} \\ + \|d(x)\|^n_{L^n(B^-(u))}. \end{aligned} \tag{27.135}$$

The last inequality can be replaced by a slightly more rough inequality

$$\phi^+(\varepsilon) \leq R_0 D_n + \frac{1}{n-1}[1 + n\|d(x)\|^n_{L^n(B^-(u))}]. \tag{27.136}$$

Inequality (27.136) is very convenient for deriving the desired lower estimate for $u(x_0)$.

The function $\gamma = \phi^+(\varepsilon) \in C^1(0, +\infty)$ is strictly increasing. Therefore there exists its inverse

$$\varepsilon = (\phi^+)^{-1}(\gamma) \in C^1(\phi^+(0), \phi^+(+\infty)). \tag{27.137}$$

This inverse also is a positive and strictly increasing function in the interval $(\phi^+(0), \phi^+(+\infty))$. According to Lemma 27.4

$$\phi^+(0) = 0 \tag{27.138}$$

and

$$\phi^+(+\infty) = \int_{k^+(u)}^{+\infty} \frac{R(\eta)}{\eta} d\eta. \tag{27.139}$$

Clearly $\phi^+(+\infty)$ can be either $+\infty$ or a finite positive number. If the inequality

$$R_0 D_n + \frac{1}{n-1}[1 + n\|d(x)\|^n_{L^n(B^-(u))}] < \int_{k^+(u)}^{+\infty} \frac{R(\eta)}{\eta} d\eta \tag{27.140}$$

holds, then it is possible to derive the desired lower estimate for $u(x_0)$. Actually from (27.136) and (27.140) it follows that there exists only one finite number $\varepsilon_0^+ > 0$ such that

$$\phi^+(\varepsilon_0^+) = R_0 D_n + \frac{1}{n-1}[1 + n\|(d(x)\|^n_{L^n(B^-(u))}]. \tag{27.141}$$

Clearly

$$\varepsilon_0^+ = (\Phi^+)^{-1}\left(R_0 D_n + \frac{1}{n-1}[1 + n\|d(x)\|^n_{L^n(B^-(u))}]\right). \tag{27.142}$$

For all $\varepsilon \leq \varepsilon_0^+$ inequality (27.136) will be correct. Thus all admissible values of ε satisfy the inequality

$$\varepsilon \leq \varepsilon_0^+. \tag{27.143}$$

From the equation

$$\frac{|u(x_0)|T(f;x_0)}{k^+(u)} = (1+\varepsilon)^{\frac{n-1}{n}} \tag{27.144}$$

we obtain the following estimate

$$|u(x_0)| \leq \frac{k^+(u)}{T(f;x_0)}(1+\varepsilon_0^+)^{\frac{n-1}{n}} \tag{27.145}$$

for any point $x_0 \in B^-(u)$.

Since $u(x_0) < 0$, then from (27.145) we obtain

$$-\frac{k^+(u)}{T(f;x_0)}(1+\varepsilon_0^+)^{\frac{n-1}{n}} \leq u(x_0) < 0 \tag{27.146}$$

for all points $x_0 \in B^-(u)$. The numbers $k^+(u)$, $T(f,x_0)$ and ε_0^+ are independent on the value of $|u(x_0)|$ and on the sign of $u(x_0)$. Thus the inequality

$$-\frac{k^+(u)}{T(f;u_0)}(1+\varepsilon_0^+)^{\frac{n-1}{n}} \leq u(x_0) \tag{27.147}$$

is correct for all points $x_0 \in B$. In (27.147) we use the following notations

$$k^+(u) = (\mu_n)^{-\frac{1}{n}}[\|\varphi^+(x,0)\|_{L^n(B^-(u))} + \|c^+(x)\cdot u(x)\|_{L^n(B^-(u))}], \tag{27.148}$$

$$T(f,x_0) = \inf_{s\in[0,+\infty)} \frac{1}{s} g\left(\int_{S^{n-1}} f(s\rho_{x_0}(\theta))d\sigma_\theta\right) \tag{27.149}$$

and

$$\varepsilon_0^+ = (\phi^+)^{-1}\left[R_0 D_n + \frac{1}{n-1}(1+n\|d(x)\|^n_{L^n(B^-(u))})\right]. \tag{27.150}$$

Thus we proved the first main theorem.

Theorem 27.1. *Inequality (27.147) is a priori lower estimate for solutions $u(x) \in W_2^n(B)\cap C(\overline{B})$ of the Dirichlet problem (27.1–2), if Conditions 1, 2", 3 and inequality (27.140) are fulfilled.*

In the same way the second main theorem can be proved.

Theorem 27.2. *Let*

$$k^-(u) = (\mu_n)^{-\frac{1}{n}}[\|\varphi^-(x,0)\|_{L^n(B^+(u))} + \|c^+(x)u(x)\|_{L^n(B^+(u))}], \tag{27.151}$$

$$\gamma = \phi^-(\varepsilon) = \int_{k^-(u)(\frac{1+\varepsilon}{\varepsilon})^{(n-1)/n}}^{k^-(u)(\frac{1+\varepsilon}{\varepsilon})^{(n-1)/n}(1+\varepsilon)} \frac{R(\eta)}{\eta}d\eta, \tag{27.152}$$

$$\varepsilon_0^- = (\phi^-)^{-1}\left(R_0 D_n + \frac{1}{n-1}(1+n\|d(x)\|^n_{L^n(B^+(u))})\right) < \int_{k^-(u)}^{+\infty} \frac{R(\eta)}{\eta}d\eta, \tag{27.153}$$

where $u(x) \in W_2^n(B)\cap C(\overline{B})$ is any solution of the Dirichlet problem (26.1–2) and Conditions C.1, C.2", C.3 are fulfilled for this Dirichlet problem.

Then the upper a priori estimate

$$u(x_0) \le \frac{k^-(u)}{T(f;x_0)}(1+\varepsilon_0^-)^{\frac{n-1}{n}} \tag{27.154}$$

holds, if the inequality

$$R_0 D_n + \frac{1}{n-1}[1+n\|d(x)\|^n_{L^n(B^+(u))}] < \int_{k^-(u)}^{+\infty} \frac{R(\eta)}{\eta}d\eta \tag{27.155}$$

is fulfilled.

In many quasilinear elliptic Dirichlet problems (27.1–2), relating to Calculations of Variations, Differential Geometry and Applied Mathematics, we often use the norms of functions $\varphi^{\pm}(x,0)$, $c^{+}(x)u(x)$ and $d(x)$ in the space $L^n(B)$ instead of norms in $L^n(B^{\pm}(u))$. Such norms are more simple and more effective. But they are slightly more rough than the norms in$L^n(B^{\pm}(u))$. The main estimates in the terms of both norms are sharp. Below we present the analogue of Theorems 27.1 and 27.2 for the case of $L^n(B)$-norms. Let

$$K^{+}(u) = (\mu_n)^{-\frac{1}{n}}[\|\varphi^{+}(x,0)\|_{L^n(B)} + \|c^{+}(x)u(x)\|_{L^n(B)}], \tag{27.156}$$

$$K^{-}(u) = (\mu_n)^{-\frac{1}{n}}[\|\varphi^{-}(x,0)\|_{L^n(B)} + \|c^{+}(x)u(x)\|_{L^n(B)}], \tag{27.157}$$

$$\gamma = \psi^{+}(\varepsilon) = \int_{K^{+}(u)(\frac{1+\varepsilon}{\varepsilon})^{n-1/n}}^{K^{+}(u)(\frac{1+\varepsilon}{\varepsilon})^{n-1/n}(1+\varepsilon)} \frac{R(\eta)}{\eta}\eta \tag{27.158}$$

and

$$\gamma = \psi^{-}(\varepsilon) = \int_{K^{-}(u)(\frac{1+\varepsilon}{\varepsilon})^{n-1/n}}^{K^{-}(u)(\frac{1+\varepsilon}{\varepsilon})^{n-1/n}(1+\varepsilon)} \frac{R(\eta)}{\eta}d\eta \tag{27.159}$$

lemmas, which are similar to Lemma 27.4 can be established for functions $\psi^{\pm}(\varepsilon)$. Therefore the functions $\psi^{\pm}(\varepsilon)$ have inverses, the properties of which are similar to the properties of the inverses of $\phi^{\pm}(\varepsilon)$.

Theorem 27.3. *Let $u(x) \in W_2^n(B) \cap C(\overline{B})$ be any solution of the Dirichlet problem (27.1–2) and let Conditions 1, 2", 3 be fulfilled. Then inequalities*

$$-\frac{K^{+}(u)}{T(f;x_0)}(1+\varepsilon_0^{+})^{\frac{n-1}{n}} \le u(x_0) \le \frac{K^{-}(u)}{T(f;x_0)}(1+\varepsilon_0^{-})^{\frac{n-1}{n}} \tag{27.160}$$

hold, if the inequality

$$R_0 D_n + \frac{1}{n-1}[1 + n\|d(x)\|^{n}_{L^n(B)}] < \int_K^{+\infty} \frac{R(\eta)}{\eta}d\eta, \tag{27.161}$$

is fulfilled, where $K = \max\{K^{+}(u), K^{-}(u)\}$ and

$$\varepsilon_0^{\pm} = (\psi^{\pm})^{-1}\left(R_0 D_n + \frac{1}{n-1}[1 + n\|d(x)\|^{n}_{L^n(B)}]\right). \tag{27.162}$$

We now introduce the functional

$$\chi(u) = (\mu_n)^{-\frac{1}{n}}[\|\varphi(x,0)\|_{L^n(B)} + \|c^{+}(x)u(x)\|_{L^n(B)}]. \tag{27.163}$$

Clearly

$$K^{\pm}(u) \le \chi(u). \tag{27.164}$$

Let

$$\Gamma(\varepsilon) = \int_{\chi(u)\tau_1(\varepsilon)}^{\chi(u)\tau_u(\varepsilon)} \frac{R(\eta)}{\eta} d\eta, \tag{27.165}$$

where $\tau_1(\varepsilon) = \left(\frac{1+\varepsilon}{\varepsilon}\right)^{\frac{n-1}{n}}$, $\tau_2(\varepsilon) = \tau_1(\varepsilon)\cdot(1+\varepsilon)$. *The* C^1*-function* $\Gamma(\varepsilon)$ *is positive and strictly increasing. Clearly the inequalities*

$$\psi^{\pm}(\varepsilon) \leq \Gamma(\varepsilon) \tag{27.166}$$

hold for all $\varepsilon \in (0, +\infty)$. *Let*

$$\delta_0 = \Gamma^{-1}\left[R_0 D_n + \frac{1}{n-1}(1 + n\|d(x)\|^n_{L^n(B)})\right], \tag{27.167}$$

then

$$\delta_0 \leq \varepsilon_0^+$$

and

$$\delta_0 \leq \varepsilon_0^-$$

where ε_0^+ *and* ε_0^- *are defined by (27.162).*

We now present the following analogue of Theorem 26.3.

Theorem 27.4. *Let* $u(x) \in W_2^n(B) \cap C(\overline{B})$ *be any solution of the Dirichlet problem (27.1–2) and let Conditions 1, 2", 3 be fulfilled. Then inequalities*

$$-\frac{\chi(u)}{T(f;x_0)}(1+\delta_0)^{\frac{n-1}{n}} \leq u(x_0) \leq \frac{\chi(u)}{T(f;x_0)}(1+\delta_0)^{\frac{n-1}{n}} \tag{27.168}$$

hold for all $x_0 \in B$, *if the inequality*

$$R_0 D_n + \frac{1}{n-1}[1 + n\|d(x)\|^n_{L^n(B)}] < \int_{\chi(u)}^{+\infty} \frac{R(\eta)}{\eta} d\eta \tag{27.169}$$

is fulfilled, where $\chi(u)$ *is defined by (27.163) and* δ_0 *is defined by (27.167).*

The proof of this theorem can be given in the same way as proofs of Theorems 27.1 and 27.2.

From inequality (27.168) it follows that

$$\|c^+(x)u(x)\|_{L^n(B)} \leq \frac{(1+\delta_0)^{\frac{n-1}{n}}}{\mu_n^{1/n}} \left\|\frac{c^+(x)}{T(f,x)}\right\|_{L^n(B)} [\|\varphi(x,0)\|_{L^n(B)} + \|c^+(x)u(x)\|_{L^n(B)}],$$

if all conditions of Theorem 27.4 are fulfilled.

Below we use the following notation

$$\tau(\delta_0, n) = [1 + \delta_0]^{-\frac{n-1}{n}} \mu_n^{\frac{1}{n}}. \tag{27.170}$$

Thus the last inequality becomes

$$\|c^+(x)u(x)\|_{L^n(B)} \leq \|\varphi(x,0)\|_{L^n(B)} \left\| \frac{c^+(x)}{T(f,x)} \right\|_{L^n(B)} \left[\tau(\delta_0, n) - \left\| \frac{c^+(x)}{T(f,x)} \right\|_{L^n(B)} \right]^{-1}. \tag{27.171}$$

Finally we obtain the following theorem.

Theorem 27.5. *Let $u(x) \in W_2^n(B) \cap C(\overline{B})$ be any solution of the Dirichlet problem (27.1–2) and let Conditions 1, 2", 3 be fulfilled. We also assume that two following conditions*

$$R_0 D_n + \frac{1}{n-1}[1 + n\|d(x)\|^n_{L^n(B)}] < \int_{\chi(u)}^{+\infty} \frac{R(\eta)}{\eta} d\eta, \tag{27.69}$$

$$\left\| \frac{c^+(x)}{T(f,x)} \right\|_{L^n(B)} < \tau(\delta_0, n) \tag{27.172}$$

hold. Then the estimate

$$|u(x_0)| \leq \frac{\|\varphi(x,0)\|_{L^n(B)}}{T(f,x_0)\left[\tau(\delta_0,n) - \left\|\frac{c^+(x)}{T(f,x)}\right\|_{L^n(B)}\right]} \tag{27.173}$$

hold for all $x_0 \in B$.

Proof. From inequalities (27.168), (27.171) and (27.173) we obtain

$$|u(x_0)| \leq \frac{(1+\delta_0)^{\frac{n-1}{n}}}{\mu_n^{1/n} T(f,x_0)} \|\varphi(x,0)\|_{L^n(B)} \cdot \left[1 + \left\| \frac{c^+(x)}{T(f,x)} \right\|_{L^n(B)} \left[\tau(\delta_0,n) - \left\| \frac{c^+(x)}{T(f,x)} \right\|_{L^n(B)} \right]\right]^{-1}$$

$$= \frac{\|\varphi(x,0)\|_{L^n(B)}}{T(f,x_0)\left[\tau(\delta_0,n) - \left\|\frac{c^+(x)}{T(f,x)}\right\|_{L^n(B)}\right]}.$$

Thus the proof is completed.

Remark. If $c^+(x) = 0$ in B, then inequality (27.173) becomes

$$|u(x_0)| \le \frac{\|\varphi(x,0)\|_{L^n(B)}}{\mu_n^{1/n} \cdot T(f,x_0)}(1+\delta_0)^{\frac{n-1}{n}}. \tag{27.174}$$

Since

$$\frac{1}{T(f,x_0)} \le \text{diam } B \tag{27.175}$$

for all $\chi_0 \in B$, then from inequality (27.174) it follows that

$$|u(x_0)| \le \frac{\|\varphi(x,0)\|_{L^n(B)} \cdot \text{ diam } B}{\mu_n^{1/n}}(1+\delta_0)^{\frac{n-1}{n}}$$

for all $x_0 \in B$.

As we mentioned above (see page 520) all proofs of Theorems 27.1–5 have been obtained by the assumption that the numbers $k^\pm(u)$, $K^\pm(u)$ and $\chi(u)$ are respectively positive. We now study the case when these numbers are equal to zero. It is sufficient to consider this problem only for Theorems 27.4 and 27.5.

We introduce the following Dirichlet problem

$$\sum_{i,k=1}^{n} a_{ik}(x,u,Du)u_{ik} = \frac{n \cdot \lambda \cdot \sqrt[n]{\det(a_{ik}(x,u,Du)}}{[R(|p|)]^{1/n}} + b(x,u,p), \tag{27.176}$$

$$u|_{\partial B} = 0, \tag{27.177}$$

where $\lambda \in (-\infty, +\infty)$ is a constant and functions $a_{ik}(x,u,p)$ and $b(x,u,p)$ satisfy Conditions 1, 2", 3. Let

$$b_\lambda(x,u,p) = \frac{n \cdot \lambda \sqrt[n]{\det(a_{ik}(x,u,p))}}{[R(|p|)]^{1/n}} + b(x,u,p). \tag{27.178}$$

Let λ be a positive constant. Then from inequality (27.38) it follows that

$$\frac{b_\lambda(x,u,p)}{n\sqrt[n]{\det(a_{ik}(x,u,p))}} \le \frac{\lambda}{[R(|p|)]^{1/n}} + \frac{\varphi(x,0) - c^+(x)u + |d(x)|\ |p|}{[R(|p|)]^{1/n}} \tag{27.178}$$

at all points $(x,u,p) \in B \times R^- \times P^n$, where $b_\lambda(x,u,p) \ge 0$.

Since we investigate the case

$$\chi(u) = (\mu_n)^{-\frac{1}{n}}[\|\varphi(x,0)\|_{L^n(B)} + \|c^+(x)u(x)\|_{L^n(B)}] = 0,$$

then inequality (27.179) becomes

$$\frac{b_\lambda(x,u,p)}{n\sqrt[n]{\det(a_{ik}(x,u,p))}} \le \frac{\lambda + |d(x)|\,|p|}{[R(|p|)]^{1/n}} \tag{27.180}$$

at all points $(x,u,p) \in B \times R^- \times P^n$, where $b_\lambda(x,u,p) \ge 0$. Let

$$\chi_\lambda(u) = \lambda(\mu_n)^{-\frac{1}{n}}(\text{meas } B)^{\frac{1}{n}}. \tag{27.181}$$

Clearly for equation (27.176) the function $\chi(u)$ coincides with $\chi_\lambda(u)$. From (27.181) it follows that

$$\chi(u) = \lim_{\lambda\to 0} \chi_\lambda(u) = 0. \tag{27.182}$$

We now introduce the function

$$\Gamma(\lambda;\varepsilon) = \int_{\chi_\lambda(u)\tau_1(\varepsilon)}^{\chi_\lambda(u)\tau_2(\varepsilon)} \frac{R(\eta)}{\eta}\,d\eta \tag{27.183}$$

for any positive constant λ and $\varepsilon \in (0,+\infty)$. $\Gamma(\lambda;\varepsilon)$ is a strictly increasing positive C^1-function of $\varepsilon \in (0,+\infty)$ for any $\lambda = \text{const} > 0$. The proof of this assertion coincides with the proof of Lemma 27.4. Since

$$0 < R(\eta) \le R_0 = \text{const} < +\infty$$

for all $\eta \in (0,+\infty)$, then

$$0 < \Gamma(\lambda;\varepsilon) \le R_0 \ln(1+\varepsilon) \tag{27.184}$$

for all $\varepsilon > 0$. Hence

$$\lim_{\varepsilon\to 0} \Gamma(\lambda;\varepsilon) = 0 \tag{27.185}$$

for any $\lambda = \text{const} > 0$. Clearly

$$\lim_{\varepsilon\to+\infty} \Gamma(\lambda;\varepsilon) = \int_{\chi_\lambda(u)}^{+\infty} \frac{R(\eta)}{\eta}\,d\eta = \int_{\frac{\lambda(\text{meas } B)^{1/n}}{\mu_n^{1/n}}}^{+\infty} \frac{R(\eta)}{\eta}\,d\eta. \tag{27.186}$$

We now consider the set of all solutions $u(x) \in W_2^n(B) \cap C(\overline{B})$ of the Dirichlet problem (27.176–177) with $\lambda = \text{const} > 0$. We are concerned with a lower estimate of these solutions at the points $x_0 \in B$ such that $u(x_0) < 0$. Let either

$$R_0 D_n + \frac{1}{n-1}[1 + n\|d(x)\|^n_{L^n(B)}] < \int_{\eta_0}^{+\infty} \frac{R(\eta)}{\eta}\,d\eta, \tag{27.187}$$

if

$$\int_{\eta_0}^{+\infty} \frac{R(\eta)}{\eta}\,d\eta < +\infty;$$

or

$$R_0 D_n + \frac{1}{n-1}[1 + n\|d(x)\|^n_{L^n(B)}] < +\infty, \tag{27.188}$$

if

$$\int_{\eta_0}^{+\infty} \frac{R(\eta)}{\eta} d\eta < +\infty.$$

Where $\eta_0 = \lambda_0 \left(\frac{\text{meas } B}{\mu_n}\right)^{1/n}$ and λ_0 be some positive number. Then for any constant $\lambda \in (0, \lambda_0]$ either

$$R_0 D_n + \frac{1}{n-1}[1 + n\|d(x)\|^n_{L^n(B)}] < \int_{\frac{\lambda(\text{meas } B)^{1/n}}{\mu_n^{1/n}}}^{+\infty} \frac{R(\eta)}{\eta} d\eta, \tag{27.169}$$

if

$$\int_{\eta_0}^{+\infty} \frac{R(\eta)}{\eta} d\eta = +\infty;$$

or

$$R_0 D_n + \frac{1}{n-1}[1 + n\|d(x)\|^n_{L^n(B)}] < +\infty, \tag{27.190}$$

if

$$\int_{\frac{\lambda_0 \cdot (\text{meas } B)^{1/n}}{\mu_n^{1/n}}}^{+\infty} \frac{R(\eta)}{\eta} d\eta = +\infty.$$

Thus we can apply Theorem 27.4 to all solutions $u(x) \in W_2^n(B) \cap C(\overline{B})$ of the Dirichlet problem (27.176–177) for all $\lambda \in (0, \lambda_0]$. Then we obtain the inequality

$$u(x_0) \geq -\frac{\lambda(\text{meas } B)^{1/n}}{\mu_n^{1/n} T(f, x_0)}(1+\delta_0)^{\frac{n-1}{n}} \tag{27.191}$$

for all points $x_0 \in B$, where $\delta_\lambda > 0$ is an unique solution of the equation

$$\Gamma(\lambda; \delta_\lambda) = R_0 D_n + \frac{1}{n-1}[1 + \|d(x)\|^n_{L^n(B)}]. \tag{27.192}$$

We now study the properties of the number δ_λ as a function of $\lambda \in (0, \lambda_0]$.

Lemma 27.6. *Let the function $R(\eta)$ non-increasing in $[0, +\infty)$. Then the function δ_λ is non-decreasing in $(0, \lambda_0]$.*

Proof. It is sufficient to establish that

$$\frac{d\delta_\lambda}{d\lambda} \geq 0$$

for all $\lambda \in (0, \lambda_0]$. From (27.192) it follows that

$$\frac{\partial \Gamma(\lambda; \delta_\lambda)}{\partial \lambda} + \frac{\partial \Gamma}{\partial \delta_\lambda} \cdot \frac{d\delta_\lambda}{d\lambda} = 0.$$

From this equation and formula (27.183) we obtain

$$\frac{R\left(\chi_\lambda\cdot\left(1+\frac{1}{\delta_\lambda}\right)^{\frac{n-1}{n}}(1+\delta_\lambda)\right)}{\left(1+\frac{1}{\delta_\lambda}\right)^{\frac{n-1}{n}}(1+\delta_\lambda)}\left(\frac{\text{meas }B}{\mu_n}\right)^{1/n}$$

$$\cdot\left(1+\frac{1}{\delta_\lambda}\right)^{\frac{n-1}{n}}\cdot(1+\delta_\lambda)$$

$$-\frac{R(\chi_\lambda\cdot\left(1+\frac{1}{\delta_\lambda}\right)^{\frac{n-1}{n}}}{\left(1+\frac{1}{\delta_\lambda}\right)^{\frac{n-1}{n}}}\frac{(\text{meas }B)^{\frac{1}{n}}}{\mu_n^{1/n}}\left(1+\frac{1}{\delta_\lambda}\right)^{\frac{n-1}{n}}$$

$$+\frac{R\left(\chi_\lambda\left(1+\frac{1}{\delta_\lambda}\right)^{\frac{n-1}{n}}\right)(1+\delta_\lambda)}{\left(1+\frac{1}{\delta_\lambda}\right)^{\frac{n-1}{n}}(1+\delta_\lambda)}$$

$$\chi_\lambda\left[\frac{n-1}{n}\left(1+\frac{1}{\delta_\lambda}\right)^{-\frac{1}{n}}\cdot\frac{(1+\delta_\lambda)}{-\delta_\lambda^2}+\left(1+\frac{1}{\delta_\lambda}\right)^{\frac{n-1}{n}}\right]\frac{d\delta_\lambda}{d\lambda}$$

$$-\frac{R\left(\chi_\lambda\left(1+\frac{1}{\delta_\lambda}\right)^{\frac{n-1}{n}}\right)}{\chi_\lambda\cdot\left(1+\frac{1}{\delta_\lambda}\right)^{\frac{n-1}{n}}}\cdot\frac{n-1}{n}\cdot\left(1+\frac{1}{\delta_\lambda}\right)^{-\frac{1}{n}}\left(-\frac{1}{\delta_\lambda^2}\right)\frac{d\delta_\lambda}{d\lambda}=0,$$

where

$$\chi_\lambda=\lambda\left(\frac{\text{meas }B}{\mu_n}\right)^{1/n}.$$

From the last equation we obtain the following formula for $\frac{d\delta_\lambda}{d\lambda}$:

$$\frac{d\delta_\lambda}{d\lambda}=\frac{1}{\lambda}\left[R\left(\chi_\lambda\left(1+\frac{1}{\delta_\lambda}\right)^{\frac{n-1}{n}}\right)-R\left(\chi_\lambda\left(1+\frac{1}{\delta_\lambda}\right)^{\frac{n-1}{n}}(1+\delta_\lambda)\right)\right]$$

$$\times\left[\frac{R\left(\chi_\lambda\left(1+\frac{1}{\delta_\lambda}\right)^{\frac{n-1}{n}}(1+\delta_\lambda)\right)}{\chi_\lambda(1+\delta_\lambda)}\right.$$

$$\left.+\frac{n-1}{n}\cdot\frac{R\left(\chi_1\left(1+\frac{1}{\delta_\lambda}\right)^{\frac{n-1}{n}}\right)-R\left(\chi_\delta\left(1+\frac{1}{\delta_\lambda}\right)^{\frac{n-1}{n}}(1+\delta_\lambda)\right)}{\delta_\lambda(1+\delta_\lambda)}\right]^{-1}.$$

Thus

$$\frac{d\delta_\lambda}{d\lambda} \geq 0$$

for all $\lambda \in (0, \lambda_0]$.

Lemma 27.6 is proved completely. □

From Lemma 27.6 and inequality (27.191) it follows that

$$u(x_0) \geq -\frac{\lambda(\text{meas } B)^{1/n}}{\mu_n^{1/n} T(f, x_0)}(1+\delta_0)^{\frac{n-1}{n}}, \tag{27.193}$$

for all points $x_0 \in B$. In (27.193) $u(x) \in W_2^n(B) \cap C(\overline{B})$ be any solution of the Dirichlet problem (27.176–177) with $\lambda \in (0, \lambda_0]$. We remind that the number λ_0 is fixed (see page 535) and δ_λ is defined by the equation (27.192).

The upper estimate for $u(x_0)$ can be obtained in the similar way. Let λ be a negative constant. Then from relation (27.39) it follows that inequality

$$\frac{|b_\lambda(x,u,p)|}{n \cdot \sqrt[n]{\det(a_{ik}(x,u,p))}} \leq \frac{|\lambda| + |\varphi(x,0)| + |c^+(x)u(x)| + |d(x)|\ |p|}{[R(|p|)]^{1/n}} \tag{27.194}$$

holds at all points $(x,u,p) \in B \times R^+ \times P^n$, where $b_\lambda(x,u,p) \leq 0$. Since we investigate the case

$$\chi(u) = (\mu_n)^{-\frac{1}{n}}[\|\varphi(x,0)\|_{L^n(B)} + \|c^+(x)u(x)\|_{L^n(B)}] = 0,$$

then inequality (27.94) becomes

$$\frac{|b_\lambda(x,u,p)|}{n\sqrt[n]{\det(a_{ik}(x,u,p))}} \leq \frac{|\lambda| + |d(x)|\ |p|}{[R(|p|)]^{1/n}}. \tag{27.195}$$

We now use the techniques developed above for the lower estimate of $u(x_0)$. Finally we obtain the following estimate

$$|u(x_0)| \leq \frac{|\lambda| \text{ meas } B)^{1/n}}{\mu_n^{1/n} \cdot T(f, x_0)}(1+\delta_0)^{\frac{n-1}{n}} \tag{27.196}$$

for all solutions $u(x) \in W_2^n(B) \cap C(\overline{B})$ of the Dirichlet problem (27.176–177), if $\lambda \in [-\lambda_0, +\lambda_0] \backslash \{0\}$, where $\lambda_0 > 0$ is a sufficiently small number defined above.

We want to establish estimate (27.196) for $\lambda = 0$. Let $u(x) \in W_2^n(B) \cap C(\overline{B})$ be any solution of the Dirichlet problem

$$\sum_{i,k=1}^n a_{ik}(x,u,Du)u_{ik} = b(x,u,Du), \tag{27.1}$$

$$u|_{\partial B} = 0, \tag{27.2}$$

for which $\chi(u) = (\mu_n)^{-\frac{1}{n}}[\|\varphi(x,0)\|_{L^n(B)} + \|c^+(x)u(x)\|_{L^n(B)}] = 0$. Then the following inequality

$$0 \le \frac{b(x,u,p)}{n[\det(a_{ik}(x,u,p)]^{1/n}} \le \frac{\lambda + |d(x)|\,|p|}{[R(|p|)]^{1/n}} \tag{27.197}$$

holds at any point $(x,u,p) \in B \times R^- \times P^n$, where $b(x,u,p) \ge 0$ and λ is any positive constant. Similarly the inequality

$$\frac{|b(x,u,p)|}{n[\det(a_{ik}(x,u,p))]^{1/n}} \le \frac{|\lambda| + |d(x)| \cdot |p|}{[R(|p|)]^{1/n}} \tag{27.198}$$

holds at any point $(x,u,p) \in B \times R^+ \times P^n$, where $b(x,u,p) \le 0$ and λ is any negative number.

Let either

$$R_0 D_n + \frac{1}{n-1}[1 + n\|d(x)\|^n_{L^n(B)}] < \int_{\eta_0}^{+\infty} \frac{R(\eta)}{\eta} d\eta, \tag{27.199}$$

if

$$\int_{\eta_0}^{+\infty} \frac{R(\eta)}{\eta} d\eta < +\infty;$$

or

$$R_0 D_n + \frac{1}{n-1}[1 + n\|d(x)\|^n_{L^n(B)}] < +\infty, \tag{27.200}$$

if

$$\int_{\eta_0}^{+\infty} \frac{R(\eta)}{\eta} d\eta = +\infty,$$

where $\eta_0 = \lambda_0 \left[\frac{(\text{meas } B)}{\mu_n}\right]^{1/n}$ and λ_0 is some positive number.

Then from Inequalities (27.197–200) and Theorems 27.4 and 27.5 it follows that

$$|u(x_0)| \le \frac{|\lambda|(\text{meas } B)^{1/n}}{\mu_n^{1/n} \cdot T(f;x_0)}(1 + \delta_{\lambda_0})^{\frac{n-1}{n}}, \tag{27.201}$$

where:

1) $u(x) \in W_2^n(B) \cap C(\overline{B})$ is any solution of the Dirichlet problem (27.1–2);
2) x_0 is any point of domain B;
3) $\lambda_0 > 0$, satisfying inequalities (27.199–200) and $\delta_{\lambda_0} > 0$ is the unique root of the equation

$$\Gamma(\lambda_0, \delta_{\lambda_0}) = R_0 D_n + \frac{1}{n-1}(1 + n\|d(x)\|^n_{L^n(B)})^{\frac{n-1}{n}};$$

4) λ is any number from $[-\lambda_0, \lambda_0]\backslash\{0\}$;
5) μ_n is the volume of the n-unit ball, $0 < R_0 = \sup_{\eta\in(0,+\infty)} R(\eta) < +\infty$,

$$D_n = \begin{cases} 0 & \text{if } \ n = 2; \\ 1 + \frac{1}{2} + \frac{1}{3} + \cdots + \frac{1}{n-2}, & \text{if } \ n \geq 3; \end{cases}$$

and $T(f, x_0) \geq \frac{1}{\text{diam } B} > 0$ is defined by (27.104).

From Lemma 27.6 it follows that

$$0 \leq |u(x_0)| \leq \lim_{\lambda\to 0} \frac{|\lambda|(\text{meas } B)^{1/n}}{\mu_n^{1/n} \cdot T(f; x_0)}[1 + \delta_0]^{\frac{n-1}{n}} = 0.$$

Thus we establish that $u(x) = 0$ in B. Hence Theorems 27.1–5 are also correct for the cases $k^{\pm}(u) = 0$, $K^{\pm}(u) = 0$ and $\chi(u) = 0$ respectively.

§28. Comments to Chapter 8. The Maximum Principles in Global Problems of Differential Geometry

28.1 Comments to Chapter 8

In Chapter 8 various variants of geometric maximum principles for solutions of general quasilinear elliptic equations are presented. These solutions are of class $W_2^n(B) \cap C(\overline{B})$ and satisfy the prescribed Dirichlet data on ∂B. The principles of such kind play the significant role in modern investigations related to existence, uniqueness and stability theorems for solutions of nonlinear elliptic boundary value problems and their applications to Differential Geometry, Calculus of Variations and Mechanics.

The first version of these maximum principles was obtained by Bakelman [4], [5] in 1959. They are related to wide classes of general quasilinear elliptic equations and linear elliptic equations of the form

$$\sum_{i,k=1}^{n} a_{ik}(x)u_{ik} + c(x)u = f(x). \tag{28.1}$$

These results were applied by Bakelman [6], [9], [10] to global problems in differential geometry.

In 1963–66 Alexandrov [6], [7] obtained extended versions of maximum principles for solutions $u(x) \in W_2^n(B) \cap C(\overline{B})$ of full linear elliptic equations

$$\sum_{i,k=1}^{n} a_{ik}(x)u_{ik} + \sum_{i=1}^{n} b_i(x)u_i + c(x)u = f(x). \tag{28.2}$$

He used a few additional analytic ideas, connected with the appearance of the term $\sum_{i=1}^{n} b_i(x)u_i$, in deriving the maximum principles.

In my recent investigations wide new classes of general quasilinear elliptic equations were introduced (see Sections 26, 27). These classes contain all full linear elliptic equations as a particular case.

In Sections 26, 27 the author developed considerably geometric maximum principles for such classes of quasilinear elliptic equations. Also significant applications of these maximum principles to Differential Geometry, Calculus of Variations and Mechanics were developed by Bakelman [16], [17], [18], [19], [20], and Bakelman and Perry [1].

Applications to Calculus of Variations are presented in Sections 21–23 and Subsection 25.3; applications to Differential Geometry are presented in Section 23 and Subsections 28.2–3, finally applications to Mechanics are presented in Section 23.

Alexandrov's versions of maximum principles for full linear equations mentioned above can be deduced as corollaries of our results presented in Section 26. Therefore we did not consider the separate proofs of Alexandrov's results.

28.2 Estimates for Solutions of Quasilinear Elliptic Equations Connected with Problems of Global Geometry

This subsection is devoted to the study of problems belonging to the frontier between quasilinear elliptic equations and differential geometry. The contents of this subsection is based on the papers by Bakelman [18], [22] and by Bakelman and Kantor [1], [2].

Let E^{n+1} be the Euclidean space of dimension $n+1$ and let $x_1, x_2, \dots, x_n$, $x_{n+1} = z$ be a fixed Cartesian system of coordinates in E^{n+1}. The hyperplane $z = 0$ is denoted by E^n. We fix a domain B on E^n and assume that $\overline{B}$ is compact. Let $z(x) \in C^2(B) \cap C(\overline{B})$ be a solution of the equation

$$\sum_{i,k=1}^{n} a_{ik}(x, z, Dz)z_{ik} = b(x, z, Dz) + G(x, z, Dz), \tag{28.3}$$

where $x = (x_1, \dots, x_n) \in B$.

We assume that the function $z(x)$ satisfies the inequalities

$$-\infty < m_1 \le z(x) \le m_2 < +\infty \tag{28.4}$$

on ∂B, where m_1 and m_2 are certain constants.

Suppose, further, that functions $a_{ik}(x, z, p)$, $b(x, z, p)$ and $G(x, z, p)$ satisfy the following conditions when $x \in \overline{B}$, $z \in (\zeta_1, \zeta_2)$ and $p = (p_1, \dots, p_n) \in R^n$:

1)

$$\sum_{i,k=1}^{n} a_{ik}\xi_i\xi_k \ge 0 \tag{28.5}$$

for any $\xi = (\xi_1, \xi_2, \ldots, \xi_n) \in R^n$.

2)
$$\zeta_1 < m_1 \le m_2 < \zeta_2 \tag{28.6}$$

(the numbers ζ_1 and ζ_2 need not be finite).

We introduce the functions

$$G_1(x) = \overline{\lim_{h \to \xi_1}}\, G(x, h, 0), \tag{28.7}$$

$$G_2(x) = \underline{\lim_{h \to \xi_2}}\, G(x, h, 0). \tag{28.8}$$

It will be assumed below that the limit passages in (28.7) and (28.8) are uniform relative to $x \in \overline{B}$, i.e. for every $\varepsilon > 0$ there exists a $\beta > 0$ such that when $h \in (\zeta_1, \zeta_1 + \beta]$ or $h \in [\zeta_2 - \beta, \zeta_2)$ we have

$$G_1(x) \ge G(x, h, 0) - \varepsilon \tag{28.9}$$

or

$$G_2(x) \le G(x, h, 0) + \varepsilon \tag{28.10}$$

uniformly for all $x \in \overline{B}$.

Remark. It can be assumed by virtue of (28.6) that β can always be chosen so small that

$$\zeta_1 + \beta < m_1 \le m_2 < \zeta_2 - \beta.$$

Below we present two theorems proved by Bakelman and Kantor [1], [2].

Theorem 28.1. *Let $z(x) \in C^2(B) \cap C(\overline{B})$ be a solution of equation (28.3) and let the conditions, formulated at the beginning of Subsection 28.2, hold.*

If there exists an $\eta \in \left(0, \frac{1}{2}(\zeta_2 - \zeta_1)\right]$ such that the inequality

$$-b(x, z, 0) > G_1(x) \tag{28.11}$$

holds for all $z \in (\zeta_1, \zeta_1 + \eta]$ and $x \in \overline{B}$, then an uniform estimate from below exists for $z(x)$ in $\overline{B}$.

Proof. The proof is more easily carried out if condition (28.11) is reformulated in the following equivalent form: for any sufficiently small $\gamma > 0$ there exists an $\eta \in \left(0, \frac{1}{2}(\zeta_2 - \zeta_1)\right]$ such that the inequality

$$-b(x, z, 0) > G_1(x)$$

holds for all $z \in (\zeta_1, \zeta_1 + \eta]$ and $x \in \overline{B}$.

We are interested in the case when $\inf_{\overline{B}} z(x) < \inf_{\partial B} z(x)$. Here $\inf_{\overline{B}} z(x)$ is achieved in $\overline{B}$ at an interior point $x_0 \in B$. At this point $dz = 0$ and $d^2 z \ge 0$. Hence

$$b(x_0, z(x_0), 0) + G(x_0, z(x_0), 0) \ge 0. \tag{28.12}$$

If $z(x_0) \geq \zeta_1+\eta$, then the desired estimate has already been obtained. Suppose

$$z(x_0) < \zeta_1 + \eta.$$

Then from (28.11) and (28.12) we obtain

$$G(x_0, z(x_0), 0) - G_1(x_0) - \gamma \geq 0. \tag{28.13}$$

Let $\varepsilon = \frac{\gamma}{2}$, then there exists a $\beta > 0$ such that $\beta \leq \eta$ and for all $z \in (\zeta_1, \zeta_1 + \eta]$ and $x \in \overline{B}$ the inequality

$$G_1(x) \geq G(x, z, 0) - \frac{\gamma}{2} \tag{28.14}$$

holds. Let us prove that $z(x_0) \geq \zeta_1 + \beta$. If this as not so, (28.13) and (28.14) would imply

$$\begin{aligned} 0 &\leq G(x_0, z_0(x_0), 0) - G_1(x_0) - \gamma \\ &\leq G_1(x_0) + \frac{\gamma}{2} - G_1(x_0) - \gamma = -\frac{\gamma}{2} < 0, \end{aligned}$$

which is impossible. The theorem is proved.

Theorem 28.2. *Let $z(x) \in C^2(B) \cap C(\overline{B})$ be a solution of equation (28.3) and let the conditions, formulated at the beginning of Subsection 28.2, be fulfilled. If there exists an $\eta \in \left(0, \frac{1}{2}(\zeta_2 - \zeta_1)\right]$ such that for all $z \in [\zeta_2 - \eta, \zeta_2)$ and $x \in \overline{B}$ the inequality*

$$-b(x, z, 0) > G_2(x) \tag{28.15}$$

holds, then an uniform estimate from above holds for $z(x)$ in $\overline{B}$.

The proof of Theorem 28.2 is completely analogous to the proof of the preceding theorem.

Applications: 1) Estimates for Solutions of Variational Problems.

Let $z(x) \in C^2(B) \cap C(\overline{B})$ be a solution of $E - L$ equation for the functional

$$\int_B [F(x, z, Dz) + \varphi(x, z)]dx, \tag{28.16}$$

where the function $F(x, z, p)$ is twice continuous differentiable with respect to $x \in \overline{B}$, $z \in (\zeta_1, \zeta_2)$ and $p = (p_1, \dots, p_n) \in R^n$, the function $\varphi(x, z)$ is continuously differentiable with respect to $z \in (\zeta_1, \zeta_2)$, and in addition the form $\sum_{i,j=1}^{n} \frac{\partial^2 F}{\partial p_i \partial p_j} \xi_i \xi_j$ is positive definite.

Under these assumptions $E - L$ equation has the following form

$$\begin{aligned} \sum_{i=1}^{n} \frac{d}{dx_i}\left(\frac{\partial F}{\partial p_i}\right) - \frac{\partial F}{\partial z} - \frac{\partial \varphi}{\partial z} &\equiv \sum_{i,j=1}^{n} \frac{\partial^2 F}{\partial p_i \partial p_j} z_{ij} \\ &+ \sum_{i=1}^{n} \frac{\partial^2 F}{\partial p_i \partial x_i} + \sum_{i=1}^{n} \frac{\partial^2 F}{\partial p_i \partial z} p_i - \frac{\partial F}{\partial z} - \frac{\partial \varphi}{\partial z} = 0. \end{aligned} \tag{28.17}$$

In accordance with equation (28.1) we introduce the functions

$$G(x,z,p) = \frac{\partial F}{\partial z} - \sum_{i=1}^{n} \frac{\partial^2 F}{\partial p_i \partial z} p_i - \sum_{i=1}^{n} \frac{\partial^2 F}{\partial p_i \partial x_i}; \tag{28.18}$$

$$b(x,z) = \frac{\partial \varphi}{\partial z} \tag{29.19}$$

and with their help write down directly conditions (28.11) and (28.15) leading to uniform estimates for a solution $z(x)$ of equation (28.17).

2) Closed Starshaped Hypersurfaces in a Riemannian Space. This application is borrowed from my recent paper [22]. Let W^{n+1} be a C^∞-Riemannian space. Let x_0 be an arbitrary interior point of W^{n+1}. We consider all geodesic rays starting from x_0. These geodesics intersect geodesic hyperspheres with center x_0 orthogonally. If this construction is considered sufficiently close to the point x_0, then all pairs of geodesic rays, starting from x_0, do not have common points except x_0. We denote by $W^{n+1}_{x_0}$ a domain in W^{n+1} (except the point x_0) which is filled out by all these geodesic rays.

We now fix a geodesic hypersphere $Q^n \subset W^{n+1}_{x_0}$. Since Q^n is a C^∞-submanifold of W^{n+1}, we introduce an C^∞-atlas of coordinate charts $\{\theta^\alpha = (\theta^\alpha_1, \dots, \theta^\alpha_n)\}$ in Q^n and extend it to C^∞-atlas of semigeodesic charts $(\theta^\alpha, \rho) = (\theta^\alpha_1, \dots, \theta^\alpha_n, \rho)$ in $W^{n+1}_{x_0}$. Then in every such chart the Riemannian metric of $W^{n+1}_{x_0}$ has the following form

$$ds^2 = d\rho^2 + \sum_{i,j=1}^{n} g_{ij}(\theta^\alpha, \rho) d\theta^\alpha_i d\theta^\alpha_j. \tag{28.20}$$

Thus we can use a simpler geometric description of the Riemannian manifold $W^{n+1}_{x_0}$: Let $E^{n+1} = \{x = (x_1, \dots, x_n, x_{n+1})\}$ be a $(n+1)$-Euclidean space with a fixed system of Cartesian coordinates $x_1, x_2, \dots, x_{n+1}$. We introduce in E^{n+1} a spherical coordinate system θ, ρ, where $\theta = (\theta_1, \dots, \theta_n)$ is a point of the n-dimensional sphere S^n: $\rho = 1$ in E^{n+1}.

Let $r_1 < r_2$ be two nonnegative arbitrary numbers. The cases $r_1 = 0$ and $r_1 = +\infty$ are not excluded. We denote by $U(r_1, r_2)$ the annulus $r_1 < \rho < r_2$ in E^{n+1}. Clearly $U(r_1, r_2)$ is an open set in E^{n+1} and the origin of E^{n+1} does not belong to $U(r_1, r_2)$ for any $r_1 > 0$.

Let $g_{ij}(\theta, \rho), i,j = 1,2,\dots,n$, be C^∞-functions in $U(r_1, r_2)$. We always assume that the matrix $g_{ij}(\theta, \rho)$ is symmetric and positive definite in $U(r_1, r_2)$. Let $V^{n+1}(r_1, r_2)$ be a Riemannian manifold, which we define on $U(r_1, r_2)$ by the metric

$$ds^2 = d\rho^2 + \sum_{i,j=1}^{n} g_{ij}(\theta, \rho) d\theta_i d\theta_j. \tag{28.21}$$

Clearly the Riemannian manifold $W^{n+1}_{x_0}$ is C^∞-isometric to the Riemannian manifold $V^{n+1}(0, r)$ for an appropriate value of $r > 0$ (the case $r = +\infty$ is not excluded).

We will consider in $V^{n+1}(r_1, r_2)$ those hypersurfaces that are given by an equation of the form

$$\rho = \rho(\theta) \in C^2(S^n) \tag{28.22}$$

and satisfy the condition

$$0 < r_1 < \rho(\theta) < r_2. \tag{28.23}$$

We are concerned with the problem of existence of hypersurface (28.22), the mean curvature of which is prescribed as a point function in $U(r_1, r_2)$. This problem has a variational structure. The $E - L$ equation for this problem is as follows

$$\frac{1}{g^{1/2}}\left[\sum_{i=1}^{n} \frac{\partial}{\partial \theta_i}\left(\frac{\partial F}{\partial \rho_i}\right) - \frac{\partial F}{\partial \rho}\right] = nH(\theta, \rho), \tag{28.24}$$

where

$$F^2 = g + \sum_{i,j=1}^{n} g^{ij} \frac{\partial \rho}{\partial \theta_i} \frac{\partial \rho}{\partial \theta_j}. \tag{28.25}$$

In (28.24) and (28.25) g^{ij} is the cofactor of the element of g_{ij} of the matrix $\|g_{ij}(\theta, \rho)\|$, $g(\theta, \rho) = \det \|g_{ij}(\theta, \rho)\|$, and $H(\theta, \rho)$ is the mean curvature in $V^{n+1}(r_1, r_2)$ at a point $(\theta, \rho(\theta))$ of the desired hypersurface. Clearly $\frac{\partial \varphi(\theta, \rho)}{\partial \rho} = nH(\theta, \rho)$. By virtue of (28.18) and (28.19) we have

$$G(\theta, \rho, 0) = \frac{\partial F}{\partial \rho} = \frac{1}{2g(\theta, \rho)} \frac{\partial g(\theta, \rho)}{\partial \rho}, \tag{28.26}$$

$$b(\theta, \rho) = nH(\theta, \rho). \tag{28.27}$$

It is convenient to represent the function $H(\theta, \rho)$ in the form $H(\theta, \rho) = -\frac{1}{2ng(\theta, \rho)} \frac{\partial g(\theta, \rho)}{\partial \rho} + h(\theta, \rho)$. As is clear from the proof of Theorem 28.1, an estimate from below can be given for a solution $\rho(\theta)$ of the problem under consideration if the inequality $-b(\theta, \rho) > G(\theta, \rho, 0)$, which in our case reduces to the inequality $h(\theta, \rho) < 0$, holds for all $\theta \in S^n$ and sufficiently small ρ. Analogously, an estimate from above can be given for $\rho(\theta)$ if the inequality $h(\theta, \rho) > 0$ for all $\theta \in S^n$ and all $\rho \in (r_2 - \eta, r_2)$, where $\eta > 0$ is a sufficiently small number.

§29. The Dirichlet Problem for Quasilinear Elliptic Equations

29.1 Introduction

We consider quasilinear elliptic equations

$$\sum_{i,k=1}^{n} a_{ik}(x, u, Du) u_{ik} = b(x, u, Du), \tag{29.1}$$

where $x = (x_1, \ldots, x_n) \in \overline{B}$ and B is an open bounded domain in the Euclidean space R^n. We suppose that $a_{ik}(x, u, p)$, $i, k = 1, 2, \ldots, n$, and $b(x, u, p)$ are defined and continuously differentiable for all real values of u and p and all points $x \in \overline{B}$. We also suppose that $a_{ik}(x, u, p)$ form a symmetric positive definite matrix. (The last assumption is equivalent to ellipticity of equation (29.1).)

The classical Dirichlet problem for equation (29.1) consists in determining a function $u(x)$ such that:

a) $u(x) \in C^2(\overline{B})$;
b) $u(x)$ takes on given continuous boundary values $\varphi(x)$ on ∂B;
c) equation (29.1) is satisfied everywhere in B.

In order to establish the existence theorem of solutions of the Dirichlet problem it is required to have more information with respect to the set of solutions for auxiliary Dirichlet problems associated with the original one. Here is a typical example of such a theorem due to Serrin [4, § 1].

Let the boundary of B and the assigned boundary values $\varphi(x)$ be of class C^3. Also let τ be an arbitrary real parameter in the closed interval $[0, 1]$. Suppose there exists a constant $M \geq 0$ independent of τ, such that the conditions:

i) $v \in C^2(\overline{B})$;
ii) $v = \tau\varphi(x)$ on ∂B;
iii)

$$\sum_{i,k=1}^{n} a_{ik}(x, v, Dv) v_{ik} = b(x, v, Dv) \quad \text{in} \quad B$$

imply

$$\sup_{\overline{B}}(|v| + |Dv|) \leq M. \tag{29.2}$$

Then the Dirichlet problem for equation (29.1) has at least one solution for the given boundary values $\varphi(x)$. Moreover, this solution is twice continuously differentiable in $\overline{B}$.

The proof of this theorem involves two different sets of ideas — first, an a priori estimate concerning the equicontinuity of the gradient vector Dv, and second a homotopic approach that solutions of i), ii), iii) exist for each value of $\tau \in [0, 1]$.

For two dimensions such a program was initiated and carried by Bernstein [1–4], (1908–1912) under the assumption that solutions v are unique. Bernstein studied the Dirichlet problem for equation (29.1) in the space of analytic functions. Leray and Schauder [1], (1934) then established a general homotopic framework for the proof which can avoid Bernstein's uniqueness condition and replace his heavy considerations in the space of analytic functions by considerations on the space of Hölder functions $C^{m,\gamma}$, $m \geq 3$, $0 < \gamma < 1$. This step has given the considerable development in the Dirichlet problem for equation

(29.1). Moreover, the derivation of the requisite a priori estimates and presentation of the Bernstein and Leray and Schauder programs were significantly clarified and developed by Nirenberg [1]. Nirenberg [2] applied his results to the solution of classical global two-dimensional geometric problems by H. Weyl and H. Minkowski.

The framework for the Bernstein investigations was presented by Bakelman based on the Kantorowich–Newton method for functional equations and of results by Nirenberg [1], mentioned above.

The homotopic technique introduced by Leray and Schauder [1] remains unchanged in higher dimensions, but the a priori estimates related to the equicontinuity of Dv become significantly more difficult. These difficulties can be surmounted if only new ideas and techniques are used. For the case of divergence structure equations the a priori estimates mentioned above were obtained by a number of authors on the basis of the fundamental papers by De Giorgi [1], Nash [3], and Moser [1]. The corresponding existence theorems were obtained by Ladyzhenskaya and Ural'tzeva [1], Gilbarg [1], and others.

The theorem from Serrin's paper, mentioned above, can be derived as a consequence of results by Ladyzhenskaya and Ural'tzeva [1] and the Leray and Schauder Theorem. Here is Serrin's proof of this theorem, who uses a method due to Gilbarg [1].

Let w be an element of the Banach space $C^{1,\gamma}(\overline{B})$, where $0 < \gamma < 1$. Consider the linear elliptic equation

$$\sum_{i,j=1}^{n} A_{ij}(x)W_{ij} = B(x), \tag{29.3}$$

where $A_{ij}(x) = a_{ij}(x, w(x), Dw(x))$, $B(x) = B(x, w(x), Dw(x))$. Since the functions $A_{ij}(x)$ and $B(x)$ are Hölder continuous in the closure of B, Schauder's Theorem implies the existence of a unique solution $W(x) \in C^{2,\gamma}(\overline{B})$, which takes on the given boundary values $\varphi(x)$, $x \in \partial B$. This process clearly defines a mapping T of $C^{1,\gamma}(\overline{B})$ into bounded sets in $C^{2,\gamma}(\overline{B})$. We will show that the mapping T is completely continuous. By Arzela's Theorem it is evident that T maps bounded sets in $C^{1,\gamma}(\overline{B})$ into sequentially compact set in $C^{1,\gamma}(\overline{B})$. Thus it only remains to prove that T is a continuous mapping. Let $w \to w_0$ in $C^{1,\gamma}(\overline{B})$. We must prove that $Tw \to Tw_0$. Suppose for contradiction that this were true. Since the elements Tw are sequentially compact in $C^2(\overline{B})$ as well as in $C^{1,\gamma}(\overline{B})$, it follows that there is a sequence $\{w_m\}$ such that $w_m \to w_0$ in $C^{1,\gamma}(\overline{B})$, $W_m = Tw_m \to \widetilde{W}$ in $C^2(\overline{B})$, where $\widetilde{W} \neq Tw_0$. On the other hand, by (29.3) we obtain

$$\sum_{i,j=1}^{n} a_{ij}(x, w_m, Dw_m)\frac{\partial^2 W_m}{\partial x_i \partial x_j} = b(x, w_m, Dw_m). \tag{29.4}$$

If $m \to +\infty$, then we obtain

$$\sum_{i,j=1}^{n} a_{ij}(x, w_0, Dw_0)\frac{\partial^2 \widetilde{W}}{\partial x_i \partial x_j} = B(x, w_0, Dw_0). \tag{29.5}$$

Consequently $\widetilde{W} = Tw_0$, and from this contradiction it follows that T is completely continuous.

The desired theorem will be proved if we can show that T has a fixed point. Serrin applies the Leray–Schauder fixed point theorem in a simplified version due to Schaeffer. This version has the following statement: T will have a fixed point if there exists a constant M', independent of τ, such that

$$v = \tau \cdot Tv \tag{29.6}$$

implies $\|v\| \leq M'$. Here τ is any real number in [0,1], while the norm of v is that in the space $C^{1,\gamma}(\overline{B})$.

Let v be a solution of (29.6). By the definition of the mapping T it is evident that $v = \tau \cdot f$ on ∂B and

$$\sum_{i,k=1}^{n} a_{ik}(x, v, Dv) v_{ij} = \tau B(x, v, Dv) \quad \text{in} \quad B. \tag{29.7}$$

Hence, by the main assumption of the theorem, we obtain

$$\sup_{\overline{B}}(|v| + |Dv|) \leq M. \tag{29.8}$$

To complete the proof it must be shown that the quantity

$$\frac{|Dv(x) - Dv(y)|}{|x - y|^{\gamma}} \tag{29.9}$$

is uniformly bounded for all x, y in B and all τ in [0,1]. From (29.8) it follows that the arguments x, v, Dv in the coefficients of (29.7) are uniformly bounded. Thus, as far as the solution v is concerned, equation (29.7) may be considered uniformly elliptic and the functions a_{ik} and b are uniformly of class C^1 in their arguments. Now by a well known theorem of Ladyzhenskaya and Ural'tzeva [1], there exists a constant γ (depending on M, on the structure of equation (29.1), and on norms of the boundary and boundary data, but independent of τ) such that the quantity (29.9) is uniformly bounded. Making this choice of γ from the beginning we complete the proof of Serrin's Theorem, stated above in this subsection.

Let us finally formulate the existence program for the Dirichlet problem for equation (29.1): This program is based on the following succession of steps:

A) Determine the estimate for the maximum absolute value of the solution in $\overline{B}$.

B) Assuming the preceding step, one determines an estimate for the gradient of the solution at the boundary of B.

C) An estimate is then obtained for the gradient of the solution in the entire domain, depending on the results derived in the preceding steps.

The step A) was studied in detail by various methods in Chapters 7 and 8. Therefore in this section we are concerned with estimates for the gradient of solutions at the boundary ∂B and in the entire domain $\overline{B}$.

29.2 Estimates for the Gradient on the Boundary of ∂B. (The Method of Global Barriers)

There are a few methods for estimating the gradient of solutions of the Dirichlet problem for quasilinear elliptic equations. In this subsection we present the method of global barriers. Such barriers were constructed by Jenkins and Serrin [2] for minimal non-parametric hypersurfaces with prescribed Dirichlet boundary data. These barriers provide a complete solution of this problem. The method of global barriers was developed by Serrin [4] for solutions of general quasilinear elliptic equations. In Subsection 29.2 we present the main results of Serrin [4].

A) The Distance Function. Consider hypersurface S of class C^3 in R^n, having no self-intersections and no boundary. Let $d = d(x)$ be the function, defined as the distance from x to the hypersurface S.

Lemma 29.1. *Let the normal curvatures of S be bounded in absolute value by K. Then d is of class C^2 at all points whose distance from S is less than $1/K$.*

Proof. For points x whose distance from S is less than $1/K$, we define $y = y(x)$ to be the (unique) point on S nearest to x. Now let $\bar{x}$ be a fixed point whose distance from S is less than $1/K$. We introduce a specific coordinate frame such that the x_n axis is oriented along the normal of S at $y = y(\bar{x})$. In the neighborhood of the point $\bar{y}$ the surface S has the following representation

$$x_n = \varphi(x_1, \ldots, x_{n-1}),$$

where φ is of class C^3. Moreover, for all points x near $\bar{x}$ the following relations

$$x_i = y_i - \frac{\varphi_i(y)}{[1 + |D\varphi(y)|^2]^{1/2}} d, \quad i = 1, 2, \ldots, n-1;$$
$$x_n = \varphi(y) + \frac{d}{[1 + |D\varphi(y)|^2]^{1/2}}$$

hold, where $\varphi_i = \partial\varphi/\partial x_i$, $D\varphi$ is the gradient vector of φ, and $y = y(x)$. It should be noted that the arguments of φ, φ_i, and $D\varphi$ are $y_1, y_2, \ldots, y_{n-1}$. Relations (29.10) have the general form

$$x = f(y_1, \ldots, y_{n-1}, d), \tag{29.11}$$

where f is a function of class C^2 in its arguments. The function $d(x)$, which can be considered as arising by inversion of (29.11), will then be of class C^2

at $\bar{x}$ provided the Jacobian of transformation is non-zero at this point. It is allowable to assume that the coordinates $x_1, x_2, \ldots, x_{n-1}$ lie along the principal directions of S at the point $\bar{y}$. In these coordinates the gradient $D\varphi$ vanishes at $\bar{y}$ and the Hessian matrix $D^2\varphi$ is diagonal there. Since the diagonal elements are equal to the corresponding principal curvatures of S at $\bar{y}$, then the expression for the Jacobian at $\bar{x}$ is as follows:

$$\prod_{i=1}^{n-1}(1 - k_i d),$$

where k_i are the principal curvatures. This completes the proof.

Lemma 29.2. *In the special coordinates introduced above, we have at* $\bar{x}$

$$Dd = (0, \ldots, 0, 1)$$

and

$$D^2 d = -\begin{pmatrix} \frac{k_1}{1-k_1 d} & & \bigcirc & \\ & \ddots & & \\ \bigcirc & & \frac{k_{n-1}}{1-k_{n-1} d} & \\ & & & 0 \end{pmatrix},$$

where $k_1, \ldots, k_{n-1}$ *are the principal curvatures of* S *at* $\bar{y}$.

Proof of Lemma 29.2 follows from relations (29.10) by direct calculations (see Serrin [4, § 3]).

B) Basic Identities. Consider a bounded domain B whose boundary is of class C^3. Let $d = d(x)$ denote the distance from points $x \in \overline{B}$ to ∂B. Let $d_0 = 1/K$, where K is an absolute bound for the normal curvatures of ∂B. Finally let $0 < a = \text{const} < d_0$. For points x in the boundary strip $0 \le d \le a$, we define

$$v(x) = g(x) + h(d), \tag{29.12}$$

where g is twice continuous differentiable in $\overline{B}$, and h is continuous for $0 \le d \le a$ and twice continuously differentiable with respect to d for $0 < d < a$. We also suppose that $h' > 0$. These assumptions together with Lemma 29.1 provide that v is of class C^2 for $0 < d < a$.

We now introduce the following notation. Let

$$A = (a_{ij}) \quad \text{and} \quad Q = (q_i) \tag{29.13}$$

be two symmetric $(n \times n)$-matrices. Then the expression

$$AU = \sum_{i,j=1}^{n} a_{ij} q_{ij} \tag{29.14}$$

is called the *trace of matrices A and Q*. If $u(x)$ and $g(x)$ are functions of class C^2 in B (or $\overline{B}$) then we assume that

$$A = A(x, u, Du) = (a_{ij}(x, u, Du)) \tag{29.15}$$

and

$$Q = D^2 g = \left(\frac{\partial^2 g}{\partial x_i \partial x_j} \right), \tag{29.16}$$

where $Du = \left(\frac{\partial u}{\partial x_1}, \ldots, \frac{\partial u}{\partial x_n} \right)$ is the gradient of $u(x)$ and $D^2 g$ is the Hessian of $g(x)$.

Thus

$$A(x, u, Du)D^2 u = \sum_{i,j=1}^{n} a_{ij}(x, u, Du) u_{ij} \quad \text{and}$$

$$A(x, u, Du)D^2 g = \sum_{i,j=1}^{n} a_{ij}(x, u, Du) g_{ij}.$$

Let $A = (a_{ij})$ be a symmetric $(n \times n)$-matrix. Then the formula

$$\eta = \xi \cdot A$$

defines a linear transformation of n-dimensional Euclidean space R^n into R^n, where $\xi = (\xi_1, \ldots, \xi_n)$ and $\eta = (\eta_1, \ldots, \eta_n)$ are vectors in R^n. The quantity

$$\xi A \xi = (\xi A, \xi) = (\eta, \xi) \tag{29.17}$$

is the quadratic form in R^n. Clearly

$$\xi A \xi = (\xi A, \xi) = \sum_{i,j=1}^{n} a_{ij} \xi_i \xi_j \tag{29.18}$$

where $(\xi A, \xi)$ is the scalar product in R^n.

Let

$$L(v) = A(x, v, Dv)D^2 v - b(x, v, Dv). \tag{29.19}$$

Lemma 29.3. *For all x in the boundary strip $0 < d < a$ the following relation holds:*

$$L(v) = F\frac{h''}{h'} - Th' + AD^2 g - b. \tag{29.20}$$

The notation here has the following meaning: first,

$$A = (a_{ij}(x, v, p)), \quad b = b(x, v, p),$$

where

$$p = p_0 + h'\nu, \quad p_0 = Dg$$

($g(x)$ is the function introduced in relation (29.12)), and ν is the inner unit normal vector at the (unique) point $y(x)$ on ∂B; and second,

$$F = (p - p_0)A(p - p_0), \quad T = \sum_{i=1}^{n-1} \frac{\lambda_i A \lambda_i}{1 - k_i d} k_i, \tag{29.21}$$

where $k_1, \dots, k_{n-1}$ and $\lambda_1, \dots, \lambda_{n-1}$ are respectively the principal curvatures and the orts of principal directions of ∂B at the point $y(x)$.

Proof. Let $\bar{x}$ be an arbitrary fixed point in the boundary strip $0 < d < a$. We introduce new Cartesian coordinates (denoted with tildes) so that the $\tilde{x}_n$-axis coincides with the normal direction into B at the point $\bar{y} = y(\bar{x})$. We assume that the axes $\tilde{x}_1, \dots, \tilde{x}_{n-1}$ are aligned along the principal directions of the boundary surface ∂B at the point $\bar{y}$. According to well known orthogonal transformation relations, we obtain

$$L(v) = \tilde{A}(\tilde{x}, v, \tilde{D}v)\tilde{D}^2 v - \tilde{b}(\tilde{x}, v, \tilde{D}v),$$

where

$$\tilde{A} = SAS^{-1}, \quad \tilde{b} = b,$$

and S is the orthogonal transformation relating the original and new Cartesian coordinate systems, that is S is a $(n \times n)$-matrix which can be represented by the orthonormal frame $\lambda_1, \dots, \lambda_{n-1}, \nu$ in the following way

$$S = \begin{pmatrix} \lambda_1 \\ \vdots \\ \lambda_{n-1} \\ \nu \end{pmatrix}.$$

Using the definition of the function $v(x)$ together with Lemma 36.2, we obtain

$$L(v) = \tilde{A}\tilde{D}^2 g - \tilde{b} - \sum_{i=1}^{n-1} \frac{\tilde{A}_{ii} k_i}{1 - k_i d} h' + \tilde{A}_{nn} h'' \tag{29.22}$$

at the point $\bar{x}$. Reverting to the original coordinates, we have

$$\tilde{A}\tilde{D}^2 g = AD^2 g, \quad \tilde{b} = b,$$

and

$$\tilde{A}_{ii} = \lambda_i A \lambda_i, \quad i = 1, 2, \dots, n-1, \quad \tilde{A}_{nn} = \nu A \nu.$$

Moreover, at $\bar{x}$ the arguments of A and b obviously are $\bar{x}, v$, and $Dv = Dg + h'Dd = p_0 + h'\nu$. By inserting these relations into (29.22), and noting that $\nu = (p - p_0)/h'$, the proof is completed.

Lemma 29.4. *For x in the boundary strip $0 < d < a$ we have also*

$$L(v) = F\frac{h'' + Hh'}{(h')^2} - h' + AD^2g - b,$$

where H denotes the mean curvature of ∂B at $y(x)$,

$$G = \sum_{i=1}^{n-1} \frac{\lambda_i A \lambda_i}{1 - k_i d} k_i + (\nu A \nu) H,$$

and the remaining notation is the same as in Lemma 29.3.

This lemma is a slightly different version of Lemma 29.3, which will be useful below. To prove this lemma we add the quantity $\tilde{A}_{nn} H h'$ to the fourth term on the right-hand side of (29.22), and correspondingly subtract this quantity from the third term.

We note particularly the inequality

$$G \geq k \text{ trace } A,$$

where k stands for the minimum (signed) normal curvature of ∂B at the point $y(x)$. The proof of this inequality is as follows: clearly

$$\frac{k_i}{1 - k_i d} \geq k_i \geq k, \qquad H \geq k.$$

Hence

$$G \geq \left(\sum_{i=1}^{n-1} \lambda_i A \lambda_i + \nu A \nu \right) k.$$

The quantity in parentheses is the trace of A in the special coordinate frame $\lambda_1, \ldots, \lambda_{n-1}, \nu$. Since the trace is an orthogonal invariant, the required inequality is proved.

By the same argument we find

$$T \geq k \left(\sum_{i=1}^{n-1} \lambda_i A \lambda_i \right).$$

Since $k \geq -K$, where K is an absolute bound for the curvatures of ∂B, it follows that $T \geq -K$ trace A.

C) Basic Concepts. Global Barriers. Let B be a bounded domain, whose boundary is of class C^3. We assume that the boundary Dirichlet data is also of class C^3. Under these assumptions there exist C^3 extensions of the boundary data into the closure $\overline{B}$ of B, while conversely any C^3-function defined in $\overline{B}$ generates C^3 boundary data by restriction. Thus we may suppose without loss of generality that the boundary data is determined by a given function $f \in C^3(\overline{B})$. Let

$$c_0 = \sup_{\overline{B}} |f|, \quad c_1 = \sup_{\overline{B}} |Df|, \quad c_2 = \sup_{\overline{B}} \|D^2 f\|,$$

where the norm $\|A\|$ of a matrix A is defined to be the root square of the sum of the squares of the components of A.

Let K be an upper bound for the unsigned normal curvatures of the boundary surface ∂B, then the distance function $d = d(x)$ (see § 29.2, Subsection A) is also of class C^2 in the boundary strip $0 \leq d < d_0$, where $d_0 = 1/K$.

It will be convenient to normalize equation (29.1) by requiring

$$tr A = 1. \tag{29.23}$$

This condition will be used in the statements of all further results. From ellipticity of equation (29.1) it follows that the eigenvalues of A must lie between zero and one. Thus

$$0 < \xi A \xi < 1 \quad \text{for} \quad |\xi| = 1. \tag{29.24}$$

Below the following invariant functions will be considered:

$$\begin{aligned} C &= C(x,u,p) = \frac{B(x,u,p)}{|p|}, \quad |p| \neq 0; \\ E &= E(x,u,p) = p \cdot A(x,u,p) \cdot p; \\ F &= F(x,u,p) = (p - p_0) \cdot A(x,u,p) \cdot (p - p_0) \end{aligned} \tag{29.25}$$

(see notation in (29.25) in § 29.2, Subsection B).

The invariant E was introduced in 1912 by S.N. Bernstein [4]. Two other invariants were introduced in 1969 by J. Serrin [4]. In particular, Bernstein considered equations in two independent variables satisfying the condition

$$m_1 \cdot |p|^\ell \leq E \leq m_2 \cdot |p|^\ell \qquad |p| \geq 1, \tag{29.26}$$

where m_1 and m_2 are positive constants and 1 is a real number. It is clear that $\ell \leq 2$. The number $g = 2 - \ell$ was called by Bernstein the *genre* of equation (29.1).

Let a, M be positive constants and let N be the neighborhood $0 \leq d < a$ of ∂B. A *global barrier* is a function v in $C^2(N)$ which satisfies the condition

$$L(v + b) \leq 0 \quad \text{in} \quad N \tag{29.27}$$

for all positive constants b, and has the form

$$v(x) = f(x) + h(x),$$

where h is continuous in the closure of N and satisfies the conditions

$$h = 0 \quad \text{when} \quad d = 0, \quad h = M \quad \text{when} \quad d = a.$$

In (29.27) we use notation (29.19).

A set of global barrier functions, defined for all $M > 0$, will be called a *global barrier family*.

Lemma 29.5. *Let $u \in C^2(\overline{B})$ be a solution of the Dirichlet problem in B for equation (29.1). Suppose that $u \leq m$ and that there exits a global barrier function corresponding to $M = m + c_0$. Then*

$$\frac{\partial u}{\partial n} \leq L \quad \text{on} \quad \partial B$$

for every direction n into B. Here the bound L depends only on the global barrier function.

Proof. Let $v = f + h$ be the global barrier function. Clearly $u = v$ on ∂B. Since $c_0 = \sup_{\overline{B}} |f|$, when $u \leq v$ when $d = a$. Thus $u \leq v$ in N. Since both u and v are differentiable on ∂B,

$$\frac{\partial u}{\partial n} \leq \frac{\partial v}{\partial n} \quad \text{on} \quad \partial B$$

for every direction $\vec{n}$ into B. Setting $L = \sup_B |Dv|$ completes the proof.

D) Regularly Elliptic Equations. According to Serrin [4, § 8] equation (29.1) is called *m-regularly elliptic* if the inequality

$$\frac{1 + |C|}{E} \leq \phi(|p|) \tag{29.28}$$

holds for $|u| \leq m, |p| \neq 0$, where $\phi(\rho)$, $0 < \rho < +\infty$, is a decreasing continuous function satisfying the condition

$$\int_\alpha^{+\infty} \frac{d\rho}{\rho^2 \phi(\rho)} = +\infty \tag{29.29}$$

for all $\alpha > 0$.

Examples:

1. Suppose that $E \geq \mu |p|$ for $|p| \geq 1$ and that for all $m > 0$

$$\frac{|B|}{E} \leq \gamma(m) \ln(1 + |p|). \tag{29.30}$$

Then it is possible to take

$$\phi(\rho) = \frac{\mu^{-1} + \gamma \ln(1 + \rho)}{\rho}.$$

Clearly equation (29.1) is regularly elliptic for each m. If in particular equation (29.1) has a genre $g \leq 1$, then it is regularly elliptic provided that (29.30) holds. An even more special case occurs for uniformly elliptic equations, these being regularly elliptic provided that

$$|B| \leq \gamma(m) \cdot |p|^2 \ln(1 + |p|)$$

for $|u| \leq m$ and $|p| \geq 1$.

2. A non-uniformly elliptic equation

$$(1 + u_x^2)u_{xx} + 2u_x u_y u_{xy} + (1 - u_y^2)u_{yy} = 0$$

is regularly elliptic, because the genre of this two-dimensional equation is zero.

3. If an equation has genre $g \geq 1$ then it can not be regularly elliptic. In particular, neither the minimal surface equation nor the equation for surfaces of constant mean curvature are regularly elliptic.

Lemma 29.6. *Suppose that* $|p_0| \leq c$. *Then*

$$\frac{1}{2}E - 2c^2 \leq F \leq 2E + 2c^2.$$

Proof. Since A is positive definite and has unit trace, then

$$|\xi A \zeta| \leq \frac{1}{2}(\xi A \xi + \zeta A \zeta) \leq \frac{1}{2}(|\xi|^2 + \zeta A \zeta),$$

for any vectors ξ and ζ. Setting $\xi = Ap$ and $\zeta = p$ then yields $|Ap|^2 \leq pAp$. Now

$$F = (p - p_0)A(p - p_0) = pAp - 2p_0 Ap + p_0 A p_0.$$

For $|p_0| \leq c$ we have obviously $|p_0 Ap| \leq c \cdot |Ap| \leq c \cdot \sqrt{(pAp)}$. Hence, recalling the definition of E, we obtain

$$E - 2c\sqrt{E} \leq F \leq E + 2c\sqrt{E} + c^2.$$

The required conclusion is now an immediate consequence of Cauchy's inequality.

Theorem 29.1 (Serrin [4, § 8]). *Let* $u \in C^2(\overline{B})$ *be a solution of the Dirichlet problem (29.1–2) in* B. *Assume that* $|u| \leq m$ *in* $\overline{B}$. *Then if (29.1) is* m*-regularly elliptic we have*

$$|Du| \leq L \quad \text{on} \quad \partial B,$$

where L *depends only on* c_0, c_1, c_2, m, K, *and the function* ϕ. *(The numbers* K, c_0, c_1, c_2 *are defined in Subsection C) of §29.2.)*

Proof. First we assume that (29.28) holds without the added restriction $|u| \leq m$; that is, we assume

$$\frac{1 + |C|}{E} \leq \phi(|p|) \tag{29.31}$$

for all $x \in \overline{B}$, all real u, and all $p \neq 0$. This assumption will be removed at the final part of the proof.

First of all we construct an appropriate global barrier family. Let

$$v(x) = f(x) + h(d), \qquad 0 \le d \le a,$$

where $a < d_0$. We suppose that h is twice continuously differentiable with respect to d and that

$$h(0) = 0, \quad h(a) = M, \quad h'(d) \ge \alpha, \tag{29.32}$$

where α is a positive constant whose value will be specified later. In order to evaluate $L(v+b)$ we may apply Lemma 29.1. Since $T \ge -K$, this yields

$$L(v+b) \le Fh''/(h')^2 + Kh' + AD^2 f - B, \tag{29.33}$$

where the arguments of A, B and F being $x, v+b$, and $p = p_0 + \nu h'$, where $p_0 = Df$.

Now $E \ge \frac{1}{\phi}$, and ϕ necessarily tends to zero as $|p|$ tends to infinity (because integral (29.29) diverges). Therefore there exists a constant α_1, depending only on c_1 and ϕ, such that

$$E \ge 8c_1^2 \quad \text{for} \quad |p| \ge \alpha_1.$$

We now set

$$\alpha = \max(c_1 + \alpha_1, MK, 1).$$

Since $p = p_0 + \nu \cdot h'$ and $|p_0| \le c_1$, then

$$\alpha_1 \le h' - c_1 \le |p| \le 2h'.$$

We now can estimate the coefficients A, B and F in the inequality for $L(v+b)$. Since eigenvalues of A lie between 0 and 1, we obtain

$$|AD^2 f| \le \|D^2 f\| \le c_2.$$

Next

$$|B| \le |p| \cdot |C| \le 2h' \cdot |C|,$$

and finally using Lemma 29.6 and the fact that $E \ge 8c_1^2$, we obtain

$$F \ge \frac{1}{2}E - 2c_1^2 \ge \frac{1}{4}E.$$

Inserting these estimates into (29.33) yields

$$\begin{aligned} L(v+b) &\le Fh'\left\{\frac{h''}{(h')^3} + 4\frac{c_2 + K + 2|C|}{E}\right\} \\ &\le Fh'\left\{\frac{h''}{(h')^3} + c\phi(|p|)\right\} \le (c = 4c_2 + 4K + 8) \\ &\le Fh'\left\{\frac{h''}{(h')^3} + c\phi(h' - c_1)\right\}, \end{aligned}$$

the last step holding because ϕ is a decreasing function of $|p|$. We now choose h' so that expression in braces vanishes and so that conditions (29.32) hold. In order to do this, define β by means of the relation

$$cM = \int_\alpha^\beta \frac{d\rho}{\rho^2\phi(\rho - c_1)}$$

$\left(\text{this is possible since } \int_\alpha^{+\infty} \frac{d}{\rho^2\phi(\rho-c_1)} = +\infty\right)$. Next, let h and d be defined by

$$ch = \int_\rho^\beta \frac{d\rho}{\rho^2\phi(\rho - c_1)}, \quad cd = \int_\rho^\beta \frac{d\rho}{\rho^3\phi(\rho - c_1)},$$

where $\alpha \leq \rho \leq \beta$. Since $h' = \rho \geq \alpha$ then the quantity in braces vanishes. Moreover

$$h(0) = 0, \quad h(a) = M,$$

where

$$a = \frac{1}{c}\int_\alpha^\beta \frac{d\rho}{\rho^3\phi(\rho - c_1)},$$

(note that $a < \frac{M}{\alpha} \leq d_0$).

Thus we have constructed for each $M > 0$ a global barrier function with $a < d_0$. By Lemma 29.5 it follows that

$$\frac{\partial u}{\partial n} \geq -L \quad \text{on} \quad \partial B,$$

where L depends only on the global barrier function corresponding to $M = m + c_0$. The latter depends, moreover, only on the quantities listed in the statement of the Theorem 29.1. Replacing u with $-u$ in the equation leaves the construction unaltered. Hence we also have

$$\frac{\partial u}{\partial n} \geq -L \quad \text{on} \quad \partial B,$$

completing the proof of the theorem subject to the initial assumption (29.31). If (29.31) does not hold, we consider a new equation with coefficient matrix A defined by

$$A(x,u,p) = \begin{cases} A(x,-m,p) & \text{if} \quad u \leq -m, \\ A(x,u,p) & \text{if} \quad -m \leq u \leq m, \\ A(x,u,p) & \text{if} \quad u > m, \end{cases}$$

and with a similarly defined inhomogeneous term B. Clearly u is also a solution of the equation $\hat{A}D^2 = \widehat{B}$, and this fact together with the evident relation

$$\frac{1 + |C|}{E} \leq \phi(|p|), \qquad |p| \neq 0,$$

allows us to repeat the previous proof in the same way. Theorem 29.1 is proved.

E) Boundary Estimates Depending on Curvature. In this subsection a priori estimate for $\frac{\partial u}{\partial n}$ on ∂B will be considered if equation (29.1) is not regularly elliptic. For such equations it turns out that the curvatures of the boundary hypersurface must be restricted in order to obtain the desired estimates.

The following conditions on the asymptotic behaviour of the invariants A and C will be considered. Below we assume that

$$A(x,u,p) = A_0(x,\sigma) + O(1), \qquad \sigma = \frac{p}{|p|}, \tag{29.34}$$

and

$$C(x,u,p) = C_0(x,u,\sigma) + O(1), \qquad \sigma = \frac{p}{|p|}, \tag{29.35}$$

as $|p| \to \infty$, where $A_0(x,\sigma)$ and $C_0(x,u,\sigma)$ are continuously differentiable functions of their arguments and where

$$\frac{\partial C_0}{\partial u} \geq 0. \tag{29.36}$$

Below in relations (29.34–36) *order terms are assumed to be uniform for* (x,u) *in any compact set, unless otherwise stated.* Condition (29.36) will be tacitly assumed whenever we deal with the asymptotic relation (29.35).

The matrix A_0 can be used to introduce a generalized mean curvature of the boundary hypersurface ∂B. Let y be a point on ∂B and let ν be the unit inner normal at y. We denote by $k_1,\ldots,k_{n-1}$ and $\lambda_1,\ldots,\lambda_{n-1}$ respectively the principal curvatures and principal directions of ∂B at y. We now set

$$A = A(y) = A_0(y,\nu)$$

and define

$$\mathcal{H} = \mathcal{H}(y) = \sum_{i=1}^{n-1}(\lambda_i A\lambda_i)k_i + (\nu A\nu)H,$$

where H is the ordinary mean curvature of ∂B at y. The function which arises upon replacing ν by $-\nu$ in this formula will be denoted by $T(y)$. It is worth emphasizing that $\mathcal{H}$ and T are *averages* of the principal curvature *) and both quantities are orthogonal invariants, exactly as is the case with the mean curvature. Moreover, if the matrix A is proportional on ∂B to the Euclidean metric tensor, then $\mathcal{H} = T = H$. This condition automatically holds in two

---

*) Actually $\mathcal{H} = \sum_i a_i k_i$, $i = 1,2,\ldots,n-1$, where $a_i \geq 0$, $\sum_i a_i = 1$ with a similar relation holding for T. Two last conditions are a direct consequence of two facts, a) the numbers $(\lambda_i A\lambda_i)$ and $\nu A\nu$ are the diagonal elements of the matrix A in the coordinate frame $\lambda_1,\ldots,\lambda_{n-1}\nu$, and b) the matrix A_0 necessarily has unit trace and is nonnegative definite.

dimensions, so that $\mathcal{H} = T = k$ in this case, where k is the curvature of the boundary curve.

Example. We illustrate the construction of A_0, C_0, and T by means of the equation for hypersurfaces of prescribed mean curvature

$$(1+|Du|^2)\Delta u - \sum_{i,k=1}^{n} u_i u_k u_{ik} = n\varphi(x)(1+|Du|^2)^{3/2}.$$

After normalizing to unit trace, we find

$$A_0 = \frac{I - \sigma\sigma}{n-1}, \quad C_0 = \frac{n}{n-1}\varphi;$$

clearly the error terms in (29.34) and (29.35) are of degree -2 and $\lambda_i A \lambda_i = (n-1)^{-1}$ and $\nu A \nu = 0$. Therefore $\mathcal{H} = T = H$.

Theorem 29.2 (Serrin). *Let $u \in C^2(\overline{B})$ be a solution of the Dirichlet (29.1–2) in $\overline{B}$. Assume that $|u| \le m$ in $\overline{B}$. If (29.34) and (29.35) hold, and if both conditions*

$$\mathcal{H} > -C_0(y, f, \nu), \quad T > C_0(y, f, -\nu) \tag{29.37}$$

are satisfied at each point y of the boundary hypersurface, then

$$|Du| \le L \quad \text{on} \quad \partial B,$$

where L depends only on $c_0, c_1 c_2, m, K$, a lower bound for the differences in (29.37), bounds for the error terms in (29.34) and (29.35), and C^1 norms of the matrix A_0 and the function C_0.

Proof. Without loss of generality we can assume that the error terms in (29.34) and (29.35), as well as the norms of A_0 and C_0, are uniformly bounded with respect to u. This is based on the arguments presented in the final step of the proof of Theorem 29.1. As before, the proof will be carried out by constructing an appropriate global barrier family and applying Lemma 29.5.

Let

$$v(x) = f(x) + h(d), \qquad 0 \le d \le a,$$

where $a < d_0$. It is further assumed that h is twice continuously differentiable with respect to d and that

$$h(0) = 0, \quad h(a) = M, \quad h'(d) \ge \alpha, \tag{29.38}$$

where a positive constant α will be determined later. By Lemma 29.4

$$L(v+b) = F\frac{h'' + Hh'}{(h')^2} - Gh' + AD^2 f - B,$$

where the arguments of A, B and F are $x, v+b$, and $p = p_0 + \nu h'$.

Let $\theta > 0$ be chosen so that

$$\mathcal{H} > -C_0(y, f, \nu) + 5\theta$$

at each point of the boundary hypersurface (this can be done since the terms in (29.37) are continuous functions of y).

An easy estimate yields $|p| \geq \alpha - c_1$ and

$$\begin{aligned}|A(x, v+b, p) - A_0(x, \nu)| &\leq |A(x, v+b, p) - A_0(x, \sigma)| \\ &\quad + \frac{2c_1}{|p|} \sup \left| \frac{\partial A_0}{\partial \sigma} \right|.\end{aligned}$$

Moreover, if $y = y(x)$ denotes the point on the boundary nearest x, we obtain

$$\|A_0(x, \nu) - A_0(y, \nu)\| \leq d \sup \left\| \frac{\partial A_0}{\partial x} \right\|.$$

Thus when α is suitably large and d is sufficiently small, the difference

$$A(x, v+b, p) - A(y)$$

becomes arbitrarily small. In particular, for $\alpha \geq \alpha_1$ and $d \leq a_1$,

$$G \geq \sum_{i=1}^{n-1} (\lambda_i A \lambda_i) k_i + (\nu A \nu) H \geq \mathcal{H} - \theta. \tag{29.39}$$

In (29.39) a_1 and α_1 depend only on θ, c_1, K, bounds for the error term $\|A - A_0\|$, and a bound of the C^1-norm of A_0. Similarly for α suitably large we obtain

$$C(x, v+b, p) \geq C_0(x, v+b, \nu) - \theta \geq C_0(x, f(x), \nu) - \theta$$

because C_0 is increasing in its second argument. Since $|Df| \leq c_1$ we may assume further that

$$C_0(x, f(x), \nu) \geq C_0(y, f(y), \nu) - \theta \quad \text{for} \quad d \leq a_2.$$

This leads to the required estimate

$$\begin{aligned}B(x, v+b, p) = |p| \cdot C(x, v+b, p) &\geq |p|[C_0(y, f, \nu) - 2\theta] \\ &\geq h'\{[C_0(y, f, \nu) - 3\theta]\}\end{aligned} \tag{29.40}$$

provided $\alpha \geq \alpha_2$ and $d \leq a_2$. Now let

$$\alpha = \max(\alpha_1, \alpha_2, M/a_1, M/a_2, MK, c_2\theta).$$

Then according to (29.39) and (29.40) we have (provided $a < \min(a_1, a_2, d_0)$)

$$L(v+b) \le F\frac{h'' + Hh'}{(h')^2} + h'[-\mathcal{H} - C_0(y, f, \nu) + 5\theta)] \le F\frac{h'' + Kh'}{(h')^2},$$

where we are using the definition of θ and the fact that $H \le K$. We now set

$$h(d) = M\frac{1 - e^{-Kd}}{1 - e^{-Ka}}, \quad \text{where} \quad a = \frac{1}{K}\ln\left(1 + \frac{MK}{\alpha}\right).$$

It is easy to check that $h'' + Kh' = 0$ and that conditions (29.38) hold. Moreover, $a < \min(a_1, a_2, d_0)$, so that finally

$$L(v+b) \le 0 \quad \text{for} \quad 0 < d < a.$$

The rest of the proof is almost exactly the same as that of Theorem 29.1. The only significant change is that when u is replaced by $-u$ we must simultaneously replace $A(x, u, p)$ by $A(x, -u, -p)$ and $B(x, u, p)$ by $B(x, -u-p)$. This accounts for the second condition in (29.38), in which $\mathcal{H}$ has been replaced by T and $C_0(y, f, \nu)$ by $-C_0(y, f, -\nu)$.

The proof of Theorem 29.2 is completed.

According to the classification of Serrin [4, § 10], equation (29.1) is (c, m)-*boundedly non-linear* provided that (29.34) and (29.35) hold and

$$\frac{\|A - A_0\| + |C - C_0| + |p|^{-1}}{F} \le \phi(|p|) \tag{29.41}$$

for $|u| \le m$, $|p_0| \le c$, $|p| > c$, where $\phi(\rho)$ is a decreasing continuous function satisfying condition

$$\int^{+\infty} \frac{d\rho}{1 + \rho^2\phi(\rho)} = +\infty.$$

The equality sign can be allowed in conditions (29.37) if equation (29.1) is (c, m) boundedly non-linear. Consider some particular cases of (c, m)-boundedly non-linear quasilinear equations.

1. Suppose that $F \ge \mu(m, c) > 0$ and

$$\frac{\|A - A_0\| + |C - C_0|}{F} \le \gamma(m, c)\frac{\ln(1 + |p|)}{|p|}$$

when $|p| \ge 1 + c$. Then we can take

$$\phi(|p|) = \frac{\mu - 1 + \gamma\ln(1 + \rho)}{\rho}, \quad \rho \ge 1 + c$$

and (29.1) is boundedly non-linear for each m and c.

If equation (29.1) has a well-defined genre $g, 1 < g < 2$, then it is boundedly non-linear provided

$$\|A - A_0\| + |C - C_0| \leq \gamma(m) \cdot |p|^{1-g} \quad \text{for} \quad |u| \leq m, |p| \geq 1.$$

(Lemma 29.6 implies $F \approx |p|^{2-g}$ for large p). Conversely, if $g > 2$ then equation (29.1) can not be boundedly non-linear.

2. The case where equation (29.1) has genre 2 is particularly interesting in view of the large number of interesting examples, including the minimal hypersurfaces equation and the equation for hypersurfaces with constant mean curvature. The following lemma states a sufficient condition for an equation of genre 2 to be boundedly non-linear.

Lemma 29.7. *Assume that equation (29.1) has genre 2 and that*

$$A - A_0 = o(|p|^{-1}), \quad C - C_0 = O(|p|^{-1})$$

as p tends to infinity. Then equation (29.1) is boundedly non-linear. If for large p the least eigenvalue of A is bounded below by a positive multiple of $|p|^{-2}$, then it is possible to replace $o(|p|^{-1})$ by $O(|p|^{-1})$ if we assume that $A - A_0 = O(|p|^{-1})$.

This lemma was established by Serrin [4].

The main result related to boundedly nonlinear equations (29.1) is as follows.

Theorem 29.3 (Serrin). *Let $u \in C^2(\overline{B})$ be a solution of the Dirichlet problem (29.1–2) in $\overline{B}$. Assume that $|u| \leq m$ in $\overline{B}$ and that (29.1) is (c_1, m)-boundedly non-linear. Then if both conditions*

$$\mathcal{H} \geq -C_0(y, f, \nu), \quad T \geq C_0(y, f, -\nu) \tag{29.42}$$

hold at each point of the boundary hypersurface, the following estimate

$$|Du| \leq L \quad \text{on} \quad \partial B$$

can be obtained, where L depends only on c_0, c_1, c_2, m, K, the function $\phi(\rho)$, and C^1 norms of the functions A_0 and C_0.

We refer the reader to Serrin [4], pages 439–440 for the proof of this theorem. Theorem 29.3 slightly extends the conclusion of Theorem 29.2.

As we mentioned above the minimal surface equation is boundedly non-linear; moreover in this case $\mathcal{H} = T = H$, where H is the mean curvature of the boundary hypersurface. Thus from Theorem 29.3 it follows that step B in the existence program (see page 614) can be carried out if

$$H \geq 0 \tag{29.43}$$

at each point of the boundary hypersurface ∂B. Now step A of the same program is trivial for the minimal surface equation, and step C follows from the fact that the derivatives of a solution satisfy the maximum principle.

Hence the Dirichlet problem for the minimal surface equation in a C^3 domain B is solvable for arbitrary prescribed C^3 boundary data provided that $H \geq 0$ at each boundary point.

This result was proved by H. Jenkins and J. Serrin [2].

29.3 Estimates of the Gradient of Solutions on the Boundary. (The Method of Convex Majorants)

In this subsection we present another method of estimating of the gradient of solutions of the Dirichlet problem for general quasilinear elliptic equations. This method is based on geometric ideas related to the theory of convex bodies and to the theory of generalized convex (concave) solutions for the Monge–Ampere equations. This method was developed by Bakelman [8], [10].

a) Twisting. Let B be a domain of R^n that is bounded by a smooth closed hypersurface ∂B. Let S be the graph of $h(x) \in C^{2,\beta}(\partial B)$, where $0 < \beta \leq 1$. Clearly S an $(n-1)$-dimensional surface in R^{n+1}. Let X be any point of S and let x be its projection on R^n. Denote by T and t, respectively, the tangent planes to S at X and to ∂B at x. Clearly t is the projection of T. Finally, let Q^- be a hyperplane passing through T and leaving T above. We suppose that Q^- is given by equation

$$z = \sum_{i=1}^{n} a_i x_i + b.$$

We define the *lower twisting* of S at X as the number

$$\mu_L(X,S) = \inf \sum_{i=1}^{n} a_i^2, \tag{29.44}$$

where the infimum is taken over all such hyperplanes Q^-. Similarly, by means of hyperplanes Q^+ passing through T and leaving S from below, we define the *upper twisting* $\mu_U(X,S)$ of S at X. The number

$$M_L(S) = \sup_{X \in S} \mu_L(X,S), \tag{29.45}$$

$$M_U(S) = \sup_{X \in S} \mu_U(X,S) \tag{29.46}$$

are called, respectively, the *lower* and *upper twistings* of S.

b) Suffice Conditions for Upper Estimates of Twistings. In what follows we suppose that the convex surface ∂B satisfies the following conditions:

1) ∂B is of class $C^{2,\beta}$ $(0 < \beta \leq 1)$, that is, at each point $x \in \partial B$ there is a neighborhood G_x that is determined in local coordinates $u_1, \dots, u_n$ by an equation

$$u_n = f(u_1, \dots, u_{n-1}) \in C^{2,\beta},$$

where the axes $u_1, \ldots, u_{n-1}$ lie in the tangent plane to ∂B at x and the u_n-axis points along the inward normal $\overline{m}$ of ∂B at x.

2) All the principal curvatures $k_1, \ldots, k_{n-1}$ at any point $x \in \partial B$ are not less than a constant $\tau_0 > 0$.

The following theorem holds:

Theorem 29.4. *The quantities $\mu_L(X, S)$ and $\mu_U(X, S)$ are bounded above, depending only on τ_0, $\|h\|_{2,\beta}$, $\|\partial B\|_{C^1}$, $\|G_x\|_{2,\beta}$, where $\beta \in (0, 1]$ is arbitrary and x is any point of ∂B.*

Theorem 29.5. *The quantities $M_L(S)$ and $M_U(S)$ are bounded above, depending only on τ_0, $\|h\|_{2,\beta}$, $\|\partial B\|_{2,\beta}$, where $\beta \in (0, 1]$ is arbitrary.*

Theorem 29.5 is an immediate corollary of Theorem 29.4, a detailed proof of which was given by Bakelman [8]. We omit it here. Note Bakelman [8] established that we can choose for G_x any domain $G \subset \partial B$ containing a point x which projects one-to-one onto the tangent plane t to ∂B at x inside an arbitrary small ball in t.

c) Upper and Lower Minorants of $R(|p|)$. Estimates of the Normal Derivative of Solutions for the Dirichlet Problem (29.1–2). Suppose that one the space P^n with the Cartesian coordinates $p_1, \ldots, p_n$ we are given a continuous strictly positive function $R(|p|)$. Further, let $M \geq 0$ be given. We construct from $R(|p|)$ a new function $N(p, R, M) = \inf R(|p|)$, where the infimum is taken over the ball of radius $\sqrt{M}$ and centre at $p = (p_1, \ldots, p_n)$. Since $R(|p|)$ is a function of the length of the vector p, it is clear that $N(p, R, M)$ is a function of $|p|$ rather than of p; we shall emphasize this by writing $N(|p|, R, M)$ rather than $N(p, R, M)$.

The number M is always related to the upper and lower twisting of the boundary of the required solution. In this connection we use the following notation:

$$\begin{aligned} N_L(|p|, R) &= N(|p|, R, M_L(S)), \\ N_U(|p|, R) &= N(|p|, R, M_u(S)). \end{aligned} \tag{29.47}$$

We call the first and the second function in (29.47) the *lower* and *upper minorants of* $R(|p|)$.

Let B be a bounded convex domain in R^n satisfying conditions 1) and 2) of § 29.3, b. We consider in $\overline{B}$ the Dirichlet problem (29.1–2), where $h(x) \in C^{2,\beta}(\partial B)$. Regarding equation (29.1) we assume that it satisfies the following conditions:

1) For any real vector $\xi = (\xi_1, \xi_2, \ldots, \xi_n)$

$$\sum_{i,k=1}^{n} a_{ik}(x, z, p)\xi_i\xi_k > 0 \tag{29.48}$$

if $|\xi| > 0$, where $x \in \overline{B}$, $z \in (-\infty, +\infty)$, $p \in P^n$. Condition (29.48) insures that equation (29.1) is elliptic and throughout $\overline{B} \times R \times P^n$ we have

$$\det(a_{ik}(x,z,p)) > 0. \tag{29.49}$$

2) Throughout $\overline{B} \times R \times P^n$, for arbitrary fixed $x \in \overline{B}$ and $p \in P^n$, the function

$$b(x,z,p)[\det(a_{ik}(x,z,p)]^{-1/n} \tag{29.50}$$

does not decrease with $z \in R$; furthermore, for any two numbers m, M with $m \leq M$

$$b(x,M,p)[\det(a_{ik}(x,M,p))]^{-1/n} \leq \frac{\varphi_{+,M}}{R_{1,M}(|p|)}, \tag{29.51}$$

$$b(x,m,p)[\det(a_{ik}(x,m,p))]^{-1/n} \geq -\frac{\varphi_{-,m}}{R_{2,m}(|p|)} \tag{29.52}$$

suppose that $R_{1,M}(|p|)$, $R_{2,m}(|p|)$, $\varphi_{+,M}$, $\varphi_{-,m}$ satisfy the conditions

a) $R_{1,M}(|p|) > 0$ and $R_{2,m}(|p|) > 0$ are locally n-th power summable in P^n;

b) $\varphi_{+,M} \geq 0$, $\varphi_{-,m} \geq 0$ are constants depending on M and m respectively. In applications a simpler version of condition 2) is often encountered: namely it suffices that

2') for all $x \in \overline{B}$, $z \in R$, $p \in P^n$ the following inequalities

$$-\frac{\varphi_-}{R_2(|p|)} \leq \frac{b(x,z,p)}{[\det(a_{ik}(x,z,p))]^{1/n}} \leq \frac{\varphi_+}{R_1(|p|)} \tag{29.53}$$

hold, where $R_i(p) > 0$ $(i = 1,2)$ are locally n-th power summable in P^n and $\varphi_+ \geq 0$, $\varphi_- \geq 0$ are constants.

Next we introduce the numbers

$$h_1 = \sup_{\partial B} h(x), \quad h_2(x) = \inf_{\partial B} h(x). \tag{29.54}$$

By Theorem 29.5 the surface S constructed from the boundary condition

$$z|_{\partial B} = h(x) \in C^{2,\beta}(\partial B) \tag{29.55}$$

has finite lower and upper twistings $M_L(S)$ and $M_U(S)$. We construct the lower minorant $N_L(p, R^n_{h_1})$ from the function $R_{h_1}(p)$ and from the lower twisting $M_L(S)$; similarly we form the upper majorant $N_U(|p|, R^n_{h_2})$ from $R^n_{h_2}(|p|)$ and $M_U(S)$.

Suppose now that $z(x) \in C^2(\overline{B})$ is a solution of the Dirichlet problem (29.1), (29.55). Let $X \in S$ be any point, and let $x \in B$ be its projection. Then through the tangent plane T to S at X we can find planes Q_1 and Q_2 leaving the surface S from below and above, respectively. If

$$Q_1:\ z = (a^1, x) + b^2, \quad Q_2:\ z = (a^2, x) + b_2$$

are the equation of these planes, then it follows from the definitions of $M_U(S)$ and $M_L(S)$ that Q_1 and Q_2 can be chosen so that

$$|a^1| \le \sqrt{M_U(S)}, \quad |a^2| \le \sqrt{M_L(S)}. \tag{29.56}$$

We denote by K_x the ball of radius $1/\tau_0$ in R^n containing $\overline{B}$ and tangent to ∂B at x. We consider in K_x the two Dirichlet problems

$$N_U(|Dv^{(2)}|, R^n_{h_2}) \det(v^{(2)}_{ij}) = \frac{1}{n^n}(-\varphi_{h_2})^n, \tag{29.57}$$
$$v^{(2)}|_{\partial K_x} = 0,$$
$$N_L(|Dv^{(1)}|, R^n_{h_1}) \det(v^{(1)}_{ij}) = \frac{1}{n^n}(\varphi_{h_1})^n, \tag{29.58}$$
$$v^{(1)}|_{\partial K_x} = 0,$$

where

$$\varphi_{+,h_1} = \varphi_{h_1} = \text{const} \ge 0 \quad \text{and}$$
$$\varphi_{-,h_2} = \varphi_{h_2} = \text{const} \ge 0.$$

It is proved in § 11, Chapter 3 that the boundary value problems (29.57) and (29.58) have unique solutions in the class of functions that are convex on the side $z > 0$, $z < 0$, respectively, provided that the following conditions are satisfied:

$$\frac{1}{n^n}\varphi^n_{h_2} \text{ mes } K_x < \int_{P^n} N_U(|p|, R^n_{h_2})dp, \tag{29.59}$$
$$\frac{1}{n^n}\varphi^n_{h_1} \text{ mes } K_x < \int_{P^n} N_L(|p|, R^n_{h_1})dp. \tag{29.60}$$

It is clear that $v^{(1)}(x)$ and $v^{(2)}(x)$ are surfaces of revolution. Furthermore, the location of the meridian $y(\rho)$, $0 \le \rho \le r$ of these surfaces, for instance $v^{(1)}$, reduces to the solution of the equation with separated variables

$$N_L\left(\left|\frac{dy}{d\rho}\right|, R^n_{h_1}\right)\left(\frac{dy}{d\rho}\right)^{n-1}\frac{d^2y}{d\rho^2} = \frac{1}{n^n}(\varphi^n_{h_1})^{n-1},$$

satisfying the boundary conditions

$$y|_{\rho=r} = 0, \quad \left.\frac{dy}{d\rho}\right|_{\rho=0} = 0.$$

Since N_U, N_L are continuous, we verify immediately that $v^{(1)}$, $v^{(2)} \in C^2(\overline{K}_x)$. The normal images of $v^{(1)}(x)$ and $v^{(2)}(x)$ are balls U_1 and U_2 in P^n with centre at $p = 0$. It follows from (29.59–60) and the relations

$$\int_{U_2} N_U(|p|, R^n_{h_2})dp = \frac{1}{n^n}\varphi^n_{h_2} \text{ mes } K_x,$$
$$\int_{U_1} N_L(|p|, R^n_{h_1})dp = \frac{1}{n^n}\varphi^n_{h_1} \text{ mes } K_x$$

that

$$\left[\sum_{i=1}^n [v_i^{(2)}]^2\right]^{1/2}\Bigg|_{\partial K_x} = F\left(N_U, \frac{1}{n^n}\varphi^n_{h_2} \text{ mes } K_x\right) < +\infty, \tag{29.61}$$

$$\left[\sum_{i=1}^n [v_i^{(1)}]^2\right]^{1/2}\Bigg|_{\partial K_x} = F\left(N_L, \frac{1}{n^n}\varphi^n_{h_1} \text{ mes } K_x\right) < +\infty \tag{29.62}$$

because the expressions at the left sides of (29.61) and (29.62) are the radii of the balls U_2 and U_1, respectively, where $F(N_U, t)$ is the inverse for the strictly increasing continuous function

$$t = \int_{|p|<\rho} N_U(|p|, R^n_{h_2})dp.$$

Next we introduce the functions

$$w^{(1)} = v^{(1)} + \sum_{i=1}^n a_i^{(1)} x_i + b^{(1)},$$
$$w^{(2)} = v^{(2)} + \sum_{i=1}^n a_i^{(2)} x_i + b^{(2)}.$$

We now show that for all $x \in \overline{B}$ the inequalities

$$w^{(2)}(x) \geq z(x) \geq w^{(1)}(x) \tag{29.63}$$

hold. First of all the construction of Q_1 and Q_2 and the condition that $v^{(1)}(x)$ and $v^{(2)}(x)$ vanish on ∂K_x provide that throughout ∂B the inequalities

$$w^{(1)}|_{\partial B} \leq z|_{\partial B} \leq w^{(2)}|_{\partial B} \tag{29.64}$$

hold. Let $\psi_2(x)$ and $\psi_1(x)$ be convex functions spanning $z(x)$ from below and above respectively, where $z(x)$ is a solution of the Dirichlet problem (29.1), (29.55). Then for any Borel set $e \subset B$ we have

$$\begin{aligned}\int_{\chi_2(e)} R^n_{h_2}(|p|)dp &= \int_{\chi_2(e\cap M_2)} R^n_2(|p|)dp \qquad (29.65)\\ &= \int_{e\cap M_2} R^n_{h_2}(|Dz|)\det(z_{ij})dx \le \frac{1}{n^n}\cdot\varphi^n_{h_2}\int_{e\cap M_2} 1\cdot dx\\ &\le \frac{1}{n^n}\varphi^n_{h_2}\cdot \text{mes } e.\end{aligned}$$

Similarly

$$\int_{\chi_1(e)} R^n_{h_1}(|p|)dp \le \frac{1}{n^n}\varphi^n_{h_1}\cdot \text{mes } e \qquad (29.66)$$

for any Borel set $e \subset B$. Furthermore, we have for all Borel sets $e \subset B$

$$\begin{aligned}\int_{\chi_{w^{(2)}}(e)} R^n_{h_2}(|p|)dp &= \int_e R^n_{h_2}(|Dw^{(2)}|)\det(w^{(2)}_{ij})dx \qquad (29.67)\\ &= \int_e R^n_{h_2}(|Dv^{(2)}+a^{(2)}|)\det(v^{(2)}_{ij})dx\\ &\ge \int_e N_L(|Dv^{(2)}|, R^n_{h_2})\det(v^{(2)}_{ij})dx\\ &\ge \frac{1}{n^n}\varphi^n_{h_2}\cdot \text{mes } e.\end{aligned}$$

Similarly

$$\int_{\chi_{w^{(1)}}(e)} R^n_{h_1}(|p|)dp \ge \frac{1}{n^n}\varphi^n_{h_1}\cdot \text{mes } e. \qquad (29.68)$$

Now by construction we have

$$\begin{aligned}w^{(1)}(x)|_{\partial B} &\le \psi_1(x)|_{\partial B} \le z(x)|_{\partial B} \qquad (29.69)\\ &\le \psi_2(x)|_{\partial B} \le w^{(2)}(x)|_{\partial B}.\end{aligned}$$

Therefore from inequalities (29.65–69) and results obtained in § 10, it follows that inequalities (29.63) are valid in the closed domain $B \cup \partial B$. Thus we can obtain from (29.61–62) and (29.6) the following estimates at any $x \in \partial B$:

$$\begin{aligned}-F\left(N_U, \mu_n\tau_0^{-n}\frac{1}{n^n}\varphi^n_{h_2}\right) - \sqrt{M_U(S)} &\le \frac{\partial z}{\partial n} \qquad (29.70)\\ &\le F\left(N_L, \mu_n\tau_0^{-n}\frac{1}{n^n}\varphi^n_{h_1}\right) + \sqrt{M_L(S)},\end{aligned}$$

where μ_n is the volume of the n-dimensional unit ball in R^n. In (29.70) we use the relation mes $K_x = \mu_n\tau_0^{-n}$. Also from (29.59–60) and inequalities between functions $w^{(1)}(x)$, $z(x)$ and $w^{(2)}(x)$ in $\overline{B} = B \cup \partial B$ we obtain the following estimates for $z(x)$ in $\overline{B}$:

$$\begin{aligned} h_2 - &\left[\sqrt{M_L(S)} + F\left(N_L, \mu_n \tau_0^{-n} \frac{1}{n^n} \varphi_{n_1}^n\right)\right] \operatorname{diam} B \\ &\leq z(x) \\ &\leq h_1 + \left[\sqrt{M_U(S)} + F\left(N_U, \mu_n \tau_0^{-n} \frac{1}{n^n} \varphi_{h_2}^n\right)\right] \operatorname{diam} B. \end{aligned} \tag{29.71}$$

Thus the following theorem is proved.

Theorem 29.6. *Let B be a domain in R^n bounded by a closed convex hypersurface. Suppose that:*

1) $B \in C^{2,\beta}$ and at each $x \in \partial B$ the normal curvatures of this hypersurface are bounded below by a constant $\tau_0 > 0$;

2) the equation (29.1) satisfies conditions 1) and 2) stated above. Consider for the equation

$$\sum_{i,k=1}^{n} a_{ik}(x, z, Dz) z_{ik} = b(x, z, Dz) \tag{29.72}$$

the Dirichlet problem on $\overline{B}$ with the boundary condition

$$z|_{\partial B} = h(x) \in C^{2,\beta}(\partial B). \tag{29.73}$$

Set

$$h_1 = M = \sup_{\partial B} h(x), \quad h_2 = m = \inf_{\partial B} h(x).$$

Suppose finally that

$$\frac{1}{n^n} \mu_n \tau_0^{-n} \cdot \varphi_{h_1}^n < \int_{P^n} N_L(|p|, R_{h_1}^n) dp, \tag{29.74}$$

$$\frac{1}{n^n} \mu_n \tau_0^{-n} \cdot \varphi_{h_2}^n < \int_{P^n} N_U(|p|, R_{h_2}^n) dp. \tag{29.75}$$

Then estimates (29.71) and (29.70) hold, respectively, for any solution $z(x) \in C^2(B \cup \partial B)$ of the Dirichlet problem (29.72–73) in $\overline{B} = B \cup \partial B$ and for its normal derivative $\frac{\partial z}{\partial n}$ on ∂B in the direction of the outer normal n of ∂B.

Theorem 29.6'. *Suppose that all conditions of Theorem 29.6 hold except for the fact that for functions a_{ik} and b condition 2) is replaced by condition 2'). Suppose also that*

$$\frac{1}{n^n} \mu_n \tau_0^{-n} \cdot \varphi_{+}^n < \int_{P^n} N_L(|p|, R_{h_1}^n) dp, \tag{29.76}$$

$$\frac{1}{n^n} \mu_n \tau_0^{-n} \cdot \varphi_{-}^n < \int_{P^n} N_U(|p|, R_{h_2}^n) dp. \tag{29.77}$$

Then for any solution $z(x) \in C^2(\overline{B})$ of problem (29.72–73) in $\overline{B}$ and $\frac{\partial z}{\partial n}\big|_{\partial B}$ the estimates (29.71) and (29.70), respectively, hold with the following modifications: replace $\varphi^n_{h_1}$ and $\varphi^n_{h_2}$ by φ^n_+ and φ^n_-, and $R_{h_1}(|p|)$ and $R_{h_2}(|p|)$ by $R_1(|p|)$ and $R_2(|p|)$ respectively.

The proof of Theorem 29.6' is verbatim the same as that of Theorem 29.6 and we omit it.

Example. For the mean curvature equation in the space R^n problem (29.72–73) takes the form

$$(1+|\nabla z|^2)\cdot \Delta z - \sum_{i,k=1}^{n} z_i z_k z_{ik} = nH(x)(1+|\nabla z|^2)^{3/2},$$

$$z|_{\partial B} = h(x) \in C^{2,\beta}(\partial B). \tag{29.78}$$

Let

$$H_+ = \max\{\sup_{\overline{B}} H(x), \} \quad H_- = \max\{-\inf_{\overline{B}} H(x), 0\}.$$

Then

$$\frac{b(x,z,p)}{[\det(a_{ik}(x,z,p)]^{1/n}} = nH(x)(1+|p|^2)^{\frac{n+2}{2n}}.$$

Hence the following inequalities hold

$$-nH_-(x)(1+|p|^2)^{\frac{n+2}{2n}} \le \frac{b(x,z,p)}{[\det(a_{ik}(x,z,p)]^{1/n}} \le nH_+(1+|p|^2)^{\frac{n+2}{2n}}.$$

Now let $z(x) \in C^2(\overline{B})$ be a solution of the boundary value problem (29.78–79). Set

$$Q_-^2 = M_U(S), \quad Q_+^2 = M_L(S).$$

Then evidently

$$N_U(|p|, R_2^n) = [1+(|p|+Q_-)^2]^{-\frac{n+2}{2}},$$

$$N_L(|p|, R_1^n) = [1+(|p|+Q_+)^2]^{\frac{-n+2}{2}}.$$

We now introduce certain auxiliary functions. The first one of them is defined by

$$T_n(Q) = n\int_0^{+\infty} \frac{p^{n-1}dp}{[1+(p+Q)^2]^{(n+2)/2}},$$

for all $Q \ge 0$. Clearly $T_n(Q)$ strictly decreases as Q increases and $T_n(0) = 1$. (Note that $T_n(Q)$ is expressible terms of elementary functions, but we do not give the expression because it is cumbersome). Next we consider the function

$$f_n(r,Q) = \int_{|p|\le r} \frac{|p|^{n-1}d|p|}{[1+(|p|+Q^2)]^{(n+2)/2}}$$

for $r \geq 0$ and $Q \geq 0$. Clearly $f_n(+\infty, Q) = \mu_n \cdot T_n(Q)$.

For any fixed Q the function $f_n(r, Q)$ of $r \in [0, +\infty)$ is continuous and strictly increasing. Hence $f_n(r, Q)$ has on the interval $[0, \mu_n T_n(Q))$ an inverse function $F_n(t, Q)$ which is also continuous and strictly increasing; in addition

$$\lim_{t \to \mu_n T_n(Q)} F_n(t, Q) = +\infty.$$

The inequalities (29.76) and (29.77) featuring in Theorem 29.6', which when valid lead to the required estimates of $z(x)$ in $\overline{B}$ and of $\frac{\partial z}{\partial n}$ on ∂B, take in our case the following form:

$$\mu_n \tau_0^{-n} H_n^n, f_n(+\infty, Q_-), \quad \mu_n \tau_0^{-n} H_+^n < f_n(+\infty, Q_+).$$

Thus, they reduce to the relations

$$H_-^n < \tau_0^n T_n(Q), \tag{29.80}$$

$$H_+^n < \tau_0^n T_n(Q_+). \tag{29.81}$$

Now on the basis of Theorem 29.6' and assuming (29.80) and (29.81), we obtain the following estimates of $z(x)$ in $\overline{B}$:

$$\begin{aligned} h_1 - [Q_+ + F_n(\mu_n \tau_0^{-n} H_+^n, Q_+)] \operatorname{diam} B &\leq z(x) \\ &\leq h_2 + [Q_- + F_n(\mu_n \tau_0^{-n} H_-^n, Q_-)] \operatorname{diam} B \end{aligned}$$

and the estimates of $\frac{\partial z}{\partial n}$ on ∂B:

$$-F_n(\mu_n \tau_0^{-n} H_-^n, Q_-) - Q_- \leq \frac{\partial z}{\partial n} \leq F(\mu_n \tau_0^{-n} H_+^n, Q_+) + Q_+.$$

If $z|_{\partial B} = \text{const}$ (in this case, $Q_- = Q_+ = 0$), then B is an n-ball of radius $r = \frac{1}{\tau_0}$ and $H(x) = H_0 = \text{const}$, then (29.80) and (29.81) become either $H_0 < \tau_0 = \frac{1}{r}$ $(H_0 \geq 0)$ or $|H_0| < \tau_0 = \frac{1}{r}$ $(H_0 < 0)$; and as shown in Chapter 7, this is a necessary condition for the construction of a regular hypersurface with constant mean curvature H_0 whose boundary satisfies the condition

$$z|_{\partial B} = \text{const}.$$

29.4 Estimates of the Gradient of Solutions on the Boundary. (The Method of Support Hyperplanes)

Let B be the domain of the Dirichlet problem (29.72–73). In the previous Subsection 29.3 we presented geometric estimates of the normal derivative for solutions $z(x) \in C^2(\overline{B})$ of problem (29.72–73), obtained by Bakelman [8], [10].

The main idea for obtaining these estimates is to use appropriate convex generalized solutions of Monge–Ampere equations according to the mutual properties of functions $a_{ik}(x,u,p)$ and $b(x,u,p)$ stated either in Conditions 1,2 or Conditions 1,2'. We did not include in these conditions the order of growth for functions $a_{ik}(x,u,p)$ and $b(x,u,p)$ as $|p| \to +\infty$. Therefore the conditions for obtaining the desired geometric estimates (29.70) required us to impose the strict convexity of the hypersurface ∂B and to use the quantities characterizing the deviation of the boundary data (29.73) from the constant.

In the present Subsection 29.4 we study the conditions which guarantee the existence of the desired estimates for $\frac{\partial z}{\partial n}$ on ∂B under a sufficiently smooth boundary data $z|_{\partial B} = h(x)$. These conditions consist of two parts. One part involves the necessary and sufficient conditions related to the mean curvature of the hypersurface ∂B in R^n. These conditions allow us to replace the convexity of ∂B by a more weak geometric inequality between mean curvature of ∂B and some geometric invariant function at any point $x \in \partial B$. The second one is related to the order of growth for functions $a_{ik}(x,u,p)$ and $b(x,u,p)$ as $|p| \to +\infty$. The results which we present below in Subsection 29.4 were established by Bakelman [9], [10]. The final statements of them are close to results of Serrin [4, § 9] related to boundary estimates depending on curvature (see Subsection 29.2.E). For some important equations, for example the mean curvature equation in Euclidean spaces, these statements coincide. However the author's results presented in Subsection 29.4 were established independently by an absolutely different geometric method.

A) Geometric Constructions. Let M be any point of ∂B, where $\partial B \in C^{k,\beta}$ ($k \geq 2, 0 < \beta \leq 1$) is a domain of the Dirichlet problem (29.72.73). We suppose that ∂B is a closed connected hypersurface in R^n. We now choose in R^n a system of Cartesian coordinates $x_1, x_2, \ldots, x_n$ lying in the tangent plane P to ∂B at M and are directed along the principal directions of ∂B at M; the x_n-axis is directed along the inward normal of ∂B at M. The following are standard facts of differential geometry concerning local properties of ∂B at M:

1) There exists a neighborhood $U_M \subset \partial B$ of M such that

a) the projection of U_M into P is the ball V_M: $\sum_{i=1}^{n} x_i^2 < \delta^2$,

b) U_M is given in V_M by the equation

$$x_n = \varphi(x_1, \ldots, x_n) \equiv \frac{1}{2}\sum_{i=1}^{n-1} k_i x_i^2 + \varepsilon(x_1, \ldots, x_{n-1}), \tag{29.82}$$

where $\varepsilon(x_1, \ldots, x_n) \in C^{k,\beta}(V_M)$, $k \geq 2$, $0 < \beta \leq 1$ vanishes together with all its derivatives of the first and second order at $x_1 = \cdots = x_{n-1} = 0$, and the k_i are the principal normal curvatures of ∂B at M. The number $\delta > 0$ is the same for all points of ∂B; from the fact that $\partial B \in C^{k,\beta}$ it follows that there exists a constant $R_0 < +\infty$ such that uniformly in ∂B we have

$$\|\varepsilon(x_1, \ldots, x_{n-1})\|_{C^{k,\beta}(V_M)} \leq R_0. \tag{29.83}$$

2) There exists a number $\eta_0 > 0$ such that for any $M \in \partial B$ the set T_M defined by relations

$$\begin{cases} \sum_{i=1}^{n-1} x_i^2 < \delta^2, \\ \varphi(x_1, \ldots, x_{n-1}) \leq x_n \leq \varphi(x_1, \ldots, x_{n-1}) + 2\eta_0 \end{cases}$$

is contained in $B \cup \partial B$.

Let f_η be the mapping of T_M into R^n given by

$$u_1 = x_1, \quad u_2 = x_2, \ldots, x_{n-1} = x_{n-1},$$
$$u_n = x_n + \phi(x_1, \ldots, x_{n-1}) - \varphi(x_1, \ldots, x_{n-1}),$$

where $\phi = \eta \left[1 - \frac{1}{\delta}\left(\delta^2 - \sum_{j=1}^{n-1} x_j^2\right)^{1/2}\right]$ and $\eta \in (0, \eta_0]$ is arbitrary. Since $\det\left(\frac{\partial u_i}{\partial y_k}\right) = 1$, then f_η is a diffeomorphism. We now define further sets. Let $Q_{M,\eta} = f_\eta(T_M)$. It is clear that $Q_{M,\eta}$ is a trip of width 2η between two halves of congruent ellipsoids of revolution with the axis of revolution directed along the inward normal of ∂B at M. Next, $\sqcap_{\lambda,\eta} \subset Q_M$, is the set of points for which

$$\sum_{j=1}^{n-1} u_j^2 \leq \lambda^2, \tag{29.84}$$

$$\eta \left[1 - \frac{1}{\delta}\left(\delta^2 - \sum_{j=1}^{n-1} u_j^2\right)^{1/2}\right] \leq u_n \leq \left[1 - \frac{1}{\delta}(\delta^2 - \lambda^2)^{1/2}\right], \tag{29.85}$$

where $0 < \lambda \leq \delta/2$. Finally, for any $\lambda \in (0, \delta/2]$ and $\eta \in (0, \eta_0]$ we define the sets

$$G_{M,\lambda,\eta} = f_\eta^{-1}(\sqcap_{\lambda,\eta}) \tag{29.86}$$

and $T_{M,\lambda,\eta}$ consisting of all points $(x_1, \ldots, x_n) \in T_M$ for which

$$\sum_{i=1}^{n-1} x_i^2 \leq \lambda^2, \tag{29.87}$$

$$\varphi(x_1, \ldots, x_{n-1}) \leq x_n \leq \varphi(x_1, \ldots, x_{n-1}) + 2\eta. \tag{29.88}$$

Clearly $G_{M,\lambda,\eta} \subset T_{M,\lambda,\eta}$, both sets have as projections onto the plane $x_0 = 0$ the ball

$$V_{M,\lambda}\colon \sum_{i=1}^{n-1} x_i^2 \leq \lambda \tag{29.89}$$

and

$$\operatorname{diam} G_{M,\lambda,\eta} \leq \operatorname{diam} T_{M,\lambda,\eta} \leq 2[\lambda + \sup_{V_{M,\lambda}} |\varphi(x_1,\dots,x_{n-1})| + \eta).$$

It follows from (29.82–85) that simultaneously for all $M \in \partial B$ there exists a constant $R_1 < +\infty$, independent of λ and η, such that

$$\operatorname{diam} G_{M,\lambda,\eta} \leq \operatorname{diam} T_{M,\lambda,\eta} \leq 2(R_1\lambda + \eta). \tag{29.90}$$

Finally, it follows from (29.83) and the properties of $\varepsilon(x_1,\dots,x_{n-1})$ that there exists a constant $R_2 < +\infty$, independent of λ and η, such that

$$\|\varepsilon(x_1,\dots,x_{n-1})\|_{C^2(V_{M,\lambda})} \leq R_2\lambda, \tag{29.91}$$

for all $M \in \partial B$.

B) The Class H of Quasilinear Equations. We consider in $B \cup \partial B$ the quasilinear elliptic equation

$$\sum_{i,k=1}^{n} a_{ik}(x,z,Dz)z_{ik} = b(x,z,Dz). \tag{29.92}$$

If we change from one Cartesian system in R^n to another, then (29.92) is clearly transformed to the quasilinear elliptic equation

$$\sum_{i,k=}^{n} a'_{ik}(x',z,Dz)z'_{ik} = b(x',z,Dz),$$

where the relationship between the matrices (a'_{ik}) and (a_{ik}) being as follows:

$$(a'_{ik}) = U^{-1}(a_{ik})U,$$

where U denotes the matrix of the orthogonal transformation from $x_1,\dots,x_n$ to the new coordinates $x'_1,\dots,x'_n$. (In U we have not included the components of the translation vector between the origins of the old and new systems.)

Let $T_{M,\lambda,\eta}$ be the set introduced by (29.87–88). For any function $z(x) \in C^2(\overline{B})$ we set

$$\gamma_1 = z(x), \quad \gamma_2 = \frac{\partial z(x)}{\partial x_1}, \dots, \gamma_n = \frac{\partial z(x)}{\partial x_{n-1}},$$

where $x_1, x_2, \ldots, x_n$ is the coordinate system in R^n connected with M (see Subsection 29.4.A). We introduce an auxiliary Euclidean space Γ^n, whose points we denote by $\gamma = (\gamma_1, \gamma_2, \ldots, \gamma_n)$. We take the usual norm

$$\|\gamma\| = (\gamma_1^2 + \cdots + \gamma_n^2)^{1/2}.$$

Each function $z(x) \in C^1(\overline{B})$ induces a mapping $\zeta_M \colon \overline{B} \to \Gamma^n$ defined by the relations

$$\gamma_1 = z(x), \quad \gamma_2 = \frac{\partial z(x)}{\partial x_1}, \ldots, \gamma_n = \frac{\partial z(x)}{\partial x_{n-1}}.$$

Equation (29.92) induces on $T_M \times \Gamma^n \times (-\infty, +\infty)$ the following functions

$$K_M(x, \gamma, p_n) = \frac{\sum_{i=1}^{n-1} a_{ii}(x_1, \ldots, x_n, \gamma_1, \ldots, \gamma_n, p_n) k_i(M)}{\sum_{i=1}^{n-1} a_{ii}(x_1, \ldots, x_n, \gamma_1, \ldots, x_n, p_n)} \tag{29.93}$$

$$N_M(x, \gamma, p_n) = \frac{|b(x_1, \ldots, x_n, \gamma_1, \ldots, x_n, p_n)|}{|p_n| \sum_{i=1}^{n-1} a_{ii}(x_1, \ldots, x_n, \gamma_1, \ldots, \gamma_n, p_n)}, \tag{29.94}$$

where the $k_i(M)$ are the principal normal curvatures of ∂B at the point M.

We now set

$$\mathcal{K}_M^+(x, \gamma) = \lim_{p_n \to +\infty} K_M(x, \gamma, p_n), \tag{29.95}$$

$$\mathcal{K}_M^-(x, \gamma) = \lim_{p_n \to -\infty} K_M(x, \gamma, p_n), \tag{29.96}$$

$$\mathcal{N}_M^+(x, \gamma) = \overline{\lim_{p_n \to +\infty}} N_M(x, \gamma, p_n), \tag{29.97}$$

$$\mathcal{N}_M^-(x, \gamma) = \overline{\lim_{p_n \to -\infty}} N_M(x, \gamma, p_n). \tag{29.98}$$

We say that *equation (29.92) belongs to the class* H, if there exists a pair of numbers $\tilde{\lambda} \in (0, \delta/2]$ and $\tilde{\eta} \in (0, \eta_0]$, such that for fixed $\gamma \in \Gamma^n$ the functions $\mathcal{N}_M^+(x, \gamma)$, $\mathcal{N}_M^-(x, \gamma)$ are uniformly bounded for all $M \in \partial B$ and for all points x in $T_{M, \tilde{\lambda}, \tilde{\eta}}$, the bound depending only on $\|\gamma\|$.

The equation

$$(1 + |\nabla z|^2) \cdot \Delta z - \sum_{i,k=1}^{n-1} z_i z_k z_{ik} = nH(x)(1 + |\nabla z|^2)^k, k = \text{ const},$$

belongs to the class H for $k \leq \frac{3}{2}$. We shall prove below that the mean curvature equation in the Riemannian space V^{n+1} also belongs to the class H.

C) Statement of Sufficient Conditions for Obtaining the Desired Estimates of the Normal Derivative of Solutions for Equations of the Class H. These sufficient conditions are as follows:

A) There exist numbers $\lambda \in (0, \delta/2]$ and $\eta \in (0, \eta_0]$ such that the inequality

$$\zeta(\gamma_0) = \inf_{M \in \partial B} \Big[\inf_{\substack{T_{M,\lambda,\eta} \\ \|\gamma\| < \gamma_0}} \{\mathcal{K}^{\pm}(x, \gamma) - \mathcal{N}^{\pm}(x, \gamma)\}\Big] > 0 \qquad (29.99)$$

holds for any $\gamma_0 > 0$.

B) There exist numbers $\lambda \in (0, \lambda/2]$ and $\eta \in (0, \eta_0]$ such that for all $M \in \partial B$, and for any $\gamma_0 > 0$ and $\varepsilon > 0$ there exists a number $L > 0$ such that throughout $T_{M,\lambda,\eta}$:

1) for $\|\gamma\| \leq \gamma_0$ and $p_n \geq L$

$$K_M(x, \gamma, p_n) \geq \mathcal{K}^+_M(x, \gamma) - \varepsilon, \qquad (29.100)$$
$$N_M(x, \gamma, p_n) \leq \mathcal{N}^+(x, \gamma) + \varepsilon, \qquad (29.101)$$

2) for $\|\gamma\| < \gamma_0$ and $p_n \leq -L$

$$K_M(x, \gamma, p_n) \geq \mathcal{K}^-_M(x, \gamma) - \varepsilon, \qquad (29.102)$$
$$N_M(x, \gamma, p_n) \leq \mathcal{N}^-_M(x, \gamma) + \varepsilon. \qquad (29.103)$$

We now point out sufficient conditions which for quasilinear equations ensure the validity of A) and B). We call a Cartesian coordinate system in R^n *admissible* if its origin lies in $B \cup \partial B$. We denote such a system briefly by U; in this system equation (29.92) can be written as

$$\sum_{i,k=1}^{n} a_{ik}(x, z, Dz, U) z_{ik} = b(x, z, Dz, U). \qquad (29.104)$$

Equation (29.104) induces on $B \cup \partial B \times \Gamma^n \times R$ the functions

$$K(x, \gamma, p_n, U) = \frac{\sum_{i=1}^{n-1} a_{ii}(x, \gamma, p_n, U) k_i}{\sum_{i=1}^{n-1} a_{ii}(x, \gamma, p_n, U)},$$

$$N(x, \gamma, p_n, U) = \frac{|b(x, \gamma, p_n, U)|}{|p_n| \sum_{i=1}^{n-1} a_{ii}(x, \gamma, p_n, U)}.$$

Clearly the functions $a_{ik}(x, u, p, U), b(x, u, p, U), K(x, \gamma, p_n, U),\ N(x, \gamma, p, U)$ are continuous in every admissible Cartesian system of coordinates U, if the functions $a_{ik}(x, u, p, U_0)$, $b(x, u, p, U_0)$ are continuous in some Cartesian system of coordinates U_0.

The following lemma is important in applications. This lemma is an immediate corollary of standard theorems about continuous functions on compact sets.

Lemma 29.8. *If $K(x,\gamma,p_n,U)$ and $N(x,\gamma,p_nU)$ converge uniformly with respect to $x \in B \cup \partial B$ and all admissible U, in any ball $\|\gamma\| < \gamma_0$ as $p_n \to \pm\infty$, then*

1) *equation (29.92) satisfies B),*
2) *$\mathcal{K}^{\pm}(x,\gamma,U)$ and $\mathcal{N}^{\pm}(x,\gamma,U)$ are continuous functions on the set of admissible systems of $x \in B \cup \partial B$ and γ over any ball $\|\gamma\| < \gamma_0$,*
3) *if, in addition, in every ball $\|\gamma\| \le \gamma_0$ the inequality*

$$\inf_{M\in\partial B}\Big[\inf_{\|\gamma\|\le\gamma_0}[\mathcal{K}^{\pm}(M,\gamma) - \mathcal{N}^{\pm}(M,\gamma)]\Big] = \zeta(\gamma_0) > 0 \tag{29.105}$$

holds, then condition A) holds as well.

We now examine in the example of the mean curvature equation in R^{n+1} what the conditions of Lemma 29.8, in particular (29.105), come down to.

Theorem 29.7. *Denote by $H_{n-1}(M,\partial B)$ the mean curvature of ∂B at $M \in \partial B$ along the inward normal of ∂B, and by $H_n(x) \in C(\overline{B})$ the mean curvature of the required hypersurface. If the following inequality holds for all $M \in \partial B$:*

$$H_{n-1}(M,\partial B) > \frac{n}{n-1}|H_n(M)|, \tag{29.106}$$

then the equation of the a hypersurface with prescribed mean curvature $H_n(x)$ in R^{n+1} satisfies A) and B).

Proof. The mean curvature equation in R^{n+1} is as follows:

$$\left(1+\sum_{i=1}^{n} z_i^2\right)\cdot\sum_{i=1}^{n} z_{ii} - \sum_{i,k=1}^{n} z_i z_i z_{ik} nH(x)\left(1+\sum_{i=1}^{n} z_i^2\right)^{3/2}.$$

This equation is invariant under orthogonal transformations of Cartesian coordinates in R^n. Therefore the functions featuring in Lemma 29.8 do not depend on the coordinate system U. It is easy to see that as $p_n \to \pm\infty$ the functions

$$K(x,\gamma,p_n) = \frac{\sum\limits_{i=1}^{n-1}\left(1+\sum\limits_{s=2}^{n}\gamma_s^2 - p_n^2 - \gamma_{i+1}^2\right)k_i}{\sum\limits_{i=1}^{n-1}\left(1+\sum\limits_{s=2}^{n}\gamma_s^2 + p_n^2 - \gamma_{i+1}^2\right)},$$

$$N(x,\gamma,p_n) = \frac{n|H_n(x)|\left(1+\sum\limits_{s=2}^{n}\gamma_s^2 + p_2^n\right)^{3/2}}{|p_n|\sum\limits_{i=1}^{n-1}\left(1+\sum\limits_{s=2}^{n}\gamma_s^2 + p_n^2 - \gamma_{i+1}^2\right)}$$

converge uniformly with respect to $x \in \overline{B}$ over any ball $\|\gamma\| < \gamma_0$ to the respective functions

$$\mathcal{K}^{\pm}(x,\gamma) = \frac{k_1+k_2+\cdots+k_{n-1}}{n-1} \quad \text{and}$$

$$\mathcal{N}^{\pm}(x,\gamma) = \frac{n}{n-1}|H_n(x)|.$$

Thus, by Lemma 29.8 condition B) holds. Since

$$\mathcal{K}^{\pm}_{M}(x,\gamma) = \frac{k_1(M) + \cdots + k_{n-1}(M)}{n-1} = H_{n-1}(M,B),$$
$$\mathcal{N}^{\pm}_{M}(M,\gamma) = \frac{n}{n-1}|H_n(M)|,$$

for any $M \in \partial B$, then (29.106) is the same as (29.105) in Lemma 29.8. Therefore A) also holds, and this proves Theorem 29.7.

D) The Main Theorem on Estimates of the Normal Derivative of Solutions for Quasilinear Elliptic Equations of Class H. Consider the following Dirichlet problem in $\overline{B}$:

$$\sum_{i,k=1}^{n} a_{ik}(x,z,Dz)z_{ik} = b(x,z,Dz), \tag{29.107}$$

$$z|_{\partial B} = h(x). \tag{29.108}$$

Concerning B and ∂B we assume that the coordinates of § 29.4.A) are satisfied and that $h(x)$ belongs to $C^{k,\beta}(\partial B)$ $(k \geq 2, 0 < \beta \leq 1)$. Let $z(x) \in C^2(B \cup \partial B)$ be a solution of the Dirichlet problem (29.107)–(29.108). For the rest of this section we suppose that

$$\sup_{\overline{B}} |z(x)| \leq Z_0 < +\infty, \tag{29.109}$$

where $Z_0 > 0$ is a constant. We also suppose that equation (29.107) satisfies conditions A) and B) stated in § 29.4.B). Let $\tilde{\lambda} \in (0, \delta/2]$ and $\tilde{\eta} \in (0, \eta_0]$ be as in A) and B).

We denote by S_h an $(n-1)$-dimensional surface that is given on ∂B by the explicit equation

$$z = h(x).$$

Let P_M be the tangent plane to S_h at the point $(M, h(M))$, where $M \in \partial B$. Then the equations of P_M are

$$\begin{cases} z = \displaystyle\sum_{i=1}^{n} q_i(M)x_i + h(M), \\ x_n = 0. \end{cases} \tag{29.110}$$

We associate with each point $M \in \partial B$ the point $\gamma(M) \in \Gamma^n$ defined by

$$\gamma_1 = h(M), \quad \gamma_2 = q_1(M), \ldots, \gamma_n = q_{n-1}(M).$$

Then clearly

$$\sup_{\partial B} \|\gamma(M)\| \leq \gamma_0 = Z_0 + \|h(x)\|_{C^1(\partial B)}.$$

Now $\gamma_0 > 0$ is fixed. As a consequence, also the number $\zeta_0 = \zeta(\gamma_0) > 0$ is fixed, which occurs in condition A). Next we choose numbers $\lambda_1 \in (0, \tilde{\lambda}]$ and $\eta_1 \in (0, \tilde{\eta}]$ such that

$$\eta_1 < \frac{\delta^2 \zeta_0}{10\left(\frac{2}{\sqrt{3}} + \frac{8}{3\sqrt{3}}\right)}, \quad \lambda_1 < \frac{\zeta_0}{10 \cdot R_2} \tag{29.111}$$

(see § 29.4.A) for the meaning of δ, R_2, and other notation). Note that λ_1 and η_1 are chosen to be the same for all points of ∂B. Let $M \in \partial B$ be arbitrary. We map T_{M,λ_1,η_1} into Q_{M,η_1} by means of the diffeomorphism f_{η_1}. On Q_{M,η_1} and hence also on $\sqcap_{\lambda_1,\eta_1}$ we define the function

$$v(u_1, \dots, u_n) = z(f_{\eta_1}^{-1}(u)),$$

where $z(x) \in C^2(B \cup \partial B)$ is a solution of the boundary-value problem (29.107–108). Since

$$z|_{\partial B \cap T_{M,\lambda_1,\eta_1}} = \psi(x_1, \dots, x_{n-1}, \varphi(x_1, \dots, x_{n-1})),$$

then we obtain

$$v|_{f_{\eta_1}(\partial B \cap T_{M,\lambda_1,\eta_1})} = \Psi(u_1, u_2, \dots, u_{n-1}, \Phi(u_1, u_2, \dots, u_{n-1})).$$

Also, since

$$\sum_{i=1}^{n-1} u_i^2 \leq \lambda_1^2 < \delta^2/4,$$

we have

$$\Psi(u_1, \dots, u_{n-1}, \phi(u_1, \dots, u_{n-1})) \in C^{k,\beta} \quad (k \geq 2, 0 < \beta \leq 1)$$

and in the ball $\sum_{i=1}^{n-1} u_i^2 \leq \lambda_1^2 < \tilde{\lambda}^2 < \delta^2/4$ we have the inequality

$$\begin{aligned} &\|\psi(u_1, \dots, u_{n-1}, \phi(u_1, \dots, u_{n-1}))\|_{C^{k,\beta}} \\ &\quad \leq K_{\lambda_1} \|\psi(x_1, \dots, x_{n-1}, \varphi(x_1, \dots, x_{n-1}))\|_{C^{k,\beta}} \end{aligned}$$

where the constant $K_{\lambda_1} < +\infty$ depends only on $\delta - \lambda_1 > \delta/2$.

We denote by F_{λ_1,η_1} the $(n-1)$-dimensional surface defined in $\partial \sqcap_{\lambda_1,\eta_1}$ by the equation

$$z|_{\partial \sqcap_{\lambda_1,\eta_1}} = v(u).$$

Let Q_M be the tangent plane to F_{λ_1,η_1} at the point $\overline{M}$, whose projection onto R^n is M. Then M is the pole of the lower half of the ellipsoid of revolution G:

$$u_n + \phi(u_1, u_2, \dots, u_{n-1}) = \eta_1 \left(1 - \frac{1}{\delta}\left[\delta^2 - \sum_{i=1}^{n-1} u_i^2\right]^{1/2}\right).$$

Since at the points of $G \cap \partial \sqcap_{\lambda_1,\eta_1}$ all the normal curvatures of G are strictly positive and since at all points of $\partial \sqcap_{\lambda_1,\eta_1} - G \cap \partial \sqcap_{\lambda_1,\eta_1}$ we have, by hypothesis,

$$|v(u)| \leq Z_0,$$

then it follows that the twisting of F_{λ_1,η_1} at $\overline{M}$ is finite. Hence there exist in R^{n+1} hyperplanes Q_1 and Q_2 passing through the $(n-1)$-plane Q_M and leaving the surface F_{λ_1,η_1} from below and above, respectively. Now if

$$z = \sum_{i=1}^{n-1} q_i^{(1)} u_i + h(M) \tag{29.112}$$

and

$$z = \sum_{i=1}^{n-1} q_i^{(2)} u_i + h(M) \tag{29.113}$$

are the equations of Q_1 and Q_2, then the quantities $\sum_{i=1}^{n} [q_i^{(1)}]^2$ and $\sum_{i=1}^{n} [q_i^{(2)}]^2$ admit estimates in terms of $\eta_1 \delta^2$, $Z_0 \eta_1^{-1} \left(1 - \frac{1}{\delta}[\delta^2 - \lambda_1^2]^{1/2}\right)$ and $\|\psi(u_1, u_2, \dots, u_{n-1}, \phi(u_1, \dots, u_{n-1})\|_{C^2,(V_{\lambda_1})}$, where $V\lambda_1$ is the ball

$$\sum_{i=1}^{n-1} u_i^2 \leq \lambda_1^2.$$

From this it follows that

1)

$$\left.\frac{\partial v}{\partial u_n}\right|_M \leq \left(\sum_{i=1}^{n} [q_i^{(1)}]^2\right)^{1/2}$$

provided that everywhere in $\sqcap_{\lambda_1,\eta_1}$ we have the inequality

$$h(M) + \sum_{i=1}^{n} q_i^{(1)} u_i \geq v(u_1, \dots, u_n), \tag{29.114}$$

and

2)

$$\left.\frac{\partial v}{\partial u_n}\right|_M \geq - \left(\sum_{i=1}^{n-1} [q_i^{(2)}]^2\right)^{1/2},$$

provided that in $\sqcap_{\lambda_1,\eta_1}$ we have the inequality

$$h(M) + \sum_{i=1}^{n} q_i^{(2)} u_i \leq v(u_1, \dots, u_n). \tag{29.115}$$

Since $\frac{\partial v}{\partial u_n}\Big|_M = \frac{\partial z}{\partial n}\Big|_M$, in the cases considered above we obtain in this way estimates for $\frac{\partial z}{\partial n}\Big|_M$ from below and from above. Thus, it remains to consider the case when $\sqcap_{\lambda_1,\eta_1}$ fails to satisfy either (29.114) or (29.115). Suppose for the sake of definiteness that (29.114) does not hold for $\sqcap_{\lambda_1,\eta_1}$, that is, there exist points of the graph of $v(u_1,\ldots,u_n)$ lying above Q_1 under the condition that $(u_1,\ldots,u_n) \in \sqcap_{\lambda_1,\eta_1}$. Through the $(n-1)$-plane Q_M (recall that Q_M is the tangent plane to F_{λ_1,η_1} at $\overline{M}$) we construct a *support hyperplane* Q to the graph of the function $v(u_1,\ldots,u_n)$ over $\sqcap_{\lambda_1,\eta_1}$ such that graph lies below Q. The support hyperplane Q is very important for obtaining the desired estimates of the normal derivative of the function $z(x)$. Therefore I call the method presented in § 29.4 as *the method of the support hyperplane*.

Since the boundary of the graph is the surface F_{λ_1,η_1}, which lies below Q_1, then it follows by our hypothesis that Q is tangent to the graph of $v(u_1,\ldots,u_n)$ at one point at least. This point may be either $\overline{M}$ or $\overline{M}_1$, which is projected onto an interior point of $\sqcap_{\lambda_1,\eta_1}$. In either case the tangent point of Q and the graph of $v(u_1,\ldots,u_n)$ is such that d^2v is a non-positive form at this point.

We denote by $u_0 = (u_1^0,\ldots,u_n^0)$ the projection of $\overline{M}$ or $\overline{M}_1$ onto R^n. Let

$$z = \sum_{i=1}^{n} q_i u_i + h(M_i) \tag{29.116}$$

be the equation of Q. It is not difficult to see that for $i = 1,2,\ldots,n-1$

$$\begin{aligned} q_i &= \frac{\partial v}{\partial u_i}\Big|_M = \frac{\partial \psi}{\partial u_i}\Big|_M + \frac{\partial v}{\partial u_n}\frac{\partial \phi}{\partial u_i}\Big|_M \\ &= \frac{\partial \psi}{\partial u_i}\Big|_M = \frac{\partial \psi}{\partial x_i}\Big|_M = q_i(M), \end{aligned}$$

that is, the quantities q_i $(i = 1,2,\ldots,n-1)$ are completely determined by the boundary condition (29.108). In other words, the point $\gamma(M) \in \Gamma^n$ constructed at the beginning of § 29.4.B) is, by virtue of (29.108), such that its coordinates $\gamma_2,\gamma_3,\ldots,\gamma_n$ satisfy the relations

$$\gamma_2 = q_1, \quad \gamma_3 = q_2,\ldots,\gamma_n = q_{n-1}. \tag{29.117}$$

Also, it is clear that

$$q_n \geq \frac{\partial v}{\partial u_n}\Big|_M. \tag{29.118}$$

Hence any upper estimate for q_n is also estimated for $\frac{\partial v}{\partial u_n}\Big|_M$ or, what is the same, for $\frac{\partial z}{\partial n}\Big|_M$. At the point u_0 we have

1)

$$\frac{\partial v}{\partial u_i} = q_i \qquad (i = 1,2,\ldots,n-1,n), \tag{29.119}$$

2) d^2v is a non-positive form.

If we change coordinates from $x_1, \dots, x_n$ to $u_1, \dots, u_n$ by becomes

$$\sum_{i,k=1}^{n} A_{ik} \frac{\partial^2 v}{\partial u_i \partial u_k} = \tilde{b} + \frac{\partial v}{\partial u_n} \sum_{i,k=1}^{n} \tilde{a}_{ik} \frac{\partial^2 [\phi - \varphi]}{\partial u_i \partial u_k}, \tag{29.120}$$

where $\sum\limits_{i,k=1}^{n} A_{ik}\xi_i\xi_k$ is a positive form with respect to the vector $\xi = (\xi_1, \dots, \xi_n)$,

$$\begin{aligned}\tilde{b} = b\Bigg(&u, v, \frac{\partial v}{\partial u_1} + \frac{\partial v}{\partial u_n}\frac{\partial(\phi - \varphi)}{\partial u_1}, \dots, \frac{\partial v}{\partial u_{N-1}} \\ &+ \frac{\partial v}{\partial u_n}\frac{\partial(\phi - \varphi)}{\partial u_{n-1}}, \frac{\partial v}{\partial u_n}\Bigg),\end{aligned} \tag{29.121}$$

$$\begin{aligned}\tilde{a}_{ik} = a_{ik}\Bigg(&u, v, \frac{\partial v}{\partial u_1} + \frac{\partial v}{\partial u_n}\frac{\partial(\phi - \varphi)}{\partial u_1}, \dots, \frac{\partial v}{\partial u_{n-1}} \\ &+ \frac{\partial v}{\partial u_n}\frac{\partial(\phi - \varphi)}{\partial u_{n-1}}, \frac{\partial v}{\partial u_n}\Bigg),\end{aligned} \tag{29.122}$$

for $i, k = 1, 2, \dots, n-1$. Equation (29.120) was derived in detail by Bakelman [9].

We denote by $\gamma(u_0, M)$ the point of Γ^n, whose coordinates are

$$\gamma_1 = v(u_0), \quad \gamma_2 = q_1(M), \dots, \gamma_n = q_{n-1}(M). \tag{29.123}$$

Clearly, if u_0 and M are coincident, then

$$\gamma(u_0, M) = \gamma(M).$$

Since $\sum\limits_{i,k=1}^{n} A_{ik}\xi_i\xi_k$ is a positive form, and since $d^2 v$ is a non-positive form at u_0, the following inequality must hold at the point u_0:

$$\begin{aligned}0 \geq\ &\tilde{b}(u_0, \gamma(u_0, M), q_0) \\ &+ q_n \sum_{i,k=1}^{n} \tilde{a}_{ik}(u_0, \gamma(u_0, M), q_n) \frac{\partial^2(\phi - \varphi)}{\partial u_i \partial u_k}.\end{aligned} \tag{29.124}$$

Bakelman [9] proved that (29.124) written out in full takes the following form

$$\begin{aligned}0 \geq\ &\tilde{b}(u_0, \gamma(u_0, M), q_n) \\ &+ q_n \left\{ \sum_{i=1}^{n-1} \tilde{a}_{ii}(u_0, \gamma(u_0, M), q_n) k_i(M) \right. \\ &+ \sum_{i,k=1}^{n-1} a_{ik}(u_0, \gamma(u_0, M), q_n) \varepsilon_{ik}(u_1^0, \dots, u_{n-1}^0)\end{aligned} \tag{29.125}$$

$$
-\frac{\eta_1}{\delta}\left(\delta^2-\sum_{j=1}^{n-1}(u_j^0)^2\right)^{-1/2}
$$

$$
\times\left[\sum_{i=1}^{n-1}\tilde{a}_{ii}(u_0,\gamma(u_0,M),q_n)+\frac{\sum\limits_{i,k=1}^{n-1}\tilde{a}_{ik}(u_0,\gamma(u_0,M),q_n)u_iu_k}{\delta^2-\sum\limits_{j=1}^{n}(u_j^0)^2}\right]\Bigg\}
$$

where the functions $\varepsilon_{ik}(u_1,\ldots,u_{n-1})$ have the properties mentioned in § 29.4.A). It should be noted that

$$
\|\gamma(u_0,M)\|\leq Z_0+\|\psi\|_{C^{(1)}(\partial B)}=\gamma_0, \tag{29.126}
$$

where Z_0 is defined in (29.109).

We now remark that no generality is lost in supposing that $q_n>0$, because we are looking for an upper estimate for q_n. There are five terms on the right-hand side of (29.125). We begin with an estimate of the last three. We have

$$
\left|q_n\sum_{i,k=1}^{n-1}\tilde{a}_{ik}\varepsilon_{ik}\right|=q_n\left|\sum_{i,k=1}^{n-1}\tilde{a}_{ik}\varepsilon_{ik}\right|
$$

$$
\leq q_n\sum_{i=1}^{n-1}\tilde{a}_{ii}\cdot\frac{\left(\sum\limits_{i,k=1}^{n-1}\tilde{a}_{ik}^2\right)^{1/2}}{\left(\sum\limits_{i=1}^{n}\tilde{a}_{ii}\right)}\left(\sum_{i,k=1}^{n-1}\varepsilon_{ik}^2\right)^{1/2}.
$$

Now since the form $\sum\limits_{I,k=1}^{n}\tilde{a}_{ik}\xi_i\xi_k$ is positive, then $\tilde{a}_{ik}^2\leq\tilde{a}_{ii}\tilde{a}_{ik}$ for all $i,k=1,2,\ldots,n-1$. Thus

$$
\left|q_n\sum_{i,k=1}^{n-1}\tilde{a}_{ik}\varepsilon_{ik}\right|\leq q_n\sum_{i=1}^{n-1}\tilde{a}_{ii}\left(\sum_{i,k=1}^{n-1}\varepsilon_{ik}^2\right)^{1/2}. \tag{29.127}
$$

Next

$$
\left|q_n\frac{\eta_1}{\delta}\left(\delta^2-\sum_{j=1}^{n-1}u_j^2\right)^{-1/2}\cdot\sum_{i=1}^{n}\tilde{a}_{ii}\right| \tag{29.128}
$$

$$
\leq qn\frac{2\eta_1}{\sqrt{3}\,\delta^2}\cdot\sum_{i=1}^{n}\tilde{a}_{ii},
$$

because for all $(u_1, u_2, \ldots, u_n) \in \sqcap_{\lambda_1,\eta_1}$ we have $\sum_{i=1}^{n-1} u_i^2 < \delta^2/4$.

Now the form $\sum_{i,k=1}^{n-1} \tilde{a}_{ik}\xi_i\xi_k$ is positive; therefore all its eigenvalues are non-negative, so that

$$\sum_{i,k=1}^{n-1} \tilde{a}_{ik}\xi_i\xi_k \le \sum_{i=1}^{n} \tilde{a}_{ii} \cdot \sum_{i=1}^{n-1} \xi_i^2.$$

Hence we obtain that

$$\begin{aligned} q_n \frac{\eta_1}{\delta}\left(\delta^2 - \sum_{j=1}^{n-1} u_j^2\right)^{-3/2} &\cdot \sum_{i,k=1}^{n-1} \tilde{a}_{ik}|u_i u_k| \\ &\le q_n \frac{8\eta_1}{3\sqrt{3}\,\delta^4} \cdot \sum_{j=1}^{n-1} u_j^2 \cdot \sum_{i=1}^{n-1} \tilde{a}_{ii} \\ &\le \frac{8}{3\sqrt{3}} \cdot q_n \frac{\eta_1}{\delta^2} \sum_{i=1}^{n-1} \tilde{a}_{ii}. \end{aligned} \tag{29.129}$$

Thus, from (29.126–129) it follows that

$$\begin{aligned} &\tilde{b}(u_0, \gamma(u_0, M), q_n) + q_n \sum_{i=1}^{n-1} \tilde{a}_{ii}(u_0, \gamma(u_0, M), q_n) k_i(M) \\ &- q_n \sum_{i=1}^{n-1} \tilde{a}_{ii}(u_0, \gamma(u_0, M), q_n) \\ &\left(\left[\sum_{i,k=1}^{n-1} \varepsilon_{ik}^2\right]^{1/2} + \left(\frac{2}{\sqrt{3}} + \frac{8}{3\sqrt{3}}\right)\frac{\eta_1}{\delta^2}\right) \le 0. \end{aligned} \tag{29.130}$$

We now use the fact that λ_1 and η_1 have been selected so that A) and B) hold on $\sqcap_{\lambda_1,\eta_1}$. The ball V_{M,λ_1}, which is the projection both of $\sqcap_{\lambda_1,\eta_1}$ and of T_{M,λ_1,η_1} onto the plane $x_n = 0$, satisfies the condition that

$$x_1 = u_1, \ldots, x_{n-1} = u_{n-1}$$

for all its points. Hence

$$\varepsilon_{ik}(u_1, \ldots, u_{n-1}) = \varepsilon_{ik}(x_1, \ldots, u_{n-1})$$

in V_{M,λ_1}. Using (29.99) and (29.111) we obtain the inequality

$$\frac{\tilde{b}(u_0, \gamma(u_0, M), q_n)}{q_n \sum_{i=1}^{n-1} \tilde{a}_{ii}(u_0, \gamma(u_0, M), q_n)} \tag{29.131}$$

$$+\frac{\sum_{i=1}^{n-1} \tilde{a}_{ii}(u_0, \gamma(u_0, M), q_n) k_i(M)}{\sum_{i=1}^{n-1} a_{ii}(u_0, \gamma(u_0, M), q_n)} - \frac{\zeta_0}{5} \leq 0.$$

Since $\|\gamma(u_0, M)\| \leq \gamma_0 = Z_0 + \|\psi\|_{C^{(1)}(\partial B)}$, then from B) it follows that there exists an $L > 0$ such that the inequalities

$$\frac{\sum_{i=1}^{n-1} \tilde{a}_{ii}(u_0, \gamma(u_0, M), q_n) k_i(M)}{\sum_{i=1}^{n-1} \tilde{a}_{ii}(u_0, \gamma(u_0, M), q_n)} \geq \mathcal{K}_M^+(u_0, \gamma(u_0, M)) - \frac{\zeta_0}{10}, \tag{29.132}$$

$$\frac{|\tilde{b}(u_0, \gamma(u_0, M), q_n)|}{q_n \cdot \sum_{i=1}^{n-1} a_{ii}(u_0, \gamma(u_0, M), q_n)} \leq \mathcal{N}_M^+(u_0, \gamma(u_0, M)) + \frac{\zeta_0}{10}, \tag{29.133}$$

hold for $q_n \geq L$, where $\zeta_0 = \zeta(\gamma_0)$ is defined by (29.99) and the functions $\mathcal{K}_M^+$ and $\mathcal{N}_M^+$ are defined by (29.95) and (29.97). Above we applied condition B) for the numbers γ_0 and $\zeta_0/10$. According to condition A) we have

$$\mathcal{K}_M^+(u_0, \gamma(u_0, M)) - \mathcal{N}_M^+(u_0, \gamma(u_0, M)) \geq \zeta_0. \tag{29.134}$$

From (29.131–134) it follows that

$$\zeta_0 - \frac{\zeta_0}{5} - \frac{\zeta_0}{5} \leq 0$$

or $\frac{3}{5}\zeta_0 \leq 0$, which is impossible. We conclude from (29.120) that if conditions A) and B) hold, then $q_n \leq L$, where L depends on $\gamma_0 = Z_0 + \|\psi\|_{C^{(1)}(\partial B)}$, $\zeta_0 = \zeta(\gamma_0)$, the choice of $\lambda_1 \in (0, \delta/2]$, $\eta_1 \in (0, \eta_0]$, the quantity $\|\psi\|_{C^{2,\beta}(\partial B)}$ and finally of quantities occurring in the conditions A) and B). Note that no restriction is made on $\|\psi\|_{C^{2,\beta}(\partial B)}$ other than that it should be finite.

Similarly, if the graph of the function $v(u) = z(f_\eta^{-1}(u))$ has points below Q_2, we obtain the estimate $q_n \geq -L$.

Returning now to our original construction we obtain the following main theorem on estimates of the normal derivative.

Theorem 29.8. *Let $z(x) \in C^2(B \cup \partial B)$ be a solution of the Dirichlet problem (29.107–108), and suppose that the conditions of Subsections 29.4.A) and 29.4.D) hold for ∂B and the function $\psi(x)$ prescribed on ∂B. If in addition,* *equation (29.107) satisfies conditions A) and B) and its solution $z(x)$ satisfies the inequality*

$$\sup_{B\cup\partial B} |z(x)| \leq Z_0 < +\infty, \tag{29.135}$$

then we can obtain upper and lower finite estimates of the normal derivative of $z(x)$ on ∂B which depend only on the following data: $\gamma_0 = Z_0 + \|\psi\|_{C^1(\partial B)}$, $\zeta_0 = \zeta(\gamma_0)$, the choice of $\lambda_1 \in (0, \delta/2]$, $\eta_1 \in (0, \eta_0]$, the quantities occurring in A) and B), and finally $\|\psi\|_{C^{2,\beta}(\partial B)}$.

29.5 Global Gradient Estimates for Solutions of Quasilinear Elliptic Equations

It is well known that such estimates are very important (together with other a priori estimates considered in Subsections 29.2–4 in the proof existence theorems for the Dirichlet problem (29.107–108). These estimates are based on the maximum principle method. The technique based on this principle was originally introduced by Bernstein [2] in 1910 for elliptic equations in two independent variables. Later Bernstein [4] further elaborated his method and applied it to various additional equations in two independent variables. Bernstein's method was later used by Leray [2] in his well known investigations of the Dirichlet problem for nonlinear elliptic equations for the case of two independent variables. Only in 1956 Ladyzhenskaya [2] studied elliptic equations with an arbitrary number of independent variables, but with the important restriction to uniformly positive definite coefficient matrices. In the same paper she also obtained the gradient estimates for solutions of quasilinear parabolic equations. The method of the maximum principle was also successfully applied by Oleinik and Kruzhkov [1] to obtain interior gradient estimates for uniformly parabolic equations.

More recently the maximum principle technique has been used independently by Serrin [4], [6], Ladyzhenskaya and Ural'tzeva [1], Bakelman and Kantor [1], [2], Trudinger [3], [4], Oskolkov [1], Simon [2], Lieberman [1], [2] to obtain gradient estimates for non-uniformly elliptic equations.

In this subsection we present results of Serrin [4, § 13], which are related to the following quasilinear equations for which the coefficient matrix has the representation

$$A(x, u, p) = G(x, u, p)A'(p) + G_1(x, u, p)pp, \tag{29.136}$$

where G and G_1 are scalar functions, A' is a positive definite matrix and

$$pp = \begin{pmatrix} p_1p_1 & p_1p_2 & \cdots & p_1p_n \\ p_2p_1 & p_2p_2 & \cdots & p_2p_n \\ \cdots & \cdots & \cdots & \cdots \\ \cdots & \cdots & \cdots & \cdots \\ p_np_1 & p_np_2 & \cdots & p_np_n \end{pmatrix}.$$

We assume without loss of generality that

$$trA' = 1. \tag{29.137}$$

Since A is positive definite, then $G > 0$. Representation (29.136) is available whenever A is independent of x and u. The same representation holds for the Euler–Lagrange equation associated with regular variational problems of the form

$$\delta F(x, u, \|Du\|)dx = 0, \qquad F \in C^3. \tag{29.138}$$

Below we present the proof of this statement. Let $u(x) \in C^1(\overline{B}) \cap C^2(B)$ be any solution of the elliptic Euler–Lagrange equation (29.138). We consider all terms in (29.138), which contain the second derivatives of the function $u(x)$. Let Q be the sum of all these terms. Then

$$Q = 2F' \cdot \Delta u + 4F'' \sum_{i,k=1}^{n} u_i u_k u_{ik}, \tag{29.139}$$

where the primes denote differention of F with respect to the variable $|p|^2$. From (29.139) it follows that equation (29.138) is elliptic if and only if the quadratic form

$$F'(x, u, |p|)|\xi|^2 + 2F''(x, u, |p|) \sum_{i=1}^{n} (p_i \xi_i)^2 \tag{29.140}$$

with respect to $\xi = (\xi_1, \xi_2, \dots, \xi_n) \in R^n$ is positive definite for all fixed $x \in \overline{B}$, $u \in (-\infty, +\infty)$, $p \in P^n$. Thus the ellipticity of equation (29.138) yields to the following inequalities

$$F'(x, u, |p|) > 0 \tag{29.141}$$

and

$$F'(x, u, p) + 2F''(x, u, p)|p|^2 > 0 \tag{29.142}$$

for all $x \in \overline{B}, u \in R, p \in P^n$.

Let

$$H = \frac{2|p|^2 \cdot F''}{nF'}. \tag{29.143}$$

Then from (29.143), (29.141) and (29.142) it follows that

$$\frac{1}{1+H} = \frac{nF'}{nF' + 2|p|^2 F''} > 0 \tag{29.144}$$

for all $x \in \overline{B}, u \in (-\infty, +\infty), p \in P^n$.

We now multiply the original equation (29.138) by the positive function $\frac{1}{2[nF' + 2|p|^2 F'']}$. Then we obtain an elliptic equation which is equivalent to equation (29.138). If $\overline{Q}$ is the sum of all terms of this new equation, which contain the second derivatives of $u(x)$, then from (29.144), (29.139) and (29.143) it follows that

$$\overline{Q} = \frac{1}{2[nF' + 2|\nabla u|^2 F'']} Q = \frac{nF'}{[nF' + 2|\nabla u|^2 F'']} \frac{1}{n} \Delta u$$
$$+ \frac{2|\nabla u|^2 F''}{nF'} \frac{nF'}{[nF' + 2|\nabla u|^2 F'']} \sum_{i,k=1}^{n} \frac{u_i}{|\nabla u|} \cdot \frac{u_k}{|\nabla u|} \cdot u_{ik}.$$

Thus the coefficient matrix $A(x, u, |p|)$ for the new equation has the following representation

$$A(x, u, |p|) = \frac{1}{1+H} \cdot \frac{1}{n} I_n + \frac{H}{1+H} \sigma\sigma, \tag{29.145}$$

where $\sigma = \frac{p}{|p|}$ and I_n is the unit matrix of the n-th order. Thus the matrix $A(x, u, |p|)$ satisfies relation (29.136) with the matrix $A' = \frac{1}{n} I_n$ and the functions

$$G = \frac{1}{1+H}, \quad G_1 = \frac{H}{1+H} |p|^2.$$

Clearly $tr A' = 1$. The proof is completed.

We now return to the derivation of global estimates for the gradient of solutions of quasilinear elliptic equations for which condition (29.136) holds. We first introduce a few notations. If W is a given scalar function of x, u, p, then we shall write

$$W_x = \left(\frac{\partial W}{\partial x_1}, \ldots, \frac{\partial W}{\partial x_n} \right), W_u = \frac{\partial W}{\partial u},$$
$$W_p = \left(\frac{\partial W}{\partial p_1}, \ldots, \frac{\partial W}{\partial p_n} \right).$$

We also set

$$\dot{W} = \dot{W}(x, u, p) = \sigma \cdot W_x + |p| \cdot W_u$$

for $p \neq 0$, where $\sigma = \frac{p}{|p|}$.

Let

$$T = \frac{C}{G}, \quad p \neq 0,$$

where $C = C(x, u, p) = \frac{B(x,u,p)}{|p|}$. Then the following result is proved by Serrin [4, §13].

Theorem 29.9. *Let $u(x) \in C^1(\overline{B}) \cap C^3(B)$ be a solution of equation (29.107) in B, such that $|Du| \leq L$ on ∂B. Suppose that condition (29.136) holds and that*

$$\dot{T} = T^2 \geq 0 \quad \text{for} \quad |p| \geq L'.$$

Then

$$|Du| \leq \max(L, L')$$

in B.

Proof. Set $w = |Du|^2$ and let B' be the open subset of B where $w \neq 0$. Then

$$\sum_{i,k=1}^{n} a_{ik} w_{ik} \geq 2G(\dot{T} + T^2)w + N \cdot Dw, \quad x \in B'.$$

where the arguments of A, G, T are x, u, Du, and where $N = N(x)$ is a continuous function on B. We may assume without loss of generality that $G = 1$ by dividing both sides of equation (29.107) by G. Of course the normalization condition $tr A' = 1$ is then lost, but it is not required for the proof.

With the new normalization in mind, we apply the operator $u_k \frac{\partial}{\partial x_k}$ to both sides of equation (29.107). Then we obtain

$$\sum_{k=1}^{n}\left(\sum_{i,j=1}^{n} a_{ij} u_k u_{ijk}\right) + |Du|\dot{G}_1 \sum_{i,j=1}^{n} u_i u_j u_{ij}$$
$$+ \sum_{k,\ell,=1}^{n}\left(\sum_{i,j=1}^{n} \frac{\partial a_{ij}}{\partial p_1} u_k u_{k\ell} u_{ij}\right) = |Du|\dot{b}$$
$$+ \sum_{k,\ell=1}^{n} \frac{\partial b}{\partial p_1} u_k u_{k\ell}.$$

Using the identities

$$w_i = 2\sum_{k=1}^{n} u_k u_{ki}, \qquad i = 1, 2, \ldots, n, \tag{29.146}$$

and

$$w_{ij} = 2\sum_{k=1}^{n}(u_k u_{kij} + u_{ki} u_{kj}), \qquad i, j = 1, 2, \ldots, n \tag{29.147}$$

we obtain

$$\frac{1}{2}\sum_{i,j=1}^{n} a_{ij} w_{ij} = \sum_{k=1}^{n}\left(\sum_{i,j=1}^{n} a'_{ij} u_{ik} u_{jk}\right) + |Du|\dot{b}$$
$$+ \text{linear function of } Dw.$$

Now we have

$$\left(\sum_{i,j=1}^{n} a'_{ij} u_{ij}\right)^2 \leq \sum_{k=1}^{n}\left(\sum_{i,j=1}^{n} u_{ik} u_{jk}\right) \tag{29.148}$$

(actually we introduce new coordinates so that (u_{ij}) is diagonal, then

$$\begin{aligned}\left(\sum_{i,j=1}^{n} a'_{ij}u_{ij}\right)^2 &= \left(\sum_{i=1}^{n} a'_{ii}u_{ii}\right)^2\\ &\le \left(\sum_{i=1}^{n} a'_{ii}\right)\left(\sum_{i=1}^{n} a'_{ii}u_{ii}^2\right)\\ &= \sum_{k=1}^{n}\left(\sum_{i,j=1}^{n} a'_{ij}u_{ik}u_{jk}\right)\end{aligned}$$

which proves inequality (29.148) since both sides are invariant under orthogonal transformation). Furthermore

$$\begin{aligned}\sum_{i,j=1}^{n} a'_{ij}u_j &= \sum_{i,j=1}^{n} a_{ij}u_{ij} + \text{linear function of } Dw\\ &= b + \text{linear function of } Dw. \end{aligned}\tag{29.149}$$

It is possible to eliminate $\sum_{k=1}^{n}\left[\sum_{i,j=}^{n} a'_{ij}u_{ik}u_{jk}\right]$ from the identity for $\sum_{i,j=1}^{n} a_{ij}w_{ij}$ if we use inequality (29.148) and the identity (29.149). Thus we obtain the inequality

$$\frac{1}{2}\sum_{i,j=1}^{n} a_{ij}w_{ij} \ge b^2 + |Du|\dot{b} + \text{linear function of } Dw$$

from which the required inequality follows at once.

The rest of the proof is an immediate consequence of the strong maximum principle (see Chapter 2 of this book).

In order to obtain a result related to regular variational problems (29.138) it is necessary to modify condition (29.136) in the following form. Below we suppose that condition (29.136) holds in the form

$$A(x,u,p) = G(x,u,p)A'(\sigma) + G_1(x,u,p)pp, \tag{29.150}$$

where $\sigma = \frac{p}{|p|}$. The following theorem is due to Serrin [2, § 14].

Theorem 29.10. *Suppose that $n > 2$ and that for some constant $\mu > 0$ the invariant function $E = E(x,u,p) = pA(x,u,p)p$ satisfies the conditions*

$$p\cdot E_p \le (1-\mu)E, \qquad E \ge \mu \tag{29.151}$$

at least for all sufficiently large values of p, say $|p| \ge 1$. Assume that as p tends to infinity

$$E_x, \quad E_u, \quad E_p = O(E) \tag{29.152}$$

and that condition (29.35) holds with the remainder $R = C - C_0$ satisfying

$$R, R_x, R_u, |p|R_p = O(E/|p|). \tag{29.153}$$

Finally we assume that condition (29.150) holds.

Let now $u(x) \in C^1(\overline{B}) \cap C^3(B)$ be a solution of (29.107) satisfying the conditions $|u| \le m$ in B and $|Du| \le L$ on ∂B. Then

$$|Du| \le M \quad \text{in} \quad B,$$

where M depends only on μ, m, ℓ, L, bounds for the orders terms listed above, and on the C^1 norm of the function C_0.

Remark. Conditions of the type (29.151), (29.152), (29.153) are consistent with the assumption that equation (29.107) is boundedly non-linear, and indeed one of the main applications of Theorem 29.10 will be to the equation for surfaces having prescribed mean curvature in Euclidean space R^n.

Proof of Theorem 29.10. Introduce a new dependent variable $\bar{u}$ by means of the transformation

$$u = \varphi(\bar{u}), \quad (\varphi'(\bar{u}) > 0).$$

Then $\bar{u}$ obeys the equation

$$\sum_{i,j=1}^{n} \bar{a}_{ij}\bar{u}_{ij} = \bar{b}$$

with

$$\bar{a}_{ij} = G(a'_{ij}(\bar{\sigma})) + (\varphi')^2 G_1 \cdot \overline{pp} \quad \text{and}$$
$$\bar{b} = \varphi'^{-1}(b + \omega E),$$

where the arguments of G, G_1, b and E are $x, u, p = \varphi'(\bar{u})\bar{p}$, and where

$$\omega = -\frac{\varphi''}{(\varphi')^2}.$$

Now Theorem 29.9 can be applied because $(\bar{a}_{ij})$ has the form (29.136). For this purpose, we note that

$$\overline{T} = \frac{C + \omega E/|p|}{G}.$$

Now forming the trace of (29.150) there results

$$1 = G + |p|^2 G_1$$

while also

$$\sigma A \sigma = (\sigma A' \sigma) G + |p|^2 G_1.$$

Eliminating $|p|^2 G_1$ from the last two relations and solving for G yields

$$G = \frac{1 - \sigma A \sigma}{1 \sigma A' \sigma},$$

whence using the invariant $D = \frac{C}{1-E^*}$, $E^* = E/|p|^2$ and $p \neq 0^{*)}$, we find

$$\overline{T} = (1 - \sigma A' \sigma) \frac{C + \omega E/|p|}{1 - \sigma A \sigma} = (1 - \sigma A' \sigma) \overline{D},$$

where $D = \frac{C}{1-E^*}$. (Note that $E^* < 1$, since $tr A = 1$.)

Thus, with regard to transformed equation,

$$\dot{\overline{T}} + \overline{T}^2 = (1 - \sigma A' \sigma) \dot{\overline{D}} + (1 - \sigma A' \sigma)^2 \overline{D}^2.$$

This quantity will be nonnegative provided $\dot{\overline{D}} \geq 0$, that is, provided the expression

$$I \equiv (1 - E^*) \cdot \delta \left[C + \frac{\omega}{|p|} E \right] + \left[C + \frac{\omega}{|p|} E \right] \cdot \delta E^* \tag{29.154}$$

is nonnegative.

Below one proves that inequality (29.154) holds for all sufficiently large values of $|p|$ when

$$\varphi(\bar{u}) = \frac{1}{a} \log \bar{u}$$

and a is chosen to be a suitably large constant. Clearly for this choice $\omega = a =$ constant.

Now $\delta E = \sigma E_x + |p| \cdot E_u - a|p| \cdot p \cdot E_p$, whence (29.152) and the first inequality of (29.151) imply

$$-(\widetilde{C} + (1 - \mu) a)|p| E \leq \delta E \leq C(1 + a) \cdot |p|^2 \cdot E \tag{29.155}$$

for all p sufficiently large, say $|p| \geq \alpha_1$; here the number $\widetilde{C}$ is a generic bound depending only on the order terms in (29.152) and (29.153) and the norm of the function C_0. Similarly from (29.154) and the second inequality of (29.151) it follows

$$\delta C = \delta C_0 + \delta R \geq \frac{\partial C_0}{\partial x} + \delta R \geq -\widetilde{C}(\mu^{-1} + 1 + a) E \tag{29.156}$$

provided $|p| \geq \alpha_2$.

From (29.155) and (29.156) it also follows that

$$\delta(C + aE/|p|) \geq (\mu a^2 - (1 + 2a + \mu^{-1}) \widetilde{C}) \cdot E$$

and

$$((1+\mu)a-\widetilde{C})\frac{E}{|p|} \leq E^* \leq \left(\frac{2a}{|p|}+(1-a)\widetilde{C}\right)E,$$

since $\delta E^* = |p|^{-2}\delta E + 2aE/|p|$. Since $C + aE/|p| = C_0 + (R + aE/|p|)$, then we can estimate two terms in the right-hand side separately; thus

$$C_0 \geq -\widetilde{C}, \quad R + \frac{aE}{|p|} \geq (a-\widetilde{C})\frac{E}{|p|}$$

for $|p| \geq \alpha_3$.

Let us assume a final choice of a satisfying the inequality $a > \widetilde{C}$. Using the estimates considered above, the function I defined by (29.154) then obeys

$$\begin{aligned} I \geq E[&(1-E^*)(\mu a^2 - (1+2a+\mu^{-1})\widetilde{C} \\ &+ E^*(a-\widetilde{C})((1+\mu)a-\widetilde{C}) - (2a\widetilde{C}+(1+a)\widetilde{C}^2)] \end{aligned}$$

provided $|p| \geq \max(\alpha_1, \alpha_2, \alpha_3, 1) \equiv L'$. Assuming $\mu < 1$, as is certainly allowable, it is easy to see that

$$I \geq E[\mu a^2 - (4\widetilde{C}+\widetilde{C}^2)a - (1+\mu^{-1})\cdot\widetilde{C} - \widetilde{C}^2].$$

Now choose the constant "a" greater than C so that the quantity in square brackets vanishes.

Thus it was shown that $\dot{\overline{D}} + \overline{D}^2 \geq 0$ provided $|\bar{p}| \geq L'/\min\varphi'$. Therefore by Theorem 29.9

$$|D\bar{u}| \leq \max(L, L')/\min\varphi'$$

and hence

$$|Du| \leq \frac{\max\varphi'}{\min\varphi}\max(L, L') \quad \max(L, L')\cdot e^{2am},$$

which completes the proof of Theorem 29.10.

Bibliography

Adams R.A.: [1] Sobolev Spaces, New York, Academic Press (1975).

Agmon S.: [1] Lectures on elliptic boundary value problems, Princeton, N.J., Van Nostrand (1965).

Agmon S., Douglis A. and Nirenberg L.: [1] Estimates near the boundary for solutions of elliptic partial differential equations satisfying general boundary conditions, I, II. Comm. Pure Appl. Mathematics **12**, 623–727 (1959); **17**, 35–92 (1964).

Albano S. and Gonzalez E.: [1] Rotating drops. Indiana University Mathem. Journal **32**, 687–702 (1983).

Alexandrov A.D.: [1] Intrinsic geometry of convex surfaces, GETTL, Moscow–Leningrad (1948). [2] Convex polyhedra, GETTL, Moscow–Leningrad (1950). [3] Theory of mixed volumes for convex bodies, I, II, III, IV. Mathem. Sb. USSR **2**, 947–972 (1937); **2**, 1205–1238 (1937); **3**, 27–46 (1938); **4**, 227–251 (1938). [4] Uniqueness theorems for hypersurfaces in the large, I–VI. Vestnik LGU **19**, 5–17 (1956); **7**, 5–26 (1957); **7**, 14–26 (1958); **13**, 27–34 (1958), (with Volkov Ju.A.); **19**, 5–8 (1958); **1**, 5–13 (1959). [5] Dirichlet problem for the equation $\mathrm{Det}(Z_{ij}) = \phi$, I. Vestnik LGU No. 1 **13**, 5–24 (1958). [6] Uniqueness conditions and estimates for the solution of the Dirichlet problem. Vestnik LGU **18**, 5–29 (1963). [7] Majorization of second order linear equations. Vestnik LGU **21**, 5–25 (1966). [8] Existence and uniqueness of convex surfaces with prescribed integral curvature. Dokl. Akad. Nauk USSR **35:8**, 143–147 (1942).

Allard W.K.: [1] On the first variation of varifold. Annal. of Mathem. (2) **95**, 417–419 (1972). [2] Some recent advances in the multidimensional parametric calculus of variations. Proc. Int. Congr. Mathem., Vancouver 1974, Canad. Int. Congr. Mathem. Vancouver 1974, Canad. Math. Congr. Montreal **2**, 231–236 (1975).

Almgren F.J.: [1] Three theorems on manifolds with bounded mean curvature. Bull. Amer. Math. Soc. **71**, 755–756 (1965). [2] The theory of varifolds (A variational calculus in the large for a k-dimensional area integrand), Princeton (1965). [3] Optimal isoperimetric inequalities. Bull. Amer. Mathem. Soc. **13**, 123–126 (1985).

Atkinson F.V., Peletier L.A., and Serrin J.: [1] Ground states for the prescribed mean curvature equation. The supercritical case. Proc. of the Conference NONLINEAR DIFFUSION EQUATIONS AND THEIR EQUILIBRIUM STATES, I. Mathematical Sciences Research Institute 51–74 (1988).

Aubin Th.: [1] Problemes isoperimetriques et espaces de Sobolev. J. Differential Geometry **11**, 573–598 (1976). [2] Best constants in the Sobolev imbedding theorem; the Yamabe problem. Seminar on Differential Geometry, Princeton Univ. Press, Princeton, N.J. 173–184 (1982). [3] Nonlinear analysis on manifolds, Monge–Ampere equations. Springer-Verlag (1982).

Bakelman I.J.: [1] Generalized solutions of the Monge–Ampere equations. Dokl. Akad. Nauk USSR **114**, 1143–1145 (1957). [2] On the theory of Monge–Ampere equations. Vestnik LGU No. 1, 25–38 (1958). [3] The regularity of generalized solutions of Monge–Ampere equations. Zap. Naucn. Leningrad State Pedagogical Institute **166**, 143–184 (1958). [4] The Dirichlet problem for equations of Monge–Ampere type and their n-dimensional analogues. Dokl. Akad. Nauk USSR **126**, 923–926 (1959). [5] The first boundary value problem for nonlinear elliptic equations. The Second Doctor Dissertation. Leningrad State Pedagogical Institute (University), Leningrad (1959). [6] On the theory of quasilinear elliptic equations. Sibirsk. Mathem. Journal No. 2, 179–186 (1961). [7] Geometric methods for solving elliptic equations. Moscow, Nauka (1965). [8] Hypersurfaces with prescribed mean curvature and quasilinear elliptic equations with strong nonlinearities. Mathem. Sb. **75**, 604-638 (1968). [9] Mean curvature and quasilinear elliptic equations. Sibirsk. Mathem. Journal **9**, 752–771 (1968). [10] Geometric problems in quasilinear elliptic equations. Uspekhi Mathem. Nauk

25:3, 49–112 (1970). [11] Applications of the Monge–Ampere operators to the Dirichlet problem for quasilinear elliptic equations. Annals of Mathem. Studies, Seminar on Diff. Geometry, Princeton University Press **102**, 239–258 (1982). [12] R-curvature, estimates and stability of solutions for the general elliptic equations. Journal of Diff. Equations No. 1 **43**, (1982). [13] Variational problems and elliptic Monge–Ampere equations. Journal of Diff. Geometry **18**, 669–699 (1983). [14] Notes concerning the torsion of hardening rods and its n-dimensional generalizations. IMA Preprints Series, University of Minnesota No. 208, 1–38 (1986). [15] Generalized elliptic solutions of the Dirichlet problem for n-dimensional Monge–Ampere equations. Proc of Symposia in Pure Mathem., Nonlinear Functional Analysis and its Applications, AMS Part 1 **45**, 73–102 (1986). [16] The boundary value problems for nonlinear elliptic equations and the maximum principle for Euler–Lagrange equations. Archives for Rational Mechanics and Analysis No. 3 **93**, 271–300 (1986). [17] Convex functions methods in the Dirichlet problem for Euler–Lagrange equations. Proc. of the International Conference VARIATIONAL METHODS FOR FREE SURFACE INTERFACES, Stanford University, California, Springer-Verlag 127–137 (1987). [18] Geometric concepts and methods in nonlinear elliptic Euler–Lagrange equations. Proc. of the International Conference NONLINEAR DIFFUSION EQUATIONS AND THEIR EQUILIBRIUM STATES, I. Mathem. Sciences Research Institute, Berkeley, California, Springer-Verlag 75–107 (1988). [19] The boundary value problems for n-dimensional Monge–Ampere equations; n-dimensional plasticity equation, I, II. Seminars of the College de France, Proc. of H. Brezis and J. Lions Nonlinear PDE Seminar, Volumes VIII (1988); IX (1989). [20] Geometric inequalities and existence theorems for convex generalized solutions of n-dimensional Monge-Ampere equations. in *Geometric Analysis and Nonlinear Partial Differential Equations*, I.J. Bakelman, ed., Marcel Dekker, New York (1993). [21] Variational problem connected with Monge–Ampere equations. Dokl. Akad. Nauk USSR **141**, 1011–1014 (1961). [22] A priori estimates and existence theorems for closed starshaped hypersurfaces in Riemannian spaces (preprint).

Bakelman I. and Perry W.: [1] Geometric inequalities and estimates of solutions for nonlinear Euler-Lagrange equations and applied problems. in *Geometric Analysis and Nonlinear Partial Differential Equations*, I.J. Bakelman, ed., Marcel Dekker, New York (1993).

Bakelman I.J. and Kantor B.E.: [1] Existence of a hypersurface homeomorphic to the sphere in Euclidean space with prescribed mean curvature. Geometry and Topology, Leningrad State Pedagogical Institute No. 1, 3–10 (1974). [2] Quasilinear elliptic equations related to the problems of global differential geometry. Math. Sb. **83:5**, (1973).

Bakelman I.J., Verner A.L. and Kantor B.E.: [1] Introduction to global differential geometry. Moscow, Nauka (1973).

Bakelman I.J. and Krasnoselskii M.A.: [1] Nontrivial solutions of the Dirichlet problem for the Monge–Ampere equation. Dokl. Acad. of Sc. USSR **136:1**, 161–163 (1961). [2] A solvability criterion for the two point boundary value problem. Vestnik LGU No. 13 **16**, 1011–1014 (1961).

Bambah R.P.: [1] Polar reciprocal convex bodies. Proc. Camb. Philos. Soc. **51**, 377–378 (1954).

Bernstein S.N.: [1] Sur la generalization du probleme de Dirichlet, I. Mathem. Annalen **62**, 253–271 (1906). [2] Sur la generalization du probleme de Dirichlet, II. Math. Annalen **62**, 82–136 (1910). [3] Sur les surfaces defines au moyen de leur courbure noyenne and total. Ann. Sci. Ecole Norm. Sup. **27**, 233–256 (1910). [4] Sur les equations du calcul des variations. Ann. Sci. Ecole Norm. Sup. **29** 431–485 (1912). [5] Collection of papers. Differential Equations, Calculus of Variations, Geometry Acad. of Sc. USSR Vol. 3, 1–436 (1960). (Vol. 3 contains 31 original papers published in 1903–1947.)

Bers L., John F. and Schecter M.: [1] Partial differential equations. New York, Interscience Publishers (1964).

Bol G.: [1] Beweis einer Vermutung von H. Minkowski. Abh. Math. Sem. Uni. Hamburg **15**, 37–56 (1943).

Bombieri E.: [1] Variational problems and elliptic equations. Proc. of the International Congress of Mathematicians, Vancouver 1974 **1**, 53–63 (1975).

Bombieri E., De Giorgi E. and Giusti E.: [1] Minimal cones and the Bernstein problem. Invent. Mathem. **7**, 243–268 (1968).

Bombieri E. and Giusti E.: [1] Local estimates for the gradient of non-parametric surfaces of prescribed mean curvature. Comm. Pure Appl. Math. **26**, 381–394 (1973).

Bonnesen T. and Fenchel W.: [1] Theorie der konvexen Körper, Berlin, Springer (1934).

Bourgain J. and Milman V.D.: [1] New volume ratio properties for convex symmetric bodies in R^n. Invent. Mathem. **88**, 319–340 (1987).

Brezis H.: [1] Some variational problems with lack of compactness. Proc. of Symposia in Pure Mathematics, Nonlinear Funct. Analysis and its Applications, AMS Part 1 **45**, 165–201 (1986).

Brezis H. and Coron M.: [1] Multiple solutions of H-systems and Rellich's conjecture. Comm. Pure and Appl. Mathem. **37**, 148–187 (1984). [2] Convergence of solutions of H-systems, or how to blow bubbles. Arch. Rational Mech. Anal. **89**, 21–56 (1985).

Brezis H. and Lieb E.: [1] A relation between pointwise convergence of functions and convergence of integrals. Proc. Amer. Math. Soc. **88**, 486–490 (1983).

Brezis H. and Nirenberg L.: [1] Positive solutions of nonlinear elliptic equations involving critical Sobolev exponents. Comm. Pure Appl. Mathem. **36**, 437–477 (1983).

Brezis H. and Browder F.E.: [1] Some properties of higher order Sobolev spaces. J. Mathem. Pures Appl. **61**, 245–259 (1982).

Browder F.E.: [1] On the regularity properties of solution of elliptic differential equations. Comm. Pure Appl. Math. **9**, 351–361 (1956). [2] Existence theorems for nonlinear partial differential equations. Proc. of Symposia in Pure Mathem., AMS **16**, 1–60 (1970). [3] Fixed point theory and nonlinear problems. Mathematical Heritage of Henri Poincare, Proc. Symp. Pure Mathem. **39**, 49–87 (1984).

Busemann H.: [1] Convex surfaces. Interscience Publ., New York–London (1958).

Caffarelli L.: [1] A localization property of viscosity solutions to Monge–Ampere equation and their strict convexity. Institute for Advanced Study, Princeton, N.J. 1–7 (1988) (preprint). [2] Interior $W^{2,p}$ estimates for solutions of Monge–Ampere equation. Institute for Advanced Study, Princeton, N.J. 1–20 (1988) (preprint). [3] Elliptic second order equations. The notes of a series of lectures at the Politecnico de Milano, April 1987 1–34 (1988) (preprint).

Caffarelli L. and Friedman A.: [1] The free boundary for elastic-plastic problems. Trans. Amer. Mathem. Soc. **252**, 65–97 (1979). [2] Reinforcement problems in elasto-plasticity. Rocky Mt. J. Mathem. **19**, 155–184 (1980).

Caffarelli L., Nirenberg L. and Spruck J.: [1] The Dirichlet problem for nonlinear second order elliptic equations: I. Monge–Ampere equations. Comm. Pure Appl. Mathem. **37**, 369–402 (1984); II. (jointly with J.J. Kohn) Complex Monge–Ampere and uniformly elliptic equations. Comm. Pure and Appl. Mathem. **38**, 209–252 (1985); III. Functions of the eigenvalues of the Hessian. Acta Mathem. **155**, 261–301 (1985); IV. Starshaped compact Weingarten hypersurfaces. Current topics in partial differential equations. Kinokunize Co., Tokyo 1–26 (1986); V. The Dirichlet problem for Weingarten hypersurfaces. Comm. Pure Appl. Mathem. **46**, 47–70 (1988).

Calabi E.: [1] Improper affine hyperspheres of convex type and a generalization of a theorem by K. Jörgens. Michigan Mathem. Journal **5**, 105–126 (1958) [2] Complete affine hyperspheres. Izt. Naz. Alta Mathem., Symp. Mathem. X (1972). [3] Extremal Kähler metrics. Annals of Mathem. Studies., Seminar in Diff. Geometry, Princeton University Press **102**, 259–290 (1982). [4] Examples of Bernstein problems for some nonlinear equations. AMS Symposium on Global Analysis, Berkeley **15**, 223–230 (1960).

Chang Sun-Yung Alice and Yang Paul C.: [1] Prescribing Gaussian curvature on S^2. Acta Mathem. **159**, 215–259 (1987). [2] Conformal deformation of metrics on S^2. J. Diff.

Geometry. **27**, 259–296 (1988). [3] Compactness of isospectral conformal metrics on S^3. Comment. Mathem. Helvetici **64**, 363–374 (1989). [4] The conformal deformation equation and isospectral set of conformal metrics. Contemporary Mathem., Providence, R.I., AMS **101**, 165–178 (1989).

Cheng S.Y. and Yau S.T.: [1] On the regularity of the solution of the n-dimensional Minkowski problem. Comm. Pure Appl. Mathem. **19**, 495–516 (1976). [2] Maximal space-like hypersurfaces in the Lorentz–Minkowski spaces. Annals of Mathem. **104**, 407–419 (1976). [3] On the regularity of the Monge–Ampere equation $\det(u_{ij}) = F(x, u)$. Comm. Pure and Appl. Mathem. **30**, 41–68 (1977). [4] The real Monge–Ampere equation and affine flat structures. Proc. 1980 Beijing Symposium on Differential Geometry and Differential Equations, **1**, 339–370 (1982).

Courant R. and Hilbert D.: [1] Methods of mathematical physics. Volumes I, II, Interscience (1953), (1962).

De Giorgi E.: [1] Sulla differenziabilità e l'analiticità della estremali degli integrali multipli regolari. Mem. Accad. Sci. Torino Cl. Sci. Fis. Mat. Natur. (3) **3**, 25–43 (1957).

Delanoë P.: [1] Equations du type Monge–Ampere sur la varietes Riemanniannes compactes, I, II. Funct. Analysis **40**, 358–386 (1981); **41**, 341–353 (1981).

Douglas J., Dupont T. and Serrin J.: [1] Uniqueness and comparison theorems for nonlinear elliptic equation in divergence form. Arch. Rational Mech. Analysis **42**, 157–168 (1971).

Evans L.G.: [1] A convergence theorem for solutions of nonlinear second order elliptic equations. Indiana University Mathem. Journal **27**, 875–887 (1978). [2] A second order elliptic equation with gradient constrain. Comm. PDE **4**, 555–572 (1978). [3] Classical solutions of fully nonlinear, convex, second order equations. Comm. Pure Appl. Mathem. **25** 333–363 (1982).

Evans L.G. and Spruck J.: [1] Motion of level sets by mean curvature, I, II. University of California and University of Massachusetts, preprints 1–73 (1989); 1–21 (1990).

Federer H.: [1] Geometric measure theory. Springer-Verlag (1969).

Fenchel W.: [1] Inégaliés quadratiques entre les volumes mixtes des corps convexes. C.R. Acad. Sci. Paris **203**, (1936), 647-650.

Fenchel W. and Jensen B.: [1] Mengenfunktionen und konvexe Körper. Danske Vid. Selsk. Mat.-Fys. Medd. **16:3**, 1–31 (1938).

Finn, R.: [1] Capillarity phenomena. Uspekhi Mathem. Nauk **29**, 131–152 (1974). [2] Equilibrium capillary surfaces. Springer-Verlag (1986).

Friedman A.: [1] Partial differential equations. New York: Holt, Rinehart and Winston (1969). [2] Variational Principles and free boundary problems. John Wiley & Sons (1982).

Gerhardt C.: [1] Existence, regularity, and boundary behaviour of generalized surfaces of prescribed mean curvature. Mathem. Z. **139**, 173–198 (1974). [2] Boundary value problems of prescribed mean curvature. J. Math. Pures and Appl. **58**, 75–109 (1979).

Giaquinta M.: [1] On the Dirichlet problem for surfaces of prescribed mean curvature. Manuscripta Mathem. **12**, 73–86 (1974). [2] Multiple integrals in the calculus of variations and nonlinear elliptic systems. Annals of Mathem. Studies, Princeton University Press **105**, (1983).

Gidas B., Ni W.M. and Nirenberg L.: [1] Symmetry and related topics via de maximum principle. Comm. Mathem. Phys. **68**, 209–243 (1979).

Gilbarg D.: [1] Boundary value problems for nonlinear elliptic equations in n variables. Nonlinear Problems, Madison, Wisconsin, University of Wisconsin Press 151–159 (1963).

Gilbarg D. and Serrin J.: [1] On isolated singularities of solutions of second order elliptic equations. Journal Analyse Mathem. **4**, 309–340 (1955/56).

Gilbarg D. and Trudinger N.S.: [1] Elliptic partial differential equations of second order. (Second Edition), Springer-Verlag (1983).

Giusti E.: [1] Boundary value problems for non-parametric surfaces of prescribed mean curvature. Ann. Scuola Norm. Sup. Pisa (4) **3**, 501–548 (1976). [2] On the equation of surfaces of prescribed mean curvature. Invent. Mathem. **46**, 111–137 (1978). [3] Equazioni elliptiche del secondo ordine. Bologna, Pitagora, Editrice (1978). [4] Minimal surfaces and functions of bounded variation. Boston, Birkhaüser (1983).

Gonzalez E., Massari U. and Tamanini I.: [1] Existence and regulariti for the problem of a pendent dop. Pacific J. Math. **88**, 399–420 (1980).

Groemer H.: [1] Minkowski addition and mixed volumes. Geometrie Dedicata **6**, 141–163 (1977).

Gromov M.: [1] Partial differential relations. Springer-Verlag (1986).

Gromov M. and Rochlin V.: [1] Embeddings and immersions in Riemannian geometry. Uspekhi Mathem. Nauk, XXV **5**, 3–62 (1970).

Hadwiger H.: [1] Vorlesungen über Inhalt, Oberfläche and Isoperimetrie. Springer, Berlin (1957). [2] Minkowskis Ungleichungen und nicht convexe Rotationskörper. Mathem. Nachr. **14**, 377–383.

Hadwiger H. and Ohmann D.: [1] Brunn–Minkowski Satz and Isoperimetrie. Math. Z. **66**, 1–8 (1956).

Heinz E.: [1] Über die existenz einer fläche konstanter mittlerer Krümmung bei vorgegebner Berandung. Math. Ann. **127**, 258–287 (1954). [2] On the existence problem for surfaces of bounded mean curvature. Comm. Pure Appl. Mathem. **9**, 467–470 (1956). [3] On certain nonlinear elliptic differential equations and univalent mappings. Journal Analyse Mathem. **5**, 197–272 (1956/57). [4] On elliptic Monge–Ampere equations and Weyl's embedding problem. Journal Analyse Mathem. **7**, 1–52 (1959). [5] Über die differentialungleichungen $0 < \alpha \leq rt - s^2 \leq \beta < +\infty$. Math. Z. **72**, 107–126 (1959). [6] Neue a-priori-abschätzungen für den Ortsvektor einer fläche positiver gaussher Krümmung durch ihr Linienelement. Math. Z. **74**, 129–157 (1960). [7] On Weyl's embedding problem. Journal Mathem. Mech. **11**, 421–454 (1962).

Heinz E. and Hildebrandt S.: [1] Some remarks on minimal surfaces in Riemannian manifolds. Comm. Pure Appl. Mathem. **23**, 371–377 (1970). [2] The number of branch points of surfaces of bounded mean curvature. Journal of Differential Geometry **4**, 227–235 (1970).

Hilbert D.: [1] Grundzüge einer allgemeinen Theorie der linearen Integralgleichungen. Leipzig–Berlin (1915).

Hildebrandt S.: [1] Über flächen konstanter mittlerer Krümmung. Mathem. Z. **112**, 107–144 (1969). [2] On the Plateau problem for surfaces of constant mean curvature. Comm. Pure Appl. Mathem. **23**, 97–114 (1970). [3] Maximum principles for minimal surfaces and surfaces of continuous mean curvature. Mathem. Z. **128**, 253-269 (1972). [4] Free boundary problems for surfaces of constant mean curvature. Proc. of International Conference VARIATIONAL METHODS FOR FREE SURFACE INTERFACES, Springer-Verlag 43–51 (1987).

Hildebrandt S. and Jäger W.: [1] On the regularity of surfaces with prescribed mean curvature at a free boundary. Mathem. Z. **118**, 289–308 (1970).

Hopf E.: [1] Elementare Bemerkunge über die Lösungen partieller Differentialgleichungen zweiter Ordnung von elliptischen Typen. Sber. Preuss. Akad. Wiss. **19**, 147–152 (1927).

Hopf H.: [1] Differential geometry in the large. (Seminar Lectures, New York University, 1946, and Stanford University, 1956), Lecture Notes in Mathematics No. 1000, Springer-Verlag (1983).

Ivočkina, N.M.: [1] An apriori estimate of $||u||_{C^2(\Omega)}$ for convex solutions of Monge–Ampere equations, Zap. Naucn. Sem. Leningrad Otdel. Mat. Inst. Steklov. **96**, 69–79 (1980). [2] Classical solvability of the Dirichlet problem for the Monge–Ampere equation, Zap. Naucn. Sem. Leningrad. Otdel. Mat. Inst. Steklov. **131**, 72–99 (1983).

Jenkins H. and Serrin J.: [1] Variational problems of minimal surface type 1. Archives Rational

Mech. Anal. **12**, 185–212 (1961). [2] The Dirichlet problem for the minimal surface equation in higher dimensions. Journal Reine Angew. Mathem. **229**, 170–187 (1968).

John F. and Nirenberg L.: [1] On functions of bounded mean oscillation. Comm. Pure Appl. Mathematics **14**, 415–426 (1961).

John F.: [1] Extremum problems with inequalities as subsidiary conditions. Courant Anniversary Volume, New York, Interscience 187–204 (1948).

Jörgens K.: [1] Über die Lösungen der Differentialgleichungen $rt - s^2 = 1$. Math. Ann. **127**, 130–154 (1954).

Kawohl B.: [1] Rearrangements and convexity of level sets in PDE. Lectures Notes in Mathematics, Springer-Verlag (1985).

Kazdan J.L.: [1] An isoperimetric inequality and Wiedersehen manifolds. Annals Mathem. Studies, Seminar on Differential Geometry, Princeton, N.J. **102**, 143–157 (1982). [2] Deformation to positive scalar curvature on complete manifolds. Math. Annalen **261**, 227–234 (1982). [3] Some applications of partial differential equations to problems in differential geometry. Surveys in Geometry Series, Tokyo University (1983). [4] Prescribing the curvature of a Riemannian manifold. Expository Lectures from the CBMS Regional Conference, Polytechnic Institute of New York No. 57, 1–55 (1985).

Kazdan J.L. and Warner F.W.: [1] Prescribing curvatures. Proc. Symp. Pure Mathem., AMS **27**, 219–226 (1975). [2] Remarks on some quasilinear elliptic equations. Comm. Pure Appl. Mathem. **28**, 567–597 (1975).

Kinderlehrer D. and Nirenberg L.: [1] Regularity in free boundary problems. Annali Scuola Norm. Sup. IV **2**, 373–391 (1977).

Kinderlehrer D. and Stampacchia G.: [1] An introduction to variational inequalities and their applications. New York, Academic Press (1980).

Kohn J.J. and Nirenberg, L.: [1] Degenerate elliptic-parabolic equations of second order. Pure Appl. Mathem. **20**, 797–872 (1967).

Korevaar N.J.: [1] Capillary surface convexity above convex domains. Ind. Univ. Journal **32:1**, 73–81 (1983). [2] Convexity properties of solutions to elliptic PDE's. Proc. of International Conference VARIATIONAL METHODS FOR FREE SURFACE INTERFACES, Menlo Park, CA, Springer-Verlag 115–121 (1987).

Koselev A.I.: [1] On boundedness in L^p of derivatives of solutions of elliptic differential equations. Mat. Sb. **38**, 359–372 (1956). [2] A priori estimates in L^p and generalized solutions of elliptic equations and systems. Uspekhi Math. Nauk **13:4**, 29–88 (1958).

Krasnoselskii M.A.: [1] Positive solutions of operator equations. Phismath Publ. Co., Nauka, Moscow (1962).

Krein S.G. and others: [1] Functional analysis, Moscow, Nauka, Second Edition (1972).

Krylov N.V.: [1] Nonlinear elliptic and parabolic equations of second order. Moscow, Nauka (1985). [2] Boundedly inhomogeneous elliptic and parabolic equations. Izvestia Akad. Nauk. SSSR **46**, 487–523 (1982) [Russian]. English Translation in Math. USSR Izv. **20** (1983). [3] Boundedly inhomogeneous elliptic and parabolic equations in a domain. Izvestia Akad. Nauk. SSSR **47**, 75–108 (1983) [Russian]. [4] On degenerate nonlinear elliptic equations. Mat. Sb. (N.S.) **120**, 311–330 (1983) [Russian].

Ladyzhenskaya O.A.: [1] A simple proof of solvability of classical boundary value problems and eigenvalue problem for elliptic equations. Vestnik LGU **11**, 23–29 (1955). [2] On integral estimates, convergence, approximate methods, and solutions in functionals for elliptic operators. Vestnik LGU **7**, 60–69 (1958). [3] Solution of the first boundary value problem in the large for quasilinear parabolic equations. Trudy Moscow. Mathem. Society **7**, 149–177 (1958). [4] The boundary value problems of mathematical physics. Springer-Verlag (1985).

Ladyzhenskaya O.A. and Uraltzeva N.N.: [1] Linear and quasilinear equations of elliptic type. Moscow, Nauka, Second Edition (1973).

Ladyzhenskaya O.A. and Visik M.I.: [1] Boundary value problems for partial differential equations and certain classes of operator equations. Uspekhi Math. Nauk **11:6**, 89–152 (1956).

Landis E.M.: [1] S-capacity and its applications to solutions of second order elliptic equations with discontinuous coefficients. Mathem. Sb. **76:2**, 186–213 (1968). [2] Theorems of Phragmen–Lindelöf type for solutions of elliptic equations. Dokl. Akad. of Sc. USSR **193:1**, 32–35 (1970). [3] New proof of the de Giorgi Theorem. Proc. of Moscow Mathem. Soc. **16**, 219–228 (1967). [4] Equations of the second order of elliptic and parabolic. Moscow, Nauka (1971).

Lax P.D.: [1] A Phragmen–Lindelöf Theorem in harmonic analysis and its applications in the theory of elliptic equations. Comm. Pure Appl. Math. **10:3**, 361–389 (1957). [2] On the Cauchy problem for hyperbolic equations and the differentiability of solutions of elliptic equations. Comm. Pure Appl. Mathem. **8**, 615–633 (1955).

Lee J.M. and Parker T.H.: [1] The Yamabe problem. Bull. Amer. Math. Soc. **17:1**, 37–91 (1987).

Leray J.: [1] Majoration des derivees secondes des solutions d'un probleme Dirichlet. J. Mathem. Pures et Appl. **17**, 89–104 (1938). [2] Discussion d'un probleme Dirichlet. J. Mathem. Pures et Appl. **18**, 249-284 (1939).

Leray J. and Lions J.L.: [1] Quelques results de Visik sur les problemeselliptiques non lineaires par les methodes de Minty–Browder. Bull. Soc. Mathem. France **93**, 97–107 (1965).

Leray J. and Schauder J.: [1] Topologie et equations fonctionelles. Ann. Sci. Ecole Norm. Sup. **51**, 45–78 (1934).

Levine H.A. and Weinberger H.F.: [1] Inequalities between Dirichlet and Neumann eigenvalues. Arch. Ration. Mech. Analysis **94**, 193–208 (1986).

Lewy H.: [1] Neuer Beweis des analytischen Characters der Löscungen elliptischer Differentialgleichungen. Math. Annalen **101**, 609–619 (1929); **107:804**, (1929). [2] A priori limitations for solutions of Monge–Ampere equations, I, II. Trans. Amer. Math. Soc. **37**, 417–434 (1935); **41**, 365–374 (1937). [3] On the existence of a closed convex surface realizing a given Riemannian metric. Proc. National Acad. Sci., USA **24**, (1938).

Lieb E.H.: [1] Bounds on the eigenvalues of the Laplace and Schroedinger operators. Bull. Amer. Math. Soc. **82**, 751–753 (1976). [2] The number of bound states of one-body Schroedinger operators and the Weyl problem. Proc. of Symp. in Pure Mathem., AMS 241–252 (1980).

Lieberman G.M.: [1] Solvability of quasilinear elliptic equations with nonlinear boundary conditions. Trans. Amer. Math. Soc. **273**, 753–765 (1982). [2] The Dirichlet problem for quasilinear elliptic equations with Hölder continuous boundary values. Arch. Ration. Mech. Analysis **79:4**, 305–383 (1982).

Lions J.L.: [1] Quelques methodes de resolution des problemes aux limites non-lineares. Dunod Gautier–Villars (1969).

Lions P.L.: [1] Resolution de problemes elliptiques quasilineaires. Arch. Rational Mech. Analysis **74**, 335–345 (1980) [2] Sur les equations de Monge–Ampere 1, 2. Manuscripta Mathem. **41**, 1–44 (1983); Arch. Rational Mech. Analysis (1984). [3] Optimal control of diffusion processes and Hamilton–Jacobi Bellman equations I, II, III; I, II. Comm. Part. Diff. Equations (1983), III — College de France Seminar IV, London, Pitman (1983).

Littman W. and Stampacchia G. and Weinberger H.F.: [1] Regular points for elliptic equations with discontinuous coefficients. Ann. Scuola Norm. Sup. Pisa **17**, 43–77 (1963).

Li P.: [1] Poincare inequalities on Riemannian manifolds. Ann. Math. Studies **102**, Seminar on Diff. Geometry, Princeton, N.J. 73–83 (1982).

Li P. and Yau S.T.: [1] A new conformal invariant and its application to the Wilmore conjecture and the first eigenvalue of compact surface. Invent. Mathem. **69**, 269–292 (1982).

Li P., Schoen R. and Yau S.T.: [1] On the isoperimetric inequality for minimal surfaces. Ann. Scuola Norm. Sup. Pisa, Cl. Sci. **11**, 237–284 (1984).

Lutwak E.: [1] Dual mixed volumes. Pacific J. Mathem. **58**, 531–538 (1975). [2] A general Bieberbach inequality. Math. Proc. Camp. Phil. Soc. **58**, 493–496 (1975). [3] On cross-sectional measures of polar reciprocal convex bodies. Geometriae Dedicata **5**, 78–80 (1976). [3] Mean dual and harmonic cross-sectional measures. Ann. Mathem. Pura Appl. **119**, 139–148 (1979).

Massari U. and Miranda M.: [1] Minimal surfaces of codimension 1. North–Holland, Amsterdam (1984).

Massari U.: [1] Esistenza regolarita' delle ipersuperfici di curvatura media assegnata in R^n. Arch. Rational Mech. Analysis **55**, 357–382 (1974).

Maz'ja V.G.: [1] Sobolev spaces. Springer-Verlag (1985).

Maz'ja V.G. and Saposnikov T.O.: [1] Multipliers in spaces of differentiable functions. Leningrad State University (1986).

McOwen R.: [1] Conformal metrics in R^2 with prescribed Gaussian curvature and positive total curvature. Indiana Univ. Math. J. **34**, 97–104 (1985).

Meyers N.G.: [1] An L^p-estimate for the gradient of solutions of second order elliptic divergence equations. Ann. Scuola Norm. Sup. Pisa **17**, 189–206 (1963).

Meyers N.G. and Serrin J.: [1] The exterior Dirichlet problem for second order elliptic partial differential equations. Journal Math. Mech. **9**, 513–538 (1960).

Meyers N. and Ziemer W.: [1] Integral inequalities of Poincare and Wirtinger type for BV-functions. Amer. J. Math. **99**, 1345–1360 (1977).

Mikhlin S.G.: [1] Variational methods in mathematical physics (translated from Russian). Pergamon Press Book, Macmillan Inc., New York (1964). [2] The problem of the minimum of a quadratic functional (translated from Russian). Holden Day Inc. 91965). [3] Multidimensional singular integrals and integrals equations (translated from Russian). Pergamon Press, Oxford–New York–Paris (1965). [4] The numerical performance of variationals methods (translated from Russian). Wolters–Noordhoff Groningen (1971).

Milman V.D.: [1] New proof of the theorem of Dvoretzky on sections of convex bodies. Functional Analysis i Prilozen **5**, 28–37 (1971). [2] Inequalite de Brunn–Minkowski inverse et applications a la theory locale des espaces normaes. C.R. Acad. Sci. Paris **302**, 25–28 (1986).

Milnor J.: [1] Morse theory. Princeton University Press, Princeton, N.J. (1963).

Minkowski H.: [1] Theorie der konvexen Körper. Ges. Abh. Leipzig–Berlin **2**, 131–229 (1911). [2] Volumen und Oberfläche. Math. Ann. **57** 447-495 (1903).

Miranda C.: [1] Partial differential equations of elliptic type. Second Edition, Springer-Verlag (1970).

Miranda M.: [1] A mathematical description of equilibrium surfaces. Proc. of Conference VARIATIONAL METHODS FOR FREE SURFACE INTERFACES, Menlo Park, Stanford University, CA 85–89 (1987).

Morrey C.B.: [1] Multiple integral problems in the calculus of variations and related topics. Univ. California Publ. Math. (N.C.) **1**, 1–130 (1943). [2] Second order elliptic equations in several variables and Hölder continuity. Math. Z. **72**, 146–164 (1959). [3] Multiple integrals in calculus of variations. Springer-Verlag (1966).

Moser J.: [1] A new proof of de Giorgi's theorem concerning the regularity problem for elliptic differential equations. Comm. Pure Appl. Mathem. **13**, 457–468 (1960). [2] On Harnack's theorem for elliptic differential equations. Comm. Pure Appl. Mathem. **14:3**, 577–591 (1961). [3] A Harnack inequality for parabolic differential equations. Comm. Pure and Appl. Mathem. **17:1**, 101–134 (1964).

Nash, J.: [1] C^1-isometric imbeddings. Ann. Math. **60:3**, 383–396 (1954). [2] The imbedding problem for Riemannian manifolds. Ann. Mathem. **63:1**, 20–63 (1956). [3] Continuity of solutions of parabolic and elliptic equations. Amer. Journal Mathem. **80**, 931–954 (1958).

Ni N.M.: [1] On the elliptic equation $\Delta u + Ku^{\frac{n+2}{n-2}} - 0$, its generalizations and applications to geometry. Indiana Univ. Math. J. **31**, 493-522 (1982).

Ni W.M. and Serrin J.: [1] Non-existence theorems for quasilinear partial differential equations. Rend. Circolo, Mathem. Palermo, Suppl. **5** (1985). [2] Existence and nonexistence theorems for ground state of quasilinear partial differential equations. The anomalous case. Acad. Naz. Lincei, Convegni del Lincei **77**, 231–257 (1986).

Nirenberg L.: [1] On nonlinear elliptic partial differential equations and Hölder continuity. Comm. Pure Appl. Mathem. **6**, 103–156 (1953). [2] The Weyl and Minkowski problems in differential geometry in the large. Comm. Pure Appl. **6**, 337–394 (1953). [3] Remarks on strongly elliptic partial differential equations. Comm. Pure Appl. Math. **8**, 649–675 (1955). [4] On elliptic partial differential equations. Annal. Scuola Norm. Sup. Pisa **13**, 115–162 (1959). [5] A strong maximum principle for parabolic equations. Comm. Pure Appl. Mathem. **6**, 167–177 (1953). [6] Monge–Ampere equations and some associated problems in geometry. Proc. Intern. Congress of Mathematicians, Vancouver 275–279 (1974).

Nitsche J.C.C.: [1] Vorlesungen über Minimalflächen. Springer-Verlag (1975). [2] Stationary partitioning of convex bodies. Arch. Rational Mech. Analysis **89**, 1–19 (1985).

Oleinik O.A.: [1] On properties of solutions of certain boundary value problems for equations of elliptic type. Mathem. Sb. (N.S.) **30**, 695–702 (1952).

Oleinik O.A. and Krushkov S.N.: [1] Quasilinear equations of second order with many independent variables. Uspekhi Mathem. Nauk **16:5**, 115–155 (1961).

Oleinik O.A. and Radkevich E.V.: [1] Second order equations with non-negative characteristic form. Moscow, Itogi Nauki (1971); English Translation, Providence, R.I., Amer. Math. Soc. (1973).

Oliker V.I.: [1] The boundary value Minkowski problem. The parametric case. Ann. Scuola Norm. Sup. Pisa **9:3**, 463–490 (1982). [2] Hypersurfaces in R^{n+1} with prescribed Gaussian curvature and related equations of Monge–Ampere type. Comm. in PDE **9(8)**, 807–838 (1984). [3] The problem of embedding S^n into R^{n+1} with prescribed Gauss curvature and its solution by variational method. Transactions of AMS **295**, 291–303 (1986). [4] Near radially symmetric solutions of an inverse problem in geometric optics. Inverse Problems **3**, 743–756 (1987).

Oskolkov A.P.: [1] A priori estimates for the first derivatives of solutions of the Dirichlet problem for nonuniformly elliptic quasilinear equations. Proc. Math. Inst. Steklov, Ak. Nauk USSR **102**, 105–127 (1967).

Osserman R.: [1] A survey of minimal surfaces. Nostrand Reinhold Co., N.Y. — London–Melbourne (1969). [2] The convex hull property of immersed manifolds. Journal of Differential Geometry **6**, 267–270 (1971). [3] Isoperimetric and related inequalities. Proc. Symp. Pure Mathem. (Stanford 1973), Providence, R.I., AMS **27**, 207–215 (1975) [4] The isoperimetric inequality. Bull. Amer. Math. Soc. **84**, 1182–1238 (1978). [5] Bonnesen-style isoperimetric inequality. Amer. Mathem. Monthly **86**, 1–29 (1979).

Osserman R. and Schiffer M.: [1] Doubly-connected minimal surfaces. Arch. Rational Mech. Anal. **58**, 285–307 (1975).

Peletier L.A. and Serrin J.; [1] Uniqueness of positive solutions of semilinear equations in R^n. Arch. Ration. Mech. Analysis **81**, 181–197 (1983). [2] Uniqueness of nonnegative solutions of semilinear equations in R^n. J. Diff. Equations **61** 380–397 (1986). [3] Ground states for the prescribed mean curvature equation. Proc. Amer. Math. Soc. **100**, 694–700 (1987).

Pisier G.: [1] The volume of convex bodies and Banach space geometry. Cambridge Tracts in Mathematics, Cambridge University Press **94**, (1989).

Pogorelov A.V.: [1] Bending of convex surfaces. GETTL, Moscow-Leningrad (1951). [2] Monge–Ampere equations of elliptic type. Groningen, Noordhoff, Groningen (1964). [3] On a regular solution of the n-dimensional Minkowski problem. Dokl. Akad. Nauk USSR **119**,

785–788 (1971). [4] On the regularity of generalized solutions of the equation $\det(u_{ij}) = f(x_1, \dots, x_n)$. Dokl. Akad. Nauk USSR **200**, 543–547 (1971). [5] The Dirichlet problem for the n-dimensional analogue of the Monge-Ampere equation. Dokl. Akad. Nauk USSR **201**, 790–793 (1971). [6] Extrinsic geometrie of convex surfaces. Providence, R.I., Amer. Math. Soc. (1973). [7] The Minkowski multidimensional problem. John Wiley & Sons, New York (1978).

Pohozaev S.I.: [1] Eigenfunctions of the equation $\Delta u + \lambda f(u) = 0$. Dokl. Akad. Nauk USSR **165** 36-39 (1965).

Protter M.H. and Weinberger H.F.: [1] Maximum principles in differential equations. Englewood Cliffs, N.J., Prentice Hall (1967).

Pucci C.: [1] Su una limitazione per soluzioni di equazioni ellittiche. Boll. Un. Math. Ital. (3), **21**, 228–233 (1966). [2] Operatori ellittici estremanti. Ann. Mat. Pura Appl. (4) **72**, 141–170 (1966). [3] Limitazioni per soluzioni di equazioni ellittiche. Ann. Mat. Pura Appl. (4) **74**, 15-30 (1966). [4] Maximum and minimum first eigenvalues for a class of elliptpic operators. Proc. of Amer. Mathem. Soc. No. 4 **17**, 788–795 (1966).

Pucci P. and Serrin J.: [1] A general variational identity. Indiana University Mathem. Journal **35:3**, 681-703 (1986).

Resetnjak Ju.G.: [1] Isometric coordinates in manifolds of bounded curvature. Dokl. Akad. Nauk USSR **94**, 631–633 (1954). [2] About one generalization of convex surfaces. Mathem. Sb. **40**, 381–398 (1956). [3] The investigation of two-dimensional manifolds of bounded curvature by isotermic coordinates. Proc. Sibir. Division, Akad. Nauk USSR **10**, 15–28 (1959). [4] Isotermic coordinates in manifolds of bounded curvature, I. II. Sibir. Mathem. Journal, I. **1**, 88–116 (1960); II **2**, 248–276 (1960). [5] Isoperimetric property of two-dimensional manifolds whose curvature is not more than K. Vestnik LGU **19**, 58–76 (1961). [6] Certain geometric properties of functions and mappings with generalized derivatives. Sibirsk. Math. Journal **7**, 886–919 (1966).

Rockafellar R.T.: [1] Convex analysis. Princeton, N.J. (1970).

Saks J. and Uhlenbeck K.: [1] The existence of minimal immersion of 2-spheres. Ann. Math. **113**, 1–24 (1981).

Santalo L.: [1] Integral geometry and geometric probability, Cambridge, Massachusetts (1977). [2] Un invariante afin pasa los cuerpos convexos del espacio de n-dimensiones. Port. Math. **8**, 155–161 (1949).

Schauder J.: [1] Der Fixpunktsatz in Functionalräumen. Studia Mathem. **2**, 171–180 (1930). [2] Über den Zusammenhang zwischen der Eindeutigkkeit und Lösbarkeit partieller Differentialgleichungen zweiter Ordnung vom elliptischen Typus. Math. Ann. **106**, 661–721 (1932). [3] Über das dirichletsche Problem in grossen für nicht-lineare elliptische Differentialgleichungen. Math. Z. **37**, 623–634 (1933). [4] Über lineare elliptische Differentialgleichungen zweiter Ordnung. Mathem. Z. **38**, 257–282 (1934). [5] Numerische Abschätzungen in elliptischen linearen Differentialgleichungen. Studia Mathem. **5**, 34–42 (1935).

Schoen R.: [1] Conformal deformation of a Riemannian metric to constant scalar curvature. J. Diff. Geometry **20** 479–485 (1984). [2] The existence of weak solutions with prescribed singular behavior for a conformally invariant scalar equation. Comm. Pure Appl. Mathem. **41**, 317–392 (1988).

Schoen R. and Uhlenbeck K.: [1] Boundary regularity and the Dirichlet problem for harmonic maps. J. Diff. Geometry **18**, 253–268 (1983).

Schoen R. and Yau S.T.: [1] Existence of incompressible minimal surfaces and the topology of manifolds with nonnegative scalar curvature. Ann. of Math. **110**, 127–142 (1979). [2] On the proof of the positive mass conjecture in general relativity. Comm. Math. Phys. **65**, 45–76 (1979). [3] The energy and the linear momentum of space-times in general relativity. Comm. Math. Phys. **79**, 47–51 (1981). [4] On the structure of manifolds with positive scalar curvature. Man. Math. **28**, 159–183 (1979). [5] Proof of the positive mass theorem, II. Comm. Math. Phys. **79**, 231–260 (1981). [6] Complete three-dimensional manifolds with positive Ricci curvature and scalar curvature. Annals of Math. Studies, Seminar on Diff. Geometry, Princeton University Press 209–218 (1982). [7] Conformally flat manifolds, Klein groups and scalar curvature. Invent. Mathem. **92**, 47–71 (1988).

Schultz F.: [1] Regularity theory for quasilinear elliptic systems and Monge–Ampere equations in two dimensions. Lecture Notes in Mathematics, Springer-Verlag **1445**, 1–120 (1990).

Schwartz L.: [1] Analyse Mathematique I. Hermann, Paris (1967).

Serrin J.: [1] On the Harnack inequality for linear elliptic equations. Journal Analyse Mathem. **4**, 292–308 (1955/56). [2] Local behavior of solutions of quasilinear elliptic equations. Acta Mathem. **111**, 247–302 (1964) [3] On surfaces of constant mean curvature which span a given space curve. Math. Z. **112**, 77–88 (1969). [4] The problem of Dirichlet for quasilinear elliptic differential equations with many independent variables. Philos. Trans. Roy. Soc. London series. **A-264**, 413–496 (1969). [5] Boundary curvatures and the solvability of Dirichlet's problem. Proc. Int. Congress of Mathem. Nice (1970). [6] Gradient estimates for solutions of nonlinear elliptic and parabolic equations. Contributions to Nonlinear Functional Analysis, New York, Academic Press 565–601 (1971).

Simon L.M.: [1] Global estimates of Hölder continuity for a class of divergence-form elliptic equations. Arch. Rational Mech. Analysis **56**, 253–272 (1974). [2] Boundary regularity for solutions of the non-parametric least area problem. Annal of Mathem. **103**, 429–455 (1976). [3] A Hölder estimate for quasiconformal mappings between surfaces in Euclidean space, with application to graphs having quasiconformal Gauss map. Acta Mathem. (1977). [4] Survey lectures on minimal submanifolds. Seminar on minimal submanifolds, Annals of Math. Studies Princeton University Press **103**, (1983). [5] Lectures on geometric measure theory. Proc. of Centre for Mathem. Analysis, Australian National University **3**, (1983).

Sobolev S.L.: [1] On a theorem of functional analysis. Mathem. Sb. (46) **4**, 471–497 (1938); English Transl.: Amer. Math. Soc. Transl. (2) **34**, 39–68 (1963). [2] Applications of functional analysis in mathematical physics. Leningrad, Izdat. LGU (1950). English Translation: Transl. of Mathem. Monographs, Providence, R.I., Amer. Mathem. Soc. **7**, (1963).

Spruck J.: [1] On the radius of the smallest ball containing a compact manifold of positive curvature. J. Diff. Geometry **8**, 257–258 (1973). [2] Gauss curvature estimates for surfaces of constant mean curvature. Comm. Pure Appl. Mathem. **27**, 547–557 (1974).

Stampacchia G.: [1] Problemi at contorno ellittici, con dati discotinui, dotati di soluzioni hölderiane. Ann. Math. Pura Appl. (4) **51**, 1–37 (1960). [2] Le probleme de Dirichlet pour les equations elliptiques du second ordre a coefficients discontinues. Ann. Inst. Fourier (Grenoble) **15**, fasc. **1**, 189–258 (1965). [3] Equations elliptiques du second ordre a coefficients discontinues. Seminaire de Mathem. Super., Montreal, Que: Les Presses de l'Universite de Montreal No. 16 (1966).

Struwe M.: [1] Nonuniqueness in the Plateau problem for surfaces of constant mean curvature. Arch. Ration. Mech. Analysis **93**, 135–157 (1986). [2] Large H-surfaces via the mountain-pass-lemma. Math. Annalen **270**, 441–459 (1985). [3] Free boundary problems for surfaces of constant mean curvature. Proc. International Conference VARIATIONAL METHODS FOR FREE SURFACE INTERFACE, Springer-Verlag 57–63 (1987).

Talenti G.: [1] Best constant in Sobolev inequality. Ann. Mat. Pura Appl. **110**, 353–372 (1976). [2] Elliptic equations and rearrangements. Ann. Scuola Norm. Sup. Pisa (4) **3**, 697–718 (1976). [3] Some estimates of solutions to Monge–Ampere type equations in dimension two. Ann. Scuola Norm. Pisa (4) **8**, 183–230 (1982).

Treibergs A.: [1] Entire space-like hypersurfaces of constant mean curvature in Minkowski space. Annals of Mathem. Study, Seminar on Diff. Geometry, Princeton University Press **102**, 229–238 (1982). [2] Entire space-like hypersurfaces of constant mean curvature in Minkowski space. Stanford University, Thesis of Ph.D. dissertation (1982).

Trudinger N.: [1] On Harnack type inequalities and their application to quasilinear elliptic equations. Comm. Pure Appl. Math. **20**, 721–747 (1967). [2] Remarks concerning the conformal deformation of Riemannian structures on compact manifolds. Ann. Scuola Norm. Sup. Pisa **3**, 265–274 (1968). [3] Gradient estimates and mean curvature. Math. Z. **133**, 165–175 (1973). [4] On the comparison principle for quasilinear divergence structure equation. Arch. Ration. Mech. Analysis **57**, 128–133 (1974). [5] Maximum principles for linear, nonuniformly elliptic operators with measurable coefficients. Math. Z. **156**, 291–301 (1977). [6] Fully nonlinear uniformly elliptic equations under natural structure conditions. Trans. Amer. Math. Soc. **278**, 751–769 (1983).

Uhlenbeck K.: [1] Removable singularities in Yang–Mills fields. Bull. Amer. Math. Soc. **1**, 579–581 (1979). [2] Variational problems for gauge fields. Ann. of Mathem. Studies, Seminar on Diff. Geometry, Princeton University Press **102**, 455–464 (1982). [3] Connections with L^p bounds on curvature. Comm. Mathem. Phys. (1983).

Weinberger H.F.: [1] Symmetrization in uniformly elliptic problems. Studies in mathematical analysis and related topics, Stanford California; Stanford University Press 424–428 (1962). [2] Variational methods for eigenvalue approximation. C.B.M.S., Regional Conf. Ser. #15, SIAM, Philadelphia (1975).

Wente H.: [1] Large solutions to the volume constrained Plateau problem. Arch. Ration. Mech. Anal. **75**, 59–77 (1980). [2] The differential equation $\Delta x = 2H(x_u \wedge x_v)$ with vanishing boundary values. Proc. Amer. Math. Soc. **50**, 131–137 (1975). [3] Counterexamples to a conjecture of H. Hopf. Pacific J. Mathem. **121**, 193–243 (1986).

Yamabe H.: [1] On a deformation of Riemannian structures on compact manifolds. Osaka Math. J. **12**, 21–37 (1960).

Yau S.T.: [1] Isoperimetric inequalities and the first eigenvalue of a compact Riemannian manifold. Ann. Sci. Ecole Norm Sup., Paris VIII 487–507 (1975). [2] Harmonic functions on complete Riemannian manifold. Comm. Pure Appl. Mathem. **28**, 201–228 (1975). [3] Calabi's conjecture and some new results in algebraic geometry. Proc. Nation. Acad. Sci. USA **74**, 1798–99 (1977). [4] On the Ricci curvature of a compact Kähler manifold and complex Monge–Ampere equation, I. Comm. Pure Appl. Mathem. **31**, 339–411 (1978). [5] Survey on partial differential equations in differential geometry. Annals of Mathem. Studies, Seminar on Diff. Geometry, Princeton University Press **102**, 3–71 (1982).

Index